CRUSTACEAN SEXUAL BIOLOGY

Crustacean Sexual Biology

EDITED BY
RAYMOND T. BAUER AND
JOEL W. MARTIN

Columbia University Press   New York

Columbia University Press
New York   Oxford
Copyright © 1991 Columbia University Press
All rights reserved

Library of Congress Cataloging-in-Publication Data

Crustacean sexual biology / edited by Raymond T. Bauer and Joel W. Martin.
  p.   cm.
  Includes bibliographical references and index.
  1. Crustacea—Reproduction. 2. Sex (Biology) I. Bauer, Raymond T.
II. Martin, Joel W.
QL445.2.C78   1990
593.3′0416—dc20
                                                    90-41006

ISBN 0-231-06880-8   CIP

Casebound editions of Columbia University Press books are
Smyth-sewn and printed on permanent and durable acid-free paper

Printed in the United States of America
c 10 9 8 7 6 5 4 3 2 1

CONTENTS

Preface Raymond T. Bauer and Joel W. Martin		vii
ONE	Bioluminescent Displays, Courtship, and Reproduction in Ostracodes *James G. Morin and Anne C. Cohen*	1
TWO	Instrinsic Factors Mediating Pheromone Communication in the Blue Crab, *Callinectes sapidus* *Richard A. Gleeson*	17
THREE	Patterns of Reproduction of Some Amphipod Crustaceans and Insights into the Nature of Their Stimuli *Betty Borowsky*	33
FOUR	Precopulatory Mate Guarding in Aquatic Crustacea: *Gammarus lawrencianus* as a Model System *Philip J. Dunham and Alan M. Hurshman*	50
FIVE	Variation in Reproductive Behavior in Stomatopod Crustacea *Roy L. Caldwell*	67

SIX	The Ecology of Breeding Females and the Evolution of Polygyny in *Paracerceis sculpta*, a Marine Isopod Crustacean *Stephen M. Shuster*	91
SEVEN	Anostracan Mating Behavior: A Case of Scramble-Competition Polygyny *Denton Belk*	111
EIGHT	Mating and Insemination in the American Lobster, *Homarus americanus* *S. L. Waddy and D. E. Aiken*	126
NINE	Sperm Competition and the Evolution of Mating Behavior in Brachyura, with Special Reference to Spider Crabs (Decapoda, Majidae) *Rudolf Diesel*	145
TEN	Sex Change, Mating, and Sperm Transfer in *Crangon crangon* (L.) *R. Boddeke, J. R. Bosschieter, and P. C. Goudswaard*	164
ELEVEN	Sperm Transfer and Storage Structures in Penaeoid Shrimps: A Functional and Phylogenetic Perspective *Raymond T. Bauer*	183
TWELVE	Functional and Evolutionary Aspects of the Sexual System in the Rhizocephala (Thecostraca: Cirrepedia) *Jens T. Høeg*	208
THIRTEEN	Functional Morphology and Evolution of Isopod Genitalia *George D. F. Wilson*	228
FOURTEEN	Functional Morphology of Spermatophores and Sperm Transfer in Calanoid Copepods *Pamela I. Blades-Eckelbarger*	246
FIFTEEN	The Reproductive Biology of Two Species of Remipedes *Jill Yager*	271
SIXTEEN	Structure and Chemical Content of the Spermatophores and Seminal Fluid of Reptantian Decapods *Gertrude W. Hinsch*	290
SEVENTEEN	Chemical Composition of Spermatophores in Decapod Crustaceans *T. Subramoniam*	308
EIGHTEEN	Morphological Diversity of Decapod Spermatozoa *Bruce E. Felgenhauer and Lawrence G. Abele*	322
	Contributors	343
	Systematic Index	347
	Subject Index	353

PREFACE

Crustaceans are a diverse assemblage of animals that includes not only the well-known shrimps, lobsters, and crabs but also a variety of less familiar but biologically important groups. Inhabiting primarily aquatic habitats, crustaceans might be considered the marine equivalent of their overwhelmingly terrestrial brethren, the insects. However, knowledge about the sexual biology of crustaceans is slight in comparison to what is known for insects and many other animal groups. In spite of the importance of selection pressures related to successful mating on the morphology, behavior, life histories, and phylogeny of crustaceans, many aspects of sex attraction, mating behavior, and insemination remain poorly known. A review of the literature on crustacean reproduction revealed that there was no comprehensive treatment of crustacean sexual biology but that in recent years much new exciting work has been (and is being) done in this area. Accordingly, we organized a symposium titled "Sex Attraction, Mating Behavior, and Insemination in Crustacea," cosponsored by the Crustacean Society and the Invertebrate Zoology Division of the American Society of Zoologists, with travel funds for participants provided by the Systematic Biology program of the National Science Foundation (Grant BSR 8806676). The symposium, which took place on December 30, 1988, served as a public forum and basis for this volume. We also solicited chapters from authorities on certain taxa or subject areas in order to round out the volume's coverage of crustacean sexual biology.

A series of entirely review articles on crustacean sexual biology, organized into chapters by taxa or by subject area, was not deemed desirable at this time, considering the lack of knowledge on the sexual habits and morphology of so many crustacean taxa. Instead, we asked our volume contributors, all active researchers in different areas of crustacean sexual biology, to strive to incorporate their recent research into a synthesis with past work in their field of interest. The chapters in this volume are loosely ordered from the topics of (1) sex attraction (bioluminescent displays in ostracods, chemical communication and sex recognition in crabs and amphipods), to (2) mating behavior and mating systems (stomatopods, isopods, anostracans, lobsters, brachyuran crabs, caridean shrimps), to (3) structure and function associated with insemination (penaeoid shrimps, rhizocephalan barnacles, isopods, copepods, remipedes, reptant decapods; chemical composition of decapod spermatophores; decapod sperm morphology). However, we have not formally arranged chapters into sections corresponding to these major subject areas because it is difficult in most cases to separate or to study these three aspects of sexual biology independently, given our present level of knowledge about crustacean reproduction. Thus there are varying degrees of overlap among these three topic areas in most chapters.

The details of sex attraction, mating behaviors, copulation, genitalia, spermatophores, and sperm structure are important not only for a basic understanding of processes and structures leading to insemination in crustaceans, but these basic data also serve as the raw material for proposing and testing hypotheses about various aspects of crustacean phylogeny and evolution. In several chapters, the question is asked "what is primitive or ancestral and what is derived or specialized" in terms of sexual morphologies or sexual systems. Other chapters deal with the evolution of mating systems and aspects of sexual selection, and one chapter deals specifically with sperm competition in crabs. Dramatic advances have been made in these areas of sexual biology by researchers using insect models, but development of the field is still in its infancy with regard to crustaceans. However, as with insect research, advances in these exciting theoretical areas must be preceded by basic studies on sex attraction, mating behavior, and functional morphology of copulation and insemination. We hope that the papers in this volume will stimulate the detailed work, couched in a broad evolutionary context, that is needed to heighten our understanding of crustacean sexual biology.

We owe a great deal of thanks to all the conscientious reviewers who contributed their time in making comments and suggestions on contributors' manuscripts. Our promise of anonymity prevents us from listing these reviewers individually. Our institutions (RTB: Center for Crustacean Research, University of Southwestern Louisiana; JWM: Natural History Museum of Los Angeles County) provided us with time, facilities, and travel funds associated with symposium and editorial activities. We thank Edward Lugenbeel of Columbia University

Press for assistance and advice at various points in the evolution of the volume. RTB thanks Adrian Wenner for helpful suggestions during the early stages of symposium and volume organization. Finally, we offer special thanks to Sue Martin and Lydia Bauer for supporting our efforts throughout this venture.

Raymond T. Bauer, Lafayette, Louisiana
Joel W. Martin, Los Angeles, California
November 1989

CRUSTACEAN SEXUAL BIOLOGY

ONE
Bioluminescent Displays, Courtship, and Reproduction in Ostracodes

**JAMES G. MORIN and
ANNE C. COHEN**

*Department of Biology,
University of California, Los Angeles, and
Division of Life Sciences,
Natural History Museum of Los Angeles County,
Los Angeles.*

Abstract

This paper presents an overview of our recent work on courtship behavior in luminescent cypridinids and a summary of reproductive and mating patterns in ostracodes. The Ostracoda, as a group, is large, and there is much variation in their reproductive patterns, especially between the two major groups: the podocopans and the myodocopans. The review of the reproductive patterns includes a summary of their population biology, reproductive system morphology, sexual dimorphism, and mating behavior. Male cypridinids from over 50 species in the nominal genus *Vargula* use their bioluminescence in courtship to attract females for reproduction throughout the Caribbean. Males secrete species-specific luminescent displays as trains of light pulses above reef habitats in the early evening throughout the year. Details of the copulation are unknown. Species show high local diversity, high endemism, and limited geographic distribution. We propose that sexual selection on the luminescent courtship displays and poor dispersal are responsible for their apparent rapid speciation in the Caribbean.

OSTRACODES ARE generally small (most less than 3 mm), reproduce by copulation, have determinate growth, and have a fixed number of instars. The class is usually divided into two subclasses, the Myodocopa, with about 650 marine species, and the Podocopa, with about 7,000 species from marine, freshwater, and terrestrial habitats (table 1.1). Luminescence occurs only in some Cypridinidae within the Myodocopida, which is the focus of this paper, as well as in some Halocyprididae within the Halocyprida. Much is known about ostracode population biology, life history patterns, and reproductive morphology, but little is known about the actual copulation process and even less is known about precopulatory courtship. In this paper we (1) briefly summarize the reproductive patterns in ostracodes, especially myodocopids (see also Cohen & Morin 1990), (2) summarize the limited data available on precopulatory courtship not involving bioluminescence, (3) present an overview on our recent work on the precopulatory courtship patterns in one group of myodocopids, the cypridinids, where

TABLE 1.1. Classification and number of taxa of the Ostracoda, with emphasis on the Myodocopa.

SubCl	Myodocopa		~100 genera,	~600 spp
O	Myodocopida			
SubO	Myodocopina			
SuperF	Cypridinoidea			
F	Cyprididinidae[a]		24 genera,	100 spp
SuperF	Cylindroleberidoidea			
F	Cylindroleberididae		27 genera,	100 spp
SuperF	Sarsielloidea			
F	Sarsiellidae		12 genera,	75 spp
F	Rutidermatidae		3 genera,	25 spp
F	Philomedidae		11 genera,	75 spp
O	Halocyprida			
SubO	Halocyprina			
SuperF	Halocypridoidea			
F	Halocyprididae[a]		7 genera,	150 spp
SuperF	Thaumatocypridoidea			
F	Thaumatocyprididae		3 genera,	10 spp
SubO	Cladocopina			
SuperF	Polycopoidea			
F	Polycopidae		4 genera,	40 spp
SubCl	Podocopa		~700 genera,	~7000 spp
O	Platycopida			
SubO	Platycopina			
SuperF	Cytherelloidea	1 family,	2 genera	
O	Podocopida			
SubO	Podocopina			
SuperF	Sigilloidea	1 family,	2 genera	
SuperF	Darwinuloidea	1 family,	2 genera	
SuperF	Bairdioidea	2 families,	11–15 genera	
SuperF	Cypridoidea	6 families,	164 genera	
SuperF	Cytheroidea	29 families,	500+ genera	

[a]Contains some luminescent species.

males use bioluminescence to attract females for mating, and (4) discuss the probable role of sexual selection in these mating systems.

REPRODUCTIVE PATTERNS

This section is a summary of a more extensive review of this topic by Cohen and Morin (1990). More specific references and information can be found in that paper.

Reproduction and Population Biology

Ostracodes generally reproduce sexually, but parthenogenesis is known in a few. Most Podocopa deposit eggs and most Myodocopa brood. Ostracode juveniles show gradual development, without metamorphosis, to a terminal adult stage. Depending on the taxon, there are 4–8 free-living juvenile instars. Total development time ranges from 16 days to over 3 years. Marine species tend to be perennial, and freshwater species tend to be annual. Seasonal reproduction is prevalent in species from habitats that are temporary, freshwater and/or subject to pronounced climatic changes. Nonseasonal reproduction is prevalent in at least a few species from less variable habitats (e.g., tropical, arctic, and antarctic waters). Resistant resting eggs occur in some species from habitats that are temporary or subject to severe winters. Sex ratios are biased toward adult females in many species.

Morphology Related to Reproduction

The male reproductive system, including the copulatory limbs, is usually paired. The copulatory limb (fig. 1.1A) is often large (figs. 1.1B, 1.2B, 1.2C, 1.2D), being as much as 35 percent of the body length (fig. 1.1B). Sperm are usually deposited in a seminal receptacle in the female (fig. 1.2E). Reproductive structures in females are generally paired, but the ovary-uterus system of some is separated from the vaginal-seminal receptacle system; in these cases mechanisms of sperm transfer are unknown. Myodocopids have the most simple reproductive system. In at least some male Cypridinidae, the gonads, vasa deferentia, seminal vesicles, and copulatory limbs are paired (fig. 1.2), but there is a single medial genital opening (penis) (fig. 1.1A). The copulatory limbs of myodocopids are taxonomically distinct (Cohen & Morin 1990) and may contribute to reproductive isolation. Copulatory organs of cypridinids show generic differences and, within some luminescent clades, specific differences as well. In addition, in some luminescent species, complex species-specific mating displays probably provide precopulatory reproductive isolation.

Sperm of some podocopids are among the largest in the animal kingdom, sometimes even longer than the animals themselves. Wingstrand (1988) described eight sperm types unique to higher ostracode taxa. Most myodocopids (Philomedidae, Rutidermatidae, Sarsiellidae, and Cyprinidae [which includes species that produce luminescent displays] but not Cylindroleberididae) have

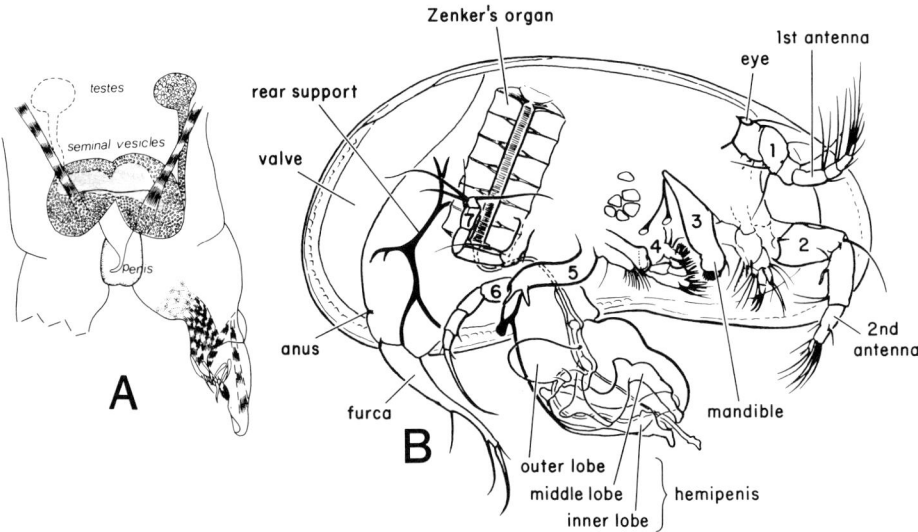

FIGURE 1.1. Male reproductive systems. A. Genitalia of *Spinacopia sandersi* (Myodocopa, Sarsiellidae): anterior view of testes, seminal vesicles, penis, and clasping limb (only left one shown). From Kornicker 1969. B. *Candona suburbana* (Podocopa, Cypridoidea) with right valve removed and left hemipenis fully erect, rotated, and in copulatory position. Outer face of hemipenis exposed to show the three lobes; limbs are numbered (1–7); sixth limb and palp of fifth limb are shown posterior to their position during copulation. Redrawn from McGregor and Kesling 1969a.

Sperm Type I (i.e., free mitochondria, a long crystalline perforatorium, and an enlarged and extended acrosomal vesicle) and a hardened spermatophore that is uniquely shaped by the external furrow on the female genital lobes during hardening (fig. 1.2E).

Most ostracodes exhibit sexual dimorphism. Sexual dimorphism generally correlates with courtship and mating, but corroborating behavioral observations are few. Lateral compound eyes, which occur only in the Myodocopida, are often larger and may have more ommatidia in males. Valve dimorphism is apparently associated with mating activities and/or brooding. Many males have some limbs specifically modified for clasping females, but clasping has been rarely observed. Suckerlike accessories occur on the first antennae of the Cyprididae and some Cladocopina. Only in the cyprididnid *Vargula hilgendorfii* have males been observed to use these suckers to grasp the female (Okada & Kato 1949). Sensory limbs (usually the first antennae) and swimming limbs (usually the second antennae) also may show dimorphism, being better developed in males. The sensory limbs may be used to find and/or court females. Finally, sexual dimorphism with uncertain function also occurs in the mandibles, fourth and fifth limbs, and furcae of many Myodocopida.

Mating Behavior

Copulation has been observed in only about 21 genera of podocopans and only three species of myodocopans. Six mating positions (two are shown in fig. 1.3)

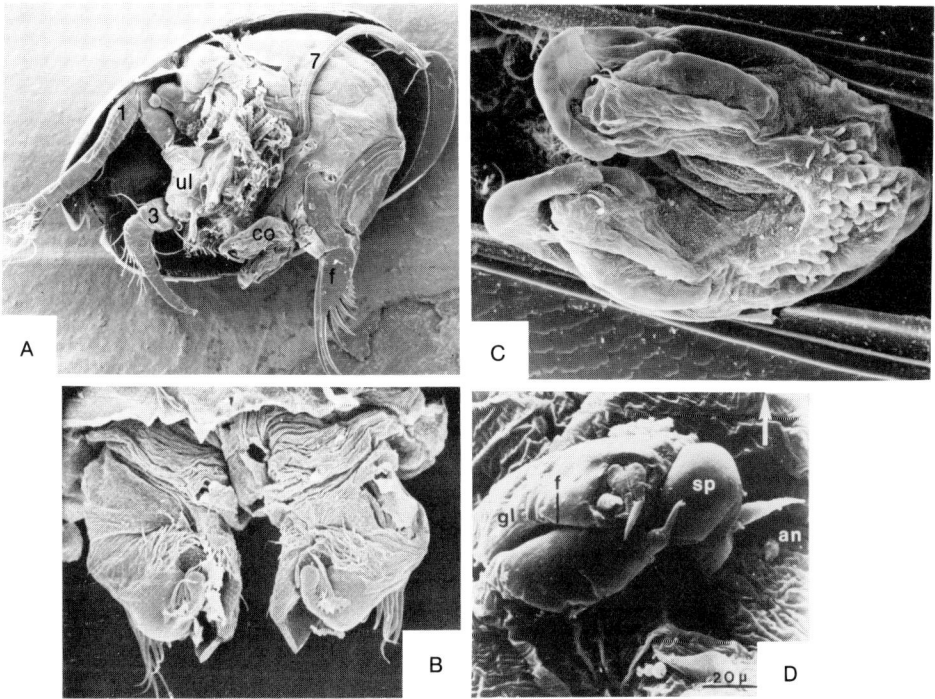

FIGURE 1.2. Scanning electronmicrographs of the reproductive limbs of cypridinid myodocopids. A. Whole lateral view of a male *Vargula kuna* (upward display type) with the left valve removed, limbs are numbered. co, copulatory limb; f, furca, ul, upper lip. B. Male copulatory limb of *Isocypridina quatuorsetae*, anterior view, from Kornicker 1975. C. Male copulatory limb of *V. contragula*. D. Female right genital lobe (gl) of *Vargula norvegica*, with spermatophore (sp) protruding from furrow (f), anterior direction shown by arrow. From Wingstrand 1988.

have been described (Cohen & Morin 1990). Copulation is rapid, usually no longer than a few seconds, but may take up to 30 mins (e.g., Okada & Kato 1949; McGregor & Kesling 1969b). Within the myodocopids mating has been observed in the cypridinid *Vargula hilgendorfii*, which copulated for more than 30 mins (Okada & Kato 1949), in the cypridinid *Cypridina dentata*, which copulated in minutes (Daniel & Jothinayagam 1979), and in two unidentified sarsiellids, which united for only a few seconds (Cohen personal observation). Behavioral observations suggest that females are monandrous with a single insemination, but a few are polyandrous, while males are polygynous, occasionally with multiple inseminations of one female (Cohen & Morin 1990).

Precopulatory courtship in which males locate or attract conspecific females must involve mechanical, chemical, or visual cues or any combination of these, if it occurs at all. Data on precopulatory behavior in the Ostracoda are sparse. In myodocopid males, structures used in searching for or attracting mates or both probably include the large and powerful second antennae used in swimming, the numerous and long sensory bristles on the first antennae, and the large eyes. We predict that precopulatory mechanical stimulation and chemical cues

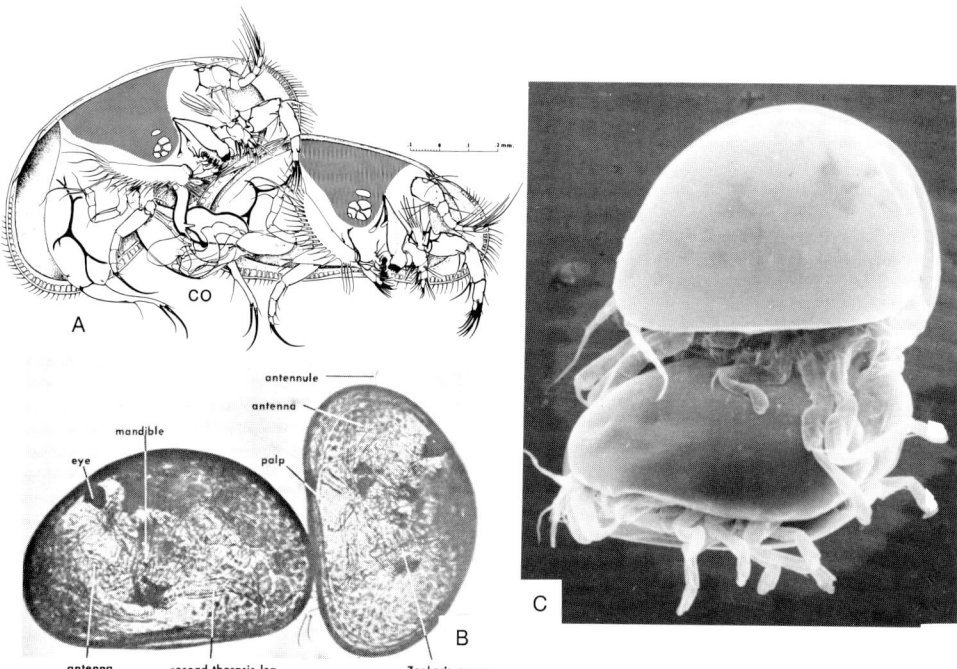

FIGURE 1.3. Copulation in some podocopids. A. *Candona suburbana* (a cypridoid) drawn with right valves removed and palp of right fifth limb of female withdrawn from proper position for clarity (see fig. 1.2B for identification of structures in male). co, copulatory limb. B. *Cypria turneri* (a cypridoid) palps or claspers of male's fifth limbs released their hold on edges of female valves when killed so that the male slid and rotated slightly down on the female carapace. Both from McGregor & Kesling 1969b. C. Scanning electron micrograph of mating entocytherids (cytheroids). From Hart and Hart 1974.

(pheromones) will prove to be a common component of ostracode courtship. Luminescent courtship displays in some cypridinids clearly involve visual cues (see below). Wherever dimorphism occurs in the size or the number of ommatidia of the lateral eyes, visual cues are probably important in courtship.

BIOLUMINESCENCE AND COURTSHIP

Bioluminescence

Luminescence is known only from some species of halocyprids and cypridinids within the Myodocopa (Herring 1978), but both groups also contain many non-luminous species. Luminescent ostracodes are strictly marine and are known from all latitudes, oceans, and depths (to at least 3,400 m) and from pelagic, demersal, and benthic habitats. In general, luminescent halocyprids tend to be pelagic while cypridinids tend to be demersal or benthic. Chemically, halocyprid and cypridinid luminescence appear to be distinct from one another (P. Herring personal communication), perhaps indicating an independent evolution of the two light-emitting systems. This possibility is further suggested by morphology: the light-emitting tissues in luminescent halocyprids are found widely distributed in the valves (Angel 1972), while luminescent cypridinids have a light organ in the upper lip (fig. 1.2A) from which luminescence is secreted outside the body (e.g., Morin 1986). Luminescent cypridinids have well-developed lateral eyes while the halocyprids are eyeless. Bioluminescence apparently serves as a mate attractant for some species of cypridinids, particularly within the greater Caribbean area (Morin 1986). In the Podocopa, none of which luminesce, relatively more is known about copulation events but little is known about precopulatory activities (courtship). The reverse is true for many of the luminescent myodocopans. Except for reports on the benthic/demersal *Vargula hilgendorfii* (Okada & Kato 1949) and the pelagic swarming *Cypridina dentata* (Daniel & Jothinayagam 1979), copulation has not been observed in luminescent species. Male swarming at the sea surface has been reported in a halocyprid (Angel 1972) and myodocopid (Daniel & Jothinayagam 1979). In fact, ostracodes are almost certainly responsible, at least in part, for the massive luminescent displays known to occur in the Indo-West Pacific (Kelly & Tett 1978; Herring & Horsman 1985); these are likely to be some sort of mating swarms. The waves of oceanic luminescence (Herring & Horsman 1985) could be visually coordinated entrainment phenomena similar to those observed in cypridinid reef displays (Morin 1986; see below). Since the early 1980s when the functions of luminescence were first precisely elucidated in cypridinids (Morin & Bermingham 1980), luminescent courtship displays have now been documented for over 50 species of cypridinids from the Caribbean (Morin 1986; Cohen & Morin 1986, 1989, in press; Morin & Cohen 1988).

Life Histories and Activity Patterns of Luminescent Cypridinids

Cypridinid species that produce luminescent courtship displays in the Caribbean Sea have fairly similar life histories and diel activity patterns. All appear to be primarily nonplanktonic species with distinct microhabitat preferences. Based on population and rearing data from a few of these species (Morin unpublished) and comparisons with other (some nonluminescent) cypridinids (Morin & Bermingham 1980; Morin 1986; Cohen & Morin 1986, 1989; Cohen 1983; Morin & Cohen 1988), Caribbean ostracodes with luminescent displays appear to share the following life history features: (1) males deposit a spermatophore that hardens in the genital furrow of the female (details of copulation are unknown); (2) sperm stored in the spermatophore fertilize the eggs as they are extruded and deposited in the brood pouch; (3) females brood from 12 to 48 eggs at a time; (4) there seems to be no brood overlap, and all embryos are of the same age and develop synchronously; (5) juveniles are released as benthic, crawl-away first stage instars, thus there is no planktonic stage anywhere in the life cycle; (6) new eggs are deposited in the brood pouch shortly after the previous brood is released (2–9 days in the nonluminescent *Skogsbergia lerneri* [Cohen 1983]); (7) embryonic development time is 2–4 weeks, and there are 5 juvenile instars and a terminal adult stage; (8) all juvenile and adult stages are capable of luminescing, but only males produce mating signals; (9) sexual characteristics become evident by the last (A-1 = fifth) juvenile stage; (10) from limited data, sex ratios of broods appear to be about equal; (11) total development time of the 5 juvenile instars is 2–3.5 months; (12) there appears to be no seasonality in their life cycles, and all stages of the life cycle are present year round; (13) adult males appear to be able to copulate many times while females may be sexually receptive and attentive to male signals only after releasing a brood or for only a brief time after their terminal molt to adulthood, at which time they copulate for the only time during their life (see below); (14) more than one brood may be produced from sperm stored during a single copulation (3 broods have been so produced in *Skogsbergia lerneri* [Cohen 1983]); (15) adults live for at least 2–6 months, and philomedid myodocopids (nontropical) live up to 3 years (Elofson 1941); (16) sexual dimorphism in secondary characteristics is distinct in all species: males have smaller and more elongate valves but larger eyes than females, and males have suckers on their first antennae for grasping the female; (17) mating displays by males are restricted to a specific period of the early evening, most often from the end of twilight for about an hour; and (18) this is the only time that males are in the water column. As details of reproduction and life histories of the various species become known, some variations in these traits will probably be found, but the overall patterns should remain.

Luminescent Courtship Displays and Sexual Selection

Nearly every evening of the year across the entire Caribbean, toward the end of twilight, shallow water cypridinid males leave their benthic haunts, enter the water column, and become demersal plankters for about an hour as they attempt to attract and mate with females. Each species produces its own stereotyped luminescence that is spatially and temporally predictable (fig. 1.4). In all cases each luminescent pulse is a discrete extracellular secretion, from glands in the upper lip, that is left behind in a stationary position (relative to the surrounding water) while the male swims rapidly onward (Morin 1986). A series of such light pulses produces a species-specific train of glowing spots in the water column. Predation is minimal on the signalers during the displays, apparently because, by doing so, the predator puts itself in danger of being consumed by its own predators who cue in on the luminescent cloud that the ostracode produces when attacked (Morin 1986).

Intramale and male-female associations appear to be similar in all signaling species. If a fine meshed net is swept through any of these displays, there are often nonsignaling males of the same species caught with the signaler; these appear to be "silent" satellite males that assemble with the signaler presumably in an attempt to sneak a copulation should a female be attracted to the display. Shining a light along the path of the display often reveals several males within a few millimeters of one another and swimming rapidly together. Only rarely are females caught in our nets. These are invariably ovigerous females without broods. Similarly, plankton tows through the display area during the display period yield high numbers of males and very few females. Juvenile stages or brooding females are almost never caught above the bottom. Furthermore, tows

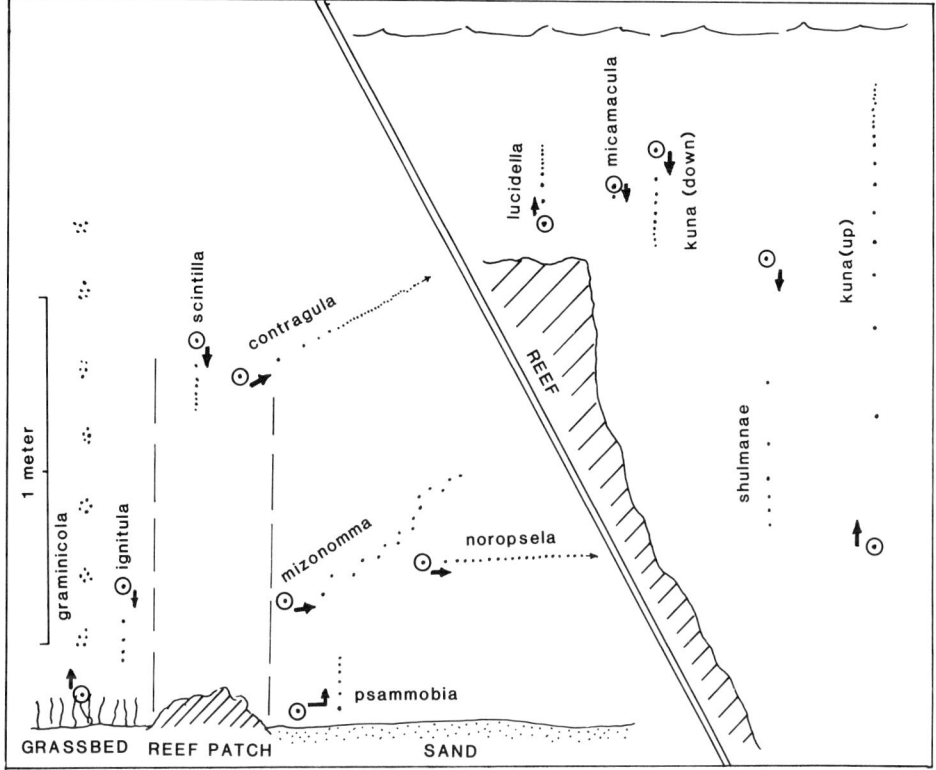

FIGURE 1.4. Spatial representation (with approximate position relative to a general habitat) of the courtship display pattern of 11 luminescent species of Caribbean cypridinid ostracodes in the nominal genus *Vargula* from Panama (see field key in Cohen & Morin 1989). Circled dot indicates first pulse in train; heavy arrow indicates direction of train; thin arrow indicates that the train continues (only in *V. noropsela* and *V. contragula*).

taken during the day or at other times of the night when they are not displaying collect few, if any, individuals of any instar or either sex. Thus during the period of luminescent displays within the water column there is a sex ratio that is strongly biased toward males.

During numerous hours of observations we have been unable to induce newly molted, virgin (nonbrooding) females to copulate with conspecific males in lighted (or unlighted) dishes in the lab, either during the day or night. They cannot be closely observed in the dark in the field. Thus, despite numerous attempts, copulation has not been observed in any of these species. We surmise that, unless conditions are appropriate (i.e., it is dark enough, the proper luminescent pattern exists, they are in the water column and not on the bottom, the female is receptive, etc.), individuals generally do not copulate.

Nonetheless, based on what is known about the life cycle of these and other related species (as indicated above) combined with the information about display activities (see below), we conclude that males are actively competing with numerous other males for copulations with females, which are only sexually receptive for a short time following their final molt to adulthood. In those species that produce luminescent displays, it is during this particular display period that males enter the courtship arena of the water column above their daytime haunts and call, using their species-specific codes, for newly molted virgin females. Once impregnated, the female is no longer sexually receptive, returns to the benthos (presumably after only a few minutes in the water column to mate), and begins the brooding process. From this one impregnation she can produce at least one brood and perhaps several. We further hypothesize that the hardened spermatophore in these and many other myodocopids may serve as a "mating plug" to prevent subsequent male impregnations. Thus she may mate only once in her life (however, at this time we lack direct evidence for this). Regardless of the number of broods produced per mating, the net effect is to skew the operational sex ratio of receptive individuals toward males, even though the actual sex ratio of the population may be about equal (i.e., the number of receptive [=all] males is about equal to the number of receptive females *plus* unreceptive females). The more broods produced by a female from a single mating, the greater will be the operational bias toward males. The operational sex ratio may be skewed further toward males if, as appears to be the case, the relatively few receptive (virgin) females are in the water column only long enough to mate, while the much more numerous and always receptive males are in the water column signaling for a much longer time (e.g., about an hour). Since we rarely find females in the vicinity of the displays, this suggests to us that females are in the water column for only a very short period of time *and* that they mate once or, at most, only a few times during their lives. A major effect of this kind of mating system is to increase the competition among males for copulations and increase the variance among males for copulation success; some males may mate often, while others seldom or never. Each female, on the other hand, would appear to have ready access to several males displaying above her. Thus variance among females for copulations should be small (all will be mated) and each has the *potential* to choose among several males. Whether females do exert a choice is not yet known, but the possibility is there. All of this adds up to a classic situation of sexual selection: many males are competing for few (choosy?) females and at least part of that competition is expressed via their luminescent displays. The basic components of this mating system are analogous to the well-known sexual selection-driven systems in many insects (see, e.g., Thornhill & Alcock 1983: Kaneshiro 1983), amphibians (e.g., Gerhardt 1982), birds (e.g., Mortin 1975), leks of various sorts (e.g., Bradbury and Gibson 1983), and other animals.

From a consideration of the biological patterns of cypridinids given above and an analysis of the luminescent displays, we hypothesize that the mating systems of these luminescent ostracodes have evolved as a result of sexual selection (both intrasexual and epigamic). Cypridinids that produce luminescent displays show the following traits that are considered evidence for sexual selection (see, e.g., Bateson 1983; Thornhill & Alcock 1983; Partridge & Halliday 1984): (1) apparent polygyny (males attempt to mate repeatedly while females do so once [or infrequently]), (2) sexual dimorphism, (3) differential investment and success in mating (males invest more in mating and show a greater variance in mating success than females), (4) differential investment in parental care (males invest little while female investment is significant), (5) the production of conspicuous mate attraction signals with particular locations and times for the displays, and (6) an operational sex ratio skewed toward males. We further hypothesize that sexual selection along with the poor dispersal capabilities at all life stages of each species has been an important factor in the sexual isolation and rapid speciation that has apparently occurred within this group in the Caribbean.

Variations in the Luminescent Display Patterns

Three major display patterns have been observed and recorded from Caribbean reef systems (Morin 1986): (1) *vertical shortening displays*, (2) *even lateral displays*, and (3) *rapid flashing displays* (fig. 1.4).

Vertical Shortening Displays. The mating display of *Vargula lucidella* from the San Blas Islands of Panama (Cohen & Morin 1989) is representative of the vertical shortening display (fig. 1.4). If the moon is not above the horizon at sunset, displays commence 44 mins after sunset (predictably within 2 min of this time), near the very end of twilight, and *only* over the shallow reef crests associated with the corals *Millepora complanata* and *Acropora palmata* at water depths of 0–2 m. The display period lasts for 45–60 mins and then ceases. Each display proceeds upward from within 5–25 cm of the reef. The display train, which is secreted by one male swimming rapidly (ca. 4–7 cm/sec), consists of a row of 4–5 (occasionally up to 7) bright points of light, each of which lasts for about 3–4 sec, secreted progressively closer together. The initial 2 pulses are about 6 cm apart and the last two about 4.5 cm apart. Occasionally this initial shortening phase is followed by a "trill" of 7–15 dim pulses, which each last about 1 sec and are spaced about 1 cm apart. The total train lasts 9–12 secs and is about 25 cm in total length (or 35 cm if there is a trill). Densities of trains over the reef are often quite high, up to about 5 displays/m^2. Displays entrain with one another so that they coincide temporally on about a 30–45 sec cycle. That is, the light from one signaling male will stimulate neighboring males to commence signaling at the same time, and they terminate at about the same time. Thus a wave of luminescent trains sweeps over the reef, remains for 10–15 secs, and is then followed by a 20–30 sec period of darkness before the next wave commences.

About two-thirds of the displaying species from the Caribbean show specific patterns that are variations of this vertical shortening display type (Morin 1986). Examples of extreme variations in vertical displays in species from the San Blas Islands of Panama (fig 1.4) include: (1) downward (e.g., *Vargula ignitula*) versus upward displays, (2) very long trains (up to 2.5 m) with widely spaced pulses (e.g., *V. kuna*) to very short trains (ca. 2–6 cm) with few, close pulses (e.g., *V. micamacula*), (3) long duration pulses (up to 11 sec) (e.g., *V. scintilla*) to short (2–3 sec) (e.g., *Vargula ignitula*), (4) many pulses per train (e.g., *V. kuna*) to few (e.g., *V. micamacula*), (5) strong entrainment between displaying males (e.g., *V. mica-*

macula) to very little (e.g., *Vargula psammobia*); (6) long display periods (e.g. *V. psammobia*) to quite short (e.g., *Vargula kuna*). All of these specific behavioral characters can be used to identify and separate the species (Cohen & Morin 1989).

Even Lateral Displays. The mating display of *Vargula noropsela* from the San Blas Islands of Panama (Cohen & Morin 1989) is representative of the even lateral display (fig. 1.4). These displays generally occur 30–100 cm above the edges of the sand-coral interfaces in spur and groove regions of coral slopes at water depths of 4–22 m. Displays commence 35–45 mins postsunset and continue for about 45 mins. The displays are lateral, either horizontal to the sand slope or slightly oblique to it. The display train from a male swimming about 3.5 cm/sec is usually many meters long, continues for many seconds, and consists of hundreds of evenly spaced (about 2 cm) pulses, each with a duration of about 2–3 secs. As new pulses are produced and older ones (5 to 10 pulses behind) fade, it gives a visual effect reminiscent of the vapor trail of a jet. Frequently 2 to 4 or more other males will begin to signal beside (about 2 cm) and in register with the initiator and then gradually diverge from him in the horizontal plane. In addition, "silent" males also often accompany the signaling males. Densities may be locally high but are generally low (6–8 displays at any one time per sand groove [ca. 3 by 8–10 m]), and there is usually weak cycling between a minute or so of displays followed by several minutes of no activity. About one-quarter of the displaying species from the Caribbean show variations of this even lateral display type (Morin 1986). Most of the species of ostracodes that produce even lateral displays are found near and display along horizontal types of habitat relief (edges of sand patches or grooves being the most common). Species with even lateral displays are found in only one of the major clades within the luminescent signaling species in the Caribbean (Cohen & Morin, in press; see below).

Rapid Flashing Displays. The mating display of *Vargula graminicola* from the San Blas Islands of Panama (Cohen & Morin 1986) is representative of the rapid flashing display (fig. 1.4). These displays occur over seagrass beds at water depths of 3–10 m. Displays commence about 45 min after sunset and continue most of the night. These remarkable displays are a series of rapid *group* flashes placed vertically upward. Each display begins within the grass bed itself as a single bright flash with a duration of about 1 sec. The second and succeeding pulses are accompanied by 2 to 40 flashes from other *non*-silent males in a pack, swimming within a few mm of each other and the leader. Swim speeds are rapid, about 10 cm/sec (or about 60 body lengths per second!). Each cluster of pulses is 15–25 cm above the previous ones, all have a duration of about 1 sec, there are about 8–12 pulse clusters per train, and each train may be as long as 2.5 m. Display densities can be very high, up to about 20 displays/m^2, and there is loose entrainment at about 60 sec intervals. The effect is dramatic: huge numbers of tight clusters of brief flashes repeatedly pulsing upward toward the sea surface in approximate synchrony with other clusters across vast expanses of grass beds. About 10 percent of the signaling species are "flashers," and most areas of the western Caribbean have one or two species that show variations of this rapid flashing. Major variations occur in the duration of the flash (some are as short as 0.250 sec), number of flashes per train, number of males signaling, and degree of entrainment. Species occur in a variety of habitats ranging from grass beds to coral reefs. Flashers are restricted to one clade within the luminescent signalers in the Caribbean (Cohen & Morin in press; see below).

Systematics and Biogeography of Ostracodes with Luminescent Displays

All of the signaling species we have discovered in the Caribbean belong to the nominal genus *Vargula*. However, our preliminary cladistic analysis, based on morphology, indicates that the genus is polyphyletic (Cohen & Morin, in press). The analysis also indicates that all the Caribbean species belong to 2 distinct monophyletic groups (clades A [with about 44 species] and F [with about 11 species]) that are distantly related. Vertical shortening displays occur in both clades while even lateral displays occur only in clade A and rapid flashing displays only in clade F. These differences infer that vertical shortening displays are the more ancestral state and evolved before the two clades diverged, while the other two signal types are more derived and evolved subsequent to the divergence in each of the two branches.

Each of the more than 50 species of signaling ostracodes we have discovered shows a high degree of habitat specificity and an apparent restricted distribution. Habitat specificity may include particular depth requirements, a minimal average amount of water movement, water clarity, substrate type (sand, coral, rubble, pavement, etc.), and particular biotic associates (e.g., particular taxa such as seagrasses, specific gorgonians or corals, etc.). The net effect of these various requirements is a highly predictable distribution of a given species within a reef system. Thus on a reef slope, for example, 4 or 5 species may co-occur, but only 2 or 3 will be displaying over the same localized reef area, and these species generally have distinctively different display patterns (e.g., upward versus downward displays, long versus short trains). Sibling species usually show very different signal patterns and/or habitat distributions (Morin & Cohen 1988; Cohen & Morin 1989).

All of the species we have found in the Caribbean appear to have limited geographic distributions; endemism is high. All species on one island or coastal area appear to be distinct in both morphology and display pattern, not only from each other but also from all other geographic locations (Cohen & Morin in press). This high local diversity and limited distribution phenomenon is likely due to isolating mechanisms produced by sexual selection of particular luminescent signals, probably by both intrasexual selection between males and epigamic selection by female choice, *and* the low dispersal characteristics of the group. Current data suggest that dispersal distances of about 50 km or less appear to be sufficient to allow gene flow, while distances of 250 km or more block it (i.e., at that distance local speciation via sexual selection exceeds rates of dispersal). The only areas where at least some species overlap apparently occurs is between (1) Curaçao and Bonaire (separated by ca. 45 km of open sea) and (2) three Belize sites and Roatan (sites with either contiguous reefs or that are within 50 km of the nearest contiguous reef). On the other hand, the Belize/Roatan area is contiguous with the San Blas area of Panama, but there is no species overlap. These areas are separated by 1500 km of coast, much of which is sand and without reef development (most of coastal Nicaragua and part of Honduras and Costa Rica). The sand and absence of reef probably act as major dispersal barriers.

How widespread around the world are these luminescent courtship displays in ostracodes? Luminescent cypridinids, but not the clades we have discovered in the Caribbean, are known from around the world, shallow to deep, boreal to equatorial. In order to detect the displays, one must be in the proper *underwater* habitat in the *dark* at the *precise time* of signaling; this is probably why the displays went unnoticed for so long. Once we knew where and when to look, we discovered signaling ostracodes on every major reef system we examined in the Caribbean; they are abundant and ubiquitous (Cohen & Morin, in press). However, based on previous and recent nocturnal underwater observations by the senior author (Morin, unpublished), we have been unable to locate any similar luminescent signaling patterns by ostracodes from: Bermuda; Cape Cod, Massachusetts; the Gulf of Aqaba in the Red Sea; Lizard, Heron, and One Tree islands on the Great Barrier Reef; Banda Islands, Indonesia; Negros and Mactan islands, Philippines; Canton Island in the Phoenix Group; Leigh, New Zealand; Isla Uvas on the Pacific coast of Panama; southern Baja, Mexico; Catalina Island and Malibu, California; and San Juan Islands, Washington. Luminescent ostracodes were found, but no signals have been seen, from the Great Barrier Reef and California sites. This lack of apparent signaling patterns elsewhere around the globe indicates either: (1) signaling is restricted to the Caribbean and has not evolved (or was lost) elsewhere or (2) it exists elsewhere, but we have not found it because (a) the species use a different luminescent code, undetected by us; (b) they display at other times of the night than the early evenings when we searched; or (c) there are restricted lunar or seasonal times for signaling, which we missed. In either case, it is clear that the types of luminescent displays described above appear to be restricted to the Caribbean. The localized nature of this phenomenon provides an interesting arena for determining more about the way that sexual selection has shaped the evolution of this intriguing group of ostracodes.

The general lessons to be learned regarding courtship and reproduction in ostracodes from our observations of luminescent sexual displays in cypridinids are that courtship activities (1) may be subtle and difficult to detect, (2) may occur on a very precise diel, lunar, or seasonal timetable and over a very short time period, and (3) may only occur in restricted habitat situations or conditions and that (4) individual interactions may occur very rapidly. If all of these phenomena are operating for a given species, they will hinder observations of the events that lead up to insemination. We can be thankful for those few species that seem not to mind the restrictive dishes, bright microscope lights, and voyeuristic scientists; those few have given us the little information we now have about copulation.

Acknowledgments

We thank L. Kornicker, R. Maddocks, and J. Martin for helpful reviews and additional information, L. Kornicker for loan of literature, and P. Baez for assistance with the Spanish literature. We thank the National History Museum of Los Angeles County for use of their facilities. Figure 1.1B was redrawn by Margaret Kowalczyk. This work was supported by the National Science Foundation (BSR89-05931).

Literature Cited

Angel, M. V. 1972. Planktonic oceanic ostracods—historical, present and future. *Proc. R. Soc. Edinb. Sect. B. (Biol. Sci)*73:219–228.
Bateson, P., ed. 1983. *Mate Choice*. New York: Cambridge University Press.
Bradbury, J. W. & R. M. Gibson. 1983. Leks and mate choice. In P. Bateson, ed., *Mate Choice*, pp. 109–138. New York: Cambridge University Press.
Cohen, A. C. 1983. Rearing and postembryonic development of the myodocopid ostracode *Skogsbergia lerneri* from coral reefs of Belize and the Bahamas. *J. Crustacean Biol.* 3:235–256.
Cohen, A. C. & J. G. Morin. 1986. Three new luminescent ostracodes of the genus *Vargula* (Myodocopida, Cyprinidae) from the San Blas region of Panama. *Contrib. Sci. (Los Angel.)*, no. 373: pp.1–23.
Cohen, A. C. & J. G. Morin. 1989. Six new luminescent ostracodes of the genus *Vargula* (Myodocopida, Cyprinidae) from the San Blas region of Panama. *J. Crustacean Biol.* 9:297–340.
Cohen, A. C.& J. G. Morin. 1990. Patterns of reproduction in Ostracoda: a review. *J. Crustacean Biol.* 10:184–211.
Cohen, A. C. & J. G. Morin. In press. Morphological relationships of bioluminescent Caribbean species of *Vargula* (Myodocopa). In R. Whatley & C. Maybury, eds., *Ostracoda and Global Events, Proceedings of the Tenth International Symposium on Ostracoda, Aberystwyth 1988.* Chichester, England: Ellis Horwood.
Daniel, A. & J. T. Jothinayagam. 1979. Observations on nocturnal swarming of the planktonic ostracod *Cypridina dentata* (Müller) for mating in the northern Arabian Sea. *Bull. Zool. Surv. India* 2:25–28.
Elofson, O. 1941. Zur Kenntnis der marinen Ostracoden Schwedens mit besonderer Berucksichtigung des Skagerraks. *Zool. Bidr. Upps.* 19:215–534. (Note: translation from German to English in 1969: *Marine Ostracoda of Sweden with Special Consideration of the Skagerrak.* Israel Program for Scientific Translations, pp. 1–286.)
Gerhardt, H. C. 1982. Sound pattern recognition in some North American treefrogs (Anura, Hylidae): Implications for mate selection. *Am. Zool.* 22:581–595.
Hart, D. G. & C. W. Hart. 1974. The ostracod family Entocytheridae. *Monogr. Acad. Nat. Sci. Phila.* 18:1–239.
Herring, P. J. 1978. Bioluminescence in invertebrates other than insects. In P. J. Herring, ed., *Bioluminescence in Action*, pp. 199–240. New York: Academic Press.
Herring, P. J. & P. Horsman. 1985. Phosphorescent wheels: Fact or fiction? *Mar. Obser.* 55:194–201.
Kaneshiro, K. Y. 1983. Sexual selection and direction of evolution in the biosystematics of Hawaiian Drosophilidae. *Annu. Rev. Entomol.* 28:161–178.
Kelly, M. G. & P. Tett. 1978. Bioluminescence in the ocean. In P. J. Herring, ed., *Bioluminescence in Action*, pp. 399–417. New York: Academic Press.
Kornicker, L. S. 1969. Morphology, ontogeny, and intraspecific variation of *Spinacopia*, a new genus of myodocopid ostracod (Sarsiellidae). *Smithson. Contrib. Zool.* 8:1–55.
Kornicker, L. S. 1975. Antarctic Ostracoda (Myodocopina). *Smithson. Contrib. Zool.* 163:1–720.
McGregor, D. L. & R. V. Kesling. 1969a. Copulatory adaptations in ostracods. Part I. Hemipenes of *Candona*. *Contrib. Mus. Paleontol. Univ. Mich.* 22:169–191.
McGregor, D. L. & R. V. Kesling. 1969b. Copulatory adaptations in ostracods. Part II. Adaptations in living ostracods. *Contrib. Mus. Paleontol. Univ. Mich.* 22:221–239.
Morin, J. G. 1986. "Firefleas" of the sea: Luminescent signaling in marine ostracode crustaceans. *Fl. Entomol.* 69:105–121.

Morin. J. G. & E. L. Bermingham. 1980. Bioluminescent patterns in a tropical ostracod. *Am. Zool.* 20:851.

Morin, J. G. & A. C. Cohen. 1988. Two new luminescent ostracodes of the genus *Vargula* (Myodocopida, Cypridinidae) from the San Blas region of Panama. *J. Crustacean Biol.* 8:620–638.

Morton, E. S. 1975. Ecological sources of selection on avian sounds. *Am. Nat.* 109:17–34.

Okada, Y. & K. Kato. 1949. Studies on luminous animals in Japan. III. Preliminary report on the life history of *Cypridina hilgendorfi*. *Bull. Biogeogr. Soc. Jpn.* 14:22–25.

Partridge, L. & T. Halliday. 1984. Mating patterns and mate choice. In J. R. Krebs & N. B. Davies, eds., *Behavioural Ecology, an Evolutionary Approach*, pp. 222–250. 2d ed. Sunderland, Mass.: Sinauer Associates.

Thornhill, R. & J. Alcock. 1983. *The Evolution of Insect Mating Systems*. Cambridge, Mass.: Harvard University Press.

Wingstrand, K. G. 1988. Comparative spermatology of the Crustacea Entomostraca. 2. Subclass Ostracoda. *Biologiske Skrifter* 32:1–149.

TWO
Intrinsic Factors Mediating Pheromone Communication in the Blue Crab, Callinectes sapidus

RICHARD A. GLEESON

The Whitney Laboratory,
University of Florida,
St. Augustine.

Abstract

A review of pheromone communication and its role in the reproductive behavior of *Callinectes sapidus* is presented. A pheromone found in the urine of pubertal females evokes a specific courtship display and cradle-carry behavior in male crabs. Males detect this pheromone via the aesthetasc sensilla on the outer flagellum of the antennules. The display-inducing pheromone appears to be absent in females that have completed the pubertal molt. Renewed release can occur in females initiating ecdysis subsequent to the pubertal molt; however, such a molt is an extremely rare event. Pubertal females appear to be preferentially attracted to waterborne odors from males. This ability to detect and locate prospective mates using chemical cues may be an important mechanism for assuring successful coupling. With eyestalk ligation/ablation a significant number of male crabs exhibit spontaneous display behavior. This effect is attributed to the loss of a hormonal factor of eyestalk origin that acts to modulate the activity of CNS "centers" controlling courtship behavior. A model for the activation and modulation of courtship behavior in the male is proposed.

THE BLUE crab, *Callinectes sapidus*, a member of the swimming crab family Portunidae, is distributed along the coastal regions of the western Atlantic Ocean from Nova Scotia to northern Argentina (Williams 1974). The female of this species will normally mate only once during her life. This copulation occurs immediately following her "maturity" or "pubertal" molt, which is also her final molt, since she enters terminal anecdysis after reaching maturity (Van Engel 1958). During the penultimate instar, the female undergoes dramatic morphological changes that prepare her for reproduction and are expressed at the time of the pubertal molt. Most prominent among these changes are the development of seminal receptacles (spermathecae) for the reception and storage of spermatophores and a transformation of the abdomen in preparation for spawning. Females in the proecdysial stage of the penultimate instar are herein referred to as "pubertal females."

Courtship and mating in *Callinectes sapidus* involve a sequence of behaviors that extends over a period of several days. Male crabs are observed to attend or to guard pubertal females that are within approximately 10 days of molting. When such a female is first encountered, the male will frequently exhibit a distinctive courtship-display behavior (Teytaud 1971). This display is characterized by an extension of the chelae toward a lateral position, extension of the walking legs (second to fourth pereiopods) such that the male is elevated high off the bottom, and rhythmic lateral waving of the swimming appendages (fifth pereiopods), which are rotated to an anterodorsal position above the carapace (fig. 2.1A). In addition, there is often an increase in the amount of water pumped through the branchial chambers; this increased pumping is associated with forward extension of the maxillipeds and rapid waving of the flagellum on each maxilliped. As a result there is a strong flow of water projected forward from the male that may be a significant component of the communication elicited by the display, e.g., by transporting chemical information to the female. The male orients the display toward the female and slowly approaches her. In response, the female either may remain passive or move toward the male, maneuvering to a position under him. The male then embraces the female and, using his first pair of walking legs, holds her beneath him in what is termed the cradle-carry position (fig. 2.1B). Laboratory observations suggest that the courtship display by the male is not an essential prelude to coupling. Males will sometimes grasp and cradle-carry females that are in close proximity without displaying at all. Furthermore, pubertal females will frequently initiate the cradle-carry themselves by approaching and repeatedly bumping against nondisplaying males. A pair of crabs may remain coupled for several days until the female undergoes her pubertal molt. At this time she is released; however, the male continues to stand over her. At the completion of the molt the pair once again assumes the cradle-carry position.

Mating is generally initiated by the female within a few minutes to an hour after she molts. With the assistance of the male, she positions herself upside-down beneath him and extends her abdomen, allowing the male to insert his paired copulatory appendages into her genital pores (fig. 2.1c). Spermatophores are deposited within the spermathecae over a 5–12 hr period after which the cradle-carry position is reestablished and maintained for two or more days. The

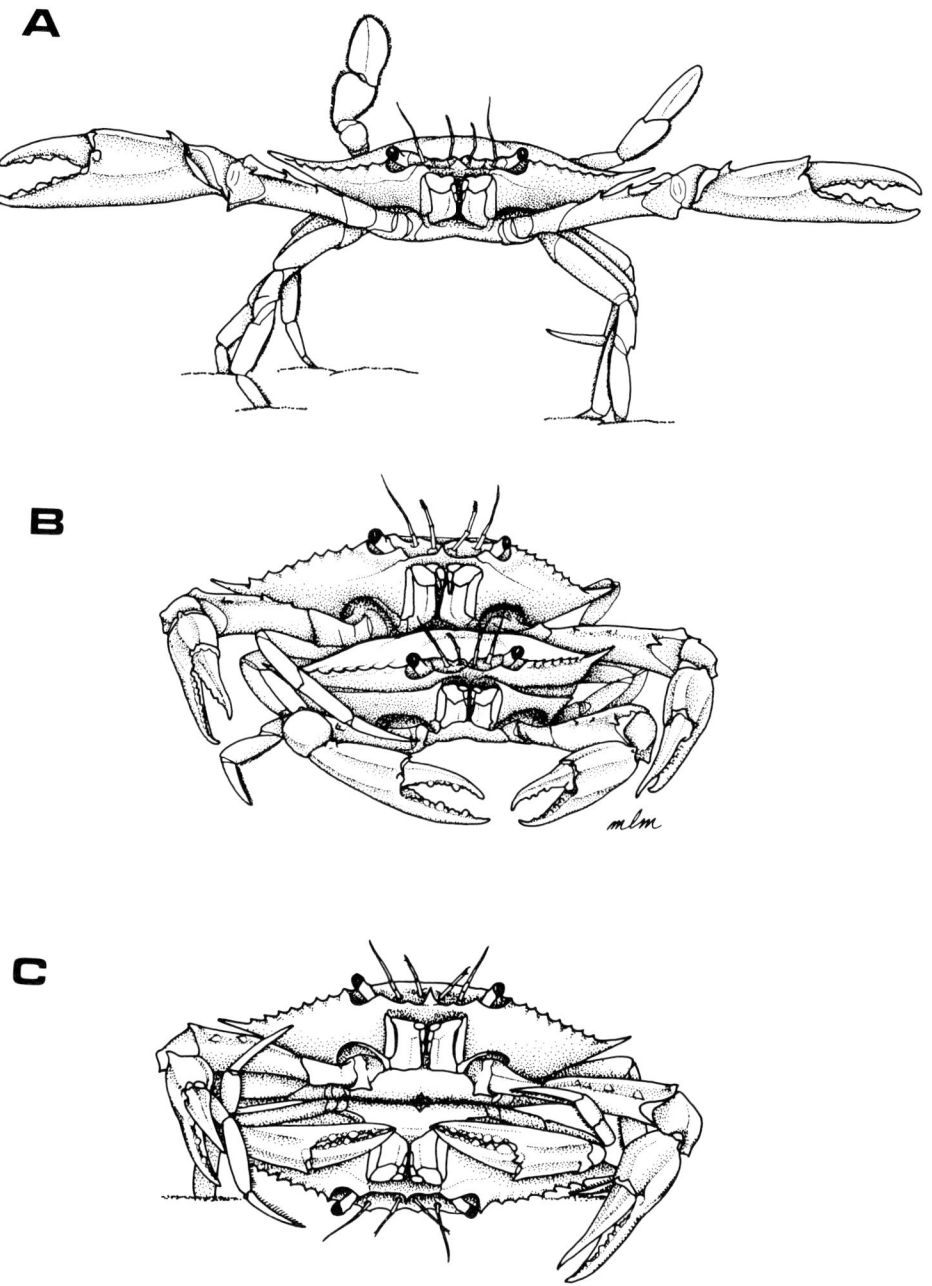

FIGURE 2.1. A. Male *Callinectes sapidus* in the courtship-display stance. B. Pubertal female in the cradle-carry position beneath a male. The female depicted is in the initial phase of her pubertal molt. C. Pair of crabs copulating. The recently molted female is upside-down beneath the male. Modified from Carr 1988.

sperm stored in the spermathecae during copulation remain viable for at least a year and are used to fertilize the two or more spawnings that the female undergoes during her lifetime (Hard 1942; Van Engel 1958).

EVIDENCE FOR PHEROMONE RELEASE BY FEMALES

Studies were performed to determine if courtship behavior (i.e., display and cradle-carry behavior) on the part of the male is triggered by a pheromone released by the pubertal female (Gleeson 1980). Water from a tank containing either pubertal females or control animals was introduced to a second tank containing "test-males." Two control groups were used: (1) mature females that were well past the pubertal molt and (2) male crabs that were of the same size and molt stage as the pubertal females. With the introduction of water from tanks containing control animals, little change in test-male behavior was observed. In contrast, when water from tanks containing pubertal females was presented to the males, a significant number of test-males began to exhibit courtship display behavior and/or attempted to cradle-carry other test-males. These findings clearly demonstrated that pubertal females release a chemical signal into the water that triggers the specific courtship behavior of the male.

The importance of visual cues in this behavior was investigated in experiments comparing the responses of males when only visual information versus visual plus chemical information from a pubertal female was available (Gleeson 1980). In this study a pubertal female was placed in a glass cylinder that was positioned within a tank containing several test-males. This allowed visual contact between the males and the female, but there was no exchange of water between the cylinder containing the female and the tank containing the males. Male activity was observed over a 10 min period. Water was then introduced into the cylinder from which it overflowed into the tank, and male behavior was again observed for five minutes. Visual cues alone did not elicit courtship behavior in males. Only with the overflow of water from the cylinder did males begin to exhibit courtship activity. The display behavior did not appear to be selectively directed toward the female; displaying males would readily orient toward other test-males and attempt to place them in a cradle-carry. These results suggested that the pheromone signal is of primary importance in initiating the male's courtship behavior and that visual cues are secondarily used to orient this behavior toward any crab in the immediate vicinity.

The urine of pubertal females was examined as a potential route of pheromone release to the environment (Gleeson 1980). For these experiments, urine samples were collected from the antennal gland ducts of pubertal females plus appropriate control animals. The two types of control animals described above were utilized as control-urine sources. Aliquots (0.1 ml) of each type of urine were sequentially introduced into tanks containing test-males and the stimulation of any courtship behavior noted over a five-minute observation period. Control urines elicited no reactions in test-males; however, urine from pubertal females stimulated courtship behavior in 15 of the 42 males tested, with a mean latency for onset of the behavior being approximately one minute. These results showed that the pheromone is indeed present in the urine of pubertal females and, consequently, may be an important route of release.

In recent experiments we have begun to examine the release of pheromone from females that have completed the pubertal molt but have not copulated (unmated, post–pubertal-molt females [UPPM-females]). Although under normal conditions, females mate immediately after the pubertal molt, laboratory observations reveal that they will copulate as late as two to four weeks after this molt and are cradle-carried immediately following copulation (Teytaud 1971; personal observation). This would suggest that the UPPM-female continues to release the pheromone until she copulates. To explore this, experiments were performed in which the induction of male display behavior was examined. Water from a tank containing UPPM-females (from 1–10 days postmolt) was introduced to tanks containing single test-male crabs, and the induction of display was monitored. Water presented under identical conditions from a tank containing pubertal females served as a benchmark for comparison of display frequency. Remarkably, of the 172 males tested, none exhibited courtship-display activity when exposed to water from tanks containing UPPM-females. In contrast, when 91 of these males were exposed to water from tanks containing pubertal females, 35 (39 percent) displayed. These findings were corroborated in tests using urine collected from UPPM-females (6–8 days postmolt). In bioassay tests with this urine, no display behavior was induced in six males, all of which consistently exhibited display activity when exposed to pubertal-female urine. Also in agreement with these results is the observation that males and UPPM-females will generally copulate without preamble when placed together in the same tank (personal observation).

Although courtship behavior in test-males can be triggered by urine collected from females immediately following the pubertal molt (Gleeson 1980), the present findings suggest that UPPM-females either do not continue to produce the display pheromone or do not release it at an effective concentration for display induction. Since males are observed to cradle-carry females after copulations that occur well past the pubertal molt, these findings also raise the possibility that the display and cradle-carry behaviors are mediated by two distinct pheromone signals. Further studies to test the induction of cradle-carry behavior by urine from UPPM-females should provide insight into these questions.

Although pheromone production/release appears to decline dramatically in UPPM-females, there is evidence that it can be released anew if the female initiates another molt. Ecdysis subsequent to the pubertal molt in the female of *Callinectes sapidus* is an extremely rare occurrence (see Olmi 1984). Although there are reports in the literature describing mature females in proecdysis, no accounts have definitively shown the release of pheromone at this time or documented the successful completion of such a molt (Hard 1942; Abbe 1974; Olmi 1984; Skinner 1985). An opportunity to examine this arose in April 1986, when a mature female in proecdysis was obtained from the St. Johns River in Florida and taken to the Whitney Laboratory for observation. When placed in a tank containing a mature male, the female approached the male and attempted to position herself beneath him. The male immediately responded by cradle-carrying the female; this position was maintained for over two hours before the female was removed and placed in a separate tank. There were no attempts to copulate during this observation period. Urine obtained from the female was bioassayed with four males, each tested individually. All four males responded strongly to the urine with courtship displays. After seven days in the laboratory, the female successfully molted, and when she was placed together with a male, copulation was initiated. The pair was immediately separated, however, and subsequent inspection of the spermathecae revealed that the female had not copulated previously.

These observations show that pheromone release can accompany a molt occurring after that of puberty. Interestingly, Truitt (in Hard 1942) has observed that females that do not copulate after the pubertal molt will initiate a second molt. If such a relationship does in fact exist, it would suggest that copulation is linked to the mechanism of terminal anecdysis in the female and may, therefore, represent a means by which females can enhance their opportunities for copulation in the event that mating is unsuccessful following the pubertal molt. This is an intriguing question that is certainly worthy of further investigation.

PHEROMONE RECEPTION IN MALES

The location of chemoreceptors mediating pheromone detection by the male was determined in a series of ablation experiments (Gleeson 1980, 1982). These studies centered on the antennules (first antennae), which are the olfactory organs in decapod crustaceans (Ache 1985). In *Callinectes sapidus* the antennule is biramous, with an inner and outer flagellum; olfactory sensilla (aesthetascs) are situated in a prominent tuft on the ventral face of the outer flagellum (fig. 2.2). To localize pheromone-sensitive chemoreceptors, various ablations of the antennules were performed in groups of male crabs and their responses to the pheromone (i.e., water from tanks containing pubertal females) compared to those of appropriate control animals. Initial experiments revealed that with bilateral removal of both flagella from the antennules, males would no longer respond to the pheromone. This effect was also observed with bilateral ablation of only the outer flagellum, whereas removal of the inner flagella had no effect on pheromone detection. These results indicated that chemosensory structures important for pheromone detection are associated with the outer flagellum of the antennule (Gleeson 1980).

Experiments then focused on the aesthetasc tuft region of the outer flagellum as a possible site of pheromone-sensitive chemoreceptors. Scanning electron microscopy studies revealed that the tuft is comprised of approximately 700 aesthetasc sensilla that originate from grooves situated distally on the ventral faces of most flagellar segments. The tuft is divided into distinct mesial and lateral halves on each segment by a central region of cuticle that has no sensilla (fig. 2.3). A second type of sensillium, which is only found on the mesial side of the tuft, arises from sockets located proximal to the aesthetasc row of each flagellar segment. From one to four of these sensilla, termed asymmetric sensilla, are found on most segments; they project across the tuft, terminating within the lateral half of the tuft (fig. 2.3). Based on these morphological findings, three

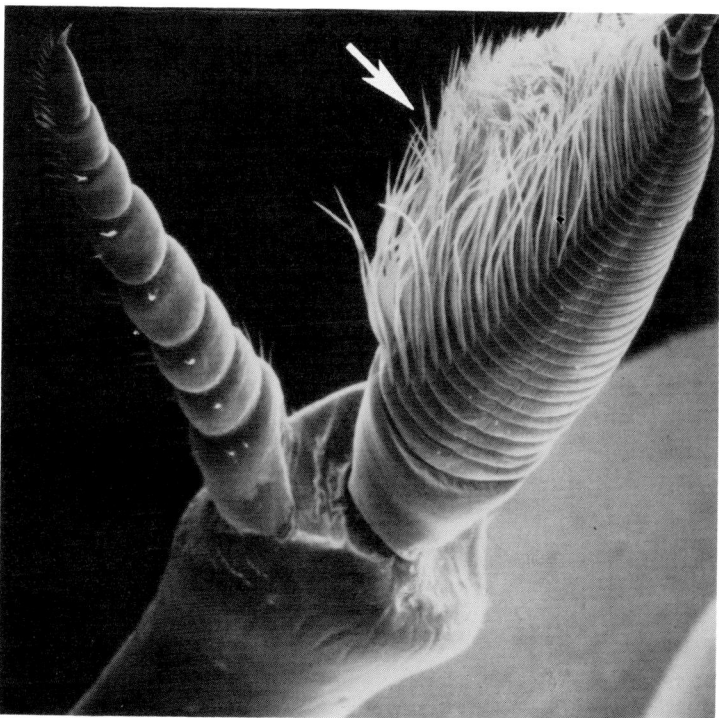

FIGURE 2.2. Scanning electron micrograph showing the inner and outer flagella of an antennule from *Callinectes sapidus* (lateral view). The tuft of aesthetasc sensilla on the outer flagellum is indicated by the arrow. The length of the outer flagellum is approximately 2 mm.

different types of lesions were bilaterally performed on the antennules of test-male crabs: (1) using fine-tipped forceps, all of the sensilla were removed from the lateral half of the tuft; (2) all sensilla were similarly removed from the mesial half of the tuft; and (3) using micro-dissecting scissors, the entire tuft was cut, resulting in the ablation of all but the basal portion of each aesthetasc. Behavioral assays demonstrated an approximately 20 percent decrease (relative to a sham control group) in the incidence of responses to the pheromone for males receiving lesions in either the mesial or lateral half of the tuft. With the entire tuft lesioned, however, there was a profound loss of response to the pheromone (Gleeson 1982). These findings are consistent with the notion that the aesthetascs are the chemosensory structures mediating pheromone detection in the male crab and suggest that the asymmetric sensilla do not play a role in this detection.

CHEMICAL PROPERTIES OF THE FEMALE PHEROMONE

The chemical nature of the pheromone of pubertal females was investigated using male crabs in a bioassay system that allowed testing small volumes of material for pheromone activity (Gleeson, Adams & Smith 1984). Urine was used

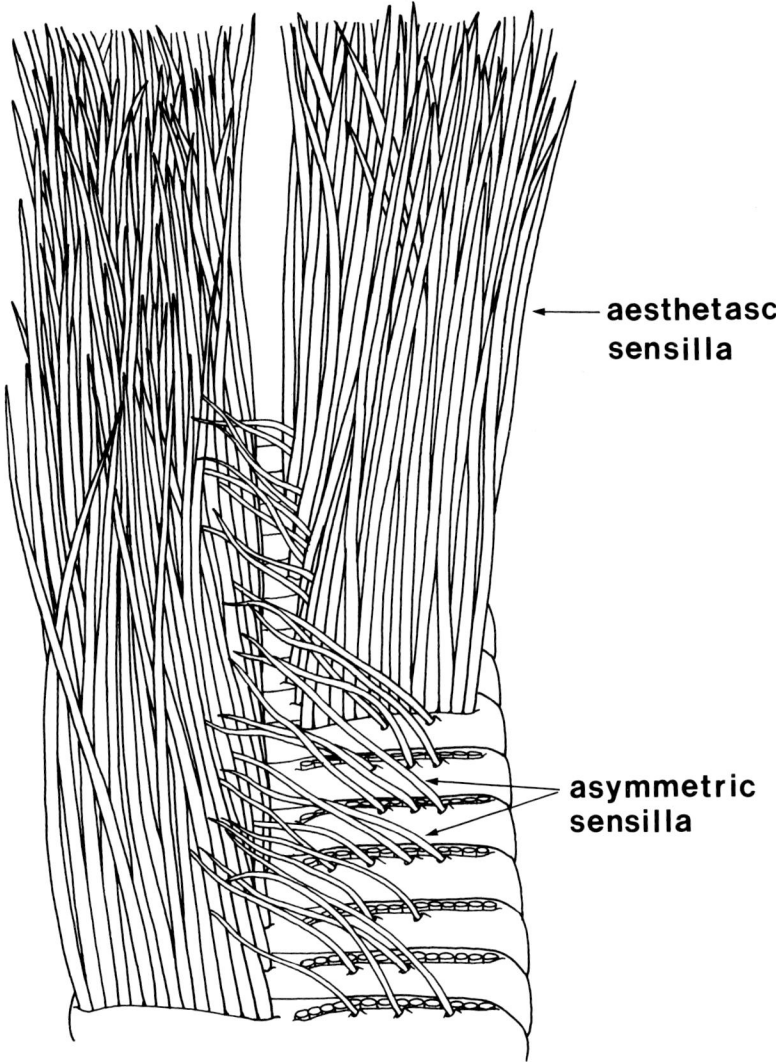

FIGURE 2.3. Portion of the outer flagellum (ventral view) of the antennule in which the aesthetasc sensilla have been removed from several segments to reveal the asymmetric sensilla. These sensilla are found only on the mesial side of the aesthetasc tuft and project laterally.

as a source of the pheromone and served as a benchmark in each bioassay for quantifying the relative activity of fractions derived from the urine. Initial studies showed that the urine can be heated to 95°C for five minutes without significant loss of activity and can be stored at −20°C for extended periods of time; it can also be lyophilized to dryness and reconstituted in water with no loss of biological activity. Sephadex gel filtration using columns calibrated with peptide molecular weight markers have indicated a molecular weight for the pheromone of between 300 and 600 daltons. Partial purification of the pheromone can be achieved by sequential ultrafiltration of the urine; since the pheromone will pass through an Amicon YC-2 filter but is retained by a YM-05 filter (molecular weight cutoffs of 1000 and 500, respectively), substances greater than 1000 and less than 500 daltons can be removed from the pheromone fraction. Further purification of the pheromone has been accomplished with high-pressure liquid chromatography using a Whatman reverse-phase preparative column (M9-ODS-3, 70:30 methanol-water solvent mixture). This procedure yields a fraction with pheromone activity equivalent to that of the urine itself and containing three components as detected by ultraviolet absorbance at 220 nm and confirmed by thin-layer chromatography (Gleeson, Adams & Smith 1984).

The potential role of the molting hormone, crustecdysone, as a pheromone in decapod crustaceans was proposed by Kittredge, Terry & Takahashi. (1971). In that study crustecdysone was reported to induce precopulatory behavior in males of *Pachygrapsus crassipes*, *Cancer antennarius*, and *C. anthonyi*. Moreover, the authors noted an apparent lack in species specificity of the pheromone released by various crabs, e.g., courtship behavior in *C. antennarius* appeared to be stimulated by *P. crassipes*. Based on these findings it was concluded that either crustecdysone is the sex pheromone for these species or that it has a structure sufficiently similar to the pheromones to mimic their actions. They subsequently proposed that the molting hormone served as a substrate for the evolution of sex pheromone communication, which initially involved leakage of

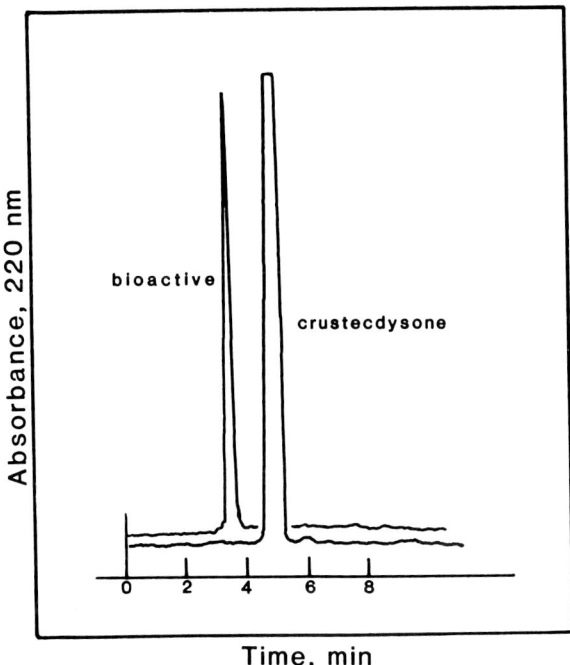

FIGURE 2.4. HPLC traces comparing the retention times for crustecdysone and the partially purified pheromone derived from the urine of pubertal *Callinectes sapidus* females. From Gleeson, Adams, & Smith 1984.

crustecdysone to the external environment with a concomitant externalization of crustecdysone receptors by males (Kittredge & Takahashi 1972; Kittredge, Terry & Takahashi 1971). The evidence supporting this hypothesis, however, is controversial (e.g., Dunham 1978; Gleeson, Adams & Smith 1984).

The molting hormone/sex pheromone concept was explored in *Callinectes sapidus* by behaviorally testing crustecdysone and by comparing the physical properties of crustecdysone with those of the partially purified pheromone derived from the urine of pubertal females (Gleeson, Adams, & Smith 1984). A comparison of the HPLC retention times for crustecdysone versus the pheromone-containing fraction from urine showed no correspondence between the two (fig. 2.4). Mass spectra using the desorption ionization technique were obtained for the HPLC-purified bioactive fraction and compared to the spectra for crustecdysone; again, no relationship between the pheromone and crustecdysone was revealed. Finally, in bioassay tests there was no evidence for the induction of courtship responses in males when exposed to crustecdysone. These results, as well as the findings of others (Atema & Gagosian 1973; Gagosian & Atema 1973; Seifert 1982), do not support an evolutionary relationship between the molting hormone and sex pheromones in crustaceans. To date, the structure of a crustacean sex pheromone has not been described.

EVIDENCE FOR AN ATTRACTANT RELEASED BY MALES

Teytaud (1971) noted that pubertal *Callinectes sapidus* females exhibit significant changes in particular behaviors (i.e., rocking and chelae waving) when presented with a visual image of a crab and simultaneously exposed to water containing male odor. These findings suggested that odor cues may be used by pubertal females in the identification of males. This notion was further supported by the observation that pubertal females appear to be attracted to males as suggested by the success of a trapping technique that is utilized commercially to obtain "peeler" crabs (crabs close to molting) in the soft-crab industry. The method involves use of a crab trap that is "baited" with live males. Interestingly, these so-called jimmy pots almost exclusively capture pubertal females.

To explore the potential chemical attraction of pubertal females to males, an apparatus was recently developed for testing the preferential attraction of crabs to point sources of waterborne odors. The test system used is schematically depicted in figure 2.5. One end of the "test"-tank is divided into two compartments by vinyl-coated wire mesh; each compartment has a funnel-shaped entry port that permits crabs to enter but very effectively prevents them from leaving. Seawater flowing through two different "odor-source" tanks (not shown), each containing 1 or 2 odor-source animals, is separately delivered to individual entry funnels (at 1800 ml/min) via tubes positioned within the tops of the funnels. Animals are tested in groups and simultaneously released at the standpipe end of the test-tank. Just prior to the beginning of each trial, the test-tank is filled with fresh seawater; water flow from the odor-source tanks is then started, followed immediately by the release of crabs in the test-tank. After a two-hour period, the number of animals moving into each compartment is determined. Each experimental condition is replicated with the odor sources reversed.

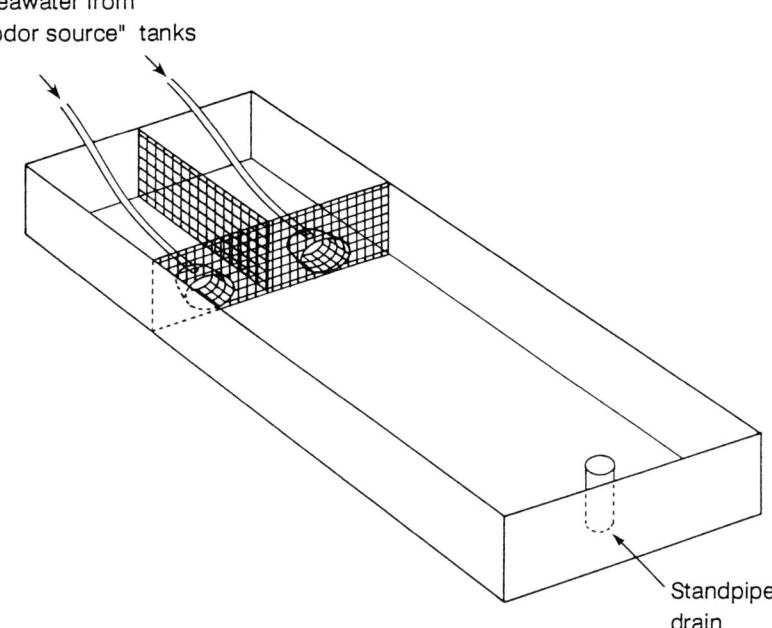

FIGURE 2.5. Schematic representation of the system used for testing the preferential attraction of crabs to point sources of waterborne odors. Test-tank dimensions are 90 x 265 cm with a water depth of 18 cm.

Initial experiments using this system suggest that pubertal females are indeed preferentially attracted to male crabs by chemical cues. The results demonstrate that pubertal females exhibit a significant preference for water from odor-source tanks containing males versus water from tanks containing either females or no animals (table 2.1). Many of these initial studies have focused on how eyestalk ligation (see Hormonal Modulation of Male Receptivity) of odor-source males affects their attractiveness. Trials comparing the relative attractiveness of males

TABLE 2.1. Preferential attraction of pubertal females to odor sources.

Experiment No.	Odor source[a]	No. Pubertal ♀♀ Attracted	χ^2 value[b]	P value
1	♂♂	86	8.4	0.0041
	♀♀	52		
2	L♂♂	45	4.5	0.0318
	L♀♀	27		
3	L♂♂	43	10.2	0.0018
	blank	18		
4	L♂♂	74	0.06	0.79
	♂♂	71		

[a] All odor-source animals were mature; L indicates animals had received bilateral eyestalk ligations.
[b] Based on an expected equal distribution of pubertal females for each odor pair.

with and without bilateral eyestalk ligations indicate that pubertal females are equally attracted by both groups of males (table 2.1, exp. 4).

Although further studies are in order, these findings suggest that pubertal females are able to utilize chemical cues (possibly a male pheromone) to locate males, an ability that could be an important mechanism for optimizing reproductive success. This chemical attraction to males may, in fact, be the basis for the selective capture of pubertal females in commercial jimmy pots. An interesting question for future study is whether females use this chemical information to discern reproductive fitness in prospective mates. For example, are females preferentially attracted to mature males or males in a receptive state? With bilateral eyestalk ligation, male crabs exhibit a high incidence of spontaneous courtship behavior (see below). Is this apparent change in receptive state detected by pubertal females via substances released from the male? Although initial results suggest that pubertal females do not discriminate between males with and without eyestalk ligations, more detailed study on this point is needed. Finally, the question of whether males utilize pheromonal cues to locate pubertal females remains to be tested.

HORMONAL MODULATION OF PHEROMONE RECEPTIVITY IN MALES

Male crabs exhibit apparent cycles in receptivity to the pheromone of pubertal females. This phenomenon was encountered during the course of experiments investigating the chemical nature of the pheromone. Routine screening of male crabs for pheromone receptivity prior to their use in bioassays revealed that there were periods when very few males responded to the pheromone, whereas at other times nearly all males were receptive. There were no obvious correlations between these peaks in male receptivity and cycles in environmental parameters, molt stage, or seasonal mating activity; however, a study that rigorously examines this periodicity is needed to explore such relationships.

There is emerging evidence that hormonal/neurohormonal factors can act on appropriate groups of neurons to alter the "behavioral state" of an organism; the behavioral output to certain stimuli is thereby biased toward a particular set of adaptive responses (e.g., Kravitz 1988; Truman & Weeks 1985). The possible role of such factors in controlling the periodic receptivity of *Callinectes sapidus* males was recently investigated (Gleeson, Adams & Smith 1987). These experiments initially targeted the androgenic glands as potential sources of a hormone modulating male-specific behavior. The androgenic glands, which are located on the subterminal regions of the deferent ducts, directly control the development and maintenance of both primary and secondary sexual characteristics in male crustaceans (Charniaux-Cotton & Payen 1985). In several crustacean species an eyestalk hormone has been shown to suppress the activity of the androgenic glands (Adiyodi & Adiyodi 1970; Adiyodi 1985). It was therefore hypothesized that if a factor from the androgenic glands acts to enhance receptivity in *C. sapidus* males, the secretion of this factor might be enhanced by eliminating the suppressing eyestalk hormone via removal or ligation of the eyestalks.

Experiments to test the effects of eyestalk ligation/ablation on the pheromone receptivity of males generated some very interesting results (Gleeson, Adams & Smith 1987). In these studies hemolymph circulation to the eyestalks was prevented by ligating each eyestalk at its base; necrosis of eyestalk tissue rapidly followed with the eyestalk frequently becoming detached after a few days. Within approximately 4–6 days after bilateral eyestalk ligation (or ablation), a significant number of male crabs began to exhibit "spontaneous display behavior" (SDB), i.e., they displayed in the absence of a pheromone signal. Bouts of display activity, sometimes continuing for several hours, were observed over a period of several days.

Responses to the pheromone were tested in spontaneously active animals by presenting pheromone-containing urine during periods between bouts of SDB. None of the animals responded to the pheromone, however, suggesting that neural pathways in the eyestalk ganglia are important links between pheromone receptors on the antennules and central nervous system (CNS) "centers" controlling courtship behavior. This view was further supported in a second group of males, all of which consistently responded to the pheromone prior to eyestalk lesions; with transection of the optic tracts connecting the eyestalk ganglia and supraesophageal ganglion or ligation of the eyestalks, these animals failed to respond to a pheromone stimulus. Together, these findings suggest that neural pathways in the eyestalk ganglia are important for transmitting or processing pheromonal information received via the antennules.

The induction of SDB following eyestalk ligation implicated the eyestalks as possible sites for the production and/or release of a hormonal factor that acts to modulate the activity of CNS "centers" (e.g., central pattern-generators) controlling courtship behavior. Since massive hypertrophy of the androgenic glands was also associated with eyestalk ligation, this suggested that the modulation of these putative "centers" might involve an androgenic hormone as well, possibly as an intermediate. An alternative hypothesis, however, was that loss of the eyestalk ganglia may remove inhibitory neural inputs to courtship "centers" located elsewhere in the CNS. To test these hypotheses, the induction of SDB was monitored in a group of males that had both optic tracts transected. With this procedure, hemolymph circulation to the eyestalks was not compromised. Of the 17 animals tested, only a single crab exhibited SDB, indicating that the loss of neural inputs from the eyestalk ganglia cannot account for the induction of SDB following eyestalk ligation, thus supporting the hypothesis that the loss of a hormonal factor is involved.

Considering the results of these studies together, a model for the activation and modulation of courtship behavior was proposed (fig. 2.6). In a receptive male, a pheromone stimulus detected via chemoreceptors on the antennules will normally activate the CNS "centers" mediating courtship activity. Neural connections in the eyestalk ganglia appear to be important components in the pathway linking these "centers" to antennular receptors. It is proposed that an eyestalk hormone (possibly a neurosecretory product from the sinus gland) can modulate the excitability of these "centers" and thereby alter the threshold for activation by a pheromone stimulus. How modulation is effected is not known; the action may be direct and/or involve one or more intermediates. For example,

30 *Richard A. Gleeson*

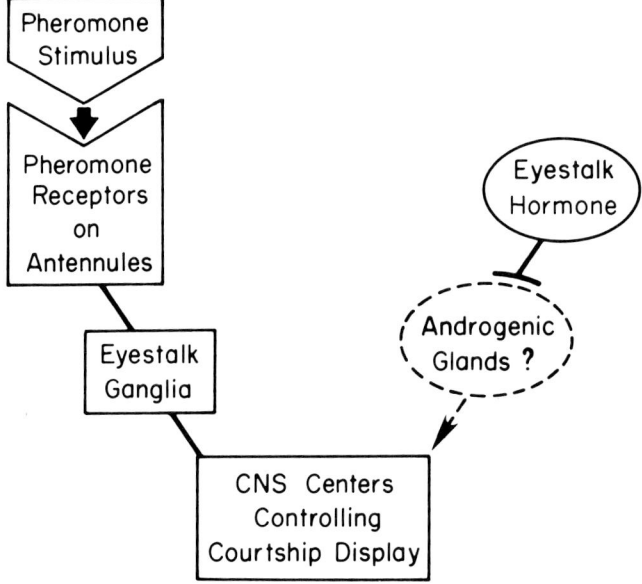

FIGURE 2.6. Hypothesized model for the activation and modulation of CNS "centers" mediating courtship behavior in *Callinectes sapidus* males. From Gleeson, Adams, & Smith 1987.

if a factor released by the androgenic glands acts to increase the excitability of courtship "centers," the eyestalk hormone may, therefore, be acting indirectly on these "centers" by modulating the androgenic glands. Studies are currently in progress to decipher the neural components that mediate courtship behavior in *Callinectes sapidus* males and to identify the neuroactive substances that influence the activity of these circuits (Wood et al. 1988a; Wood, Gleeson & Derby 1988b). Clearly, discerning the mechanisms underlying SDB should provide important insight into the processes that modulate the "behavioral state" of the male crab and contribute to an understanding of how this modulation influences the pheromonal activation of courtship behavior.

CONCLUSIONS

The pheromone released by the pubertal female of *Callinectes sapidus* plays an important role in coordinating reproduction in this species. By inducing the cradle-carry behavior of the male, the female is afforded protection from predators at the time of her pubertal molt and copulation following ecdysis is assured. The use of chemical cues by pubertal females to identify and locate prospective mates is presently less well understood, but may be an important mechanism for enhancing the prospects of successfully coupling with a male.

There are a number of intriguing questions that remain to be addressed with this pheromone system. One of the most prominent among these is the chemical nature of the female pheromone. With an elucidation of the structure, important insight can be gained on such points as its metabolic origins, the receptor processes involved in its detection, and its evolutionary relationships with other crustacean pheromone systems. Other matters of particular interest include: What neural elements are involved in controlling courtship behavior in the male? What are the hormonal/neurohormonal factors that modulate these neural circuits, and how is the modulation effected? Because of the very distinct behavior elicited in the male by the pheromone of pubertal females, *Callinectes sapidus* represents a particularly promising model system for detailed study of the processes underlying the activation and modulation of a chemically evoked behavior.

Acknowledgments

This work was supported in part by the following grants from the National Science Foundation: BNS-7523217, PCM-8104336, PCM-8038776, and BNS-8607513; and by a grant from the National Institutes of Health (5F32NSO5942-02). The illustrations were prepared by M. L. Milstead.

Literature Cited

Abbe, G. R. 1974. Second terminal molt in an adult female blue crab, *Callinectes sapidus* Rathbun. *Trans. Amer. Fish. Soc.* 103:643–644.
Ache, B. W. 1985. Chemoreception and thermoreception. In H. L. Atwood & D. C. Sandemann, eds., *The Biology of Crustacea*, vol. 3:369–398. New York: Academic Press.
Adiyodi, K. G. & R. G. Adiyodi. 1970. Endocrine control of reproduction in decapod Crustacea. *Biol. Rev. Camb. Philos. Soc.* 45:121–165.
Adiyodi, R. G. 1985. Reproduction and its control. In D. E. Bliss & L. H. Mantel, eds., *The Biology of Crustacea*, 2:147–216. Orlando: Academic Press.
Atema, J. & R. B. Gagosian. 1973. Behavioral responses of male lobsters to ecdysones. *Mar. Behav. Physiol.* 2:15–20.
Carr, W. E. S. 1988. The molecular nature of chemical stimuli in the aquatic environment. In J. Atema, R. R. Fay, A. N. Popper & W. N Tavolga, eds., *Sensory Biology of Aquatic Animals*, pp. 3–27. New York: Springer-Verlag.
Charniaux-Cotton, H. & G. Payen. 1985. Sexual differentiation. In D. E. Bliss & L. H. Mantel, eds., *The Biology of Crustacea*, 9:217–299. Orlando: Academic Press.
Dunham, P. J. 1978. Sex pheromones in Crustacea. *Biol. Rev. Camb. Philos. Soc.* 53:555–583.
Gagosian, R. B. & J. Atema. 1973. Behavioral responses of male lobsters to ecdysone metabolites. *Mar. Behav. Physiol.* 2:115–120.
Gleeson, R. A. 1980. Pheromone communication in the reproductive behavior of the blue crab, *Callinectes sapidus*. *Mar. Behav. Physiol.* 7:119–134.
Gleeson, R. A. 1982. Morphological and behavioral identification of the sensory structures mediating pheromone reception in the blue crab, *Callinectes sapidus*. *Biol. Bull. (Woods Hole)* 163:162–171.
Gleeson, R. A., M. A. Adams & A. B. Smith III. 1984. Characterization of a sex pheromone in the blue crab, *Callinectes sapidus:* Crustecdysone studies. *J. Chem. Ecol.* 10:913–921.
Gleeson, R. A., M. A. Adams & A. B. Smith III. 1987. Hormonal modulation of pheromone-mediated behavior in a crustacean. *Biol. Bull. (Woods Hole)* 172:1–9.
Hard, W. L. 1942. *Ovarian Growth and Ovulation in the Mature Blue Crab*, Callinectes sapidus Rathbun. Chesapeake Biol. Lab., Solomons, Md., publ. no. 46.
Kittredge, J. S. & F. T. Takahashi. 1972. The evolution of sex pheromone communication in the Arthropoda. *J. Theor. Biol.* 35:467–471,.
Kittredge, J. S., M. Terry & F. T. Takahashi. 1971. Sex pheromone activity of the molting

hormone, crustecdysone on male crabs *(Pachygrapsus crassipes, Cancer antennarius* and *C. anthonyi). U.S. Fish Wild. Serv. Fish. Bull.* 69:337–343.

Kravitz, E. A. 1988. Hormonal control of behavior: Amines and the biasing of behavioral output in lobsters. *Science (Washington, D.C.)* 241:1775–1781.

Olmi, E. J. 1984. An adult female blue crab, *Callinectes sapidus* Rathbun (Decapoda, Portunidae), in proecdysis. *Crustaceana (Leiden)* 46:107–109.

Seifert, P. 1982. Studies on the sex pheromone of the shore crab, *Carcinus maenas*, with special regard to ecdysone excretion. *Ophelia* 21:147–158.

Skinner, D. M. 1985. Molting and regeneration. In D. E. Bliss & L. H. Mantel, eds., *The Biology of Crustacea*, 9:43–146. Orlando: Academic Press.

Teytaud, A. R. 1971. *The Laboratory Studies of Sex Recognition in the Blue Crab*, Callinectes sapidus *Rathbun*. Sea Grant Tech. Bull. no. 15, University of Miami Sea Grant Program.

Truman, J. W. & J. C. Weeks. 1985. Activation of neuronal circuits by circulating hormones in insects. In A. I. Silverston, ed., *Model Neural Networks and Behavior*, pp. 381–399. New York: Plenum Press.

Van Engel, W. A. 1958. The blue crab and its fishery in Chesapeake Bay. Part I. Reproduction, early development, growth and migration. *Commer. Fish. Rev.* 20:6–17.

Williams, A. B. 1974. The swimming crabs of the genus *Callinectes* (Decapoda Portunidae). *U.S. Natl. Mar. Fish Serv. Fish. Bull.* 72:685–798.

Wood, D. E., B. Beltz, C. D. Derby & R. A. Gleeson. 1988a. Neural and hormonal control of a courtship behavior in the blue crab. *Am. Zool.* 24:92A.

Wood, D. E., R. A. Gleeson & C. D. Derby. 1988b. Evidence for neuromodulation of a pheromone-mediated courtship behavior in the blue crab, *Callinectes sapidus*. *Chem. Sens.* 13:747.

THREE

Patterns of Reproduction of Some Amphipod Crustaceans and Insights into the Nature of Their Stimuli

BETTY BOROWSKY

Osborn Laboratories of Marine Sciences, Brooklyn.

Abstract

Compared to the larger Crustacea, relatively little is known about the biology of gammaridean amphipods. Yet amphipods are often key members of benthic communities and merit study. Further, their ready adaptation to laboratory conditions facilitates observations of their life histories. This paper provides an overview of the group, emphasizing studies on two species in particular: *Gammarus palustris* and *Microdeutopus gryllotalpa*.

Most gammarideans are diecious. However, since there is no sperm storage and females can reproduce many times in succession, a female must be accompanied by a male each time fertilization occurs. Female reproductive behaviors may be divided into four sequential stages: Stage I, mate location, is assisted by the secretion of waterborne pheromones. Stage II, initiation of pairing, is stimulated by contact, and the specific stimulus involved is probably a contact pheromone. Stage III, pairing, continues until the female molts, and its stimulus is currently unknown. Stage IV, copulation, occurs shortly after the female molts and is probably stimulated by a contact pheromone. The majority of species studied to date share the same basic reproductive physiology and exhibit the same four behavioral stages. But the behaviors exhibited during the first three stages can differ markedly. These differences are ascribed to adaptations to different habitats.

GAMMARID AMPHIPODS are ubiquitous and often extremely abundant in benthic communities (Bousfield 1973). Yet compared with other crustacean groups, their biology is relatively little understood, even though they are particularly amenable to study in the laboratory. Amphipods are small, so their habitats can be closely mimicked and large numbers can be maintained. Many species reproduce repeatedly and frequently in the laboratory, and some tolerate harsh experimental manipulations.

This paper deals with amphipod reproductive behaviors and their stimuli. Its objectives are to review what is known to date, to suggest future directions for study, and to point out the advantages of using amphipods as model systems for the study of crustacean biology.

Results of observations of amphipod reproductive behaviors have sometimes been contradictory. In particular, there has been some controversy about whether or not waterborne pheromones are produced by amphipods. Dahl, Emanuelsson, and von Mecklenburg (1970a) found that male *Gammarus duebeni* congregated on a double mesh net separating males from females in a dish, suggesting that males were attracted to female effluents. But Hartnoll and Smith (1980) were unable to repeat Dahl et al.'s results. In another study, employing different experimental methods, Ducruet (1973) concluded that waterborne pheromones are present in *Gammarus pulex*. Males were attracted to conspecific females at all intermolt stages and also responded to females of another species *(G. fossarum)*. However, the experimental conditions confounded the effect of male and female attractants. Hartnoll and Smith (1980) were also unable to replicate these observations.

A series of experiments on two amphipods, *Gammarus palustris* and *Microdeutopus gryllotalpa*, has revealed that these two species do produce pheromones. This paper summarizes this work and relates the results to other studies. It will be shown that amphipod reproductive behaviors are not simple, as early workers supposed, and further, that there are real differences among species. The complexity of the behaviors plus differences among species may explain some of the inconsistencies previously reported.

Although their specific behaviors differ, the behaviors of both *Gammarus palustris* and *Microdeutopus gryllotalpa* occur in the same four successive stages through the female's intermolt period. At the beginning of the intermolt the female is alone. In Stage I a receptive male and receptive female find each other (mate location); Stage II involves behaviors designed to initiate pairing (pair formation); Stage III is the period during which the male and female remain together (are paired); and Stage IV is copulation.

REPRODUCTIVE BIOLOGY OF GAMMARID AMPHIPODS

Gammarids are small (between 0.5 and 1.5 cm), flattened laterally, and have pereiopods of various morphologies. Females ovulate within minutes after the parturial molt, and the eggs pass into a ventral brood pouch (called oviposition hereafter). This brood pouch, or marsupium, is a unique characteristic of the peracarid Crustacea. Since females do not store sperm, a female must be accompanied by a male at each molt so that copulation and fertilization of the current brood will occur in a timely manner. The eggs develop and hatch during the female intermolt. The offspring may remain in the brood pouch for a variable length of time but cannot remain past the female's next molt. Amphipods exhibit direct development, and eggs hatch into juveniles with the general adult morphology, although they still lack secondary sex characteristics.

In *Gammarus palustris*, as in most gammaridean species, females do not ovulate unless they copulate. Observations of *G. palustris* showed that delays between the female molt and copulation can be costly in terms of reduced fecundity. A delay of only six hours led to a 19 percent reduction in the number of juveniles produced per brood, and after 48 hrs no ovulation occurred even if copulation occurred (Borowsky 1988).

In some species, the females oviposit even if no copulation occurs (some examples: *Microdeutopus gryllotalpa*, *Elasmopus laevis*, *Jassa marmorata*, [personal observation]). However, these eggs never develop. Unfertilized broods generally disappear from the marsupium within a few days (most probably eaten by the female).

Reproduction is seasonal in most natural populations (some genera: *Gammarus* [Steele & Steele 1975], *Ampelisca* [Mills 1967], *Corophium* [Fish & Mills 1979]), but in the laboratory, under constant conditions, females of most species continue to molt and reproduce and males continue to inseminate females (personal observation of ten species; Kaestner 1970). This is true even for *Jassa marmorata*, whose males exhibit terminal ecdysis (Borowsky 1985a).

Two environmental factors that have been shown to affect reproduction in specific amphipods are temperature and day length. Intermolt periods and brood development times vary inversely with temperature (Steele & Steele 1973; Borowsky 1980a; Moore & Francis 1986).

The juveniles of all gammarideans that have been examined to date are released in the warmer months. However, the time of oviposition varies with latitude and with species. Some species oviposit in the fall and release their young in the spring; others oviposit in the spring and release their broods within a few weeks (Steele & Steele 1975).

The time of oviposition is a species-specific response to day length. Long days initiate reproduction and short days terminate it in spring and summer ovipositors (*Hyalella azteca* [DeMarch 1977], *Talitrus saltator* [Williams 1985], *Gammarus lawrencianus* [Steele & Steele 1986]). In contrast, *short* days initiate reproduction in fall ovipositors (*Gammarus setosus* [Steele & Steele 1986] and *Pontoporeia affinis* [Segerstrale 1970]). Thus the same environmental stimulus triggers opposite physiological phenomena.

Once sexual maturity is attained, the female molt, ovulatory, and reproductive behavioral cycles are linked. Since male reproductive behaviors vary with the female's intermolt, male behaviors are linked to the female cycles as well. Thus female hormones drive the reproductive activities of both sexes.

During the reproductive season ecdysterone probably acts as the "master control" hormone, which coordinates the female molt, ovarian, and behavioral cycles. The effects of ecdysterone on the molt cycle are well documented (i.e., Skinner 1985). Ecdysterone affects the ovarian cycle in amphipods in at least two ways: (1) by directly controlling folliculogenesis (Charniaux-Cotton 1975) and (2) by stimulating the secretion of Vitellogenin Stimulating Hormone, which controls vitellogenin synthesis (Junera, Martin & Meusy 1977). Finally, it affects reproductive behaviors. *Microdeutopus gryllotalpa* females injected with ecdysterone not only molt earlier but express reproductive behaviors earlier in their intermolt cycles than do the controls (Borowsky 1980b).

PATTERNS OF REPRODUCTION IN *GAMMARUS PALUSTRIS* AND *MICRODEUTOPUS GRYLLOTALPA*

Gammarus palustris

This species is ubiquitous in salt marshes along the Northwestern Atlantic coast. During the warmer months it is found on *Spartina alterniflora* culms. At low tide these animals hide in the small spaces between the roots of the plant. Both sexes apparently travel freely when their habitat is submerged. The intertidal habitat can be extremely harsh, exposed to large variations in temperature and salinity. In addition, predation may be heavy, since shore birds have free access to the culms at low tide and shallow water fish have access during periods of submersion. Thus *Gammarus palustris*'s pattern of reproduction must allow for protection against environmental stresses while satisfying the general constraints of amphipod reproductive physiology.

Gammarus palustris's pattern of reproduction is as follows: toward the end of its intermolt period, amplexus is initiated. The animals' orientation toward each other, the specific gnathopods (claws; pereiopods 1 and 2) that the male uses for attachment, and the points of attachment on the female are highly stereotyped (Borowsky 1984a). Males carry the relatively passive females about until they molt. The male releases the female briefly to permit her to shed the exuvium, then regrasps her to permit copulation. During copulation, the male holds the female perpendicularly. Intromission generally occurs two or three times, and then the male and female separate. The female ovulates and oviposits a few minutes after copulation.

Microdeutopus gryllotalpa

The reproductive behavior of *Microdeutopus gryllotalpa* differs markedly from that of *Gammarus palustris* because individuals of all ages and both sexes build residential tubes. The tubes are constructed of local debris, which is glued together with sticky, threadlike secretions exuded from the tips of the animals' first and second walking legs (pereiopods 3 and 4). The tubes function as protection against exposure (Borowsky 1989) and also serve as storage depots for food. If a food item is too large to be consumed all at once, it is glued to the tube and eaten later (personal observation). Thus *M. gryllotalpa* individuals have considerable energetic investment in their tubes and most of their activities, including reproduction, occur inside them.

Each animal lives alone in its tube and defends it against conspecifics. Immediately prior to a female's molt, this behavior changes and the female permits a male to enter and share the tube. Tube sharing continues until the female molts, when copulation occurs (Borowsky 1980c). Copulation behavior is the same in this species as in *Gammarus palustris*. After copulation the male leaves the female and travels from tube to tube until it encounters the tube of another receptive female (Borowsky 1983a).

Females may abandon their tubes and swim to a new location in search of a better habitat if the quality of the local environment declines or if the sediment is disturbed (DeWitt 1985). But they generally remain in place, constantly rebuilding and perfecting their tubes. Thus the burden of mate location falls on the males.

STIMULI FOR THE REPRODUCTIVE BEHAVIORS OF *GAMMARUS PALUSTRIS* AND *MICRODEUTOPUS GRYLLOTALPA*

Overview

Investigations into the nature of stimuli eliciting male behaviors analyzed each reproductive stage separately. By definition, Stage I stimuli must function before individuals contact each other. Thus the existence of waterborne pheromones was investigated.

In addition to waterborne pheromones, there are three other broad categories of potential stimuli for Stage II–IV behaviors, since these are expressed after contact: female morphology, contact chemicals, and female behavior. The overall plan of research was to test the effects of each of these potential stimuli on each behavioral stage in *Gammarus palustris* and *Microdeutopus gryllotalpa*. The results of these studies demonstrate that many stimuli are involved. These include waterborne pheromones, contact pheromones, and female behaviors.

Gammarus palustris

Pheromones for Mate Location. Mate location in this species involves both sexes traveling toward each other. To determine whether waterborne pheromones exist, a two-choice test apparatus based on a Y-tube was employed (Borowsky 1985b). Water was slowly dripped into each of two bottles containing different types of source animals whose secretions were being tested. Test-animals (those whose responses to source-animals were scored) were placed in another bottle. Effluents from the source-animals flowed into the two arms of a Y-tube, united at the fork, and entered through the single stem into the bottle containing test-animals. Test-animals could either remain in their bottles or travel to the bottles of either type of source-animal. The apparatus mimicked field conditions to allow animals to express their preferences as normally as possible. Since animals tend to seek shelter in small, hidden spaces, they readily entered the narrowed openings of the tubes leading from the test bottles in response to appropriate effluents (see Borowsky 1985b for diagram and details).

When tests were conducted in subdued light, animals moved actively inside the apparatus but did not exhibit preferences for different types of conspecific effluents. However, when tests were conducted in the dark, marked preferences were expressed. This supported casual field observations that suggested that animals are nocturnal.

All subsequent tests were conducted in the dark. When males were offered the choice of conspecific effluents versus plain seawater, they traveled significantly more often to the former than to the latter. They not only preferred effluents of receptive females over plain seawater, but they also preferred effluents of nonreceptive females and males over seawater. In contrast, they showed no greater response to effluents from *Gammarus mucronatus*, a closely related species, than to plain seawater.

When females were tested in a similar way, they also exhibited preferences for conspecific effluents over seawater and exhibited no preference for *Gammarus mucronatus* effluents. These observations suggested that *G. palustris* produces a species-specific pheromone to which both males and females are attracted.

In subsequent experiments, males were presented with effluents from receptive versus nonreceptive females and then from nonreceptive females versus males. Males showed a marked preference for receptive female effluents in the first tests but exhibited no preferences when presented with the choice of nonreceptive females versus males. When females' responses were tested under similar experimental conditions, it was found that females preferred males over other types of conspecifics. Nonreceptive females were employed in these tests, because receptive females are typically carried about by males and remain relatively quiescent when separated from them. Thus sex-specific attractants also exist in this species.

Stimuli for Pair Formation and Copulation. Both sexes exhibit distinctive behaviors after contact. When a test-male contacts another individual (the object), the two can separate immediately or the male can "grab" the object-animal with its claws. After grabbing, the test-male either "palpates" the object animal (the test-male passes its antennae and limbs over much of the exoskeleton of the object-animal) or drops the object-animal. During palpation, males may "brush" their pleopods rapidly against the female's body. After palpation the test-male may initiate amplexus with the object-animal, copulate with it, or drop it.

Females also behave differently when approached by a male. Their responses depend upon their own intermolt stage and are aimed at either rejecting the initiation of amplexus—at the beginning of their intermolt stage—or enhancing the occurrence of the reproductive behavior appropriate for their current intermolt stage (Borowsky & Borowsky 1987).

To test the effects of female effluents, female body (morphology and chemistry), and female behaviors separately, different types of object-animals were freeze-fixed and then presented to test-males either in water that contained female effluents or in plain seawater. The four different types of object-animals were: receptive females, nonreceptive females, newly molted females, and males. Freeze fixing eliminated behaviors while leaving the object-animals' bodies intact. Freeze fixing involves removal of an animal from its dish, blotting the excess water with a paper towel, then placing the animal on a piece of aluminum foil that is resting on a piece of dry ice. The animals are frozen within seconds and, when replaced in water, defrost within seconds.

The response of test-males to object-animals was the same in female effluent waters and in plain seawater. What mattered was the nature of the object-animal presented to the male. To determine whether the stimulus on the body was morphological or chemical, object animals that were freeze-fixed were held at 0°C for 24 hrs. Under these conditions, the female exoskeleton remains unchanged, but substances on the exoskeleton may denature. Test-males contacted and palpated these object-animals, but amplexus and copulation were almost entirely eliminated. Thus the evidence suggests that contact pheromones are the principal stimuli for amplexus and copulation and that female behaviors enhance the expression of appropriate male behaviors.

Microdeutopus gryllotalpa

Pheromones for Mate Location. An apparatus similar in design, but with smaller components, was employed to test for the existence of waterborne pheromones in *Microdeutopus gryllotalpa*. It was shown that a waterborne pheromone is secreted by receptive females and that it attracts males (Borowsky 1984b). In contrast to *Gammarus palustris*, males responded in the presence of indirect light. This suggests that males may travel to females even during the day in the field.

Also in contrast to *Gammarus palustris*, no evidence for the existence of a species-specific pheromone has been found. Males did not distinguish between male effluents and plain seawater or between nonreceptive female effluents and plain seawater (Borowsky 1984b). Female responses could not be tested because they rarely entered the arms of the Y-tube. Instead, females tended to remain fixed to a particular site in the apparatus and some of them even began to build residential tubes.

Stimuli for Pair Formation and Copulation: Overview. In addition to the three broad classes of potential female stimuli that could influence male *Gammarus palustris* behaviors—female effluents, female body, female behavior—a fourth potential stimulus, tubes, had to be considered for *Microdeutopus gryllotalpa*. Data presented below shed light on the importance of each of these on stimulating pair formation and the importance of the female's body on stimulating copulation.

Pair formation in *Microdeutopus gryllotalpa* is accompanied by a stereotyped behavioral sequence called Intermittent Pleopod Beats (IPBs; Borowsky 1983a). IPBs are only expressed by males at the initiation of tube sharing and involve a rapid fanning movement of the pleopods a few times, followed by a rest of a second or two. This may be repeated 4–5 times.

IPBs are only expressed in the presence of tubes. Females in the absence of tubes failed to elicit the behavior (Borowsky 1984b). This observation suggested the possibility that females might mark their tubes either chemically and/or structurally to indicate the state of their receptivity. However, detailed investigations have revealed no evidence for any kind of marking. A tube must be present and encounters between the male and female must occur when both are at least partially inside it, but the nature of the tube does not affect the expression of IPBs. IPBs are expressed about as often to receptive females inside their own tubes as to receptive females inside the tubes of other conspecifics or even inside the tubes of other amphipods (Borowsky 1989).

Stimuli for Pair Formation and Copulation: Preliminary Data. Three types of object-animals were tested: receptive females, nonreceptive females, and males. First, each object-animal was permitted to build its tube in an individual dish. The effects of female effluents on test-male IPBs were determined by removing animals from their dishes, exchanging the ambient seawater with fresh seawater or with water that had been in other receptive females' dishes, then replacing the animals in their tubes and observing test-males' responses. There was a significant difference in the expression of IPBs under the six different conditions (table 3.1: $\chi_5^2 = 71.92$, $p < 0.001$). Although there was a difference in expression to different types of object animals (0 of 40, 20 of 80, and 40 of 50 test-males expressed IPBs to males, nonreceptive females, and receptive females, respectively: $\chi_2^2 = 69.29$, $p < 0.001$), there was no difference in expression to object-animals in the two types of waters ($\chi_1^2 = 0.005$, $p > 0.05$). Thus ambient receptive female effluents had no detectable influence on male IPBs.

To test the effects of eliminating behavior and to test the effects of the state of the female's body, another experiment was performed. Three types of object-animals were freeze-fixed and then replaced immediately inside their tubes. During test-male encounters with object-animals, the object-animals were either held in place or moved back and forth gently inside their tubes. There was a significant difference in test-males' expression of IPBs among the six treatment classes ($\chi_5^2 = 53.80$, $p < 0.001$, table 3.2). The difference was due to whether or not the animals were moved (94 of 260 versus 40 of 160 test males expressed IPBs

TABLE 3.1. Effects of female effluents on male IPBs in *Microdeutopus gryllotalpa* (object animals untreated)

	In Seawater				In Female Effluent Water			
Object Animal	IPBs	No IPBs	N	(% IPBs)	IPBs	No IPBs	N	(% IPBs)
Male	0	20	20	(0)	0	20	20	(0)
Nonreceptive female	17	43	60	(28)	3	17	20	(15)
Receptive female	18	2	20	(90)	22	8	30	(73)

N = number of tests

TABLE 3.2. Effects of animal bodies and movements on IPBs in *Microdeutopus gryllotalpa* (object animals freeze-fixed).

Object Animal	Animals Moved				Animals Not Moved			
	IPBs	No IPBs	N	(% IPBs)	IPBs	No IPBs	N	(% IPBs)
Male	3	57	60	(5)	2	18	20	(10)
Nonreceptive female	26	54	80	(33)	4	16	20	(20)
Receptive female	65	55	120	(54)	34	86	120	(28)

N = number of tests

TABLE 3.3. Effects of animal bodies on copulations in *Microdeutopus gryllotalpa* (number of copulations of 40 tests).

Time past female molt	Untreated Controls	Freeze-Fixed
3 hours	14	9
2 days	0	0
6 days	0	1

when object animals were moved versus not moved, respectively; $\chi_1^2 = 5.17$, $p < 0.05$) and also due to the nature of the object-animal (5 of 80, 30 of 100, and 99 of 240 males expressed IPBs to males, nonreceptive females, and receptive females, respectively; $\chi_2^2 = 34.05$, $p < 0.001$).

Thus neither ambient receptive female effluents nor their tubes have a detectable effect on male IPBs. In contrast, receptive female bodies enhance the expression of IPBs, especially when they are moved (Table 3.2).

The effects of the females' bodies on stimulating copulation was examined in a similar way. Isolated females that had molted within the previous three hours were freeze-fixed and replaced in their tubes, after which test-males were introduced (one per dish). Two- and six-day postmolt females were also tested. The frequency of copulations was significantly different among these classes ($\chi_2^2 = 50.345$, $p < 0.001$, 2 and 6 days combined, table 3.3). Three-hour postmolt females copulated more often than all the other classes combined ($\chi_1^2 = 44.07$, $p < 0.001$). Males copulated with about the same number of freeze-fixed females as controls (10 or 120 versus 14 of 120; $\chi_1^2 = 0.417$, $p > 0.05$).

DISCUSSION

Reproductive Behaviors

Stage I, mate location, involves the movement of both sexes in *Gammarus palustris* but only the movement of males in *Microdeutopus gryllotalpa* (table 3.4). Stage II (pair formation) behaviors are preceded by contact in both species, but after contact the behaviors expressed by the two species are quite different.

TABLE 3.4. Reproductive behaviors of male *Gammarus palustris* (G.P.) and *Microdeutopus gryllotalpa* (M.G.).

	Behavior		Stimulus	
Stage	G.P.	M.G.	G.P.	M.G.
I. Mate location	Both sexes travel	Only males travel	Male and female pheromones	Receptive female pheromone
II. Pair formation	Contact	Contact	Visual	?
	Grab	—	?	—
	Palpation	—	?	—
	Brushing	IPBs	?	?
	Amplexus begun	Tube sharing begun	Contact pheromone likely + female behavior	Contact pheromone likely + female behavior
III. Pairing	Amplexus	Tube sharing	?	?
IV. Copulation			Contact pheromone likely + female behavior	Contact pheromone likely

Palpation and manipulation of the female's body does not occur in *M. gryllotalpa*. Instead, the male touches the female, expresses IPBs, then turns around inside the tube so its antennae extend outside an opening.

However, the description of "beating" by male *Gammarus duebeni* (Hartnoll & Smith 1980) is similar to "brushing" in *G. palustris* and to IPBs in *Microdeutopus gryllotalpa*. Further, rapid pleopod movements have been observed in some isopods during pair formation (personal observation) and in the lobster *Homarus americanus* (J. Atema personal communication). Pleopod movements enhance the flow of female pheromone over the males' chemoreceptors in the lobster, and the possibility that "brushing," "beating" and/or "IPBs" have the same function in peracarids warrants investigation.

While the Stage III pairing behaviors, amplexus and tube sharing, are different in the two species, both behaviors ensure the proximity of the male and female at the time of the female's molt, perhaps the most important requirement for maximizing the number of offspring.

Stage IV, copulation, is virtually identical in both species and in six other species (personal observation).

Amplexus is common in amphipods. In addition to *Gammarus*, some other genera that exhibit amplexus are: *Hyale, Melita* (Borowsky 1984a), *Hyalella* (Strong 1973), and *Eulimnogammarus (Marinogammarus)* (Sheader & Chia 1970). The common feature of these genera is that both males and females are free roaming. In the semiterrestrial species, amplexus is confined to the period between the female's molt and copulation (five species; Williamson 1953). These are free roaming species, but carrying a female may be more difficult in the absence of the buoyancy provided by water, so the time during which it occurs is abbreviated.

In *Microdeutopus gryllotalpa*, pairing involves tube sharing. This behavior has been found in *Lembos websteri* (Shillaker & Moore 1987), *Jassa (falcata) marmorata* and *Amphithoe valida* (Borowsky 1983b) and is consistent with observations on *Corophium volutator* and *C. arenarium* (Fish & Mills 1979) and *C. insidosum* (Nair & Anger 1979). The common features of these species are that the animals build residential tubes, females produce several broods in succession, and the females' reproduction is unsynchronized.

Other amphipods exhibit different types of pairing behaviors. For example, *Ampelisca abdita* and *A. vadorum* are tube-building species that do not have cruising males. Instead, males encounter females when both abandon their tubes and enter the water column in response to environmental stimuli (Mills 1967). Other species with similar behavior are *Neohaustorius biarticulatus* and *Haustorius canadensis* (Sameoto 1969) and *Bathyporeia pilosa* and *B. pelagica* (Fish 1975). In these species both sexes respond to the same stimulus, so males need not travel between females' tubes to seek successive mating.

Stimuli for Reproductive Behaviors

Pheromones from receptive females have been shown to influence male behaviors in many crustaceans. In some cases, the effects are dramatic: vigorous, stereotyped reproductive behaviors are expressed in the presence of receptive female pheromones in the crabs *Portunus sanguinolentus* (Ryan 1966) and *Callinectes sapidus* (Gleeson 1980). In other cases, however, the behaviors are not expressed only in a reproductive context. For example, female effluents increase the general level of activity and/or orient males' movements in the crabs *Carcinus maenas* (Seifert 1982) and *P. sanguinolentus* (Christofferson 1978) and in the shrimp *Heptacarpus paludicola* (Bauer 1979). In *Microdeutopus gryllotalpa* and *Gammarus palustris*, males travel toward receptive female effluents. But in these species all subsequent sex-specific behaviors are unaffected by receptive female effluents (table 3.4).

There had been some controversy about whether waterborne effluents should be considered pheromones if they do not elicit sex-specific behaviors. Dunham (1978, 1988) provided rigorous and logical guidelines for interpreting studies on pheromones and concluded that a chemical that induced behaviors leading, either directly or indirectly, to reproductive success should be considered a pheromone.

The receptive female pheromones of *Gammarus palustris* and *Microdeutopus gryllotalpa* probably enhance mate location in the field. Male and female *G. palustris* travel toward effluents of any conspecific, but given the opportunity, individuals travel toward effluents of the opposite sex. This makes sense in the field. Animals' responses to species-specific and to sex-specific pheromones lead to aggregations at the preferred habitat. Thus both types of pheromones enhance habitat location. Once there, the sex pheromones may assist in locating appropriate partners. In contrast, male *M. gryllotalpa* respond only to female effluents. A species-specific pheromone may not exist in this species. While a species-specific pheromone is of value to a species in which both sexes are mobile, such a pheromone may be of little value to a species in which only the males are free roaming.

There could be as many as three different waterborne pheromones in *Gammarus palustris:* a species-specific, a male, and a receptive female pheromone. A more conservative model is that only two pheromones exist. One is produced continuously by both sexes and serves as the species-specific attractant (SP). Possibly males produce more SP than females. The other is produced only by females (FP), and its titer increases as the female intermolt advances. Both sexes respond to SP, but only males respond to FP. Thus females would be attracted by SP but more to males than to other females. Males would respond to SP in the same way, but superimposed on their responses to SP would be those to FP. In the early intermolt stages no preference would be expressed, but as the female's intermolt advanced and the FP titer increased, the males' response to FP titer would be greater than their response to SP. This hypothesis remains to be tested.

Female behaviors play a role in pair formation and copulation in *Gammarus palustris* and *Microdeutopus gryllotalpa*. In *G. palustris* females curl up and rest quietly, allowing the males to attach themselves, and in *M. gryllotalpa* the females turn away from the entering male, allowing the male to remain in the tube (table 3.4). But the principal stimuli for male reproductive behaviors are factor(s) associated with the female's body. Although female gross morphology and exoskeletal structure remains the same throughout the intermolt, male responses to it change. This suggests that the stimulus, while associated with the exoskeleton, is labile and probably is not a morphological feature. Thus the weight of the evidence suggests that the factor is a contact pheromone.

Another explanation for these observations is that the stimulus is a diffusible pheromone that acts in combination with the physical presence of the female. In other words, it may be that the pheromone is not firmly fixed to the exoskeleton and diffuses into the water immediately surrounding the exoskeleton. The observations reported here cannot distinguish between a "contact pheromone" versus a "contact plus pheromone" interpretation of the data. The extensive touching that occurs during *Gammarus palustris* palpation tends to support the contact pheromone hypothesis, but both hypotheses should be tested in detail.

Contact precedes many reproductive displays in crustaceans (Salmon 1983), and contact chemoreception has been suggested for the shrimp *Heptacarpus pictus* (Bauer 1976) and several other carideans (summarized in Bauer 1976). Contact pheromones have been demonstrated in several insects (summarized in Stadler 1984). Seelinger and Schuderer (1985) not only demonstrated the existence of contact pheromones in the cockroach *Periplaneta americana;* they also succeeded in transferring the substance from receptive to nonreceptive females, thereby making the latter capable of eliciting courtship displays from males. These investigators concluded that the airborne pheromones served as male attractants and sexual motivators and that mate recognition was accomplished via the contact pheromones. The experiments conducted on *Gammarus palustris* and *Microdeutopus gryllotalpa* suggest that this may be true in those species as well.

Site of Secretion of the Waterborne Pheromone

The site of secretion of amphipod waterborne pheromones remains to be determined. However, pheromones of other Crustacea are secreted in the urine (e.g., the shore crab *Carcinus maenas* [Eales 1974], the crab *Portunus sanguinolentus* [Christofferson 1978], the blue crab *Callinectes sapidus* [Gleeson 1980], and the lobster *Homarus americanus* [McLeese et al. 1977]). A reasonable working hypothesis is that amphipods secrete pheromone in their urine also.

Site of Secretion of the Contact Pheromone

Upon contacting an object-animal, males pass their antennae and limbs all over the female's exoskeleton but do not seem to prefer a particular area. Thus I suggested that the contact pheromone is secreted by pore canals, which are distributed all over the female exoskeleton (pore canals described by Halcrow 1978). Halcrow and Bousfield (1987) disagreed, however, because pore canals are present in both sexes. While it certainly has not been established that pore canals are the site of the contact pheromone, these authors' objection is not compelling. On the same grounds it might have been objected that urine cannot contain sex-specific pheromones. Pore canals might secrete the same substances in both sexes (such as antifoulants [Halcrow 1978]) but pheromone only in females.

Dunham (1986) suggested that the female brood pouch might contain the stimulus for amplexus. This was an interesting suggestion, because the brood develops as the female intermolt period advances. Therefore, a conservative method for regulating male reproductive behaviors would be to have them respond directly to the brood. Dunham found that female *Gammarus lawrencianus* without broods were more attractive than females with broods at the same intermolt stage. To test this idea more fully, I exchanged eggs of different developmental stages between females in *G. palustris*. Males expressed reproductive behaviors most often to females in the later intermolt stages regardless of the stage of development of the broods inside their pouches. They did not amplex with females with empty brood pouches if they were in the early intermolt stages. Thus the stage of development of the brood had no detectable effect on male responses (Borowsky 1987).

Site of Perception of the Waterborne Pheromone

Studies on *Gammarus duebeni* (Dahl, Emanuelsson & von Mecklenburg 1970a,b) demonstrated that substances secreted into the water by females accumulate in males and are particularly concentrated on the males' second antennae. Females were fed H^3-labeled food and then placed in dishes with males, but the sexes were kept apart by a double mesh net. Scintillation and autoradiographic techniques demonstrated that males accumulated label. The label was about 1,000 times more concentrated in the second antennae than in the rest of the males' bodies. In addition, examination of autoradiographs showed that the label was particularly concentrated on the calceoli, a structure found only in males' second antennae in this species. These are bulblike structures, about 100 μm in diameter and with complex microstructures, that are attached by a thin stalk to the distal portion of the second antennae of male *G. duebeni* (Lincoln 1985). Since this is a sexually dimorphic characteristic and since it was assumed that males would travel to females, this was taken as indirect support for the existence of waterborne pheromones in this species. However, a contrary conclusion comes from Lincoln and Hurley (1981), who examined the microstructure of calceoli and stated that they are not chemosensors, but that they perceive some form of sound, vibration, or pressure waves.

Studies of most other crustaceans suggest that chemosensory receptors include hairlike aesthetascs distributed singly or in groups (summarized in Ache 1982). Lowry (1986) described an elaborate antennal structure present in some amphipods (and in other eucarideans) that has associated aesthetascs and that the author suggests has a chemosensory function. Dahl later (1973a,b) described hairlike structures on amphipods to which he ascribed a chemosensory function. Thus the weight of all observations suggests that the calceoli are not chemosensory. Lincoln and Hurley (1981) suggested that the autoradiographs were misleading and that, in fact, the label was concentrated on adjacent aesthetascs.

While there is no consensus on the exact site of the label, Dahl's published autoradiographs clearly show that it is concentrated on the second antennae. Further, when Ducruet (1973) ablated these antennae the males' responses to conspecifics' effluents was eliminated. Thus the chemosensors for waterborne pheromones may be on the second antennae in amphipods.

CONCLUSIONS

As noted in the opening of this paper, the nature of the stimuli for reproductive behaviors in amphipods has been the subject of some controversy. The most likely reason for this is that amphipod behaviors and stimuli are not simple: (1) there are two basic types of pheromones, waterborne and contact; (2) there may be two or more types of waterborne pheromones in some species; (3) the same type of pheromone may elicit different behaviors in different species; and (4) other, nonchemical stimuli are also important in eliciting reproductive behaviors.

The differences in the way chemical cues are employed in different, even closely related amphipods is determined, in great part, by ecology. The most effective way to explore the role of chemical stimuli in reproductive behaviors is to begin with a good understanding of the reproductive biology and the typical habitat of the species in question. Only then can experimental methods that mimic the species' habitat be devised to permit the animals to respond in as normal a way as possible when exposed to putative pheromones.

A great deal more remains to be learned about amphipod behaviors and their stimuli. First, the details of the behavior of most species have not been described. Second, the waterborne pheromone(s) remain to be characterized chemically. At present, the only information that exists is that the attractant of *Microdeutopus gryllotalpa* is a nonpolar substance (Borowsky, Augelli & Wilson 1987). Third, the contact pheromone hypothesis must be tested by chemical methods. Fourth, the stimuli that maintain pairing (Stage III) must be described. It should not be automatically assumed that they are the same as those for pair formation.

Some of the most important discoveries about crustacean reproductive physiology were learned from investigations on an amphipod (by Madame Charniaux-Cotton and her colleagues). It is possible that the use of amphipods as models to study crustacean reproductive behavior and its stimuli may prove equally fruitful.

Acknowledgments

I wish to thank R. Borowsky and D. Franz for their critical reading of the manuscript and for their invaluable suggestions.

Literature Cited

Ache, B. 1982. Chemoreception and thermoreception. In D. Bliss & L. Mantel, eds., *The Biology of the Crustacea*, 3:365–735. New York: Academic Press.
Bauer, R. T. 1976. Mating behaviour and spermatophore transfer in the shrimp *Heptacarpus pictus* (Stimpson) (Decapoda: Caridea: Hippolytidae). *J. Nat. Hist.* 10:415–440.
Bauer, R. T. 1979. Sex attraction and recognition in the caridean shrimp *Heptacarpus paludicola* (Decapoda: Hippolytidae). *Mar. Behav. Physiol.* 6:157–174.
Borowsky, B. 1980a. Reproductive patterns of three intertidal salt-marsh gammaridean amphipods. *Mar. Biol. (Berl.)* 55:327–334.
Borowsky, B. 1980b. The physiological control of reproduction in *Microdeutopus gryllotalpa* (Crustacea: Amphipoda). I. The effects of exogenous ecdysterone on the females' molt and behavioral cycles. *J. Exp. Zool.* 213:399–403.
Borowsky, B. 1980c. The pattern of tube-sharing in *Microdeutopus gryllotalpa* (Crustacea: Amphipoda). *Anim. Behav.* 28:790–794.
Borowsky, B. 1983a. Behaviors associated with tube-sharing in *Microdeutopus gryllotalpa* (Costa) (Crustacea: Amphipoda). *J. Exp. Mar. Biol. Ecol.* 68:39–51.
Borowsky, B. 1983b. Reproductive behavior of three tube-building peracarid crustaceans: The amphipods *Jassa falcata* and *Amphithoe valida* and the tanaid *Tanais cavolinii*. *Mar. Biol. (Berl).* 77:257–263.
Borowsky, B. 1984a. The use of the males' gnathopods during precopulation in some gammarid amphipods. *Crustaceana (Leiden)* 47:245–250.
Borowsky, B. 1984b. Effects of receptive females' secretions on some male reproductive behaviors in the amphipod crustacean *Microdeutopus gryllotalpa*. *Mar. Biol. (Berl.)* 84:183–187.
Borowsky, B. 1985a. Differences in reproductive behavior between two male morphs of the amphipod crustacean *Jassa falcata* Montagu. *Physio. Zool.* 58:497–502.
Borowsky, B. 1985b. The responses of the amphipod crustacean *Gammarus palustris* to conspecifics' and congenerics' secretions. *J. Chem. Ecol.* 11:1545–1552.
Borowsky, B. 1987. The effects of embryos of different developmental stages on reproductive behavior and physiology in brooding females of the amphipod crustacean *Gammarus palustris*. *Biol. Bull. (Woods Hole)* 172:155–160.
Borowsky, B. 1988. Delaying copulation in the amphipod crustacean *Gammarus palustris*: Effects on female fecundity and consequences for the frequency of amplexus. *Mar. Behav. Physiol.* 13:359–368.
Borowsky, B. 1989. The effects of residential tubes on reproduction in *Microdeutopus gryllotalpa* Costa. *J. Exp. Mar. Biol. Ecol.* 128:117–126.
Borowsky, B. & C. E. Augelli & S. R. Wilson. 1987. Towards chemical characterization of waterborne pheromone of amphipod crustacean *Microdeutopus gryllotalpa*, *J. Chem. Ecol.* 13:1673–1680.
Borowsky, B. & R. Borowsky, 1987. The reproductive behaviors of the amphipod crustacean

Gammarus palustris and some insights into the nature of their stimuli. *J. Exp. Mar. Biol. Ecol.* 107:131–144.

Bousfield, E. L. 1973. *Shallow-Water Gammaridean Amphipods of New England.* Ithaca; Cornell University Press.

Charniaux-Cotton, H. 1975. L'ovogenese et sa regulation chez les Crustaces superieurs. *Ann. Biol. Anim. Biochem. Biophys.* 15:715–724.

Christofferson, J. P. 1978. Evidence for the controlled release of a crustacean sex pheromone. *J. Chem. Ecol.* 4:633–639.

Dahl, E. 1973a. Antennal sensory hairs in talitrid amphipods. *Acta Zool. (Stockh.)* 54:161–171.

Dahl, E. 1973b. Presumed chemosensory hairs in talitrid amphipods (Crustacea). *Entomol. Scand.* 4:171–180.

Dahl, E., H. Emanuelsson & C. von Mecklenburg. 1970a. Pheromone reception in the male of the amphipod *Gammarus duebeni* Lilljeborg. *Oikos* 21:42–47.

Dahl, E., H. Emanuelsson & C. von Mecklenburg. 1970b. Pheromone transport and reception in an amphipod. *Science (Washington, D.C.)* 170:739–740.

DeMarch, B. G. E. 1977. Effects of photoperiod and temperature on induction and termination of the reproductive resting stage in the freshwater amphipod *Hyalella azteca. Can. J. Zool.* 55:1595–1600.

DeWitt, T. 1985. The Behavior and Ecology of Migration and Colonization in the Epibenthic, Tubicolous Amphipod *Microdeutopus gryllotalpa.* Ph.D. dissertation, State University of New York at Stony Brook.

Ducruet, J. 1973. Interattraction chez les gammares du groupe pulex (Crustaces Amphipodes). *Rev. Comp. Animal* 7:313–322.

Dunham, P. 1978. Sex pheromones in crustacea. *Biol. Rev. Camb. Philos. Soc.* 53:555–583.

Dunham, P. 1986. Mate-guarding in amphipods: A role for brood pouch stimuli. *Biol. Bull. (Woods Hole)* 170:526–531.

Dunham, P. 1988. Pheromones and behavior in Crustacea. In H. Lauffer & R. Downer, eds., *Endocrinology of Selected Invertebrate Types,* pp. 375–392. New York: Alan R. Liss.

Eales, A. J. 1974. Sex pheromone in the shore crab *Carcinus maenas* and the site of its release from females. *Mar. Behav. Physiol.* 2:345–355.

Fish, J. D. 1975. Development, hatching, and brood size in *Bathyporeia pilosa* and *Bathyporeia pelagica. J. Mar. Biol. Assoc. U. K.* 55:357–368.

Fish, J. D. & A. Mills. 1979. The reproductive biology of *Corophium volutator* and *Corophium arenarium* (Crustacea: Amphipoda). *J. Mar. Biol. Assoc. U. K.* 59:355–368.

Gleeson, R. A. 1980. Pheromone communication in the reproductive behavior of the blue crab *Callinectes sapidus. Mar. Behav. Physiol.* 7:119–134.

Halcrow, K. 1978. Modified pore canals in the cuticle of *Gammarus* (Crustacea: Amphipoda); a study by scanning and transmission electron microscopy. *Cell Tissue. Res.* 10:267–276.

Halcrow, K. & E. Bousfield. 1987. Scanning electron microscopy of surface microstructures of some gammaridean amphipod crustaceans. *J. Crustacean Biol.* 7:274–287.

Hartnoll, R. G. & S. M. Smith. 1980. An experimental study of sex discrimination and pair formation in *Gammarus duebeni* (Amphipoda). *Crustaceana (Leiden)* 38:253–264.

Junera, H., C. Zerbib, M. Martin & J. J. Meusy 1977. Evidence for the control of vitellogenin synthesis by an ovarian hormone in *Orchestia gammarella* (Pallas), Crustacea: Amphipoda. *Gen. Comp. Endocrinol.* 31:457–462.

Kaestner, A. 1970. *Invertebrate Zoology, Crustacea.* New York: Wiley.

Lincoln, R. J. 1985. Morphology of a calceolus, an antennal receptor of gammaridean amphipoda (Crustacea). *J. Nat. Hist.* 19:921–927.

Lincoln, R. J. & D. E. Hurley. 1981. The calceolus, a sensory structure of gammaridean amphipods (Amphipoda: Gammaridea). *Bull. Br. Mus. (Nat. Hist.) Zool.* 40:103–116.

Lowry, J. K. 1986. The callynophore, a eucaridean/peracaridean sensory organ prevalent among the Amphipoda (Crustacea). *Zool. Scr.* 15:333–349.

McLeese, D. W., R. L. Spraggins, A. K. Bose & B. N. Pramanik. 1977. Chemical and behavioral studies of the sex attractant of the lobster *Homarus americanus. Mar. Behav. Physiol.* 4:219–232.

Mills, E. L. 1967. The biology of an ampeliscid amphipod crustacean sibling species pair. *J. Fish. Res. Board Can.* 24:305–355.

Moore, P. G. & C. H. Francis. 1986. Notes on breeding periodicty and sex ratio of *Orchestia gammarellus* (Pallas) (Crustacea: Amphipoda) at Millport, Scotland. *J. Exp. Mar. Biol. Ecol.* 95:203–209.

Nair, K. K. C. & K. Anger. 1979. Life cycle of *Corophium insidosum* (Crustacea: Amphipoda) in laboratory culture. *Helgol. Wiss. Meeresunters.* 32:279–294.

Ryan, E. P. 1966. Pheromone: Evidence in a decapod crustacean. *Science (Washington, D.C.)* 151:340–341.

Salmon, M. 1983. Courtship, mating systems, and sexual selection in decapods. In S. Rebach & D. W. Dunham, eds., *Studies in Adaptation, the Behavior of Higher Crustacea*, pp. 143–170. New York: Wiley.

Sameoto, D. D. 1969. Comparative ecology, life histories and behavior of interidal sand-burrowing amphipods (Crustacea: Haustoriidae) at Cape Cod. *J. Fish Res. Board Can.* 26:361–388.

Seelinger, G. & B. Schuderer. 1985. Release of male courtship display in *Periplaneta americana*: Evidence for female contact sex pheromones. *Anim. Behav.* 33:599–607.

Segerstrale, S. G. 1970. Light control of the reproductive cycle of *Pontoporeia affinis* Lindstrom (Crustacea: Amphipoda). *J. Exp. Mar. Biol. Ecol.* 5:272–275.

Seifert, P. 1982. Studies on the sex pheromone of the shore crab, *Carcinus maenas*, with special regard to ecdysterone secretion. *Ophelia* 21:147–158.

Sheader, M. & F. S. Chia. 1970. Development, fecundity and brooding behavior of the amphipod *Marinogammarus obtusatus*. *J. Mar. Biol. Assoc. U.K.* 50:1079–1099.

Shillaker, R. O. & P. G. Moore. 1987. Tube-emergence behavior in the amphipods *Lembos websteri* Bate and *Corophium bonnellii* Milne Edwards. *J. Exp. Mar. Biol. Ecol.* 111:231–241.

Skinner, D. M. 1985. Molting and regeneration. In D. Bliss & L. Mantel, eds., *The Biology of the Crustacea*. Vol. 9., Integument, Pigments, and Hormonal Processes, pp. 43–146. New York: Academic Press.

Stadler, E. 1984. Contact Chemoreception. In William J. Bell & Ring T. Carde, eds., *Chemical Ecology of Insects*, pp. 3–35. London: Chapman and Hall.

Steele, V. J. & D. H. Steele. 1973. The biology of *Gammarus* in the northwestern Atlantic. VII. The duration of embryonic development in five species at various temperatures. *Can. J. Zool.* 51:995–999.

Steele, V. J. & D. H. Steele. 1975. The biology of *Gammarus* in the northwestern Atlantic. XI. Comparison and discussion *Can. J. Zool.* 53:1116–1126.

Steele, V. J. & D. H. Steele. 1986. The influence of photoperiod on the timing of reproductive cycles in *Gammarus* species (Crustacea, Amphipoda). *Am. Zool.* 26:459–467.

Strong, D. R. 1973. Amphipod amplexus, the significance of ecotypic variation. *Ecology* 54:1383–1388.

Williams, J. A. 1985. The role of photoperiod in the initiation of breeding and brood development in the amphipod *Tallitrus saltator* Montagu. *J. Exp. Mar. Biol. Ecol.* 86:59–72.

Williamson, D. I. 1953. Mating behavior in the Talitridae (Amphipoda). *Br. J. Anim. Behav.* 1:83.

FOUR
Precopulatory Mate Guarding in Aquatic Crustacea: Gammarus lawrencianus as a Model System

PHILIP J. DUNHAM and ALAN M. HURSHMAN

*Department of Psychology,
Dalhousie University,
Halifax, Nova Scotia.*

Abstract

In many species of Crustacea, males and females pair together for a brief period prior to copulation. The mating behavior of the amphipod *Gammarus lawrencianus* is a prototypical example of this precopulatory mate guarding strategy. The present chapter discusses both the *proximate* and *ultimate* causal mechanisms proposed to explain the precopulatory mate guarding phenomena in Crustacea and reviews recent evidence concerned with these mechanisms in amphipods.

Choice of Model Organism

Gammarus lawrencianus is a euryhaline amphipod found in tidal waters of the small streams that flow into estuaries along the northeastern coast of North America (Steele & Steele 1970). During a breeding season that is most productive from March to October, the males and females pair briefly prior to each female molt, after which they copulate. Following sperm transfer, the female releases eggs into a specialized ventral brood pouch where they are fertilized and develop for a number of days until they are released as juveniles. This *precopulatory mate guarding strategy*, which functionally isolates the female from other males during the guarding period, has been observed in a number of different vertebrate and invertebrate groups, but it seems to be particularly widespread among aquatic Crustacea (Ridley 1983).

The mate guarding behavior of *Gammarus lawrencianus* is highly stereotyped. When a male comes into contact with an acceptable free swimming female a few days prior to her molt, he pulls her close to his ventral surface with their longitudinal axes at right angles to each other and actively palpates her with various appendages. After the period of palpation, the female is either released or rotated into a more efficient carrying position with the female underneath the male and their body axes parallel. The pair swims together in this precopula position until the female molt and copulation, after which they separate.

We have adopted *Gammarus lawrencianus* as a model for research on precopulatory mate guarding for several reasons. First, we agree with Wickler and Seibt (1981) in suggesting that Crustacea, in general, are a valuable group in which to analyze mating systems. Although this group has been largely ignored in classic accounts of animal social systems (e.g., Wilson 1975; Emlen & Oring 1977), their reproductive behaviors rival those of the more frequently studied social insects in diversity, and many closely related crustacean species have adapted their mating strategies to the widest possible range of habitats. Second, among the crustaceans using this reproductive strategy, the mate guarding behavior of *G. lawrencianus* (and gammarid amphipods in general) is prototypical. The pattern of contacting a female, palpating her briefly, and then physically carrying her until molt and copulation has been observed across a large number of different species (Dunham 1978; Ridley 1983; Salmon 1983). Although there are substantial variations on the theme (e.g., shelter sharing prior to copulation), a better understanding of this prototypical pattern should also facilitate research on the exceptions.

Gammarus lawrencianus is also easily cultured and maintained in the laboratory under seminatural conditions. Under optimal photoperiods and temperature, we observe generation times of 30 to 40 days and reproductive cycles that range between 9 to 12 days. Gravel bottom tanks with either flow-through or unfiltered aerated seawater are sufficient to maintain a very large population of animals at all stages of development. Individual animals or male-female pairs can also be kept for several weeks in relatively small containers where their reproductive cycles and precopulatory mate guarding behavior are easily monitored on a daily basis.

As the Danish physiologist August Krogh (1929) once suggested, "for a large number of problems there will be some animal of choice, or a few such animals, on which it can be most conveniently studied." When considered together, we would argue that the combination of substantive and practical characteristics described above makes *Gammarus lawrencianus* a particularly convenient and valuable model for studying precopulatory mate guarding.

Levels of Analysis

Attempts to understand precopulatory mate guarding fall into two categories based on the level of explanation sought. One level of analysis examines the phenomenon in terms of *ultimate causal mechanisms*. The goal is to elucidate the adaptive value and selective consequences of precopulatory mate guarding in evolutonary terms (e.g., Trivers 1972; Parker 1974; Manning 1980; Wickler & Seibt 1981; Ridley 1983; Grafen & Ridley 1983).

This evolutionary causal analysis can be contrasted with attempts to understand the *proximate causal mechanisms*. The goal at the proximate level is to identify the necessary and sufficient conditions under which precopulatory mate guarding behavior will occur and elucidate the physiological and behavioral mechanisms that mediate the process (e.g., Dunham 1978; Salmon 1983; Atema 1985).

These ultimate and proximate levels of analysis tend to generate very different kinds of research activity. The former usually make inferences retrospectively from comparative data while the latter are more susceptible to the logic of analytic, experimental manipulations (Dunham 1988b). Our own research on *Gammarus lawrencianus* has been primarily concerned with the proximate causal mechanisms that control mate guarding, although we recognize that the two levels of explanation are not mutually exclusive. Progress is always most satisfying when knowledge gained at one level interacts with theory and data at other levels, and some of the data to be discussed later will illustrate this kind of interaction.

Two Generic Assumptions

Several assumptions are frequently encountered in explanations of precopulatory mate guarding at both the ultimate and proximate levels of analysis. Our research has direct implications for at least two of these assumptions. First, at the proximate level, one of the most commonly encountered notions is the *female sex pheromone hypothesis*, which suggests that the female releases a chemical signal late in her reproductive cycle that attracts the male and elicits mate guarding behavior. Kittredge and Takahashi (1971), for example, suggested that primordial Crustacea externalized a receptor site and released the molting hormone into the external environment at about the same point in their evolution. The reproductive advantage of these two events in terms of finding a mate permitted them to be fixed in the genome with the molting hormone serving a dual function as a sex pheromone (cf. McLeese 1970; Atema & Gagosian 1973; Gagosian & Atema 1973; Hammoud, Compte & Ducruet 1975; Dunham 1979; Seifert 1982, Gleeson, Adams & Smith 1984).

Two corollary assumptions are implicit in most versions of the sex pheromone hypothesis. One is the *passive female behavior assumption*, which suggests that the female is a passive behavioral agent in the mate guarding sequence (i.e., she simply releases an attractant); the other is the *single stimulus assumption*, which suggests that all aspects of the mate guarding and copulation sequence are controlled by a single chemical stimulus. The reader is referred to Dunham

(1978, 1988a) for a more detailed review of the female sex pheromone hypothesis and data associated with it.

The second generic assumption is the *constant male threshold hypothesis*. This notion emerges most explicitly when mate guarding is discussed at the level of ultimate causes. For example, Parker (1974) and, more recently, Ridley (1983) both conceptualize precopulatory mate guarding as a time investment strategy that permits males to maximize their reproductive success when mating is restricted to a relatively brief period during the female's reproductive cycle. Given these time constraints, a randomly searching male encounters acceptable females at a very low rate and male intrasexual competition for acceptable females is high. Ridley (1983) argues that, under these conditions, sexual selection favors a mutant that is able to assess the female's reproductive status and is willing to invest time guarding her until copulation. Species with male biased sex ratios and/or very restricted opportunities to copulate are assumed to guard females for longer periods of time than species in which the female resource is more plentiful. As Grafen and Ridley (1983) indicated in their discussion of this time investment strategy, "the evolutionary variable of interest is the male's criterion, which is a point in the female's cycle, after which the male will guard the female." (p. 552). For a given species, an evolutionarily stable criterion is assured to evolve that maximizes the male's reproductive success (i.e., a criterion that cannot be bettered by any mutant male in the population using a different criterion).

The general purpose of the present chapter is to review the research relevant to these assumptions with an emphasis on the subgroup Amphipoda and with particular reference to our own studies of the amphipod *Gammarus lawrencianus*.

THE SEX PHEROMONE HYPOTHESIS: EVIDENCE IN AMPHIPODA

The evidence for a sex pheromone in the Amphipoda mirrors the inconsistency encountered in the more general literature on sex pheromones in Crustacea (see Dunham 1978, 1988a). Although males in several different amphipod species can make the required discrimination between females early and late in the female reproductive cycle (e.g., Birkhead & Clarkson 1980; Hartnoll & Smith 1980; Borowsky 1984; Hunte, Myers & Doyle 1985; Dunham, Alexander & Hurshman 1986; the evidence for a sex pheromone that attracts a mate from a distance and/ or on physical contact has not been conclusive.

Several early studies reported that the premolt female *Gammarus duebeni* releases a substance that attracts males from a distance (Dahl, Emanuelsson & von Mecklenburg 1970a,b). However, other research on two different gammarid species *(G. pulex* and *Gammarus fossarum)*, using different behavioral assays, found no evidence that males could be attracted from a distance (Ducruet 1973; Hammoud, Compte & Ducruet 1975). In a subsequent, very thorough series of experiments on *G. duebeni*, Hartnoll and Smith (1980) failed to replicate *any* of the previously reported positive results for a sex pheromone in amphipods using the same assays that had been employed in the earlier literature.

More recently, Borowsky (1984) demonstrated that males of the tubiculous amphipod *Microdeutopus gryllotalpa* will selectively choose the arm of an olfactometer containing water from premolt female conspecifics over an arm containing nonreceptive intermolt females. Control experiments indicated that the males did not exhibit a preference for water flowing from intermolt males over untreated seawater. In a second paper, Borowsky (1985) examined the preferences of the intertidal marine amphipod *Gammarus palustris* in a similar choice procedure. In this species, she reported that: (1) both males and females prefer water flowing from a conspecific (male or female) over untreated seawater; (2) males prefer water flowing from receptive premolt females over water from nonreceptive intermolt females; and (3) females prefer water flowing from males over water flowing from receptive females. Hence in this species there appeared to be a general attraction to water from conspecifics regardless of sex and a preference for water that increases the probability of encountering an "appropriate" mate (e.g., helps the male find a female near molt and helps the female find a male). It should also be noted that these tests were conducted during the dark phase of the daily light-dark cycle. Although this variable was not systematically investigated across all choice conditions, Borowsky suggested that testing during the dark phase may be a necessary condition to observe the preferences she reported.

It has been difficult to determine why evidence for the sex pheromone assumption is so inconsistent and contradictory within this group of Crustacea. Across different experiments, there are differences in the species studied, in the assays employed, in sampling procedures, and in the background laboratory conditions. Given this confusion, our approach has been to use several different behavioral assays with the same species *(Gammarus lawrencianus)* and the same background laboratory conditions to test the simple assumption that a premolt female releases secretions capable of influencing any aspect of the male's behavior from a distance. We assume that a better understanding of one prototypical species will help guide subsequent research of a more comparative nature. A brief digression to describe our general methods will facilitate discussion of the evidence we have obtained thus far.

RESEARCH METHODS

Breeding Cultures and Maintenance

Animals used in our experiments are randomly sampled from large breeding cultures maintained in tanks of aerated seawater. The amphipods are fed daily with Tetramin fish food, bits of mussel, and squash flakes. Algae *(Fucus)*, gravel, assorted shells, and larger rocks are provided to simulate the animals' natural environment in the breeding tanks.

When a specific experimental procedure requires information about the reproductive behavior and/or physiology of the males and females, precopula pairs are sampled from the breeding culture and maintained as a pair in individual 400 ml plastic cups with bits of *Fucus* for cover. Daily records of their reproductive status are maintained (e.g., free swimming, mate guarding, copulation, molt), and when necessary, examination of the ovaries and/or embryo development can be used to provide more precise information on the female's reproductive status (cf. Steele & Steele 1969). Maintaining pairs in isolated containers permits us to select males or females with a known reproductive history for an experimental procedure (e.g., comparison of a male's behavior 24 hrs postcopulation to a male 72 hrs postcopulation). A pair is usually maintained for one complete breeding cycle prior to participation in an experimental procedure.

The 3-3 Precopula Assay

In the 3-3 precopula behavioral assay, a male and female are isolated from each other for three minutes in two 50 ml cups of seawater. At the end of three minutes, the two cups of seawater containing the animals are emptied into a larger, opaque plastic, 6 cm diameter, circular test arena. The interactions between the sexes are recorded on videotape during a 3 min test period. Our experience with this assay over a wide range of conditions indicates that a male and an acceptable female will display precopulatory mate guarding within 3 mins. If not, a longer test seldom changes the outcome.

The videotape records of the behavioral interactions are coded by an observer trained to identify various behaviors but naive with reference to the hypothesis under investigation and the condition of the animals in the test. The observer makes multiple passes through the videotape and on each pass records specified responses by depressing an event recorder button for as long as the behavior continues. A microcomputer is used to integrate the input from the event recorder across different passes, reconstruct a real-time image of all events recorded, and store the image on magnetic disk for subsequent analysis. The behavioral events that we record most frequently, and their definitions, are presented in table 4.1.

Some of the advantages of this assay are: (1) a male and female with known reproductive histories can be placed together in the arena (e.g., a female that has been in precopula for 48 hrs can be separated from her partner and placed in the

TABLE 4.1. Definitions of the behaviors typically measured in the 3-3 precopula assay.

Behavior	Operational Definition
Contact	Any physical touch between animals
Palpation	Female held at 90-degree angle to the longitudinal male axes
Precopula	Female held in a parallel position underneath male
Male activity	Time the male is swimming actively around the test arena
Female activity	Time the female is swimming actively around the test arena

arena with a strange male that has copulated with his partner 24 hrs previously); (2) the stimulus conditions in the arena can be varied (e.g., pairs can be tested in water taken from females late in their molt cycle); and most important, (3) we can assess the effects of various conditions on different *components* of the behavioral sequence leading to precopulatory mate guarding (e.g., the number of contacts, the duration of palpation, etc.).

The Olfactometer Assay

A traditional olfactometer assay (Kennedy 1977) based specifically on the apparatus described by Borowsky (1985) has also been developed to measure the responses of *Gammarus lawrencianus* to chemical stimulation. The apparatus is essentially a Y-shaped water maze. The stem of the Y consists of a stem tank (25 m long × 9 cm wide × 7.5 cm deep). Animals can leave the stem tank by swimming into an 8 mm diameter clear plastic tube positioned 1 cm above the tank floor (a small pebble placed at the exit hole encourages departure from the stem tank). Passage through the stem tube (17 cm long) places the animal at the choice point of the maze (a Y intersection in the tubing). Each arm of the Y tube (17 cm long) terminates with entry into respective left and right arm tanks. Each arm tank is 27 cm long × 19 cm wide × 5 cm deep. The water that flows into each arm tank first passes through a pretreatment tank (325 ml volume). The pretreatment tanks are used to isolate the substances (e.g., premolt females) to be introduced into each arm tank.

Untreated seawater, flowing at the rate of 150 ml/hr enters each pretreatment container, flows from each pretreatment container into its respective arm tank, then down the arms of the Y into the stem, and enters the stem tank through the exit hole. Animals in the stem tank have two choices. They can remain in the stem tank or depart through the exit hole; next they can swim up one arm or the other of the Y maze and enter the arm tanks. Natural cover in the form of three small pebbles and a mussel shell is provided in each arm tank; the animals cannot enter the isolated pretreatment containers that supply the arm tanks. The measures of interest in the assay are: (1) whether or not animals depart from the stem tank and (2) whether they discriminate between the substances presented at the choice point and selectively enter one arm tank or the other.

THE SEX PHEROMONE HYPOTHESIS

Evidence in *Gammarus lawrencianus*

Our first experiment examined *specific behavioral components* of the interactions between males and females during the 3-3 precopula assay described earlier (Dunham, Alexander & Hurshman 1986). The rationale was simple. If females are releasing a pheromone late in their reproductive cycle, a female late in her cycle should attract a male in the test arena more rapidly than a female early in her cycle. The pattern of results obtained in the 3-3 precopula assay was very clear. When the latency to the first contact initiated by the male in the test arena is used as a measure of female attraction, there is no significant difference between females known to be early and late in their reproductive cycle. The measure provided no evidence for chemical influence from a distance. However, once contacted, the females late in their reproductive cycle are palpated for longer time and are more likely to be guarded by the male. These data offer no evidence for distance attraction and permit only the conclusion that males can discriminate between females early and late in the reproductive cycle upon contact. The nature of the stimulation that permits the discrimination on contact is yet to be identified, although more will be said about this issue later in the discussion.

A second experiment was designed to determine if secretions from premolt receptive females will influence any aspect of the male's precopulatory mate guarding behavior. A perforated false bottom was placed over the floor of the test arena during the 3-3 precopula assay. Premolt females placed in the 50 ml space below the false bottom were isolated from the arena above, but water flowed freely through the perforations during the behavioral observations.

Using the 3-3 precopula assay, interactions between *unacceptable* females (3 days postcopulation) and males also known to be 3 days postcopulation were observed in the test arena above the false bottom. The interactions were observed in two different conditions. In Group P, 15 male-female pairs were videotaped while the compartment below the false floor contained 20 premolt, *acceptable* females known to have been in precopula for at least 3 days prior to being placed in the compartment. In Group C, 15 male-female pairs were videotaped while the compartment below the false floor was empty (i.e., contained only untreated control seawater). The question of interest was whether secretions from the receptive females under the perforated false floor would modify any component of the behavioral interactions between the male and unacceptable female above the floor (i.e., whether the unacceptable female would be treated as more acceptable in the presence of the secretions).

The analysis of videotaped interactions included measures of (1) the frequency of precopulatory mate guarding responses observed in each treatment group, (2) the average latency to the first physical contact between the male and female, (3) the total number of physical contacts observed during the test session, (4) the latency to the first palpation response observed, (5) the number of palpation bouts initiated, (6) the average duration of the palpation bouts, (7) the total duration of precopulatory mate guarding behavior observed, (8) the total amount of time the male was swimming and resting during the test session, (9) the average duration of the male's swimming bouts, (10) the total amount of time the female was swimming and resting during the session, and (11) the average duration of the female's swimming bouts.

Stated very simply, the results of this extensive behavioral analysis were negative on all dimensions measured. Two pairs of animals established precopulatory mate guarding in Group C and one pair in Group P. No other measure of behavioral interactions during the assay revealed statistically reliable differences between the two treatment groups.

Our third approach employed the olfactometer apparatus described earlier. One experiment has been completed to date. In each replication of this experiment, 50 male-female *precopula* pairs were randomly sampled from the breeding culture and each pair was separated. The 50 females were immediately placed in one of the two pretreatment tanks that flow into either the left or right arm of the Y maze. Water flowed through the apparatus for a period of one hour to stabilize the system, and then the 50 males were placed in the stem tank from which they could enter the stem of the Y maze.

All testing sessions were conducted between 0900 and 1400 hrs during the light period of the LD cycle. Each test was three hours in duration, after which the males remaining in the stem tank and the males in each arm tank were counted. Six replications of the experiment were completed, with target females placed in the left-arm pretreatment container for three replications and in the right-arm pretreatment container for three replications (i.e., target position counterbalanced). Between replications, the apparatus was washed in dechlorinated water and air dried.

It should be noted that males were free to retrace any path and "correct" their choices during the three-hour assay. Casual observations of the animals during the procedure indicated that they seldom left an arm tank once they reached it, but they did occasionally retrace their path to the choice point of the Y while they were in the tubing connecting the various tanks.

The results of this assay are presented in table 4.2. These data, which are collapsed across the six replications of the procedure, indicated that males departing from the start tank showed a preference for the arm tank containing water flowing from receptive females. A hierarchical log linear analysis using a backward elimination method to analyze for target preference, target position, and their interaction revealed that the preference for the receptive female water was highly significant (L.R. ratio chi square = 23.33, $p < .0001$)), and the target preference by position interaction was reliable (L.R. ratio chi square = 12.96 $p < .0003$). The significant interaction indicates that the preference for the receptive female water was most pronounced when it flowed from the arm tank on the left side of the apparatus. This position bias does not, however, alter the conclusion that males preferred the female water. We suspect an imbalance in flow rate on that side of the apparatus is the most likely explanation for the position bias.

Although these results suggest that acceptable premolt females can attract males from a distance, considerably more work is necessary to elucidate the nature of the process. Both *males and females* should be tested with a wider variety of substances (e.g., nonreceptive female water, male water) under a

---

TABLE 4.2. Frequency of male choices when receptive females are on either the left side or right side of the olfactometer. Data are collapsed over six replications with target position counterbalanced (statistics reported in text).

	Target Position	
	Left side	*Right side*
Receptive female	93	67
Untreated seawater	29	56

number of different conditions (e.g., time of testing, water temperature, photoperiod) in order to determine the boundary conditions of the effects. This work is currently in progress.

When one considers the results obtained across these three different behavioral assays, an interesting pattern emerges. Water from premolt females has a direct effect on the male's behavior only in that assay in which a water current is used to deliver the test stimulus. In all our variations on the 3-3 precopula assay (no water currents present), water associated with premolt females had no observable effect on any aspect of the male's behavior. Alternatively, in the olfactometer a water current that is probably less saturated with the test stimulus (animals per unit volume) had a definite effect on the male's behavior (i.e., they swam upstream and selected the stream containing receptive female secretions).

This pattern of results can also explain some of the inconsistencies in the existing literature. The early research that Hartnoll and Smith (1980) were unable to replicate also lacked a water current for delivering the test stimuli. The two reports of positive evidence (Borowsky 1984, 1985) employed assays in which the stimuli were presented in a water current.

Although comparisons across experiments require caution, both the data in the existing literature on amphipods and our research with *Gammarus lawrencianus* suggest that a test stimulus in a water current is a necessary condition to observe chemical influence from a distance. As a working hypothesis, the available data suggest that a combination of water current and the chemical signal causes the animal to orient and swim upstream (e.g., a rheotropism [Loeb 1918:132]). Understanding other details of the mechanism (e.g., how the behavior changes when the signal is lost) will obviously require additional research. If one substitutes air currents (anemotaxis) for water currents (rheotaxis) the literature on the flying insects provides many models for further research on amphipods in the aquatic environment (Kennedy 1977). At the minimum, the role of water currents deserves explicit attention in future research on the sex pheromone assumption in aquatic Crustacea.

The Passive Female Assumption

As noted earlier, one corollary of the female sex pheromone hypothesis is the assumption that the female is a passive behavioral agent in the mate guarding interaction. Its only function is to release the attractant. Although we have not yet examined the validity of this corollary assumption in our research on *Gammarus lawrencianus*, there is a growing body of evidence on other species indicating that this assumption should be reconsidered.

Warner (1976) describes an interesting method used by Chesapeake Bay "watermen" to catch soft-shelled (recently molted) female crabs *(Callinectes sapidus)*. In the spring, the fishermen bait their traps with large male "Jimmies" to selectively catch recently molted female "peelers." The success of this method suggests that the female may be *actively* seeking dominant males shortly after the female's pubertal molt and that she can be attracted from considerable distances to the male bait (Cheung 1966).

Similarly, in his research on the mating behavior of the American lobster,

Homarus americanus, Atema (1986) observed these animals in large seminatural tanks and monitored them in the field over a period of several years. Among other things, these heroic efforts revealed, contrary to the passive female assumption, that premolt females actively "forage" for dominant males and share the male's shelter during molt and copulation.

Finally, Borowsky's (1985) olfactometer research on the amphipod *Gammarus palustris* also suggests that females in this group play an active part in the mate guarding process. *Both males and females* swam upstream toward conspecifics of either sex. Although the females were less active then males, those swimming upstream preferred water associated with conspecific males over water from conspecific females.

Considered together, these observations question the traditional passive female assumption as a corollary of the sex pheromone hypothesis. It is also the case that any convincing evidence for active female choice in the mate guarding process has important implications for both ultimate and proximate causal explanations of precopulatory mate guarding. At the level of ultimate causes, such data suggest that *both* female choice and male intrasexual competition may be components of sexual selection in these species. At the level of proximate causes, the data require a mechanism more elaborate than the traditional female sex pheromone hypothesis. At the least, hypotheses about a male sex pheromone and/or a bidirectional aggregation pheromone become viable alternative mechanisms.

The Single Stimulus Assumption

The second corollary to the female sex pheromone hypothesis is the single stimulus assumption. Perhaps more an error of omission than design, most versions of the female sex pheromone hypothesis place substantial demands on a single channel of chemical communication. The pheromone is assumed to attract the male from a distance, elicit the mate guarding behavior, maintain the male's guarding behavior, and finally elicit the copulatory response of the male.

Our knowledge of chemical communication in the social insects (Shorey & McKelvey 1977) indicates that nature seldom builds a social system that is totally dependent on a single channel of communication. It seems more reasonable to suggest that attraction from a distance may be only the first stage in a more complex sequence of signals involved in the final mate guarding decision.

We have some initial evidence that suggests a second source of information mediates the interactions between male and female amphipods once the animals make physical contact (Dunham 1986). The rationale for this experiment was as follows.

Gammarus lawrencianus females are seldom guarded immediately after copulation, when the brood pouch is full of large, recently fertilized, yolk-filled eggs (Stage A embryos [Steele & Steele 1969]). Alternatively, a female with an *empty* brood pouch signals a short investment time for the male (all embryos have hatched and the female molt is imminent). In order to determine if encountering young embryos in the brood pouch during *palpation* might inhibit the male's mate guarding behavior, we experimentally flushed young embryos from the brood pouch of unguarded females early in their reproductive cycle. The flushed females were then tested with a male in the precopula test arena. Eight of 15 previously unacceptable females were guarded after their brood pouches had been flushed. This compared to 1 of 15 females exposed to a sham flushing procedure. These data suggest that the brood pouch contents become involved in the mate guarding decision process during palpation. Previously unacceptable females can be made acceptable by altering the contents of their brood pouch.

While these data are interesting and suggest that a second source of stimulation influences mate guarding behavior during the palpation sequence, they should be viewed with caution. The argument would have been more definitive if we had also tested a group in which young embryos were experimentally inserted into the empty brood pouch of an acceptable premolt female. The hypothesis predicts that males would fail to guard the previously acceptable female who is now carrying young embryos.

It should also be noted that Borowsky has personally communicated an inability to replicate the effects of these brood pouch manipulations in *Gammarus palustris*. We do not know, at present, whether species differences or procedural differences account for the discrepancy, but a replication and extension of our initial results are obviously necessary.

CONSTANT MALE THRESHOLD ASSUMPTION: EVIDENCE IN *GAMMARUS LAWRENCIANUS*

Explanations of precopulatory mate guarding in aquatic Crustacea at both the ultimate and proximate levels of analysis imply that the male maintains a threshold for female stimulation below which the female is not acceptable for mate guarding. As noted earlier, at the ultimate level of causal analysis the threshold for a particular species is assumed to be an evolutionarily stable strategy that will maximize fitness; deviations from the threshold in either direction reduce fitness (Ridley 1983). At the proximate level of analysis, implicit in the female pheromone assumption is the notion that females will only be acceptable for mate guarding when the pheromone is secreted in quantities that exceed the male's threshold. Substances associated with the late stages of the female's reproductive cycle (e.g., molting hormones) are proposed as pheromones precisely because they signal a short time investment for the male (Dunham 1978, 1988a). As such, both levels of explanation assume that the male is a constant in the equation, with differences in the female's reproductive status accounting for all of the variance in the male's mate guarding behavior.

We first questioned the constant male threshold assumption during an experiment that examined the mate guarding behavior of males after two different male postcopulation intervals (Dunham, Alexander & Hurshman 1986: exp. 3). In this experiment, 40 precopula pairs were sampled from the breeding culture and observed in individual containers through one complete reproductive cycle. Twenty of the males were isolated from their female partners on the first day the pair established mate guarding, and these males were held in isolation for an additional 24 hrs. Twenty other males were similarly separated from their female partners and held in isolation for an additional 72 hrs. Each of the males were subsequently placed in the 3-3 precopula test arena with a strange female known to be acceptable for mate guarding (i.e., she was in the first day of precopula with a different male). These differences in postcopulation intervals (an additional 24 versus 72 hrs) did not have any effect on the *probability* that the males would guard the strange female during the test. These differences in postcopulation interval did, however, produce qualitative differences in the male-female interactions. For example, it took the males in the 72 hr group less time to establish initial contact with the females in the test arena, and this group also tended to palpate the females for a longer period prior to establishing a mate guarding relationship with the female. These data suggest that the male's behavior changes as the time since his last copulation increases.

In a more recent extension of the above-described research, Dunham, Hurshman, and Gavin (1989) sampled 45 pairs of *Gammarus lawrencianus* from a breeding culture and isolated the males immediately following copulation with their female partner. One group of 15 males was isolated for 24 hrs following copulation (M24), a second group was isolated for 240 hrs (M240) and a third group for 600 hrs (M600).

After the designated isolation period, each male was tested in the 3–3 precopula assay with a strange female. The female target was two days postcopulation and known to be *unacceptable* for mate guarding when she was with her previous partner. In this experiment, *none* of the males in Group M24 attempted to guard the unacceptable female targets, but *all* of the males in Groups M240 and M600 established a mate guarding relationship with previously unacceptable female targets during the three-minute test session.

Table 4.3 presents other qualitative differences in the mate guarding behavior of these males during the 3–3 precopula assay. As the data in table 4.3 indicate, the longer the interval since the male's last copulation, the more active the male is during the assay, the longer the male palpates the female, the shorter the latency to the first palpation response, and the shorter the latency to the precopula response. Collectively, the data suggest, contrary to the *constant male*

TABLE 4.3. Means and F values for various behaviors measured during the 3-3 precopula assay in the three male treatment groups (M = males, following numbers represent hours) presented with unacceptable females. Temporal measures are in seconds. Adapted from Dunham, Hurshman, and Gavin 1989 with permission.

Behaviors	M24	M240	M600	F	Sig.
Latency to first contact	6.1	21.8	8.5	<1	—
Latency to palpation	135.1	28.6	12.6	21.6	.0001
Latency to precopula	180.0	61.4	43.0	41.3	.0001
Duration palpation	1.9	13.3	16.3	7.7	.001
Duration precopula	0.0	60.9	101.1	8.5	.001
Frequency contacts	8.5	3.5	1.7	5.5	.01
Proportion male activity	0.2	0.2	0.7	17.6	.0001

threshold assumption, that the male's threshold for precopulatory mate guarding behavior will vary with the male's reproductive history. A 48 hr postcopulation female is below threshold after a 24 hr male postcopulation interval, but above threshold after a 240 hr male postcopulation interval.

These results have important implications for both ultimate and proximate causal explanations of precopulatory mate guarding behavior. With respect to ultimate causes the data suggest that the notion of an evolutionarily stable threshold that has evolved to maximize fitness is inappropriate. Instead, there appears to be considerable plasticity in the male's threshold for mate guarding, and his current criterion is determined by some aspect of his *ontogenetic* reproductive history.

The question of primary interest for future research on male threshold assumption is to determine which aspect of the male's reproductive history directly influences the male's threshold. We have just started a series of experiments to answer this question. The initial data are somewhat surprising.

Notice first that our research thus far on the constant male threshold assumption confounds two dimensions of the male's reproductive history. Males isolated for 24, 240, and 600 hrs not only have different postcopulation intervals; they have also been isolated for different periods of time without any type of social contact. The change in the male threshold could be produced by either factor (i.e., time since last copulation or time since last social contact with a conspecific).

Using the same basic rationale and behavioral assay described earlier, we tested males in the 3–3 precopula test arena with an unacceptable female after the males had been isolated (1) *alone* for 240 hrs, (2) *with another different male* each day for 240 hrs, (3) with a *different unacceptable female* each day who had just copulated. Our initial results indicate that *contact with an unacceptable female* accounts for most of the variance in the male's threshold. If intermittent contact with an unacceptable female is permitted, the male maintains his typically "high" threshold for guarding females *late* in their reproductive cycle. The *time since the male's last copulation* alone does not appear to affect the male's threshold at the postcopulation intervals we have investigated thus far.

When thinking about proximate causal mechanisms, it should be recalled that our socially isolated males will eventually guard females very early in the female reproductive cycle. This observation also directly questions several specific versions of the female sex pheromone hypothesis. If, as these data indicate, males will eventually guard an unacceptable female (two days *postmolt*), the substances that are uniquely associated with the molt would not appear to be necessary for mate guarding to be elicited. Both the *molting hormone hypothesis* (Kittredge & Takahashi 1971) and the more recently proposed *arthropodin hypothesis* (Dunham 1988a) make this assumption and are consequently questioned by the male threshold data. The fact that the male's threshold permits a response from stimuli very early and very late in the female's reproductive cycle leads one to search, instead, for a female stimulus dimension that carries information about the stage of the female's reproductive cycle *throughout* the cycle.

CONCLUSIONS

Evidence obtained thus far on *Gammarus lawrencianus* as a model system of precopulatory mate guarding suggests that we should reconsider at least two of the basic assumptions about the ultimate and proximate causal mechanisms that underlie this mating strategy. Consider first the *female sex pheromone assumption*.

Our data on *Gammarus lawrencianus* are consistent with the assumption that a chemical stimulus does influence the social behavior of conspecifics from a distance. As noted earlier, some of the inconsistencies in the existing literature on chemical communication among amphipods that have undermined this conclusion in the past can be resolved by a comparison of positive and negative results obtained across different behavioral assays. Our data indicate that differential orientation and attraction will be observed only when the stimulus is presented in a water current. The pattern of positive and negative results in the earlier literature is also predictable from this procedural difference.

Although the evidence for a pheromone communication system in amphipods is now accumulating across several different species, there are also converging lines of research from our laboratory and others that suggest that these chemical signals are not uniquely associated with *premolt females* as the traditional hypotheses would imply. Borowsky's (1985) data on *Gammarus palustris* indicate that *both* males and females are capable of attracting conspecifics in a choice olfactometer, and similar conclusions can be drawn about other groups of Crustacea (e.g., Warner 1976; Atema 1986). In addition, our research with *G. lawrencianus* on the effects of the male postcopulation interval indicates that *postmolt* females early in their reproductive cycle are also capable of eliciting mate guarding behavior. Hence the premolt female condition may be sufficient to elicit mate guarding behavior from the male, but it is apparently not necessary.

Our data on *Gammarus lawrencianus*, which revealed that the brood pouch contents may play a role in the mate guarding decision, also suggest a modification of the sex pheromone hypothesis and define a problem for future research. It is unlikely that this mating strategy is mediated by a single channel of stimulation (i.e., a pheromone). Indeed, the general notion that various sources of stimulation are involved at several different stages of a behavioral sequence that culminates in copulation would seem to be a more reasonable model for future research. The stimulus conditions that bring the animals together need not be the same as those that maintain the precopulatory mate guarding behavior or those that eventually elicit copulation.

Finally, consider the data that indicate that the male's threshold for accepting a female varies with his reproductive history (i.e., the male threshold effect). This discovery also raises several important questions for future research. We do not at present understand why the intermittent contact with an unacceptable female keeps the male's threshold high, but we are considering several models of the process. It is interesting, for example, to compare the effects we observe in the male threshold experiments to the dynamics of intercellular pulsatile hormonal systems in mammals (see Goldbeter 1988 for an overview). In these hormonal systems, the threshold of target cells is dependent on the *temporal pattern* of pulsed hormones that reach the target cell. Most systems are tuned to an optimal temporal pattern. If the target cells are pulsed at a rate higher or lower than the optimal pattern, the target cell threshold changes. If one thinks of the free swimming male amphipod as a target cell and the females encountered as pulsatile hormones, the male's threshold appears to behave in our experiments in much the same manner as the threshold of target cells in pulsatile hormonal systems. Although we are not proposing that the social interactions of aquatic Crustacea are the precursors of these pulsatile hormonal communication systems (cf. Haldane 1955), we would suggest that there is some heuristic value in using the parallel to generate future research on the male threshold phenomenon. Indeed, the next step in our own research will be to systematically examine the effects of both *increases* and *decreases* in the rate at which males are "ontogenetically pulsed" by a female conspecific to determine how far the parallel can be extended.

Acknowledgments

This research was supported by Grant A0194 from the Natural Sciences and Engineering Research Council of Canada.

Literature Cited

Atema, J. 1985. Chemoreception in the sea: Adaptations of chemoreceptors and behavior to aquatic stimulus conditions. *Soc. Exper. Biol. Symp.* 39:387–423.

Atema, J. 1986. Review of sexual selection and chemical communication in the lobster, *Homarus americanus. Can. J. Fish. Aquat. Sci.* 43:2283–2390.

Atema, J. & R. Gagosian. 1973. Behavioral responses of lobsters to ecdysones. *Mar. Behav. Physiol.* 2:15–20.

Birkhead, T. R. & K. Clarkson 1980. mate selection and precopulatory mate guarding in *Gammarus pulex. Tierpsychol.* 52:365–380.

Borowsky, B. 1984. Effects of receptive females' secretions on some male reproductive behaviors in the amphipod crustacean, *Microdeutopus gryllotalpa. Mar. Biol. (Berl.)* 84:183–187.

Borowsky, B. 1985. Responses of the amphipod crustacean *Gammarus palustris* to waterborne secretions of conspecifics and congenerics. *J. Chem. Ecol.* 11:1545–1552.

Cheung, T. S. 1966. An observed act of copulation in the shore crab *Carcinus maenas. Crustaceana (Leiden)* 11:107–108.

Dahl, E., H. Emanuelsson & C. Von Mecklenburg. 1970a. Pheromone transport and reception in an amphipod. *Science (Washington, D.C.)* 170:739–740.

Dahl, E., H. Emanuelsson & C. Von Mecklenburg. 1970b. Pheromone reception in the males of the amphipod *Gammarus duebeni* (Lilljeborg). *Oikos* 21:42–47.

Ducruet, J. 1973. Inter attraction chez les gammares du group pulex. *Rev. Comp. Anim.* 7:313–322.

Dunham, P. J. 1978. Sex pheromones in Crustacea. *Biol. Rev. Camb. Philos. Soc.* 53:555–583.

Dunham, P. J. 1979. Mating in the American lobster: Stage of moult cycle and sex pheromone. *Mar. Behav. Physiol.* 5:209–214.

Dunham, P. J. 1986. Mate guarding in amphipods: A role for brood pouch stimuli. *Biol. Bull. (Woods Hole)* 170:526–531.

Dunham, P. J. 1988a. Pheromones and behavior in Crustacea. In H. Laufer and R. Downer, eds., *Endocrinology of Selected Invertebrate Types*, pp. 375–392. New York: Alan R. Liss.

Dunham, P. J. 1988b. *Research Methods in Psychology*. New York: Harper & Row.
Dunham, P. J., T. Alexander & A. Hurshman. 1986. Precopulatory mate guarding in an amphipod, *Gammarus lawrencianus*, Bousfield. *Anim. Behav.* 34:1680–1686.
Dunham, P. J., A. Hurshman & C. Gavin. 1989. Precopulatory mate guarding in an amphipod *Gammarus lawrencianus:* Effects of the male post-copulation interval. *Mar. Behav. Physiol.* 14:181–187.
Emlen, S. T. & L. W. Oring. 1977. Ecology, sexual selection and the evolution of mating systems. *Science (Washington, D.C.)* 197:215–223.
Gagosian, R. & J. Atema 1973. Behavioral responses of male lobsters to ecdysone metabolites. *Mar. Behav. Physiol.* 2:115–120.
Gleeson, R., M. Adams & A. Smith. 1984. Characterization of a sex pheromone in the blue crab, *Callinectes sapidus:* Crustecdysone studies. *J. Chem. Ecol.* 10:913–921.
Goldbeter, A. 1988. Periodic signalling as an optimal mode of intracellular communication. *N.I.P.S.* 3:103–105.
Grafen, A. & M. Ridley. 1983. A model of mate guarding. *J. Theor. Biol.* 102:549–567.
Haldane, J. 1955. Animal communication and the origin of human language. *Sci. Prog.* 43:385–401.
Hammoud, W., J. Compte & J. Ducruet. 1975. Recherche d'une substance sexuellement attractive chez les gammares du group pulex. *Crustaceana (Leiden)* 28:152–157.
Hartnoll, R. G. & S. M. Smith. 1980. An experimental study of sex discrimination and pair formation in *Gammarus duebeni* (Amphipoda). *Crustaceana (Leiden)* 38:253–264.
Hunte, W., R. Myers & R. Doyle. 1985. The effect of past investment on mating decisions in an amphipod: Implications for models of decision making. *Anim. Behav.* 33:366–372.
Kennedy, J. S. 1977. Behaviorally discriminating assays of attractants and repellents. In H. Shorey & J. McKelvey, eds., *Chemical Control of Insect Behavior*, pp. 215–229. New York: Wiley.
Kittredge, J. S. & F. T. Takahashi. 1971. The evolution of sex pheromone communication in the Arthropoda. *J. Theor. Biol.* 35:467–471.
Krogh, A. 1929. Progress of physiology. *Am. J. Physiol.* 90:243–251.
Loeb, J. 1918. *Forced Movements, Tropisms, and Animal Conduct*. Philadelphia: Lippincott.
McLeese, D. W. 1970. Detection of dissolved substances by the American lobster, *Homarus americanus*, and olfactory attraction between lobsters. *J. Fish. Res. Board. Can.* 27:1371–1378.
Manning, J. T. 1980. Sex ratio and optimal male time investment strategies in *Asellus aquaticus* and *A. meridianus*. *Behaviour* 74:264–273.
Parker, G. A. 1974. Courtship persistence and female-guarding as male time investment strategies. *Behaviour* 48:157–184.
Ridley, M. 1983. *The Explanation of Organic Diversity*. Oxford: Clarendon Press.
Salmon, M. 1983. Courtship, mating systems and sexual selection in Decapoda. In S. Rebach & D. Dunham, eds., *Studies in Adaptation*, pp. 143–162. New York: Wiley.
Shorey, H. H. & J. J. McKelvey. 1977. *Chemical Control of Insect Behavior*. New York: Wiley.
Siefert, P. 1982. Studies on the sex pheromone of the shore crab *Carcinus maenas* with special regard to ecdysone secretion. *Ophelia* 21:147–158.
Steele, D. H. & V. J. Steele. 1969. The biology of *Gammarus* (Crustacea, Amphipoda) in the northwestern Atlantic. I. *Gammarus duebeni*. *Can. J. Zool.* 47:235–244.
Steele, D. H. & V. J. Steele. 1970. The biology of Gammarus (Crustacea, Amphipoda) in the northwestern Atlantic. IV *Gammarus lawrencianus* Bousfield. *Can. J. Zool.* 48:1261–1267.
Trivers, R. L. 1972. Parental investment and sexual selection. In B. Campbell, ed., *Sexual Selection and the Descent of Man*. pp. 1871–1971. Chicago: Aldine.
Warner, W. W. 1976. *Beautiful Swimmers: Watermen Crabs and the Chesapeake Bay*. New York: Penguin.
Wickler, W. & U. Seibt. 1981. Monogamy in Crustacea and man. *Z. Tierpsychol.* 57:215–234.
Wilson, E. O. 1975. *Sociobiology*. Cambridge: Harvard University Press.

FIVE
Variation in Reproductive Behavior in Stomatopod Crustacea

ROY L. CALDWELL

Department of Integrative Biology,
University of California at Berkeley.

Abstract

Reproductive behaviors are shaped by a variety of physical and biological constraints. Almost all stomatopods live in some type of refuge, but species differ greatly in the type of cavity or burrow occupied, in the time and effort it takes to acquire and maintain a dwelling, in the risk of losing their home to a competitor, and in the danger of being attacked by predators when away from the safety of their refuge. In stomatopods, these factors have had a major influence on the evolution of mate location, courtship, and maternal behaviors. Here I discuss the mating systems of several species of stomatopods and the selective forces that have shaped these reproductive patterns.

LIKE ALL behaviors, those involved in courtship and reproduction are shaped by a variety of physical and biological constraints. The same selective forces that produce highly aggressive, territorial species may require novel solutions to problems such as mate location and intraspecific aggression. Conditions that favor more gregarious life-styles consequently produce more frequent male-female encounters and, in turn, may promote the evolution of such phenomena as sperm competition or mate choice. One of the first considerations of the ecological bases for the evolution of social systems was Crook's classic study of weaver birds (Crook 1964). He related predator defenses and the temporal and spacial distribution of food to the variety of social systems found in these birds. In a more general treatment of the evolution of mating systems, Emlen and Oring (1977) proposed that resource pattern, distribution, and quality dictate the pattern of spacing in a variety of taxa. In turn, mating systems evolved to function within the constraints of the spatial and temporal organization of the populations in question. Using similar approaches to study reproductive organization in Crustacea, Salmon (1983) analyzed courtship and mating systems in decapods, and Christy and Salmon (1984) stressed understanding the underlying ecology of species when interpreting the evolution of mating systems in fiddler crabs.

A second major determinant of the evolution of mating systems is parental investment in offspring and the relative contributions of males and females. Costs incurred by each sex in mate location and selection, gamete production, and nurturing offspring will impact on the evolution of courtship and reproductive patterns (Trivers 1972). Costs include not only actual energy expenditure but also risk of predation, time lost, general wear and tear, and anything else that reduces the probability of future reproduction. The importance of understanding such costs is evident in Knowlton's (1980) now classic study of monogamy in the snapping shrimp, *Alpheus armatus*.

A myriad of other factors can affect the evolution of courtship and reproductive patterns, such as the need to produce offspring at a particular time and/or location, specific requirements of the young, etc. Understanding phyletic constraints is also critical. Developmental and life history characteristics such as the method of fertilization, type of offspring produced (i.e., oviparity or viviparity), and larval form are not easily modified (for a review of such constraints in decapods, see Hazlett 1983). To understand the functions, origins, and evolutionary trajectories of courtship and reproductive behaviors, we must be able to identify the biological foundations on which these behaviors are built and the selective forces that shape them.

Here I explore the diversity of reproductive behavior found in stomatopod Crustacea and attempt to attribute interspecific differences to various morphological and ecological factors. While the basic mode of reproduction in stomatopods is fairly conservative (all known species have separate sexes, and females are oviparous), the assortment of mating systems is considerable, ranging from monogamy to promiscuity, from mate guarding and forced copulation to mate choice and pair bonding. Some species of stomatopods exhibit behaviors unusual for marine Crustacea, such as paternal care, male choice, and extended care of offspring.

This paper cannot be an exhaustive survey. Reproduction has been described in fewer than 10 percent of the approximately 500 known species of stomatopods, and most of these come from two groups, the Squilloidea and the Gonodactyloidea. I will draw heavily on observations my students and I have made on reproduction in more than 30 species of stomatopods.

BASIC STOMATOPOD BIOLOGY

What Are Stomatopods?

The Stomatopoda, the only living order of the subclass Hoplocarida, are a small group of carnivorous marine Crustacea. Early hoplocarids, akin to the primitive Aeschronectida, diverged from primitive malacostracan stock 400 million years ago. They were filter feeders or sifted detritus from soft bottoms, using thoracopods all similar in form and subequal in size. By the Late Devonian, the appearance of the Paleostomatopoda suggests that the stomatopod lineage was moving toward a more predatory nature. Their long, slender subchelate second through fifth thoracopods, subequal in size, almost certainly were used to seize and tear at other creatures (Schram 1979). The first true Stomatopoda (Suborder Archaeostomatopodea) appear during the Upper Mississippian. While the second through fifth thoracopods (frequently called maxillipeds) were still subequal in size, their subchelate form was better developed, suggesting a more active predatory life-style (Schram 1986).

During the Upper Jurassic, a dramatic change occurred, profoundly influencing the evolution of all stomatopods yet to come. In the living suborder Unipeltata, the second thoracopods became enlarged and occur today as powerful raptorial appendages, capable of reaching out over a considerable distance to seize prey. The ischiomerus is very large, housing the muscles that drive the raptorial strike. The carpus is reduced and acts as a trigger mechanism that releases the rapid forward extension of the propodus and dactylus. These two distal segments, which form the subchelate part of the appendage, are both long and are frequently armed with spines and teeth to facilitate prey capture. All modern stomatopods possess these highly modified and potent weapons.

Spearers and Smashers

Functionally, stomatopods can be divided into two groups, spearers and smashers, based on the morphology and use of their raptorial appendages (Caldwell & Dingle 1975, 1976). Spearing is the probable primitive condition and occurs in all four superfamilies. The dactylus of the second thoracopod is long and slender and is armed with two or more barbs. Spearers tend to specialize on soft-bodied prey, such as shrimp and fish, which they grasp or impale with a lightening quick extension of the raptorial appendages. The strike is analogous to, but much faster than, that of a praying mantis, giving the stomatopods their common name of mantis shrimp.

The raptorial appendage of smashers is derived from that of spearers and evolved independently in several different lineages, although most living representatives are in the superfamilies Gonodactyloidea and Lysiosquilloidea. In smashers, the heel of the dactylus is inflated, the dactylus is usually shorter, and it may be unarmed. This type of appendage is specialized for breaking apart armored prey, such as snails and crabs. The strike is delivered with the dactylus folded tightly against the propodus, the point of impact being the heavily calcifed dactylus heel. Some spearers will attack weakly armored prey, such as thin-shelled clams, by striking them with a folded dactylus (*Oratosquilla oratoria* [Hamano & Matusuura 1986], *Squilla aculeata* [Caldwell, personal observation]), and it is easy to hypothesize how the smashing mode of attack evolved. The impact of the strike of a smasher is much greater than that of a spearer. The ability to land crushing blows not only allows smashers to take more heavily armored prey, it also provides them with a more potent weapon that can be employed for defense and in intra- and interspecific agonistic encounters.

The evolution toward smasher or spearer specialization is intertwined with numerous morphological, physiological, behavioral, and ecological features. Prey abundance and distribution, habitat, the availability of suitable refuges, competitors, and predators all may play a role in driving the evolution of raptorial appendage morphology and use. In turn, the type of appendage possessed opens up opportunities for selection on traits such as feeding specialization, social interactions, and habitat use. Certain suites of characters are commonly associated with convergent smashing or spearing specializations, and most probably also lead to the diversity of reproductive patterns described here. Differences between spearers and smashers have been reviewed by Caldwell and Dingle (1975, 1976) and Reaka and Manning (1981, 1987). Here I mention only some of the more prominent aspects of this divergence as they relate to reproduction.

Most spearers live in self-excavated burrows in sand or mud. Almost all smashers occupy preexisting cavities in hard substrata such as rock, coral, and calcareous algae. Cavities in hard substrata are often in short supply, with a variety of organisms, including fish, crabs, snapping shrimp, octopuses, and other stomatopods, vigorously competing for them (Reaka 1985; Steger 1987). The lethal strike of the smashing stomatopods is more effective against these competitors. Self-excavated burrows are less likely to be limiting for spearers.

In conjunction with the potency of the raptorial appendage and the level of competition for home sites, smashers have evolved more intense and sophisticated aggressive behaviors than have spearers. (This is certainly true for the gonodactyloids, although the behavior of the small smashing lysiosquilloids is unknown.) Even within a functional group, there is typically a positive correlation between the scarcity of cavities or the amount of energy that goes into constructing a burrow and the type and intensity of aggression displayed in its defense (Caldwell & Dingle 1976; Dingle 1983).

Type of raptorial appendage, habitat, and aggressive behavior also are associated with sensory development in stomatopods. Smashers typically are diurnal and live in clear water. They tend to have acute vision with some groups possessing sophisticated color processing (Manning, Schiff & Abbott 1984; Marshall 1988; Cronin & Marshall 1989). Spearers are more likely to be nocturnal and/or live in murky environments associated with soft substrata. Their eyes are better adapted for detection of prey at short distances in dim light. With respect to chemical senses, several gonodactylids have been shown to identify other individual stomatopods by odor and use this ability to mediate aggressive encounters (Caldwell 1979, 1982, 1985). In the few spearers I have examined for this ability *(Pseudosquilla ciliata, Squilla empusa)*, it was not found (Caldwell unpublished).

STOMATOPOD REPRODUCTIVE BIOLOGY

The basic reproductive biology of stomatopods is conservative and is similar in smashers and spearers. Sexes are separate, there is little or no sexual dimorphism, fertilization is internal, females produce heavily yolked eggs, and there is at least a short period of maternal care of the eggs while they are in the female's cavity or burrow. However, the reproductive biology of very few stomatopods is even partially known, and we may discover exceptions to the general description that follows.

Reproductive Morphology and Physiology

In males, paired testes are contained entirely within the abdomen, extending from the fourth segment to the telson, where they fuse. Separate vasa deferentia extend anteriorly from the testes to the eighth thoracomere, where they enter long processes called genital papillae or gonopods (Deecaraman & Subramoniam 1980a; Schram 1986). These tubes protrude and hang down from the medial surface of the precoxae of the last pair of walking legs. They often have sclerotized tips that appear adapted for insertion into the female's gonopores. Just prior to copulation, the gonopods of *Pseudosquilla ciliata* become more rigid (Caldwell unpublished). Deecaraman and Subramoniam (1983b) also mention that the gonopods of *Squilla holoschista* become erect during copulation, so this may be a general phenomenon. Mature sperm are round, vesicular, and nonflagellated (Komai 1920, Deecaraman & Subramoniam 1980a). The vasa deferentia carry the sperm forward and embed them in a mucopolysaccharide cement. This forms a sperm cord that is ejaculated into the female (Deecaraman & Subramoniam 1983b).

Paired male accessory glands lie in the thorax and also open through the gonopods. They have openings separate from the genital orifices and deliver a proteinaceous fluid during copulation. This secretion probably serves to break down the sperm cords inside the female, releasing sperm. There also is evidence that protein from the male's accessory glands is taken up by the female's ovaries (Deecaraman & Subramoniam 1980a, 1983b).

In females, paired ovaries lie below the pericardial sinus and above the digestive glands, extending from thorax to telson, where they fuse. The oviducts arise anteriorly and enter a single median pouch above the sixth thoracic sternite. This opens to the outside through a highly sclerotized medial "figure-8" structure on the rib of the sixth thoracic sternite that receives the tips of the male's gonopods. Little is known concerning sperm storage in the female's seminal vestibule or exactly where fertilization occurs. In at least some species, viable sperm are stored for up to several weeks (*Pseudoquilla ciliata* [Hatziolos & Caldwell 1983], *Oratosquilla oratoria* [Hamano 1988], *Odontodactylus scyllarus*, *Gonodactylus bredini* [Caldwell, unpublished]).

Reproductively mature females possess cement glands lying just under the sixth, seventh, and eighth thoracic sternites. These glands develop in synchrony with the ovaries (Do Chi 1975; Deecaraman & Subramoniam 1983a; Caldwell

1986) and provide the cement used to hold the eggs in a mass. Their packed, milky-white appearance when fully developed, clearly visible through the integument, provides a convenient means of determining when females are ready to spawn. In nonreproductive females, the glands are barely visible. Deecaraman and Subramoniam (1980b, 1983a, 1983c) report that cement glands in *Squilla holoschista* contain two types of secretory cells that produce glycoprotein and mucopolysaccharide. These products are released simultaneously to agglutinate the eggs.

The endocrine control of reproduction is poorly understood in stomatopods. Charniaux-Cotton (1960) reports that males have androgenic glands, lying along the vas deferens, that are involved in sexual differentiation. In females, Deecaraman and Subramoniam (1983c) examined the endocrine control of ovarian and cement gland development in *Squilla holoschista* and found a typical malacostracan pattern. Eyestalk ablation, removing the sinus glands and x-organs, promoted rapid development of the ovaries and cement glands. This suggests the presence of a gonad inhibiting hormone. Injection of various CNS extracts stimulated ovarian and cement gland development, implying a CNS source of a gonad stimulating hormone or a gonad inhibiting hormone binding agent.

Unlike many malacostracans, reproduction in stomatopods is not coupled to molting. Females typically mate and spawn when they are in an intermolt condition (Reaka 1976). In fact, reproduction and molting are mutually exclusive. Newly molted females are not reproductively developed, and brooding females do not molt. This is clearly demonstrated in *Gonodactylus bredini*. My students and I collected over 50,000 animals at one location in Panama and recorded only a single brooding female molt. No newly molted female laid eggs within two weeks after capture. In this species, molting and reproduction have a lunar periodicity, both occurring at the full moon. A female that molts one month will not reproduce until at least the next full moon (Caldwell 1986 personal observation).

In males, the relationship between molting and reproduction is poorly understood. In *Pseudosquilla ciliata*, testicular resorption occurs during molting (Hatziolos & Caldwell 1983), although it is not known if this occurs among other stomatopod species. In Panamanian *Gonodactylus bredini*, the molting of sexually mature males peaks a week after the full moon, a week later than for females. This allows males the opportunity to locate and pair with receptive females before they must physiologically commit to a molt (Caldwell personal observation).

In *Gonodactylus bredini* females, the presence of males may accelerate reproductive development. In a pilot study, I placed six adult females in a 0.7 m diameter tank with running seawater, unlimited snails as food, and six cavities. In three identical tanks I placed two females and four males. Prior to the experiment, each female was marked and her reproductive condition scored on the basis of cement gland development (Caldwell 1986). All females had moderate reproductive development (i.e., with thin, translucent strands of glandular tissue). Five days later, the females were removed and their cement glands checked. None of the six females maintained without males had changed, but all of the females housed with males showed increased cement gland development (opaque, white glands), indicating that they were ready to spawn (sign test, $p < .05$). While these females were not observed to determine if they paired or mated with the males, this experiment does suggest that reproductive development is under some social influence.

Tropical stomatopods typically breed year-round (Reaka 1979a), although seasonal variation in the proportion and/or size of individuals spawning may exist (Caldwell 1986). More temperate species usually breed seasonally (Morgan 1980; Greenwood & Williams 1984). Even different populations of the same species may vary with respect to seasonality. In the Bay of Panama, *Gonodactylus bahiahondensis* breed year-round, except for a variable period in February and March, when cold upwellings occur. In the Gulf of Chiraqui, 400 km to the west, where temperatures are more stable, they breed year-round (Caldwell, unpublished). Seasonal breeding appears most constrained by low water temperature, although other factors, such as range of tidal fluctuation or food availability, may be important in the tropics.

Life History

Sexual Maturation. The size and age at sexual maturation are known for only a few species (see Reaka 1979a). In general, males begin developing gonopods soon after metamorphosis from postlarval to adult form. For example, gonopod buds are visible in most species of *Gonodactylus* when they are 10 to 12 mm in total length and probably less than two months past settlement. The external gonopores of females develop later. In *Gonodactylus*, this typically occurs when juveniles are 16 to 18 mm in length and 4 or 5 months past settling (Caldwell, unpublished). Age of sexual maturation is more difficult to pin down. In a Panamanian population of *G. bredini*, males have mature sperm and will copulate at approximately 24 mm total length (7 or 8 months postsettlement). I have recorded females as small as 24 mm in length with eggs, but size of reproduction varies with season. During early wet season (June through August), females under 30 mm are not found with eggs, and typically, only females larger than 36 mm breed. During the late wet season and extending into the dry season (October through April), many females under 30 mm are found with eggs (Caldwell 1986). These sizes translate to females initiating breeding from as early as 9 to as late as 16 months, depending on the season (Steger & Caldwell, unpublished). As a rule of thumb, female stomatopods begin breeding at approximately 40 percent maximum body length for the species (Reaka 1979a; Caldwell, unpublished). This may not apply to some very small species, such as *Nannosquilla*, that may delay reproduction.

Mating. Courtship and mating systems are quite varied in stomatopods and will be discussed below. However, the final stages of courtship, mounting, and copulation are remarkably similar in the 8 families (Squillidae, Harpiosquillidae, Lysiosquillidae, Nannosquillidae, Odontodactylidae, Pseudosquillidae, Protosquillidae, and Gonodactylidae) in which I have observed mating. The following general description of mating is derived from these unpublished observations plus the few published reports of copulation in the literature (Serene 1954;

Dingle & Caldwell 1972; Hatziolos & Caldwell 1983; Deecaraman & Subramoniam 1983b; Hamano 1988).

Typically, the final stages of courtship include mutual antennulation (stroking with the first antennae), often of the partner's head, telson, and ventral thorax. This is followed by the male and/or female pushing his and/or her head under the other animal and the male then mounting the female. This often occurs near the female's telson, but the male may grasp her dorsum at any location, oriented toward either her head or telson. He then proceeds to work his way forward, using his third, fourth, and fifth maxillipeds to cling to the female. If the male was oriented posteriorly, he attempts to turn around when he reaches the female's telson and then moves anteriorly. Once the male reaches the female's thorax, he hooks the dactyli of his maxillipeds under the edges of her carapace and extends the large second maxillipeds to further stabilize his position. At the same time, the male uses his first maxillipeds to scratch the female's anterior carapace and rostrum. At this point the male attempts to partially turn the female on her side, apposing their ventral thoraxes. If the male fails to properly position the female, he often tries to turn her from the other side or remounts. Once the male succeeds in placing his erect gonopods near the female's gonopores, he gains insertion by strong single thoraco-abdominal thrusts. When intromission occurs, the male passes the sperm cords into the female's seminal vestibule through a rapid series of copulatory thrusts that may last for several seconds. Once the male is finished thrusting, the female begins to struggle and the union is broken. In some species, remating may occur after a few minutes.

Spawning and Brooding. Spawning females typically lie slightly curled on their side or back while the eggs and cement are extruded through the gonopores. The female reaches back with her first maxillipeds and uses them to form the eggs into a loose mass against the ventral thorax, walking legs, and first pleopods. When all the eggs are extruded, the female then uses her maxillipeds to knead the mass into a thick disk. Final shaping takes place over the next few hours while the mass is still sticky (Hamano & Matsuura 1984; Caldwell unpublished). The shape of egg masses ranges from thin, circular sheets only two to four eggs thick in squillids and lysiosquillids to near circular balls in some gonodactylids. One notable exception to this pattern occurs in *Nannosquilla*, a small sandburrowing lysiosquilloid. At least some species in this genus cement eggs to the side of the burrow (Rathbun 1910; Manning 1967, 1979; Watanabe personal communication).

Females of most species of stomatopods spawn and brood their eggs alone, although in a few cases males remain with females during spawning and even brooding (see below). In all known cases, females remain with their eggs, in the burrow or cavity, at least until hatching. As well as defending the brood site, females aerate the eggs by the physical beating of their pleopods, circulating water through the cavity or burrow. More direct care is also provided by frequently palpating or picking up the egg mass with the maxillipeds, manipulating and cleaning it, and eating stray or undeveloped eggs. Without such care, egg masses typically foul and degenerate (Reaka & Manning 1981; Hamano & Matsuura 1984; Caldwell, unpublished). On two occasions, I observed female

Squilla aculeata apparently assist the hatching of their young. Both females were captured in the field with large, disc-shaped egg masses. They were held in aquaria for several days with their eggs and were frequently observed picking them up. when the eggs began to hatch, the females spread the masses into their full diameter and slowly and repeatedly flapped them as one would a rug. The result was a massive release of larvae. When most had hatched, the females discarded, and then ate, the matrix material (Caldwell unpublished).

Larvae and Postlarvae. Giesbrecht (1910), in his classic study of Mediterranean stomatopod larvae, established the foundation for what is currently known of their development. Pyne (1972) surveyed the literature and identified only 32 species of stomatopods for which larvae had been positively identified. Since then, many more stomatopod species have been described, but larval forms have been attributed to a very few. Despite the cannibalistic and aggressive tendencies of stomatopod larvae, a few researchers have successfully reared gonodactylids and squillids from hatching to postlarvae. Much of our knowledge comes from these studies. The general picture that has emerged is that the cavity dwelling gonodactylids have relatively large yolk-rich eggs and three propelagic larval stages that remain in the cavity with the female. There are four pelagic stages that feed in the plankton for one to two months before settling as postlarvae (Dingle 1969; Dingle & Caldwell 1972; Manning & Provenzano 1963; Provenzano & Manning 1978; Morgan & Goy 1987). The burrowing lysiosquillids and squillids generally have smaller eggs and fewer propelagic stages that remain in the burrow, usually only two. In *Squilla aculeata*, larvae become photopositive and leave the burrow 48 hours after hatching (Caldwell unpublished). Greenwood and Williams (1984) report that the lysiosquillid *Heterosquilla tricarinata* has only one such stage. On the other hand, the larvae of these groups typically remain in the plankton longer, have a greater number of pelagic stages, nine being the typical number, and settle as larger postlarvae (Alikuhni 1950; Gohar & Al-Kholy 1957; Pyne 1972; Reaka 1986; Hamano & Matsuura 1987a). Unfortunately, we know so little about the eggs and larvae of other groups that few generalizations can be made about the type of home they occupy or their life history. Reaka (1979a) and Reaka and Manning (1980, 1981, 1987) discuss divergence in life history strategies with respect to dispersal and rates of evolution.

The survival of larvae in burrows, as opposed to cavities in hard substrata, and the defensibility of larvae while in these homesites would seem to warrant further study. We know practically nothing about the micro-environment inside stomatopod burrows and cavities. Females of at least some burrowing species plug the entrance when brooding young (*Oratosquilla oratoria* [Hamano & Matsuura 1987b], *Pseudosquilla ciliata* [Hatziolos 1979], *Lysiosquilla glabriuscula, L. sulcata, Squilla parva, S. aculeata, Meiosquilla dawsoni, M. swetti* [Caldwell, unpublished]). While the same is true of cavity dwelling gonodactyloids (Caldwell & Dingle 1972; Montgomery & Caldwell 1984), given the type of substrate and the size and shape of the brooding chambers, low oxygen tensions and/or the buildup of metabolic wastes might present more of a problem for burrowers. This could contribute to the evolution of smaller larvae and shorter retention times in the burrow.

Differences in the ability of burrowing, as opposed to cavity living, stomatopods to defend and transport eggs and propelagic larvae might also affect the types of eggs and larvae produced. While we have almost no information on the persistence of individuals in burrows and cavities, one study on a brooding gonodactylid found it remarkably resistant to eviction (Montgomery & Caldwell 1984). Less aggressive, spearing burrowers may be subject to more frequent eviction or, because of the nature of the substrata they inhabit, destruction of their burrows. Stomatopods can carry their egg masses in their maxillipeds and have been observed in the field to successfully move them to another cavity (*Gonodactylus bredini, G. oerstedii*; Caldwell unpublished), but there is considerable risk. A female gonodactylid can search through a complex environment and even fight for a cavity while carrying her egg mass, but a burrower must traverse relatively open terrain and may have to excavate a new burrow while protecting her eggs. This would be most difficult, since the same appendages are needed to hold the eggs and dig. Once hatched, large numbers of larvae cannot be transported and loss of the burrow or cavity would probably spell their doom. Greater risk while in the brood chamber may select for fewer and/or shorter propelagic stages in spearers.

MATING SYSTEMS

Because of the relative ease with which cavity dwelling stomatopods can be collected from rocks and coral and maintained and observed in the laboratory, we know more about their reproductive behavior than we do about that of species that live in burrows. For the remainder of this paper I will briefly summarize what is known about the mating systems of a typical cavity living, smashing gonodactylid, *Gonodactylus bredini*, and a burrowing, spearing squillid, *Oratosquilla oratoria*. I will then conclude by describing the mating systems of a few exceptional species for which I have data and attempt to determine what factors have caused them to diverge from more typical mating patterns.

Reproduction in *Oratosquilla oratoria*

Commonly known as the Japanese mantis shrimp, *Oratosquilla oratoria* has been studied extensively by Tatsuo Hamano and his colleagues. This species grows to 15 cm in total length and is commercially important in Japan. From trawling catches Hamano has obtained detailed information on growth, reproductive condition, and population structure throughout the year. This species also survives well in the laboratory, living in artificial plastic burrows. This has permitted observation of feeding, mating, spawning, and brooding behavior.

Common in the muddy bays and inlets of Japan, *Oratosquilla oratoria*, like most squillids, lives in U-shaped burrows that it excavates. The burrows are about six times the length of the resident. Both entrances are open, but one is usually larger than the other (Matsuura & Hamano 1984). Diet consists mostly of various shrimps and clams, although fish and annelids also were found in their stomachs. Interestingly, large females of reproductive size were more likely to take clams (Hamano & Matsuura 1986). At Hakata Bay, animals settle out of the plankton in August and September and live three or four years. There is little or no growth during the winter when water temperatures are below 15°C. Females probably do not breed until their second year, when they have a carapace length of 20–25 mm. Breeding is highly seasonal, and females with ripe ovaries are found from May to August. In the laboratory some females were able to lay two clutches during this period, and this may occur in the field, particularly in large third and fourth year females that tend to breed earlier in the year (Hamano, Morrissy & Matsuura 1987). Eggs are about 0.45 mm in diameter, and clutch size is positively correlated with body length, ranging from 30,000 to more than 100,000 eggs. Time to hatching is dependent on water temperature and varies from approximately 22 days at 20°C to 8 days at 27°C. The larvae remain in the burrow with the female during the first two propelagic stages (2 or 3 days) and leave as third stage larvae when their yolk is exhausted and they are approximately 3 mm in total length. In the plankton, they usually pass through nine larval stages and settle out as 16–17 mm juveniles one to two months later (Hamano & Matsuura 1987a). However, it is not uncommon to find supernumerary larval instars that have extended their time in the plankton (Hamano & Matsuura 1987c).

While mating by *Oratosquilla oratoria* has not been observed in the field, laboratory observations suggest that this species is promiscuous, with mating probably occurring on the surface or in the burrow entrance. Females captured at the beginning of the spawning season and isolated from males all contained viable sperm and laid fertile eggs up to two months later. When males and females were held together in the laboratory, courtship and copulation proceeded through the typical sequence of antennulation, mounting, scratching, and thrusting. Following copulation, the male and female did not interact and the male did not occupy the female's burrow (Hamano 1988). Mating in the open with no cohabitation in the burrow is probably typical for burrowing squillids, although I know of no published accounts of mating observed in the field. On the Pacific coast of Panama I have witnessed *Squilla parva* mating at night during a low tide when the water on the mud flat was only a few centimeters deep. Pairs courted and copulated in the open and then quickly separated, males and females returning to their own burrows.

Reproduction in *Gonodactylus bredini*

Gonodactylus bredini is a moderate-sized (50–70 mm) stomatopod found throughout the Caribbean, north to the Carolina coast and Bermuda. Individuals occupy cavities in various hard substrata such as rock, coral rubble, calcareous algae, shells, and even such unlikely items as seeds and bottles. Individuals are typically found in the low intertidal or just subtidally, although at some locations *G. bredini* may occur down to 50 m (Manning 1969; Caldwell 1988). Like most *Gonodactylus*, *G. bredini* is a generalist, feeding on various snails, crabs, and hermit crabs (Caldwell, Shuster & Roderick 1989, Caldwell & Childress 1990). In Bermuda, *G. bredini* reproduces only in the summer, but in the Caribbean, where I have studied it most extensively, populations breed year-round

(Dingle 1969; Dingle & Caldwell 1972; Caldwell 1986). Postlarvae recruit from the plankton at 6–10 mm of length, molt to juveniles within a few days, and begin occupying cavities. Animals achieve sexual maturity by the end of their first year (see above) and probably live no more than 4 or 5 years.

Reproduction in the coastal Panamanian population of *Gonodactylus bredini* displays a strong lunar periodicity. Females develop their ovaries and cement glands one to two weeks prior to the full moon and oviposit at the full moon. Unless a female has developed ovaries, she will not mate and acts aggressively toward males. A few days prior to the full moon, pairs form in cavities and mating occurs (fig. 5.1). There are two ways that males and females come together. In some cases, males leave their burrows and search for receptive females. however, even if the male locates a female ready to mate, she typically will not admit him to her cavity. The male must fight his way in. If he is sufficiently strong and/or persistent to gain entry, the female usually mates and remains with him in the cavity. This means that when a male pairs with a female, he will typically be as large as, or slightly larger than, his mate. (Since cavity entrances are usually only slightly larger in diameter than the resident, it is unlikely that a male much larger than a female could physically enter her

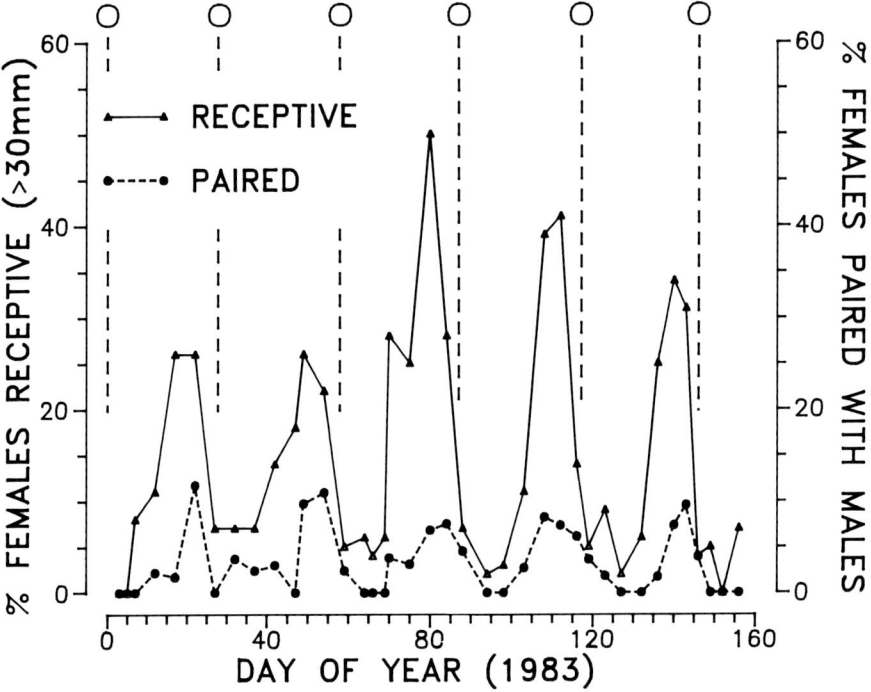

FIGURE 5.1. Percentage of females in collections of *Gonodactylus bredini* from Isla Mina, Panama, that were receptive or that had paired. Data are from 1654 females greater than 25 mm total body length collected during the first 160 days of 1983. Animals were obtained by breaking apart randomly collected pieces of coral rubble. Each collection had a minimum of 50 adult stomatopods. Receptivity in females was determined by the presence of developed cement glands.

cavity.) However, it is costly for males to leave their cavities to search for females because of the risk of predation and the fact that their own cavities may be occupied while they are away.

Alternatively, receptive females may go looking for a male. When they find one, he usually admits the female without hesitation. males do not seem to be able to tell if a female is receptive. They also admit nonreceptive females, who then attempt to evict them. Once a receptive female enters a male's cavity, they mate and pair. Since males are not aggressive toward approaching females, pairs may form even if the female is somewhat larger or considerably smaller than the male. While the majority of pairs captured in the field are closely matched, discrepancies are not normally distributed. There are more large male, small female pairs than would be expected, especially considering that in larger size classes males are rare due to the risks associated with breeding (see below). Females exhibit some choice when they are residents and have a strong positional advantage. They are less choosy and will accept a wider range of males when they must search for a mate. The precise dynamics of who searches for whom and when are not understood, but as the time for spawning approaches, more receptive females are found away from their cavities, suggesting that they can no longer wait for a male to come to them. Also, at times other than just prior to a full moon, males are less willing to admit a female into their cavity, presumably because she is less likely to be receptive (Caldwell 1986).

Once a pair forms, the male is vigilant and defends the cavity against both male and female intruders. While males are reluctant to leave the cavity to court another female, females will mate with a different male if he is successful in evicting her mate. This rarely happens, however, since guarding males are extremely effective in their defense (Shuster & Caldwell 1989). A number of factors favor male guarding behavior. First, while we know nothing about sperm precedence in stomatopods, it seems reasonable to assume that a male's paternity would be compromised should his mate subsequently copulate with another male. This also may explain why males copulate repeatedly with their partners. Second, because of his strong positional advantage, the male is usually able to sequester the female and prevent her from mating with other males. Third, female breeding synchrony makes it less likely that the male will find another receptive female should he leave (Knowlton 1979). Breeding synchrony could be driven by the need to release larvae at an appropriate time in the lunar tidal cycle, as a by-product of synchronized molting that is favored for physical or biological reasons (Reaka 1976, 1979b) or as a means of reducing the cost to males and/or females of searching for mates. Finally, because the cost of searching for receptive females is extremely high due to the risk of predation, once a female is located and mated it is probably better to remain with her and protect one's investment until she spawns than it would be to leave and search for other females while risking loss of paternity. It is also to the female's benefit to pair with a male. He defends the cavity not only against other males but also against any intruder (females or other species of stomatopods, fish, octopus, etc.) that might cost the female a site to brood her eggs.

After the female spawns, the male leaves the cavity within a few hours and does not return. However, should he encounter her again during the next month while she is brooding eggs or larvae that he probably fathered, he recognizes his former mate and will not attempt to evict her (Caldwell, personal observation). Females that are brooding or that are not receptive frequently give a specific display to males that approach their cavity. The display involves a rapid whirling and pulsing of the maxillipeds and appears to signal an unwillingness to mate. The display is rarely directed at other females but is somewhat effective in causing males to leave the area without challenging the resident.

Because males leave the cavity when their mate spawns, each breeding episode requires that a male locate a new cavity. Observations and trapping in the field show that large numbers of males are moving through a population immediately following the full moon. Presumably, these are males that have just left their mates (Caldwell 1986). In populations where cavities are scarce and/or predation is intense, this exacts a heavy toll on males. In Panama, we find that at the age of first reproduction, the sex ratio is 1:1. However, after two years of breeding, large males (>40 mm) are outnumbered by females at least two to one (Jackson et al. 1989). This may help explain why some large females search for males, and it may make it possible for large males to remain in their cavities until they are at least assured of successfully breeding. Still, each time a male breeds, he must ultimately run the gauntlet of finding a new cavity.

Small males, under about 35 mm, appear to engage in an alternative mating tactic. If they see a female moving near their cavity, they frequently will rush out and chase and attempt to grasp the female and copulate. Usually the female is not receptive and will flee, attack, or attempt to beat the male back to his cavity and take it over. While many males lose their cavities in this way, I have seen several small males successfully mate with a female in the open. I have not seen large males attempt this ploy. Because cavities are not as limiting for small animals (Steger 1987), these males may be able to locate new homes more easily and therefore have less to lose than a large male attempting the same tactic.

Another, albeit infrequent, mating tactic that I have observed in *Gonodactylus bredini* occurs when males discover a recently molted female. After molting, stomatopods are defenseless and cannibalism frequently occurs (Steger & Caldwell 1983). However, during staged laboratory interactions between recently molted (less than 24 hrs) females and same-sized intermolt males, I have recorded several instances when the male seized the female and forced a copulation. If the female was outside the cavity, she was released after mating. If the female was resident in a cavity, she was evicted following copulation and the male retained the cavity. Although the female will not spawn for at least a month, sperm can be stored for at least that long (see above). Apparently, the chance that the female will use his sperm to fertilize her eggs is worth more to the male than is the female's nutritional value, were the male to eat part of her. A newly molted male trapped in a cavity by another male or female is usually killed and partially eaten.

Reproduction in *Pseudosquilla ciliata*

One of the most widely distributed stomatopods is *Pseudosquilla ciliata*, occurring circumtropically except for the eastern Pacific. A burrowing, spearing species, it is typically found in shallow seagrass beds or on coral rubble and sand flats (Kinzie 1968; Hatziolos 1979). What makes mating in this species so unusual is the promiscuous nature of the females and the reticence of males to mate with some females. While males and females may engage in elaborate courtship, males frequently must be "forced" to mate by the female. The female grasps the male's telson and holds on until he tires, turns, mounts, and copulates. Following copulation, the male and female may continue to court for a few minutes and may even remate, but then they separate. Females have been observed to mate with several males in one day, and they will copulate at any stage in their reproductive cycle. Hatziolos and Caldwell (1983) demonstrated that males are usually coy and readily mate only with larger females.

It is not clear why males should be reluctant to mate with smaller females. There is little risk involved in mating, either from predators or from the female. The male does not give up his burrow to mate. And it seems unlikely that sperm are limiting, since males have been observed to mate with different females after only a few minutes. As in some insects, males may pass some limiting nutrient to the female during copulation. Deecaraman and Subramoniam (1983b) suggested that in *Squilla holoschista* male accessory sex gland secretion is passed to the female during copulation and later appears in the ovary. Another explanation might be a venereal agent that decreases the probability of subsequent fertilizations by the male. While there is no evidence supporting either hypothesis in *Pseudosquilla ciliata*, in either case, males would do better mating with the largest possible females because they produce many more eggs.

Cavity Dwelling Spearers

There are a few species of spearers known to live in preexisting cavities rather than in the more typical burrow. While we know very little concerning the mating systems of these species, the few fragments of information available are instructive. Reaka and Manning (1981) report finding a cohabitating pair of the squillid *Meiosquilla tricarinata*, although they do not report the reproductive condition of the female. On the Atlantic coast of Panama I have collected pairs of a similar cavity dwelling species, *M. lebouri*, and in every case but one the female had developed cement glands indicating that she was ready to spawn. In the one exception, the female had just spawned and the male was still in the cavity with her. These observations suggest a mating system in cavity living *Meiosquilla* similar to that seen in most gonodactylids. It seems likely that the same factors associated with cavity living that promote mate guarding in *Gonodactylus*, namely competition for cavities, high search costs, and defensibility of the female, are also involved in *Meiosquilla*.

However, at least some burrowing *Meiosquilla* also form pairs. While working on a mud flat in the Bay of Panama from March through July, I collected 96 *M. swetti* from burrows, including two pairs. In both cases, the pairs were taken during the full moon and the females had developed cement glands and ovaries. Similarly, of 272 *M. dawsoni* collected from burrows, 12 pairs were found. Again, the paired females were reproductively developed and 11 of the pairs were found during a full moon, the other during a new moon. Species of this genus tend to be found in burrows in gravelly mud or under rocks as well as in cavities. Such living sites are relatively stable and may be costly to construct. Laboratory studies on *M. dawsoni* and *M. swetti* show that both species are aggressive and attempt to evict other individuals from burrows (Dingle 1983). Pairing in this genus may be associated with their occupation of defensible homesites. *Pseudosquilla* sp. (similar to *P. ornata*) is a dwarf species (postlarvae settle at 17 to 20 mm; females produce eggs at 22 mm) found in cavities in coral rubble in the Society Islands. While studying it on Moorea, I was able to record the details of courtship and brooding (Caldwell, personal observation). Its reproductive behavior is a mixture of that described above for the larger, burrowing *P. ciliata* and cavity dwelling gonodactylids. Courtship and copulation usually take place in the open in front of the male's cavity. Females approach the entrance, then turn and move slowly away. The male then comes out of his cavity and follows. While the courtship displays are elaborate and similar to those seen in *P. ciliata*, males do not appear particularly coy. Copulation usually occurs twice, males mounting once from each side, with a few minutes of continued "dancing" in between. After the second copulation, the male moves back to his cavity, followed by the female. The pair remains in the cavity for only a few hours before the male leaves. Out of 145 animals captured in the field, only one pair (male 24 mm, female 27 mm with cement gland development) sharing a cavity was collected. In contrast to *Gonodactylus*, the male leaves the female in the cavity before she lays eggs. In fact, copulation and spawning were not strongly correlated in this species, although females with cement gland development were more likely to mate. Females are promiscuous and mated with up to three males in quick succession if they were not allowed to enter the males' cavities. I suspect that females trade copulations for cavities in this species and that the multiple matings and short period of cohabitation are male tactics involved in sperm competition, although much work remains to be done to test this hypothesis.

Monogamy in *Lysiosquilla*

Lysiosquilla is a genus of burrowing spearers found primarily on sandy substrates. Most species are quite large, and one, *L. maculata*, obtains the greatest total length (39 cm) of any stomatopod. Burrows are U-shaped and are remarkable for their size. Caldwell (1988) reports that the burrow of one *L. tredecimdentata* was 10 m in length and 10 cm in diameter. *Lysiosquilla* are specialized ambush predators and hunt from their burrow entrances. During the day, the burrow is capped with a sand-mucus membrane, with only a small hole in the center. The animal sits in the entrance with its camouflaged eyes and antennules protruding through the opening. Passing fish are impaled by the strike of the raptorial appendages, which slash through the burrow cap.

Adults of each species of *Lysiosquilla* for which we have data occupy burrows as single, male-female pairs: *L. maculata* (Borradaile 1898; Serene 1954; Caldwell unpublished), *L. tredecimdentata*. (Caldwell 1988), *L. sulcata* (Steger & Steger-Bemis in press), and *L. panamica* and *L. glabriuscula* (Caldwell unpublished). That *Lysiosquilla* occur in pairs first came to my attention when I found an article by Sather (1966) describing how Bajau fisherman of Sabah catch *Lysio-*

squilla with snares. He reports that after catching a "mantis-prawn," the fisherman always reset the snare to catch the stomatopod's mate.

Males and females are sexually dimorphic. Males have larger raptorial appendages and eyes that seem to be associated with their doing most of the hunting from the burrow entrance. Females are provisioned by the males (Steger, personal communication; Caldwell unpublished). Pairs appear to form prior to or at the time of sexual maturity. The smallest pairs of *Lysiosquilla glabriuscula* and *L. panamica* that I captured were 80–90 mm in total length. The females had gonopores, but the ovaries were not developed. Steger and Steger-Bemis (in press) report the same situation for *L. sulcata* from Moorea. The pairs stay together for months, if not years, and may remain in the same burrow for life. During molting, and probably brooding, the burrows are sealed with thick plugs, both animals remaining inside. When females die or are removed from the burrow, the males typically leave. However, if the male is lost, the female may recruit a new mate, often smaller than her previous partner (Steger & Steger-Bemis personal communication).

Long-term monogamy is relatively rare in invertebrates (Wickler & Siebt 1981) and has not been reported in other stomatopods. Why should it occur in this group? I suspect the answer lies in the cost of constructing the burrow and the risk of moving form one burrow to another. *Lysiosquilla* are strongly adapted for living in burrows. They have lost much of their body armor and have legs better suited for holding position in a burrow than for walking. In fact, they are clumsy on the surface and are easy prey for predators. Given the open sand substrates on which they are usually found, moving from one burrow to another must be perilous. Also, as mentioned above, their burrows are extremely large and represent a considerable investment. Adult *L. sulcata* are unable to construct a new burrow because they lack the tremendous amounts of mucus needed to stabilize the shifting sands in which they live. Burrows may be excavated by postlarvae and then continually expanded (Steger, Steger-Bemis & Caldwell, personal observation). The risks of predation and/or loss of a burrow while periodically searching for a mate may be so great that monogamy is favored.

Female Choice in *Haptosquilla trispinosa*

Haptosquilla trispinosa is a small (35 mm) Indo-Pacific gonodactyloid that lives subtidally on open sandy plains in cavities in coralline algae, shells, and coral rubble. While it is similar in size and feeding habits to many *Gonodactylus* species that also occupy similar habitats, its reproductive behavior is very different. Females brood eggs and larvae as in *Gonodactylus*, but I have never found a pair cohabiting. Also, in contrast to *Gonodactylus*, *H. trispinosa* females will mate repeatedly at any stage in the reproductive cycle, including while brooding eggs and larvae and just prior to a molt. But perhaps the most unusual aspect of reproduction in *H. trispinosa* is that females exercise considerable choice over which males they will accept as mates and generally only copulate with males considerably smaller than themselves.

Courtship is complex in *Haptosquilla trispinosa*. When a female detects a male, she immediately blocks the entrance of her cavity with her telson (fig.

5.2A). If the male does not immediately approach, she will periodically look out, but then quickly replugs the entrance. Females also threaten approaching males, lunging partway out of the entrance while spreading the raptorial appendages. When the male nears the entrance plugged by the female's telson, he antennulates the opening and then grasps her telson with his maxillipeds, scratching it with the brush-tipped first maxillipeds. After a few seconds, he backs up one or two body lengths along the edge of the cavity and waits for the female to reappear. If she looks out, he may start a series of raptorial appendage pumping displays, rapidly moving them in and out laterally while simultaneously pulsing the other maxillipeds. If the female replugs the entrance, he again approaches and resumes scratching. After several bouts of such behavior, sometimes lasting up to 30 mins, if the female is receptive, she comes partially out of the cavity, extending just far enough to expose her gonopores, and remains motionless. The male rapidly approaches, grasps her head and carapace, scratches her rostrum with his first maxillipeds, turns her on her side, mounts, achieves insertion, and gives a rapid series of copulatory thrusts (fig. 5.2B). Copulation is much more rapid than in other stomatopods, lasting only one or two seconds. At this point the female invariably pulls back into the cavity, often while threatening, and the male flees. If the male does not leave immediately, he is attacked. The larger the male, the less likely the female is to unblock the entrance and look out and the more likely she is to threaten and strike. When I presented each of 43 females with a male her own size and a male 15 percent smaller (males presented 12 hrs apart, order balanced), a significantly greater number of females mated with the smaller males (McNemar Test, $N = 22$, $p < .02$).

Again, the most likely explanation for this behavior is cavity limitation and the cost of locating mates. Most suitable cavities in the habitat I studied were occupied. Individuals were often several meters apart. Moving across the open sand plain in a world filled with predatory fish is dangerous. In over 100 hrs of diving between dawn and dusk and 8 hrs at night I did not a single *Haptosquilla trispinosa* out of its cavity, except to make short sallies to capture prey. Animals released in the open were quickly eaten by fish. A stomatopod without a cavity in this environment has a very short life expectancy.

The costs of locating new cavities has lead to extreme competition for living space. Contests for cavities are intense and frequently last several minutes. At the same time, the telson of *Haptosquilla trispinosa* is highly modified for blocking the entrance. Residents have a strong positional advantage and are not easily displaced, as long as the intruder remains outside. When large males approach females in cavities, they frequently attempt to evict them rather than court, suggesting that cavities are more valuable than matings. In the study involving the 43 females discussed above, 12 females were evicted by same-sized males while none were evicted by smaller males (McNemar Test, $N = 12$, $p < .01$). The risk of eviction probably explains why females mate at the entrance with smaller, less dangerous males and do not give up their positional advantage. Also, because of the risks associated with finding mates and the infrequent movement of males, females that copulate at any opportunity, as long as the males pose little threat of evicting them, are more likely to have sperm available when they are ready to reproduce.

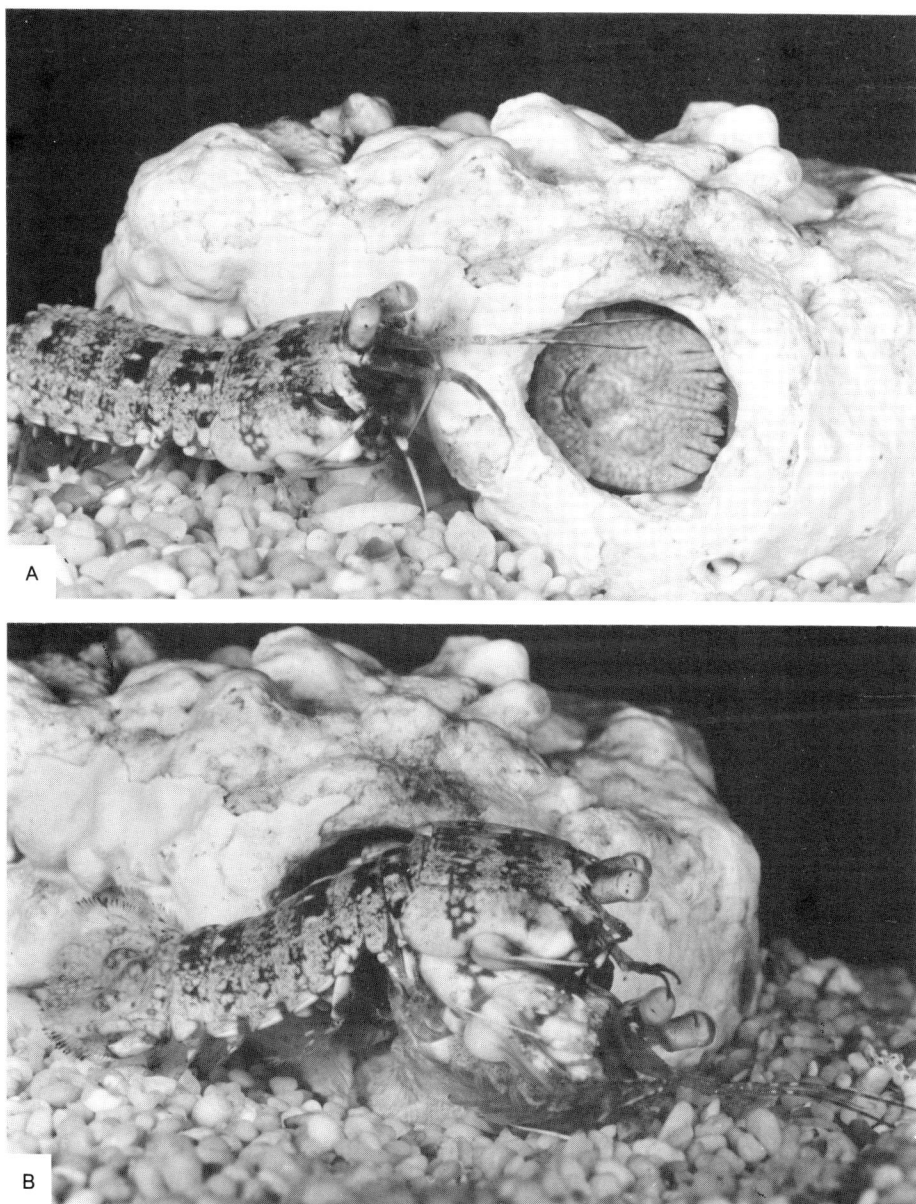

FIGURE 5.2. A. Mating in *Haptosquilla trispinosa*. During the initial stages of courtship, a female blocks the entrance to her cavity using her armored telson. The animal on the left is a 30 mm courting male. B. After 12 mins of courtship by the male, the female extends her body halfway out of the cavity entrance, exposing her gonopores. The male mounts, grasps the female by her carapace, and achieves intromission.

CONCLUSIONS

My major objective in this paper has been to demonstrate the diversity of mating systems that has been uncovered in stomatopods, considering the relatively few species that have been examined, and to tentatively point to those physical and biological factors that appear to play a role in the evolution of these systems. As is true for many aspects of stomatopod biology, the form of raptorial appendage often appears to be associated with the type of mating system a species has evolved. Since the raptorial appendages often reflect the type of habitat occupied, predatory specializations, social interactions, and perhaps most important, the kind of burrow or cavity occupied, this is not surprising. The same selective forces that shape these coevolved characters will obviously also influence the evolution of mating systems.

Mating Systems

Burrowers seem much more likely to mate promiscuously on the surface, and there is little evidence of mate guarding. This may reflect the more ephemeral nature of burrows, their poorer defensibility, or physical constraints that preclude sharing of burrows due to space and/or metabolic considerations. The major exception to this is found in the genus *Lysiosquilla*, whose large, stable, and expensive burrows, coupled with severe risks of locating mates, favor long-term monogamy. However, I must point out that we have very little information on any species of burrowing stomatopod with respect to these parameters. This should prove a fruitful area for further study.

Preexisting cavities in rock or coral generally provide more defensible and permanent homes, making postcopulatory mate guarding a viable tactic for species that live in cavities and have the ability to defend them. This appears to have evolved in several groups, including some spearing species, that occupy holes. At the same time, intense competition for cavities appears to have set the stage for the evolution of alternative mating tactics such as forced copulation and extra-cavity pursuit of females by small males. It may even have led to females using the prospect of mating as a competitive ploy to gain access to males' cavities and thus bypassing the males' positional advantage. And in some species, such as *Haptosquilla trispinosa*, cavities may be so limiting that mate guarding does not occur because neither member of the pair is willing to give up a cavity once they have possession of it.

Care of Eggs and Larvae

Again, the use of burrows and cavities appears to dictate the extent of the care provided by the female to her eggs and larvae. In burrowing species, egg size is generally small and developmental time relatively short. After hatching, larvae usually enter the plankton after a couple of molts. This is in contrast to many cavity living species that invest in large, yolk-laden eggs that have longer developmental times and that produce larvae that remain for a longer period (three propelagic stages in Gonodactylidae) in the breeding cavity. It seems reasonable to postulate that the more transient nature of burrows, the inability of residents to defend them, and possible physical and metabolic problems associated with the burrow environment select for shorter periods in the burrow by eggs and larvae.

In all known cases, except in the genus *Lysiosquilla*, the female assumes sole responsibility for the care of the eggs and larvae. Since there is little sexual dimorphism in stomatopods and females can defend their residence as effectively as can a male, the male is likely to gain little by remaining with the eggs after they are laid. In fact, because of tight quarters, his presence could actually be harmful because of mechanical disturbance, lowered oxygen titers, or the release of metabolic wastes. It is not known what role male *Lysiosquilla* take in caring for eggs and larvae. At the very least they are probably involved in the sealing of the burrow during brooding and defending the entrance should it be uncovered. The relatively large size of the burrow system may reduce the chance that the presence of the male will harm the brood.

As more stomatopod species are studied, a more comprehensive picture of the processes that have led to this diversity of reproductive behavior should develop. For now, there are more questions than answers, but I hope that I have convinced the reader that the problems are tractable and that the avenues open for exploration are many.

Acknowledgments

I would like to thank my colleague, Dr. Hugh Dingle, and my former students, Drs. Marjorie Reaka, Ken Evans, Marea Hatziolos, Ilze Berzins, George Roderick, Eldridge Adams, and Stephen Shuster, who have contributed to our understanding of stomatopod behavior. Funds were partially supplied by NSF BNS 80-23414 and BNS 85-17573.

Literature Cited

Alikuhni, K. H. 1950. Observations on some larval and post-larval stomatopods. *J. Bombay Nat. Hist. Soc.* 49:101–109.

Borradaile, L. A. 1898. On some crustaceans from the South Pacific. Part 1. Stomatopoda. *Proc. Zool. Soc. Lond.*, no. 3, pp. 32–38.

Caldwell, R. L. 1979. Cavity occupation and defensive behavior in the stomatopod, *Gonodactylus festae:* Evidence for chemically mediated individual recognition. *Anim. Behav.* 27:194–201.

Caldwell, R. L. 1982. Interspecific chemically mediated recognition in two competing stomatopods. *Mar. Behav. Physiol.* 8:189–197.

Caldwell, R. L. 1985. A test of individual recognition in the stomatopod *Gonodactylus festae*. *Anim. Behav.* 33:101–106.

Caldwell, R. L. 1986. Withholding information on sexual condition as a competitive mechanism. In L. C. Drickamer, ed., *Behavior as a Factor in the Dynamics of Animal Populations*, pp. 83–88. Toulouse: Private publisher.

Caldwell, R. L. 1988. Interspecific interactions among selected intertidal stomatopods. In G. Chelazzi & M. Vannini, eds., *Behavioral Adaptation to Intertidal Life*, pp. 371–385. New York: Plenum Press.

Caldwell, R. L. & M. J. Childress. In press. Prey selection and processing in a stomatopod crustacean. In R. N. Hughes, ed., *Behavioral Mechanisms of Food Selection*. NATO ASI Series G, Ecological Sciences, V. 20, pp. 143–164. Berlin: Springer-Verlag.
Caldwell, R. L., & H. Dingle. 1975. Ecology and evolution of agonistic behavior in stomatopods. *Naturwissenshaften* 62:214–222.
Caldwell, R. L. & H. Dingle. 1976. Stomatopods. *Sci. Am.* 234:80–89.
Caldwell, R. L., S. Shuster & G. Roderick. 1989. Studies of predation by *Gonodactylus bredini*. In E. A. Ferrero, ed., *Biology of Stomatopods*, Selected Symposia and Monographs U.Z.I., 3, pp. 117–132. Modena, Italy: Mucchi Editore.
Charniaux-Cotton, H. 1960. La glande androgene du crustace stomatopode *Squilla mantis*. *Bull. Soc. Zool. Fr.* 85:110–114.
Christy, J. H. & M. Salmon. 1984. Ecology and evolution of mating systems of fiddler crabs (genus *Uca*). *Biol. Rev. Camb. Philos. Soc.* 59:483–509.
Cronin, T. W. & N. J. Marshall. 1989. A retina with at least ten spectral types of photoreceptors in a mantis shrimp. *Nature (Lond.)* 339:137–140.
Crook, J. H. 1964. The evolution of social organization and visual communication in the weaver birds (Ploceinae). *Behaviour Supplement* 10:1–178.
Deecaraman, M. & T. Subramoniam. 1980a. Male reproductive tract and accessory glands of a stomatopod, *Squilla holoschista*. *Int. J. Invertebr. Reprod.* 2:175–188.
Deecaraman, M. & T. Subramoniam. 1980b. Cement gland activity in *Squilla holoschista* (Crustacea: Stomatopoda). In T. Subramoniam & S. Varadarajan, eds., *Progress in Invertebrate Reproduction and Aquaculture*, pp. 68–76. Madras, Proc. of the First All India Sym. on Invert. Reproduction.
Deecaraman, M. & T. Subramoniam. 1983a. Synchronous development of the ovary and the female accessory sex glands of a crustacean, *Squilla holoschista*. *Proc. Indian Acad. Sci. Anim. Sci.* 92:179–184.
Deecaraman, M. & T. Subramoniam. 1983b. Mating and its effect on female reproductive physiology with special reference to the fate of male accessory sex gland secretion in the stomatopod *Squilla holoschista*. *Mar. Biol. (Berl.)* 77:161–170.
Deecaraman, M. & T. Subramoniam. 1983c. Endocrine regulation of ovarian maturation and cement glands activity in a stomatopod crustacean, *Squilla holoschista*. *Proc. Indian Acad. Sci. Anim. Sci.* 92:399–408.
Dingle, H. 1969. Ontogenetic changes in phototaxis and thigmokinesis in stomatopod larvae. *Crustaceana (Leiden)* 16:108–110.
Dingle, H. 1983. Strategies of agonistic behavior in Crustacea. In S. Rebach and D. W. Dunham, eds., *Studies in Adaptation: The Behavior of Higher Crustacea*, pp. 85–111. New York: Wiley.
Dingle, H. and R. L. Caldwell. 1972. Reproduction and maternal behavior of the mantis shrimp *Gonodactylus bredini* Manning (Crustacea: Stomatopoda). *Biol. Bull. (Woods Hole)* 142:417–426.
Do Chi, T. 1975. Biometrié de la reproduction de Squilla mantis (L) (Crustacé: Stomatopode) dans le golfe d'Aigues-Mortes (Mediterranée Nord-Occidentale). *Publ. Staz. Zool. Napoli* 39 (Supp.):114–139.
Emlen, S. T. & L. W. Oring. 1977. Ecology, sexual selection, and the evolution of mating systems. *Science* 197:215–223.
Giesbrecht, W. 1910. Stomatopoden. *Erster Theil. Fauna Flora Neapel Monogr.* 33:vii, 239.
Gohar, H. A. F. & A. A. Al-Kholy. 1957. The larval stages of three stomatopod Crustacea. *Publ. Mar. Biol. Stn. Ghardaqa* 9:85–130.
Greenwood, J. G. & B. G. Williams. 1984. Larval and early post-larval stages in the abbreviated development of *Hetersquilla tricarinata* (Claus, 1871) (Crustacea: Stomatopoda). *J. Plankton Res.* 6:615–635.
Hamano, T. 1988. Mating behavior of *Oratosquilla oratoria* (De Haan, 1844) (Crustacea: Stomatopoda). *J. Crustacean Biol.* 8:239–244.
Hamano, T. & S. Matsuura. 1984. Egg laying and egg mass nursing behavior in the Japanese mantis shrimp. *Bull. Jpn. Soc. Sci. Fish.* 50:1969–1973.
Hamano, T. & S. Matsuura. 1986. Food habits of the Japanese mantis shrimp in the benthic community of Hakata Bay, *Bull. Jpn. Soc. Sci. Fish.* 52:787–794.
Hamano, T. & S. Matsuura. 1987a. Egg size, duration of incubation, and larval development of the Japanese mantis shrimp in the laboratory. *Nippon Suisan Gakkaishi* 53:23–39.
Hamano, T. & S. Matsuura. 1987b. Sex ratio of the Japanese mantis shrimp in Hakata Bay. *Nippon Suisan Gakkaishi* 53:2279.
Hamano, T. & S. Matsuura. 1987c. Delayed metamorphosis of the Japanese mantis shrimp in nature. *Nippon Suisan Gakkaishi* 53:167.
Hamano, T., N. M. Morrissy & S. Matsuura. 1987. Ecological information on *Oratosquilla*

oratoria (Stomatopoda, Crustacea) with an attempt to estimate the annual settlement date from growth parameters. *J. Shimonoseki Univ. Fish.* 36:9–27.

Hatziolos, M. E. 1979. Ecological correlates of aggression and courtship in the stomatopod *Pseudosquilla ciliata.* Ph.D. dissertation, University of California, Berkeley (unpubl.).

Hatziolos, M. E. & R. L. Caldwell. 1983. Role reversal in courtship in the stomatopod *Pseudosquilla ciliata* (Crustacea). *Anim. Behav.* 31:1077–1087.

Hazlett, B. A 1983. Parental behavior in decapod Crustacea. In S. Rebach & D. W. Dunham, eds., *Studies in Adaptation: The Behavior of Higher Crustacea,* pp. 171–193. New York: Wiley.

Jackson, J. B., J. D. Cubit, B. D. Keller, V. Batista, K. Burns, H. M. Caffey, R. L. Caldwell, S. D. Garrity, C. D. Getter, C. Gonzalez, H. M. Guzman, K. W. Kaufman, A. H. Knap, S. C. Levings, M. J. Marshall, R. Steger, R. Thompson & E. Weil. 1989. Ecological effects of a major oil spill on Panamanian coastal marine communities. *Science (Washington, D.C.)* 243:37–44.

Kinzie, R. A., III. 1968. The ecology of the replacement of *Pseudosquilla ciliata* (Fabricius) by *Gonodactylus falcatus* (Forskal) (Crustacea; Stomatopoda) recently introduced into the Hawaiian Islands. *Pac. Sci.* 22:464–475.

Knowlton, N. 1979. Reproductive synchrony, parental investment, and the evolutionary dynamics of sexual selection. *Anim. Behav.* 27:1022–1033.

Knowlton, N. 1980. Sexual selection and dimorphism in two demes of a symbiotic, pair-bonding snapping shrimp. *Evolution* 34:161–173.

Komai, T. 1920. Spermatogenesis of *Squilla oratoria. J. Morphol.* 34:307–333.

Manning, R. B. 1967. *Nannosquilla anomala,* a new stomatopod from Californa. *Proc. Biol. Soc. Wash.* 80:147–150.

Manning, R. B. 1969. Stomatopod Crustacea of the Western Atlantic. *Stud. Trop. Oceanogr. Inst. Mar. Sci. Univ. Miami* 8:1–380.

Manning, R. B. 1979. *Nannosquilla vasquezi,* a new stomatopod crustacean from the Atlantic coast of Panama. *Proc. Biol. Soc. Wash.* 92:380–383.

Manning, R. B. & A. J. Provenzano, Jr. 1963. Studies on development of stomatopod Crustacea. I. Early larval stages of *Gonodactylus oerstedii* Hansen. *Bull. Mar. Sci.* 28:297–315.

Manning, R. B., H. Schiff & B. C. Abbott. 1984. Cornea shape and surface structure in some stomatopod Crustacea. *J. Crustacean Biol.* 4:502–513.

Marshall, N. J. 1988. A unique colour and polarization vision system in mantis shrimps. *Nature (Lond.)* 333:557–560.

Matsuura, S. & T. Hamano. 1984. Selection for artificial burrows by the Japanese mantis shrimp with some notes on natural burrows. *Bull Jpn. Soc. Sci. Fish.* 50:1963–1968.

Montgomery, E. L. & R. L. Caldwell. 1984. Aggressive brood defense by females in the stomatopod *Gonodactylus bredini. Behav. Ecol. Sociobiol.* 14:247–251.

Morgan, S. G. 1980. Aspects of larval ecology of *Squilla empusa* in Chesapeake Bay. *U.S. Natl. Mar. Fish. Serv. Fish. Bull* 78:693–700.

Morgan, S. G. & J. W. Goy. 1987. Reproduction and larval development of the mantis shrimp *Gonodactylus bredini* (Crustacea: Stomatopoda) maintained in the laboratory. *J. Crustacean Biol.* 7:595–618.

Provenzano, A. J., Jr. & R. B. Manning. 1978. Studies on the development of stomatopod Crustacea. II. The later larval stages of *Gonodactylus oerstedii* Hansen reared in the laboratory. *Bull. Mar. Sci.* 28:297–315.

Pyne, R. R. 1972. Larval development and behaviour of the mantis shrimp *Squilla armata.* Milne Edwards (Crustacea: Stomatopoda). *J. R. Soc. N. Z.* 2:121–146.

Rathbun, M. J. 1910. The stalk-eyed Curstacea of Peru and the adjacent coast. *Proc. U.S. Natl. Mus.* 38:531–620.

Reaka, M. L. 1976. Lunar and tidal periodicity of molting and reproduction in stomatopod Crustacea: A selfish herd hypothesis. *Biol. Bull. (Woods Hole)* 150:468–490.

Reaka, M. L. 1979a. The evolutionary ecology of life history patterns in stomatopod Crustacea. In S. Stancyk, ed., *Reproductive Ecology of Marine Invertebrates,* pp. 235–260. Columbia, South Carolina: Belle W. Baruch Library in Marine Sciences, University of South Carolina Press.

Reaka, M. L. 1979b. Patterns of molting frequencies in coral-dwelling stomatopod Crustacea. *Biol. Bull. (Woods Hole)* 156:328–342.

Reaka, M. L. 1985. Interactions between fishes and motile benthic invertebrates on reefs: The significance of motility vs. defensive adaptations. *Proc. Fifth Int. Coral Reef Cong., Tahiti* 5:439–444.

Reaka, M. L. 1986. Biogeographical patterns of body size in stomatopod Crustacea: Ecological and evolutionary consequences. In R. H. Gore & K. L. Heck, eds., *Crustacean Biogeography,* pp. 209–235. Rotterdam: Balkema Press.

Reaka, M. L. & R. B. Manning. 1980. The distributional ecology and zoogeographical relationships of shallow water stomatopod Crustacea from the Pacific coast of Costa Rica. *Smithson. Contrib. Mar. Sci.* 7:1–29.

Reaka, M. L. & R. B. Manning. 1981. The behavior of stomatopod Crustacea, and its relationship to rates of evolution. *J. Crustacean Biol.* 1:309–327.

Reaka, M. L. & R. B. Manning. 1987. The significance of body size, dispersal potential, and habitat for rates of morphological evolution in stomatopod Crustacea. *Smithson. Contrib. Zool.* 448:1–46.

Salmon, M. 1982. Courtship, mating systems, and sexual selection in decapods. In S. Rebach & D. W. Dunham, eds., *Studies in Adaptation: The Behavior of Higher Crustacea*, pp. 143–169. New York: Wiley.

Sather, C. A. 1966. A Bajau prawn snare. *Sabah Soc. J.* 3:42–45.

Schram, F. R. 1979. The genus *Archaeocaris* and a general review of the Palaeostomatopoda. *Trans. San Diego Soc. Nat. Hist.* 19:57–66.

Schram, F. R. 1986. *Crustacea.* New York: Oxford University Press.

Serene, R. 1954. Observations biologiques sur les stomatopodes. *Mem. Inst. Oceangr. Nhatrang* 8:1–93.

Shuster, S. M. & R. L. Caldwell. 1989. Male defense of the breeding cavity and factors affecting the persistence of breeding pairs in the stomatopod, *Gonodactylus bredini* (Manning). *Ethology* 82:192–207.

Steger, R. 1987. Effects of refuges and recruitment on gonodactylid stomatopods, a guild of mobile prey. *Ecology* 68:1520–1533.

Steger, R. & R. L. Caldwell. 1983. Intraspecific deception by bluffing: A defense strategy of newly molted stomatopods (Arthropoda: Crustacea). *Science* 221:558–560.

Steger, R. & B. Steger-Bemis. In press. Abundance and distribution of piscivorous mantis shrimps around Moorea, French Polynesia. *Proc. Sixth Int. Coral Reef Cong.*

Trivers, R. L. 1972. Parental investment and sexual selection. In B. Campbell, ed., *Sexual Selection and the Descent of Man*, pp. 136–179. Chicago: Aldine Press.

Wickler, W. & U. Siebt. 1981. Monogamy in Crustacea and man. *Z. Tierpsychol.* 57:215–234.

SIX

The Ecology of Breeding Females and the Evolution of Polygyny In Paracerceis sculpta, *A Marine Isopod Crustacean*

STEPHEN M. SHUSTER

*Department of Ecology and Evolution,
University of Chicago.*

Abstract

Recent schemes for mating system classification are based on distinctive patterns of male behavior associated with mate control. As a result, much of mating system research concerns the analysis of male reproductive adaptations. Despite the assumption, moreover, that female breeding ecology exerts a fundamental influence on the evolution of male reproductive behavior, mating system analyses routinely examine patterns of female availability *after* male activities have been described. Such analyses provide little direct information about the process by which mating systems evolve. A more balanced approach to mating system research begins with investigation of biotic and abiotic factors that limit female fecundity and thus circumscribe female distribution and abundance. Possible male responses to these female-imposed constraints may then be defined and empirically tested, and the process by which mating systems evolve may thus be more clearly identified. To illustrate the latter approach, this chapter describes aspects of female reproductive ecology that affect the availability of sexually receptive females and thus influence the evolution of polygyny in *Paracerceis sculpta*, a marine isopod crustacean.

EMLEN AND Oring (1977) proposed the first ecological classification of animal mating systems. The assumptions of this scheme were: (1) male reproduction in most species is limited by the availability of breeding females and (2) as males compete for access to females, sexual selection favors adaptations that permit males to monopolize potential mates. Emlen and Oring argued that the degree to which males are capable of monopolizing multiple females (i.e., becoming polygynous) is determined by the distribution and abundance of resources crucial to female reproduction, the consequent distribution and abundance of sexually receptive females, and variation in male ability to monopolize the above distributions.

Emlen and Oring (1977) suggested that if resources or females are clumped in space and time and if young are not dependent on male parental care, polygynous mating systems are likely to evolve. In contrast, dependent young and dispersed distributions of resources or females are expected to favor monogamous mating systems. The authors therefore recommended that tests of their hypothesis involve examination of the relationships between resource/female distributions and male breeding behavior.

With their first assumption, Emlen and Oring (1977) identified female availability as central to mating system evolution. The mating systems they subsequently defined, however, were categorized by male activities associated with mate monopolization. Evidently because data on male behavior are more easily obtained than data on female availability, the Emlen and Oring (1977) hypothesis stimulated an explosion of mating system research emphasizing the activities of breeding males. To this day, a majority of mating system analyses, including studies concerned with female mate selection, primarily address the adaptive significance of particular patterns of male reproductive behavior (reviews in Bradbury & Veherencamp 1977; Borgia 1979; Wickler & Seibt 1981; Wittenberger 1981; Oring 1982; Salmon 1983; Thornhill & Alcock 1983; Bradbury 1985).

While this male-oriented approach is indeed successful in categorizing animal breeding assemblages and does demonstrate the diversity of male adaptations possible in response to sexual selection, the strong emphasis on male behavior obscures the fundamental influence of female availability on mating system evolution. Rather than serving as passive receptacles for the ejaculates of monopolizing males, females and their requirements for reproductive success *define* the arena for male-male competition. Furthermore, if the distribution and abundance of females truly does circumscribe male behavior, it is appropriate to begin mating system research with analyses that explain why reproductive females occur where and when they do. This approach minimizes the possibility of bias in testing the Emlen & Oring (1977) hypothesis.

To illustrate this point of view, I will begin by briefly summarizing the breeding biology of *Paracerceis sculpta*, a marine isopod crustacean inhabiting the northern Gulf of California (Brusca 1980). I will next identify biotic and abiotic factors likely to generate variance in relative female fitness and explain how female life history characteristics arising from such selection can determine the types of resources reproductive females require, thus influencing patterns of female distribution and abundance. I will then discuss male reproductive adaptations that appear likely to evolve in response to these patterns of female availability. Finally, I will describe variation in male reproductive behavior and morphology, and I will identify ways in which male attributes may reciprocally influence female reproductive success.

A BRIEF SUMMARY OF THE REPRODUCTIVE BIOLOGY OF *PARACERCEIS SCULPTA*

Like many members of the family Sphaeromatidae, *Paacerceis sculpta* exhibits conspicuous sexual dimorphism (Hansen 1905; Hurley & Jansen 1977; Iverson 1982). Most males are larger than females and are equipped with robust pleotelsons and elongated uropods (fig. 6.1). This species is nocturnal, and both sexes spend nearly all of their prereproductive lives feeding on subtidal coralline algae (Shuster 1986). Following their adult molt, males with elongated uropods (α-males [Shuster 1987a]) swim from algae to the midintertidal zone, where they establish breeding territories in the spongocoels of the calcareous sponge, *Leucetta losangelensis* (Shuster 1986, 1987b). Alpha-males situate themselves in spongocoels with their uropods and telsons protruding from the spongocoel osculum.

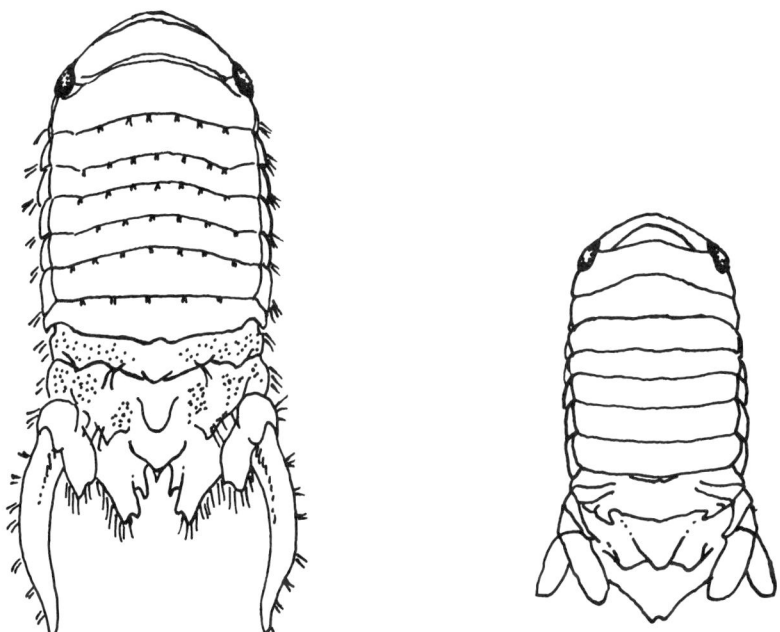

FIGURE 6.1. Sexual dimorphism in *Paracerceis sculpta*. Alpha-male (left) and female (right). Elongated structures projecting from the posterior end of the male are uropods. Horizontal bar represents 1 mm. Redrawn from Brusca 1980.

Sexually mature females leave algae before their final molt and swim to sponges, where they are attracted to spongocoels containing α-males and other breeding females. Following a brief courtship (Shuster 1986, 1990), females enter spongocoels, molt, and mate. Females deposit embryos into ventral brood pouches (Shuster 1986) and remain within spongocoels during gestation. As successive females enter spongocoels and become gravid, individual α-males may accumulate harems of as many as 19 females (Shuster 1987a). Three to four weeks after females become gravid, young isopods (mancas) emerge from brood pouches, disperse from the spongocoel, and settle on coralline algae to feed. Females reproduce once in their lives and die soon after parturition (i.e. females are semelparous [Shuster 1986, 1990]).

Males in this species are polymorphic (fig. 6.2 [Shuster 1987a]). Alpha-males comprise approximately 82 percent of the male population, approximately 4 percent of the male population consists of β-males, and the remaining 14 percent of the male population consists of γ-males. Beta-males are smaller than α-males, lack elongated uropods and rugose pleotelsons, and resemble sexually mature females in external morphology. Beta-males mimic female courtship behavior and enter spongocoels containing reproductive females by apparent deception of resident α-males. Gamma-males are smaller than β-males, also lack pleotelsonic modifications and use their rapid movements and tiny proportions to slip around the bodies of resident α-males and into spongocoels. Gamma-males, like β-males, prefer spongocoels containing sexually receptive females (Shuster 1987a, 1989a).

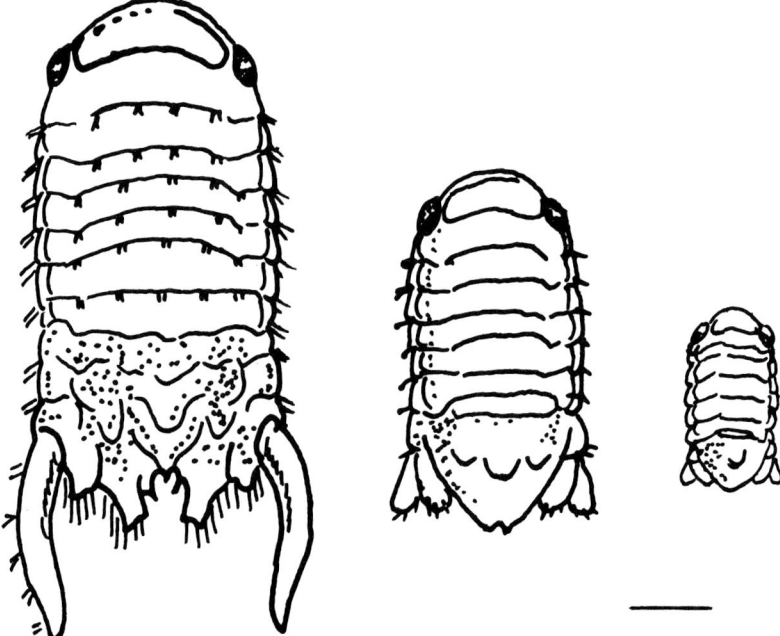

FIGURE 6.2. The relative body sizes of the three male morphs in *Paracerceis sculpta*. Left to right, α-male, β-male, γ-male. Horizontal bar represents 1 mm. Redrawn from Shuster 1990.

When reared under laboratory conditions, the three male morphs exhibit distinct life history trajectories: γ-males mature in approximately four weeks, β-males mature in approximately six weeks, and α-males mature in approximately eight weeks (Shuster 1986). The three male morphs show heritable differences in growth rates (Baitoo, Shuster & Sassaman 1988). Furthermore, the electromorphs of the enzyme phosphoglucomutase (PGM) exhibit Mendelian segregation with adult male morphology, suggesting close linkage between this locus and genetic factors influencing male phenotype (Shuster & Sassaman unpublished data). Male polymorphism in this species thus appears substantially controlled by a simple genetic mechanism.

FACTORS INFLUENCING THE TIMING, FREQUENCY, AND LOCATION OF FEMALE REPRODUCTION

Female Life History Evolution

The relationship between environmental variation and individual fitness for most characteristics under natural selection can be described by a Gaussian fitness function (Wright 1968), i.e., fitness is usually greatest under a limited range of conditions and tapers off with environmental extremes. Provided that environmental conditions do not fluctuate wildly, phenotypes under Gaussian fitness functions should eventually become normally distributed. Life history characters in most animal populations appear to fit this description (Stearns 1977).

Natural selection simultaneously acts on phenotypic variation in female growth and maturation rates, resistance to predators and disease, the ability to select and process food, the timing and frequency of reproduction, the ability to locate adequate breeding sites, and the quality of parental care. If the function describing female fitness in each of these contexts is Gaussian, the combined effects of these environmental factors generate a multidimensional space that describes relative female fitness (Lande & Arnold 1983; Wade & Kalisz, 1990). This space defines the conditions and resources females require for successful reproduction. In turn, these requirements directly influence when, where, how many, and how long sexually receptive females will be present in a particular environment.

These patterns of female availability define the nature and opportunity for sexual selection on males for a given species (Emlen & Oring 1977; Wade 1987). If receptive females are present in the environment at a particular time and place, a fitness function is imposed on males whose shape is determined by the frequency distribution of females at the preferred site. Thus males able to locate, remain in, or control areas preferred by the majority of females will mate more successfully than males unable to do so (review in Thornhill & Alcock 1983). An exhaustive survey of factors influencing female life history evolution in *Paracerceis sculpta* will not be attempted in this chapter. Certain factors are identifiable, however, and appear especially important in determining the timing, location, and frequency with which sexually receptive females become available to males.

The Distribution and Abundance of Food

The primary food sources for *Paracerceis sculpta* are the branching and encrusting forms of red coralline algae (*Amphiroa* and *Corallina* spp.). These genera occur throughout the year in lower intertidal and subtidal zones in the northern Gulf of California (Dawson 1966). Intertidal transects oriented perpendicular to shore and censused two to four times per year indicate that coralline algal abundance fluctuates seasonally, evidently in response to nutrient availability and water temperature (Turk-Boyer & Boyer, personal communication). In general, algal abundance is relatively high in the spring and reaches its lowest level in the late summer (Shuster 1986). Although algal censuses were too few in 1983–1984 to permit statistical analysis, the relative abundances of *P. sculpta* mancas collected from algae appear to follow the general pattern of algal abundance (fig. 6.3A).

The abundance of adults in spongocoels peaks in the spring, following the period of maximum manca abundance (fig. 6.3A). The relative abundance of mature females fluctuates widely, however, and the operational sex ratio (the number of sexually mature males: the number of sexually receptive females) may shift abruptly within a few weeks (Shuster 1989A; fig. 6.3B). Sexually mature females occur in spongocoels throughout the year, indicating that isopod generations overlap (Shuster 1986). However, while total isopod abundance appears related to coralline algae abundance, the relative abundance of reproductive females does not appear to be directly related to the relative availability of food.

Water Temperature and Disturbance

Fluctuations in surface water temperature in the northern Gulf of California are considerable (Thomson & Lehner 1976). Although no correlation between water temperature and receptive female abundance is apparent between October 1983–September 1984 (Spearman's $r=0.227$, $P>0.05$, $N=12$), samples collected over a longer duration (October 1983–November 1985) indicate that sexually mature females become relatively more abundant as surface waters warm in the summer (the operational sex ratio is negatively correlated with sea surface temperature; $r=-0.212$, $F_{[1,127]}=5.96$, $P=0.016$; Shuster 1986). Differential effects of water temperature on male and female maturation characteristics and on mechanisms of sex determination are known to occur in Crustacea (Charniaux-Cotton & Payen 1982; Ginsberger-Vogel & Charniaux-Cotton 1982). Whether seasonal changes in water temperature induce shifts in this species' sex ratio, however, is unclear.

Storms and wave action negatively affect the abundance of breeding isopods. During even brief periods of high surf, waterborne sand can abrade large numbers of calcareous sponges from the reef. The dramatic rise in the operational sex ratio in May 1984 (fig. 6.3B) was caused in part by the loss of numerous sponges containing receptive females during one such storm (Shuster 1986). Calcareous sponges such as *Leucetta* seem particularly susceptible to abrasion during the summer, and sponge availability decreases markedly between June

Polygyny in *Paracerceis sculpta* 97

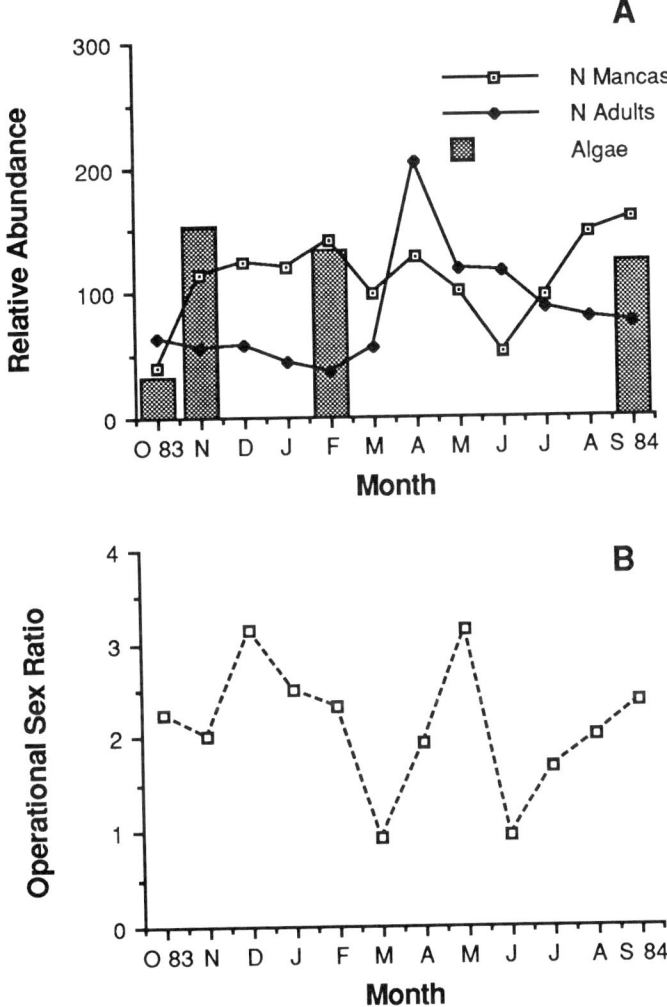

FIGURE 6.3. A. The relative abundances of coralline algae, *Paracerceis sculpta* mancas on coralline algae, and adult isopods collected in spongocoels between October 1983 and September 1984 near Puerto Peñasco, Sonora, Mexico. Algae abundances represent the number of point samples containing coralline algae along a 140 m transect oriented perpendicular to shore and censused at low tide. Twenty samples were randomly selected and examined for algae in each 10 m portion of the transect (total samples = 280 for each transect; Turk-Boyer & Boyer unpublished data). Manca abundances represent the number of individuals collected in 15 randomly selected sweep samples along a 100 m transect oriented parallel to shore. The transect was located in the lower intertidal zone and censused at low tide once each month (details in Shuster 1986). Adult abundances represent the number of isopods inhabiting sponges within 15 randomly selected 0.25 m plots along a 100 m transect oriented parallel to shore. The transect was located in the midintertidal zone and censused once each month. B. The operational sex ratio (N adult males/N sexually receptive females) calculated for samples of adults collected as described in A.

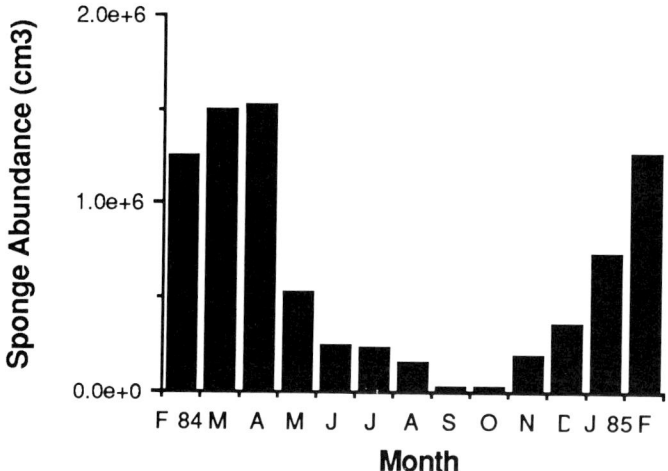

FIGURE 6.4. The relative abundance of *Leucetta losangelensis* sponges collected from 15 randomly selected 0.25 m plots along the 100 m transect described in figure 6.3 between February 1984 and February 1985. Sponge abundance represented in millions of cm³; $1.0^e + 6 = 1 \times 10^6$.

and September (Fig. 6.4; Shuster 1986). Perhaps because of the decreased availability of sponges during the late summer, the number of adult isopods decreases at this time and remains relatively low through the winter (fig. 6.3A; Shuster 1986).

The Distribution and Abundance of Breeding Sites

Sponges used as breeding habitat by isopods are located primarily in the midintertidal zone under boulders in permanent tide pools (Shuster 1986). Discriminate analysis of the physical characteristics of spongocoels (in the fall of 1983 when sponges were relatively rare [fig. 6.5A] and in the spring of 1984 when sponges were relatively abundant [fig. 6.5B]) indicates that occupied and unoccupied spongocoels are morphologically distinguishable. Discriminant functions in both samples loaded most heavily on osculum diameter (table 6.1), suggesting a minimum osculum diameter necessary for occupancy by isopods. While many spongocoels are clearly uninhabitable, the distributions of discriminate functions for occupied and unoccupied spongocoels overlap considerably, both in the spring and in the fall. This overlap suggests that even when sponges are relatively rare, there is no clear shortage of spongocoels that could serve as reproductive habitat for isopods.

Predation

Evidence of Predation. Isopods comprise a major proportion of the diets of Gulf of California reef fish (Thomson & Lehner 1976). Moreover, preliminary feeding trials indicate that female *Paracerceis sculpta* are readily taken as food by at least two common intertidal fish species (triggerfish [*Balistes* sp.] and damselfish

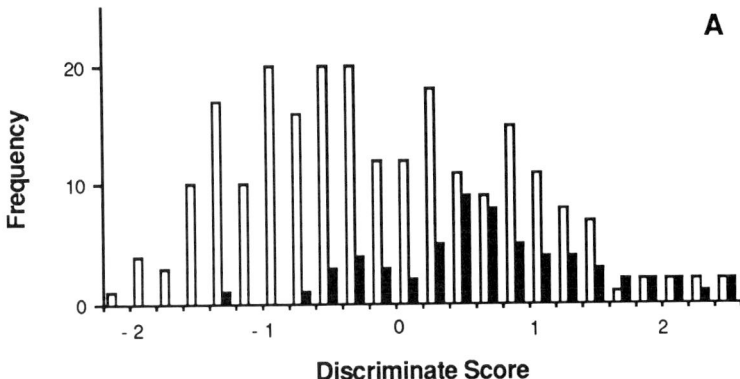
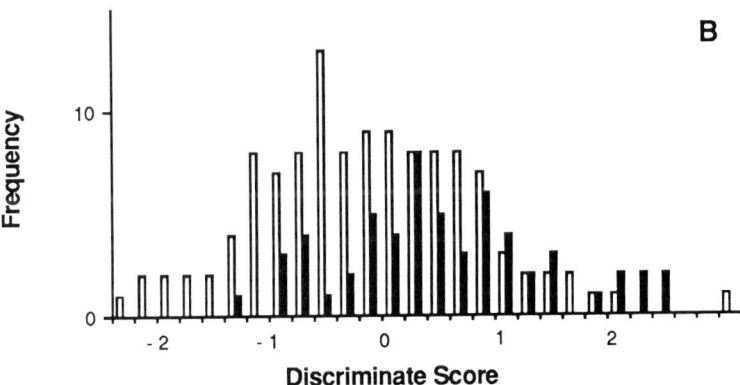

FIGURE 6.5. Discriminate analyses of spongocoels occupied and unoccupied by isopods in A. Fall 1983, Mahalanobis' distance = 0.68, F = 10.68, P < 0.001; and in B. Spring 1984, Mahalanobis' distance = 0.77, F = 12.25, P < 0.001. Methods and variables included in this analysis summarized in table 6.1. Black bars represent discriminate scores for occupied spongocoels; white bars represent discriminate scores for unoccupied spongocoels.

[*Abudefduf* sp.]). *P. sculpta* is cryptic on coralline algae, and individuals exhibit a wide range of white, purple, and red pigmentation patterns that resemble calcareous and other coralline algae encrustations (Shuster 1989a). Similar color polymorphisms are widespread among coastal isopods exposed to predation by fish (review in Hedgecock, Tracey & Nelson 1982). Moreover, young *P. sculpta* detect and orient away from objects probing through algae. Young seagrass shrimp use similar behavior to avoid their visual predators (*Tozeuma* sp. Main 1987).

The Influence of Predation on Female Activity Cycles. Nocturnality is a well-documented response of aquatic invertebrates to selection by visual predators

TABLE 6.1. Standardized canonical discriminant function coefficients from discriminant analysis of occupied and unoccupied spongocoels collected near Puerto Peñasco, Sonora, Mexico in Fall 1983 and Spring 1984. Sponge dimensions were measured to the nearest 0.05 mm; isopods were removed from spongocoels, sexed, and measured to the nearest 0.05 mm; and spongocoels were injected with melted paraffin wax. Osculum diameter and spongocoel length were measured on the wax casts to the nearest 0.05 mm, and casts were weighed to the nearest 0.0001 g; weight is proportional to volume. All values were log-transformed to meet assumptions for normality before analysis.

Variable	Fall 1983	Spring 1984
Osculum diameter	0.9845	0.8079
Spongocoel weight	0.3187	−0.2446
Spongocoel length	0.4410	0.4424
Sponge volume	0.1005	0.0156

(McLaren 1963; Zaret 1980; Gliwicz 1986). Not surprisingly, the onset of activity at sunset by sexually mature females in *Paracerceis sculpta* is pronounced (fig. 6.6A; Shuster 1986). This pattern persists across seasons as daylength changes (Shuster 1986), and females become active later during spring tides than during neap tides, evidently delaying their activities to compensate for periods of intertidal exposure.

The Influence of Predation on the Association of Females with Sponges. Isopods commonly use sponges as breeding habitat (Holdich 1976; Delaney 1984; N. Upton personal communication), evidently because sponges contain toxins or spicules that make them difficult for predators to process. The association of *Paracerceis sculpta* females with spongocoels is reasonable in this context, as the cuticles of gravid females are quite soft and females are relatively sluggish when brooding young. *Leucetta* sponges are present throughout most of the year in the northern Gulf of California (fig. 6.4); thus refugia provided by spongocoels are predictably available. Sponges circulate water through their oscula and are typically located in tidal channels or other areas with frequent water movement. Desiccation of brooding females within spongocoels is therefore unlikely. Furthermore, if more than one female occupies the same cavity, as occurs frequently in this species, the possibility of oxygen stress is minimized in a spongocoel, whereas a nonliving cavity might become anoxic. *Leucetta* spongocoels thus provide female isopods with safe, predictably available, and well-aerated reproductive habitat.

The Influence of Predation on the Evolution of Female Semelparity. Despite the apparent suitability of spongocoels as breeding sites, their use for this purpose by isopods necessitates migration from feeding sites in algae to sponges higher in the intertidal zone (Shuster 1989b). The risk of predation in a single such trip appears considerable. The probability of successful multiple trips from algae to sponges and back to algae to feed (as would be necessary if females attempted to rear successive broods) seems small. While males, once established in spongo-

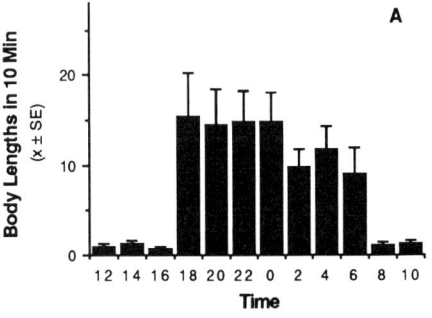
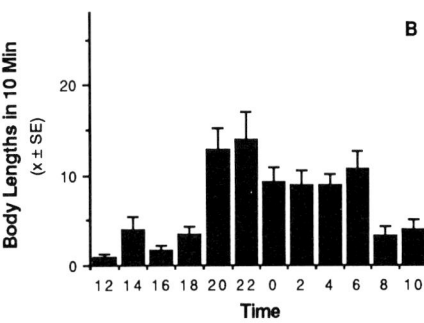
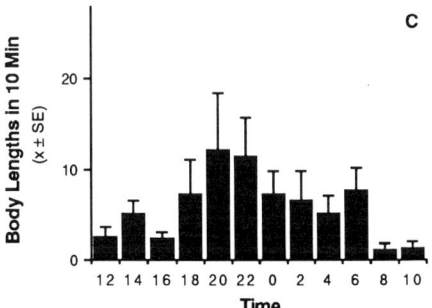
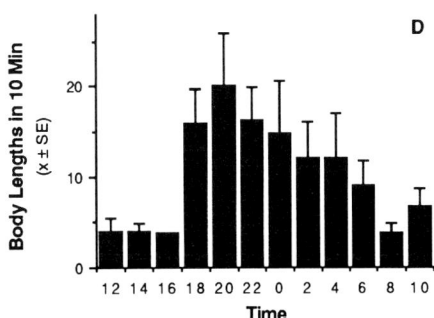

FIGURE 6.6. Twenty-four–hour activity cycles for A. unmolted adult females, N = 51; B. α-males, N = 38; C. β-males, N = 13; D. γ-males, N = 24 in January 1985. Individuals were placed into separate 7 cm³ containers suspended in flowing seawater. At 2 hr intervals for 24 hrs, the number of body lengths each isopod traveled in 10 mins was recorded. Histogram bars represent mean number of body lengths (+ SE) traveled in 10 mins.

coels, may sire multiple broods, the risks associated with repeated reproductive episodes could favor females that invest their lifetime reproductive effort in a single brood of young (Bell 1980).

Such obligate semelparity occurs in two of the five sphaeromatid subfamilies: the Sphaeromatinae and the Dynameninae. Females in 6 of 15 sphaeromatine genera and in 10 of 23 dynamenine genera (21 percent of all sphaeromatid genera, N=77; Harrison 1984) lose functional mouthparts during their reproductive molts and die soon after parturition. Detailed life history data are available for only 2 of these 16 semelparous sphaeromatid genera: *Dynamene* (Holdich 1968, 1971) and *Paracerceis* (Shuster 1986, 1989b). While females in these two genera must migrate from feeding sites to remote breeding sites to reproduce, females in all 18 iteroparous genera for which life history data are known breed and feed in the same locations (Shuster unpublished data).

Obligate semelparity similar to that observed in the Sphaeromatidae is widespread among gnathiidean isopods. All species for which life history data are known have remote feeding and reproductive habitats (Holdich & Harrison 1980; Upton 1987). The life histories of semelparous sphaeromatines are presently uninvestigated (Ridley 1983; Harrison 1984). Thus these genera could provide an interesting test of the hypothesis that obligate semelparity in isopods evolves in response to predator-imposed selection on females traveling between feeding and reproductive habitats.

Patterns of Female Mate Discrimination and Sexual Receptivity

Female Discrimination of Reproductive Habitat Characteristics. The distribution of harem sizes among α-males differs significantly from that expected by chance; single α-males and highly polygynous α-males occur more frequently than expected (Shuster 1987a). Multiple regression analyses were conducted to determine whether variation in the physical characteristics of sponges or of α-males explains significant variation in harem size (Shuster 1986, 1987b). Significant relationships suggest characteristics of reproductive habitat or males that may be important in female mate choice (Emlen & Oring 1977; Pleszczynska 1978).

Male body length was the only independent variable that consistently explained significant variance in harem size (table 6.2). Sponge characteristics, including total sponge volume, osculum diameter spongocoel depth, spongocoel volume, and the location of spongocoels on sponges, did not contribute significantly to the regression (Shuster 1987b). Thus females do not appear to use these sponge characteristics to discriminate among potential breeding sites.

Female Discrimination of Male Characteristics. The relationship between α-male body length and harem size suggests that females prefer large α-males as mates. However, when given a choice between α-males differing by 10 percent in total body length, females showed no preference for large males (Shuster 1990; bi-

---

TABLE 6.2. Results of multiple regression to determine the influence of α-male body length (MLENGTH), osculum diameter (OSCDIA), spongocoel length (SPCLL), spongocoel weight (SPCLWT), and sponge volume (SPV) on the harem size of α-males, Fall 1983. Methods as in table 6.1.

ANOV	df	Sum of Squares	F	P
Regression	5	14.9956	22.1680	<0.0001
Residual	46	6.2234		

$R^2 = 0.7067$

Contribution of each variable to the regression:

Variable	R^2	R^2 Change	F	P
MLENGTH	0.6457	0.6457	15.3052	<0.001
OSCDIA	0.6983	0.0526	2.3911	>0.100
SPCLWT	0.7053	0.0069		
SPCLL	0.7066	0.0014		
SPV	0.7067	0.0007		

nomial test, $P > 0.25$, $N = 70$). Furthermore, direct observations indicated that females do not compare individual males; females simply courted the first male they encountered and usually entered that spongocoel.

Females also showed no tendency to discriminate α-male age (newly molted α-males versus α-males 30 days postmolt) or α-male uropod condition (missing a uropod or intact, $P > 0.25$, $N = 20$ for both tests, Shuster 1990). Females did, however, detect the presence of aggregations of breeding females within spongocoels. When given a choice between spongocoels containing three gravid females and spongocoels containing no gravid females, females preferred the established spongocoels in 12 of 16 trials (Binomial test, $P = 0.04$; Shuster 1990).

Female Sexual Receptivity. Females become sexually receptive within five days after entering spongocoels (Shuster 1989b). Receptivity begins when females shed the posterior portion of their cuticle and expose genital pores at the base of each fifth walking leg. Females can delay this molt if males are not present, an ability perhaps associated with the fact that a female's search for a breeding site may take several days. Females remain in half-molted condition for approximately 24 hrs (median = 24.0 hrs, range 6.0–50 hrs, $N = 40$), are receptive to any nearby male during this time, and were observed to mate up to six times within 2 hrs in the laboratory (Shuster 1989b). Receptivity ends when females shed the anterior half of their cuticle. The duration of receptivity seems unaffected by the presence of α-males (U-test, $P > 0.45$, $N = 40$; Shuster 1989b).

Multiple Insemination and Sperm Mixing. Females in this species possess no sperm storage organs. Males deposit sperm directly into a simple oviduct that leads from each vagina to the dorsal ovaries. Microscopic examination of the oviducts of singly inseminated females reveals sperm entwined in a loosely associated ball. Females inseminated several times have larger sperm masses of the same form. Separate ejaculates, however, are indistinguishable (Shuster 1989b).

In addition to this physical suggestion of sperm mixing, the proportions of progeny sired by genetically marked α-, β-, and γ-males in sperm competition

TABLE 6.3. Fecundities of females with single or multiple inseminations. Singly inseminated females were retained in 225 ml cups containing seawater and were allowed to mate once with their designated male. Females inseminated more than once were randomly selected from among 45 females that had been retained in cups with a male of the designated type for the duration of their receptivity. All of these females were observed to mate at least twice.

	Single Matings					Multiple Matings			
Male	Female	Live young	Undev.	N	Male	Female	Live young	Undev.	N
748β	117	49	12	61	952β	374	58	2	60
986β	050	72	18	90	354β	201	49	0	49
793α	161	4	86	90	256α	293	97	0	97
		125	116	241			204	2	206

experiments suggest that multiple insemination is common and sperm precedence is minimal in this species (Shuster 1989a). Multiple insemination may even be a necessity for females, as females mated only once experienced decreased fecundity compared to females inseminated several times (table 6.3). In many insect species in which multiple insemination is the rule, females possess reproductive tracts that encourage sperm mixing (Walker 1980).

The Evolution of Patterns of Female Discrimination and Sexual Receptivity. Risks incurred during mate selection are thought to diminish opportunities for females to discriminate among potential mates (Kirkpatrick 1987). Females in this species are clearly exposed to predators when searching for spongocoels. Thus the degree to which females may move among breeding sites, discriminating male and sponge characteristics, is limited. Rather than discriminating particular characteristics of spongocoels or males, females may instead use the scent of other females in spongocoels as an indication of breeding site quality. This tactic would permit females to move toward established breeding sites with minimal exposure to predation (Shuster 1990).

The use of chemical cues by female to locate breeding sites increases the probability that more than one female will find the same spongocoel. Since α-males guard spongocoels in this species, this type of cueing by females enhances the potential for polygyny (Emlen & Oring 1977). Females must mate, however, while in half-molted condition or lose their only opportunity to reproduce (Shuster 1989b). Thus if guarding α-males are successful in excluding other α-males, accumulations of females within spongocoels could decrease an individual female's chances of insemination.

Such circumstances could favor females that (1) remain receptive for durations longer than necessary to accomplish a single mating and/or (2) possess reproductive tracts that encourage sperm competition. These adaptations could increase the incentive for resident α-males to mate with females more than once, as well as encourage matings by other males (Walker 1980), thus decreasing a female's risk of infertility.

THE INFLUENCE OF PATTERNS OF FEMALE AVAILABILITY ON MALE REPRODUCTIVE BEHAVIOR AND MORPHOLOGY

The Opportunity for Sexual Selection

Recent mating system analyses have used variance in reproductive success to estimate the intensity of sexual selection on males (Emlen & Oring 1977; Wickler & Seibt 1981; Oring 1982; Thornhill & Alcock 1983). Wade (1987) has argued that the intensity of male sexual selection is better expressed as the *opportunity for sexual selection*, I, which is equal to the variance in male reproductive success divided by the square of its mean. This statistic measures variance in *relative* fitness with respect to the male character under sexual selection, permitting prediction of the maximum change possible in the phenotypic distribution of a character within one generation. The predicted response represents the upper limit for change because only the process of selection is considered; heritable genetic variation associated with the character is not. Thus while large statistical variance in male reproductive success may indicate that some males are very successful in acquiring mates, variance alone says nothing about how sexual selection may affect male phenotypes. By defining the upper limit for evolutionary change by sexual selection, I, the opportunity for sexual selection, provides this information.

Factors Influencing the Opportunity for Sexual Selection on Males

Changes in the Operational Sex Ratio

Predictions. Variance in male reproductive success depends on the relative abundance of females (i.e., the operational sex ratio) *as well as* their distribution in space and time (Oring 1982). An additional advantage provided by the opportunity for sexual selection is that both of these variables are incorporated into a single expression (Wade 1987). Operational sex ratios skewed toward males (>1.0) generate a large opportunity for sexual selection if the distribution of females permits a few males to obtain most of the available mates. In such circumstances, variance in male reproductive success is high and the square of mean male success is relatively low. If the distribution of females is dispersed, the opportunity for sexual selection decreases because more males obtain mates. Nevertheless, the opportunity for sexual selection in such circumstances can remain relatively high if females are rare.

Operational sex ratios equal to or less than 1.0 decrease the opportunity for sexual selection considerably if females are dispersed in space and time. If, however, as in *Paracerceis sculpta*, females prefer particular breeding sites and a few males acquire very large harems, the squared mean of male reproductive success may remain relatively small while variance in male success becomes very large. The opportunity for sexual selection in such circumstances can thus become enormous.

Male Responses. In *Paracerceis sculpta*, receptive females tend to aggregate in particular spongocoels even when the operational sex ratio is skewed toward males (Shuster 1986). The population frequencies of β- and γ-males are relatively low when females are rare; thus the opportunity for sexual selection is probably high among α-males at these times. Large body size and elongated uropods among α-males may have evolved under such conditions.

When females become more abundant, harems can become relatively large ($N > 19$; Shuster 1987a). At these times, the frequency of β- and γ-males increases, evidently because β- and γ-males locate and prefer to enter spongocoels containing sexually receptive females (Shuster 1986). Within spongocoels containing α-males, γ-males, and sexually receptive females, the operational sex ratio affects relative male fertilization success (Shuster 1989a). When few receptive females are present, α-males seem capable of guarding females that enter their spongocoel and siring nearly all progeny. As the number of receptive females in the spongocoel increases, however, the fertilization success of γ-males increases at the expense of α-male success. The fertilization success of β-males in spongocoels containing an α-male and his harem is more variable but also appears affected by female density (Shuster 1989a).

Periodic fluctuations in the abundance of sexually receptive females appears to result in frequency dependent selection in spongocoels. When females are rare, reproduction by α-males seems favored; when females are more abundant, females aggregate in spongocoels and the reproductive success of γ-males, and possibly by β-males, increases considerably (Shuster 1989a). At such times sexual selection may cease to favor male adaptations associated with physical combat and begin to favor male adaptations associated with invading spongocoels. The maturation rates of γ- and β-males do not appear sufficiently rapid to permit these males to capitalize on short-term fluctuations in the operational sex ratio (Shuster 1986). Rapid developmental rates, however, may allow cohorts of these males to enter the breeding population faster than cohorts of α-males. These conditions and perhaps other yet unknown factors are evidently sufficient to maintain a genetic polymorphism among males in this population (Shuster 1989a).

Female Activity Patterns and the Association of Females with Sponges

Predictions. A preference by females for particular breeding sites can favor males that develop rapidly, locate the reproductive habitat early, and "anticipate" the arrival of females. Such males are likely to outcompete slower developing, randomly searching individuals (Parker 1978). Similar reasoning predicts that males should begin their daily activities somewhat earlier than females. Early males may obtain better locations to encounter females than males that become active later (Thornhill & Alcock 1983). With respect to *Paracerceis sculpta*, males are thus expected to begin their nocturnal activity earlier than females and to settle on sponges before females arrive.

Male Responses. *Paracerceis sculpta* males are indeed nocturnal. Thus selection against daylight activity appears to have operated similarly on both sexes. The activity periods of females (fig. 6.6A), however, are more rigidly confined to evening hours than those of any male type (fig. 6.6B–D). Strong predation pressure should disfavor daylight activity by males, but anticipation of female activity could, as predicted (Parker 1978), increase mating opportunities for males. While more detailed examination of activity cycle differences among the sexes and among males is needed, the attenuated activity patterns of males compared to females in this species are consistent with Parker's hypothesis.

As expected, α-males establish themselves in spongocoels and wait for females to arrive (Shuster 1986). Alpha-males are not, however, simply stationing themselves at any particular spongocoel. When given a choice, α-males prefer spongocoels containing gravid females over spongocoels that are empty (Shuster 1990). Alpha-males also permit gravid (i.e., unreceptive) females to enter their spongocoels (Shuster 1986) and tend to fight more aggressively over occupied spongocoels than over unoccupied spongocoels.

In light only of male competition for mates, α-male combat over spongocoels containing unreceptive females makes little sense. Alpha-males should be unconcerned with spongocoels containing females mated by other males and should exclude gravid females attempting to enter their spongocoels; such females offer no fertilizable ova. Sexually receptive females, however, are attracted to spongocoels containing gravid females (Shuster 1990). This creates an enormous opportunity for certain males to acquire multiple mates. Alpha-males capable of monopolizing spongocoels containing females are likely to mate, not only with receptive females presently in their harem but also with females attracted to that site.

The Pattern of Female Mate and Reproductive Habitat Discrimination

Predictions. *Paracerceis sculpta* females are attracted to established breeding aggregations, but females do not appear to discriminate the physical characteristics of sponges or of males (Shuster 1986, 1987b, 1990). Large body size and elongated uropods in α-males may have arisen primarily in response to male-male competition (Shuster 1987b). If so, larger α-males should be more successful in combat than smaller α-males. Furthermore, while removal of α-male uropods may not affect patterns of female choice, such modification should affect an α-male's ability to fight with other α-males.

Male Responses. Alpha-males compete for access to breeding sites, and, as expected, large α-males and α-males with intact uropods are more successful at spongocoel takeover than smaller α-males or α-males with one uropod missing (Shuster 1986, 1990). Alpha-males use their body size and uropods in offensive and defensive maneuvers during combat over spongocoels (Shuster 1986). However, while large body size and intact uropods appear useful in spongocoel takeover, α-males with one uropod missing are as successful at defending their spongocoels as intact α-males when residents and intruders are equal in size (Shuster 1986). Spongocoels evidently provide a positional advantage that may protect residents with damaged uropods from eviction.

The Pattern of Female Sexual Receptivity

Predictions. Multiple mating by females and sperm mixing within the oviduct are expected to favor male adaptations associated with sperm competition (Parker 1970). Two possible outcomes of such conditions are: (1) if each successful mating confers some fitness to the mating male, but sperm precedence is incomplete, males should copulate briefly with females and not attempt postcopulatory guarding (Parker 1974); (2) if males are capable of retaining females and can mate repeatedly, males may guard females for the duration of their receptivity and copulate often, in attempts to overwhelm the ejaculates of males that have stolen matings (Shuster & Caldwell 1989). In either case, the potential for sperm competition is high and males are likely to possess the capacity for prodigious sperm production (Parker 1970, 1974).

Male Responses. Females capable of remaining sexually receptive for one to several days may increase their chances for successful insemination by resident α-males. An extended period of receptivity, however, provides an opportunity for invasion of the mating system by alternative male mating strategies (Shuster 1989a). Males capable of invading spongocoels containing multiple receptive females could experience considerable fertilization success once inside. Large α-males may be capable of excluding smaller α-males from spongocoels, but very small males, not engaging in physical combat, might successfully enter spongocoels if they avoid physical combat altogether. Female mimicry and sneaker behavior in this context are well known in other animal species (review in Shuster 1987a).

In the laboratory, α-males tend to retain females after copulation, while β- and γ-males mate very rapidly and abandon females immediately (Shuster 1989b). Beta-males possess seminal vesicles that are proportionally several times larger than those of α-males, and the seminal vesicles of γ-males nearly fill the body cavity (Shuster 1987a). The details of sperm competition in this species are poorly known, but strategies of mate guarding and sperm competition appear to vary with differences in adult male morphology.

SUMMARY AND CONCLUDING REMARKS

The reproductive behavior and morphology of *Paracerceis sculpta* males appears to have arisen in response to patterns of receptive female availability. The availability of females is evidently the result of physical conditions imposed by an intertidal existence, food and breeding habitat availability, predation, and male-male competition. While analysis of male behavior and morphology alone permits classification of this polygynous mating system and does provide insight into the mechanics of competitive interactions among males, certain male characteristics such as male polymorphism, sexual dimorphism, sperm competition, and patterns of male mate and habitat selection are better understood if examined after the details of female breeding biology are described.

The labor required to obtain these details may be prohibitive for many species. Yet even species not suited for field or laboratory manipulation may be more productively examined if the sequence with which data are collected is chosen with care. Mating system analyses that begin with descriptions of male behavior and end by examining female availability, unless combined with careful manipulation experiments (e.g., Plesczscynska 1978), do not test the Emlen and Oring (1977) hypothesis; they merely describe the behavior of males and females. An unbiased understanding of how male reproductive behavior may arise and persist requires exploration of the context in which mating systems may evolve. To clearly understand this context we must first understand the reproductive biology of females.

Acknowledgments

I thank R. Bauer, M. Wade, D. Clayton, M. Kooda, and two anonymous referees for their comments and criticism on earlier drafts of this manuscript. The Center for the Study of Deserts and Oceans in Puerto Peñasco and the Environmental

Research Laboratory in Tucson, Arizona, provided logistical support for my research in Mexico. Financial assistance was provided by the Theodore Roosevelt Fund, the University Regents' Fellowship, the Alice Galloway Fund, the Departments of Zoology and Genetics at the University of California, Berkeley, NSF dissertation improvement grant OCE-8401067, and NSF grant BSR 87-00112.

Literature Cited

Baitoo, H., S. M. Shuster & C. A. Sassaman. 1988. Polymorphic male molting and growth schedules in a marine isopod crustacean. *Am. Zool.* 28:134A.
Bell, G. 1980. The costs of reproduction and their consequences. *Am. Nat.* 112:45–76.
Borgia, G. 1979. Sexual selection and the evolution of mating systems. In M. S. Blum & N. A. Blum, eds., *Sexual Selection and Reproductive Competition in Insects*, pp. 19–80. New York: Academic Press.
Bradbury, J. W. 1985. Contrasts between insects and vertebrates in the evolution of male display, female choice and lek mating. In B. Holldobler & M. Lindauer, eds., *Fortschritte der Zoologie: Experimental Behavioral Ecology*, pp. 273–288. Stuttgart: G. Fisher Verlag.
Bradbury, J. W. & S. L. Veherencamp. 1977. Social organization and foraging in emballonurid bats. III. Mating systems. *Behav. Ecol. Sociobiol.* 2:1–17.
Brusca, R. C. 1980. *Common Intertidal Invertebrates of the Gulf of California*. Tucson: University of Arizona Press.
Charniaux-Cotton, H. & G. Payen. 1982. Sexual differentiation. In D. E. Bliss & L. H. Mantel, eds., *The Biology of Crustacea*. Vol. 9, *Integument, Pigments and Hormonal Processes*, pp. 217–299. New York: Academic Press.
Dawson, E. Y. 1966. *Marine Algae in the Vicinity of Puerto Peñasco, Sonora, Mexico*. Gulf of California Field Guide Series. Tucson: University of Arizona Press.
Delaney, P. M. 1984. Isopods of the genus *Excorallana* Stebbing, 1904, from the Gulf of California, Mexico (Crustacea Isopods Corallanidae). *Bull. Mar. Sci.* 34:1–20.
Emlen, S. T. & L. W. Oring. 1977. Ecology, sexual selection and the evolution of mating systems. *Science (Washington, D.C.)* 197:215–223.
Ginsberger-Vogel, T. & H. Charniaux-Cotton. 1982. Sex determination. In L. G Abele, ed., *The Biology of Crustacea*. Vol. 2, *Embryology, Morphology and Genetics*, pp. 257–283. New York: Academic Press.
Gliwicz, Z. M. 1986. A lunar cycle in zooplankton. *Ecology* 67:883–897.
Hansen, J. J. 1905. On the propagation, structure and classification of the family Sphaeromatidae. *Q. J. Microsc. Sci.* 49:69–135.
Harrison, K. 1984. The morphology of the sphaeromatid brood pouch (Crustacea: Isopoda: Sphaeromatidae). *Zool. J. Linn. Soc.* 82:363–407.
Hedgecock, D., M. L. Tracey & K. Nelson. 1982. Genetics. In L. G. Abele, ed., *The Biology of Crustacea*. Vol. 2, *Embryology, Morphology and Genetics*, pp. 284–403. New York: Academic Press.
Holdich, D. M. 1968. Reproduction, growth and bionomics of *Dynamene bidentata* (Crustacea: Isopoda). *J. Zool. (Lond.)* 156:136–153.
Holdich, D. M. 1971. Changes in the physiology, structure and histochemistry occurring during the life-history of the sexually dimorphic isopod *Dynamene bidentata* (Crustacea: Peracarida). *Mar. Biol. (Berl.)* 8:35–47.
Holdich, D. M. 1976. A comparison of the ecology and life cycles of two species of littoral isopod. *J. Exp. Mar. Biol. Ecol.* 24:133–149.
Holdich, D. M. & K. Harrison. 1980. The crustacean isopod genus *Gnathia* Leach from Queensland waters with descriptions of nine new species. *Aust. J. Mar. Freshwater Res.* 31:215–240.
Hurley, D. E. & K. P. Jansen. 1977. The marine fauna of New Zealand: Family Sphaeromatidae (Crustacea: Isopoda: Flabellifera). *Mem. N. Z. Oceanogr. Inst.* 63:1–95.
Iverson, E. W. 1982. Revision of the isopod family Sphaeromatidae (Crustacea: Isopoda: Flabellifera). I. Subfamily names with diagnosis and key. *J. Crustacean Biol.* 2:248–256.
Kirkpatrick, M. 1987. The evolutionary forces acting on female mating preferences in polygynous animals. In J. W. Bradbury & M. B. Andersson, eds., *Sexual Selection: Testing the Alternatives*, pp. 67–82. New York: Wiley.

Lande, R. & S. J. Arnold. 1983. The measurement of selection on correlated characters. *Evolution* 37:1210–1226.

McLaren, I. A. 1963. Effects of temperature on growth of zooplankton and the adaptive value of vertical migration. *J. Fish. Res. Board Can.* 20:685–767.

Main, K. 1987. Predator avoidance in seagrass meadows: Prey behavior, microhabitat selection and cryptic coloration. *Ecology* 68:170–180.

Oring, L. W. 1982. Avian mating systems. In D. S. Farmer & J. R. King, eds., *Avian Biology*, 6:1–92. New York: Academic Press.

Parker, G. A. 1970. Sperm competition and its evolutionary consequences in insects. *Biol. Rev. Camb. Philos. Soc.* 45:525–567.

Parker, G. A. 1974. Courtship persistence and female guarding as male time investment strategies. *Behaviour* 48:157–184.

Parker, G. A. 1978. Evolution of competitive mate searching. *Annu. Rev. Entomol.* 23:173–196.

Pleszczynska, W. K. 1978. Microgeographic prediction of polygyny in the Lark Bunting. *Science (Washington, D.C.)* 201:935–937.

Ridley, M. 1983. *The Explanation of Organic Diversity: The Comparative Method and Adaptations for Mating*. Oxford: Oxford Science Publications.

Salmon, M. 1983. Courtship, mating systems and sexual selection in decapods. In S. Rebach & D. W. Dunham, eds., *Studies in Adaptation: The Behavior of Higher Crustacea*, pp. 143–169. New York: Wiley.

Shuster, S. M. 1986. *The Reproductive Biology of* Paracerceis sculpta *(Crustacea: Isopoda)*. Ph.D. dissertation, University of California, Berkeley.

Shuster, S. M. 1987a. Alternative reproductive behaviors: Three discrete male morphs in *Paracerceis sculpta*, an intertidal isopod from the northern Gulf of California. *J. Crustacean Biol.* 7:318–327.

Shuster, S. M. 1987b. Male body size, not reproductive habitat characteristics predicts polygyny in a sexually dimorphic intertidal isopod crustacean, *Paracerceis sculpta* (Crustacea: Isopoda). In G. Malagrino & H. Santoyo. eds., *Mem. V Simp. Biol. Mar. Univ. Auton. Baja California Sur*, 24–26 Octubre de 1984, La Paz B.C.S., Mexico, pp. 71–80.

Shuster, S. M. 1989a. Male alternative reproductive strategies in a marine isopod crustacean (*Paracerceis sculpta*): Use of genetic markers to measure differences in fertilization success among α-, β- and γ-males. *Evolution.* 43:1683–1698.

Shuster, S. M. 1989b. Female sexual receptivity associated with molting and differences in copulatory behavior among the three male morphs in *Paracerceis sculpta*, (Crustacea: Isopoda). *Biol. Bull. (Woods Hole)* 177:331–337.

Shuster, S. M. 1990. Courtship and female mate selection in the marine isopod crustacean, *Paracerceis sculpta*. *Anim. Behav.* 40:390–399.

Shuster, S. M. & R. L. Caldwell. 1989. Male defense of the breeding cavity and factors affecting the persistence of breeding pairs in the stomatopod, *Gonodactylus bredini* (Crustacea: Hoplocarida). *Ethology* 82:192–207.

Stearns, S. C. 1977. The evolution of life history traits: A critique of the theory and a review of the data. *Annu. Rev. Ecol. Syst.* 8:145–71.

Thomson, D. A. & C. E. Lehner. 1976. Resilience of a rocky intertidal fish community in a physically unstable environment. *J. Exp. Mar. Biol. Ecol.* 22:1–29.

Thornhill, R. & J. Alcock. 1983. *The Evolution of Insect Mating Systems*. Cambridge, Massachusetts: Harvard University Press.

Upton, J. P. D. 1987. Asynchronous male and female life cycles in the sexually dimorphic, harem-forming isopod. *Paragnathia formica* (Crustacea: Isopoda). *J. Zool. (Lond.)* 212:677–690.

Wade, M. J. 1987. Measuring sexual selection. In J. W. Bradbury & M. B. Andersson, eds., *Sexual Selection: Testing the Alternatives*, pp. 197–207. New York: Wiley.

Wade, M. J. & S. Kalisz. 1990. The causes of natural selection. *Evolution.*

Walker, W. F. 1980. Sperm utilization strategies in non-social insects. *Am. Nat.* 115:780–799.

Wickler, W. & U. Seibt. 1981. Monogamy in crustacea and man. *Z. Tierpsychol.* 57:215–234.

Wittenberger, J. F. 1981. *Animal Social Behavior*. Boston, Massachusetts: Duxbury Press.

Wright, S. 1968. *The Evolution and Genetics of Populations*. Vol. 1, *Genetic and Biometric Foundations*. Chicago: University of Chicago Press.

Zaret, T. M. 1980. *Predation and Freshwater Communities*. New Haven, Connecticut: Yale University Press.

SEVEN
Anostracan Mating Behavior: A Case of Scramble-Competition Polygyny

DENTON BELK

Biology Department,
Our Lady of the Lake University of San Antonio, Texas.

Abstract

Anostracan males avoid aggressive encounters, concentrating their reproductive efforts on finding, courting, and inseminating receptive females in what amounts to a competitive scramble to mate the most females and thus contribute the most offspring to the next generation. Anostracan females obtain sperm from a different male for each clutch, thus increasing the genetic variability of their offspring. Variability is likely to be important for a female producing eggs that must hatch in the uncertainties of ephemeral aquatic habitats. Anostracan males use imprecise visual cues to guide their mating behavior. Anostracan females use tactile cues provided largely by outgrowths of the male's second antennae in choosing mates. Evolution of the species-specific characteristics of these often elaborate male courtship structures seems to be controlled by allopatric genetic mechanisms unrelated to antihybridization mechanisms and driven by sexual selection by female choice.

INFORMATION ON mating behavior in the Anostraca is available for only 10 of the more than 200 described species. These 10 represent examples from only 7 of the 23 genera currently accepted. The first observations on mating behavior in anostracans are summarized by Baird (1850). Mathias (1937) reviewed the subject and discussed duration of amplexus, copulation time, and mating posture. Kuenen (1939) and Lochhead (1941) presented useful information on *Artemia*. Moore and Ogren (1962) reported observations on mating behavior in *Eubranchipus moorei* (misidentified as *Eubranchipus holmani*) and discussed their findings in relation to those of other workers. The only mating behavior studies making use of experimental approaches are those of Pearse (1913), Wiman (1981), and Belk (1984). Thus the generalizations I offer in this paper are based on a very limited sample of Anostraca. However, given the morphological similarity of reproductive structures in the Anostraca, I consider these generalizations to be reasonable. They should also serve as valuable starting points for future research.

Based on the information at hand, male anostracans are continually on the prowl, searching for females. Males of *Artemia* and the unrelated *Artemiopsis stefanssoni* seem to spend more time in amplexus and less in searching (Lochhead 1941; Daborn 1977). Relying on imprecise visual cues, male anostracans orient toward any promising object (see below). They will attempt to clasp and mate with females already carrying fertilized, shelled eggs and even females of other species. Anostracan females exercise choice of mates by not accepting every male that clasps them, even when in a receptive condition (Gissler 1883; Pearse 1913; Lochhead 1941; Belk 1984). A female is considered in receptive condition when she has mature, unfertilized eggs in the lateral pouches of the oviducts and an empty ovisac.

The general mating pattern that emerges is one in which males avoid aggression and simply search for receptive females. Alcock (1980) classified this type of mating system as scramble-competition polygyny. The male-female association in these polygynous crustaceans, when considered in relation to the probable freedom that females have in mate choice, fits the Type 1 category of Borgia (1979) as presented by Thornhill and Alcock (1983): "Females acquire resources on their own because their mates provide nothing, and thus females are free to choose among all available males on the basis of phenotype differences reflecting relative genetic quality."

The concept of scramble-competition polygyny helps us understand mating behavior in the context of anostracan ecology and in the context of how individual males behave in an effort to leave more offspring than their rivals. Since anostracan mating behavior cannot be appreciated without knowledge of the reproductive structures involved, I begin my treatment of the subject at hand with a consideration of male and female reproductive structures. Readers wishing to review general anostracan morphology may consult McLaughlin (1980).

REPRODUCTIVE STRUCTURES

Male

Males have paired penes located on the ventral surface of a genital region formed by the partial fusion of the two anteriormost abdominal segments. Each penis is divisible into a basal and an apical part. The basal parts are rigid or semirigid, variously shaped, and typically integrated into the architecture of the genital region, with definable boundaries. The apical parts are carried retracted within the genital segments. They are everted during copulation and are probably the only parts of the penis to enter the female. The apical parts may have few to no spines or other outgrowths as in Artemiidae and Linderiellidae, have two dentate warts located distally as in Branchinectidae, or have spines or other outgrowths of various development and arrangement as found in the other families.

The penes are not highly divergent among members of the same genus. Their morphology is of only limited value in providing species-specific characters. The only species differences established to date are associated with the basal parts of the penes in the family Chirocephalidae (Brtek 1966, 1967). The claim by two workers that differences exist in the microspination on the terminal spine of the eversible penes in three *Eubranchipus* species is not correct. Van Cleave and Hogan (1931) misinterpreted duplicate serrated patterns in the hypodermis that had separated from the exoskeletons of the terminal spines in their material as representing additional rows of microspines. In general, apical penes exhibit little or no variability in clearly defined genera as, for example, *Branchinecta* with more than 20 described species and *Streptocephalus* with about 50 described species. The long, tubular apical penis in *Streptocephalus* has a lateral and a medial row of flat, triangular spines (Belk 1973:figs. 1–2). Recently I found the first exception to this: I observed a third row along the dorsal surface in specimens of *S. dendyi* from Cape Point Reserve, Republic of South Africa.

The highly divergent and species-specific characters of anostracan males are most commonly associated with the second antennae. These antennae are large, muscular structures directed ventrally from lateral positions at the anterior end of the head. They are usually two-jointed structures that function as claspers during precopulatory courtship and during copulation. In members of the family Polyartemiidae, the antennae are unjointed, unitary structures. Schrehardt (1987) carefully documented development in *Artemia* using SEM. He demonstrated clearly that the adult antenna develops from the exopodite of the larval antenna and that its two-part configuration results from development of a secondary hinge-joint. Earlier workers considered the distal segment to be derived from the larval exopodite and the basal segment from the larval protopodite in a number of genera, including *Artemia* (Linder 1941; Šrámek-Hušek, Straškraba & Brtek 1962:22; Baqai 1963). Schrehardt's work suggests that the question of segment origins in these important mating structures needs to be reinvestigated for the Anostraca in general.

The morphology of male anostracan antennae takes on great variety in the form of outgrowths ranging from surface texturing, spines, knobs, and other

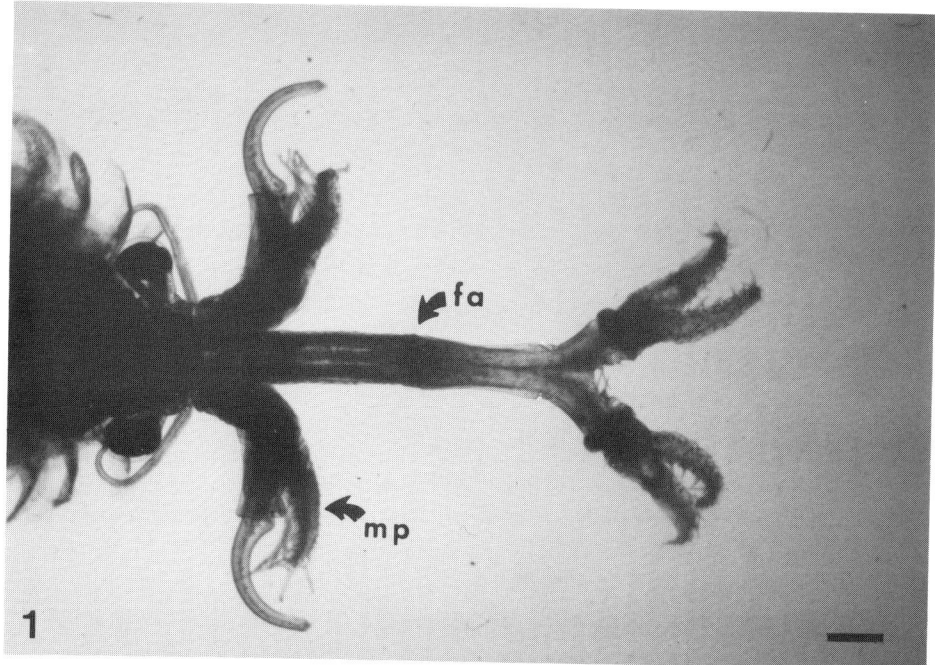

FIGURE 7.1. Anterodorsal view of *Branchinella kugenumaensis* (Ishikawa). Medial processes (mp) originate from the medial surface of the basal joints of the antennae near their distal ends. The trunk of the frontal appendage (fa) inserts on the anteromedial surface of the head near the proximal ends of the antennae. Scale = 1.0 mm.

processes to the elaborate medial processes, antennal appendages, and frontal appendages (figs. 7.1, 7.2). These undoubtedly function in providing tactile cues during what amounts to premating courtship when a male clasps a female and brings them into intimate contact with her body. Belk (1984) demonstrated experimentally that females of *Eubranchipus serratus* consider the antennal appendage in choosing among males. Judging from this evidence, evolution of the various morphological features of the antennae is probably best understood as resulting from sexual selection by female choice (see Eberhard 1985).

A number of other male anatomical features exhibit modifications that most likely play a role as tactile cues in courtship, since they are positioned so that they contact the female during amplexus and are species-specific in their morphology. Their evolution, too, has probably been the result of sexual selection by female choice. The unique knob on the anterior end of the labrum in *Eubranchipus bundyi* is a good example. The unusual species-specific divergent outgrowths of the endopods of the first three legs found in all but one species of *Dendrocephalus* are another (Pereira & Belk 1987). Spines occur at various places on the abdomen of a few species in several genera belonging to different families. There is no apparent phylogenetic pattern to the presence or absence of abdominal spines. They too may have a role in courtship.

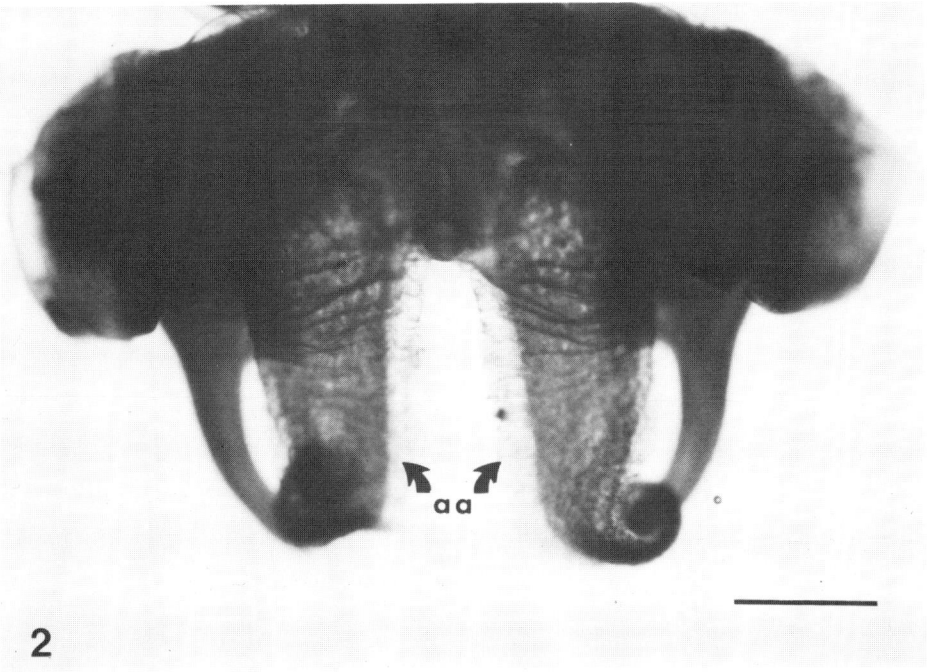

FIGURE 7.2. Anterodorsal view of *Eubranchipus bundyi* Forbes. Antennal appendages (aa) originate from the anterior surface of the basal joints of the antennae near their proximal ends. Scale = 1.0 mm.

Female

In female anostracans a brood pouch arises from the ventral surface of the genital segments. The median region of the brood pouch contains a unitary ovisac. A lateral oviducal pouch attaches to each side of the ovisac. These two lateral pouches function as storage areas for ripe eggs passing from the ovaries down the oviducts. Shell glands that secrete the tertiary envelope material, which covers fertilized eggs, attach to the ovisac at several points. In all but one species, the ovisac opens to the outside at the posterior end of the brood pouch through a gonopore. Opening and closing of the gonopore is controlled by a dorsal flap. In *Artemiopsis stefanssoni* the gonopore region is devoid of any opening and fertilized eggs are apparently retained in the ovisac throughout the life of the female (Daborn 1977). I examined specimens of *Artemiopsis stefanssoni* and observed a very small pore with what appeared to be a tube leading from it to the ovisac. The pore, which is much smaller than one shelled egg diameter, is located anterior and ventral to the sealed gonopore region. This may be an insemination pore and duct. The terminal spine on the eversible penes of *Artemiopsis* has a groove in it (Brtek 1966:plate IX) and is of a size that would fit into the pore. Additional study is needed to test this hypothesis.

Variations in brood pouch morphology have not received the broad comparative study they deserve. Nevertheless, it is not unusual for anostracan students to note brood pouch differences in their species descriptions. In identifying American species of *Branchinecta*, brood pouch morphology is diagnostic in some species but for others the brood pouches appear alike. Mathias (1937) pointed out that brood pouch length has an effect on mating posture. One cannot help but wonder what, it any, mating function the lateral outpocketings on the brood pouch of *Branchinecta campestris* serve (Lynch 1960). I observed a red stripe along the midventral region of the brood pouch in *Streptocephalus* species I collected in India. I have never seen such a mark on American streptocephalids. The stripe disappears in preserved specimens, making females from India indistinguishable from those from America.

MATING BEHAVIOR

The general pattern of mating in anostracans includes detection, orientation, station taking, clasping (amplexus), intromission, copulation, and disengagement (modified from Wiman 1981). Detection may be the result of visual, chemical, or vibrational cues. In observations of *Eubranchipus serratus* discussed below, vision alone is shown to be sufficient. However, many anostracan habitats are turbid some of the time, and others are typically turbid. In these situations, detection by visual cues may be aided or replaced by use of chemical or vibrational cues or both. However, there are no studies dealing with the use of chemicals or vibrations in mate detection. Orientation involves the male moving toward the female. In station taking, the male positions his head just below, or in the case of *Streptocephalus* just above, the female's genital region. He typically aligns his body with hers and follows her every move. Clasping involves quick movement by the male during which he grasps the female with his antennae between her last pair of legs and brood pouch. All male anostracans observed so far, even *Streptocephalus*, clasp the female from below. The pair may remain in amplexus for only a few seconds as in *Streptocephalus*, up to a minute or more as in *Eubranchipus*, five minutes to one hour as in *Branchipus*, or hours to days as in *Artemia*. Illustrations of amplexus are available in Pearse (1913), Mathias (1937), Moore and Ogren (1962), and Belk (1984). Intromission involves the male moving his genital region around one side or the other of the female's body and inserting one of his penes into her ovisac. Several attempts may precede intromission. Copulation may last from a few seconds to several minutes. In the best studied species, *Eubranchipus serratus*, disengagement results from actions initiated by the female (Pearse 1913; Belk 1984). Gissler (1883), Moore and Ogren (1962), and Wiman (1981) suggested that the males effected disengagement in the species they studied. Kuenen (1939) reported that although female *Artemia* may "struggle vigorously" they do not succeed in escaping amplexus. Thus it may prove to be the case that males terminate copulation and control amplexus in some species. This has important implications for the operation of sexual selection and thus evolution in the Anostraca (see Eberhard 1985).

I made some simple observations demonstrating that vision can play a key role for males in detection, orientation, station taking, and clasping. For these observations, I placed *Eubranchipus serratus* females in 50 mm deep by 100 mm diameter flat-bottomed Pyrex dishes floating in an aquarium populated with both males and females of *E. serratus*. Males in the aquarium repeatedly detected and oriented toward females in the dishes. When approaching from the bottom of a dish, a male would position his head below the dorsal surface of the contained female's genital segments and follow her every move, as is typical of station taking behavior. In a number of instances, a station taking male attempted to clasp one of the contained females only to strike and bounce off the bottom of the dish.

Some additional studies on *Eubranchipus serratus* also point to an important role for vision in anostracan reproductive behavior. Using two of the dishes described above, I placed five receptive females in each dish. The two dishes were in a water bath that kept them at 5–6°C. At 2030 hrs on 17 April 1975, I added five males to each dish and immediately turned off the lights. The anostracans remained in the dark in a windowless basement room until 0900 hrs the next day, when I checked them for evidence of mating. I found none. No semen was visible in the ovisacs of any of the 10 females. All still carried unfertilized eggs in their lateral pouches. The results were very different when I repeated the experiment adding only one male to each of two dishes containing five receptive females and leaving the lights on. In this case, all 10 females were mated, as evidenced by movement of eggs from the lateral pouches into the ovisacs and the fact that these eggs were shelled two days later. This second experiment started at 1030 hrs on 29 April 1975 and ended with removal of the two males at 2030 hrs the same day.

The cues used by male anostracans in detection, orientation, and station taking do not seem to be very precise. Wiman (1981) observed that *Streptocephalus* males in laboratory cultures oriented toward *Branchinecta* half their size and *Thamnocephalus* twice their size. Using observation chambers with males and females of *S. mackini* present in equal numbers, Wiman found about 70 percent of all orientation movements were directed toward other males. He considered this was most likely a result of males encountering other males more frequently than females, since males more actively cruise their habitat. There was, however, a statistically significant difference between the number of male-male and male-female orientation events that were followed by station taking. The *S. mackini* males in Wiman's study assumed station taking in 65 percent of encounters with other males as compared to 90 percent of cases when they initially oriented toward a female. Thus station taking seems to involve more selectivity than detection and orientation.

The cues used by male anostracans in deciding when to clasp also seem imprecise. Female anostracans carrying fertilized eggs in their ovisacs and belonging to species that do not spend long periods in amplexus, like *Artemia*, actively avoid mating attempts by males. The males, notwithstanding consistent rejection by such females, continue to attempt mating them (Gissler 1883; Pearse 1913; Mathias 1937; Wiman 1981). To check for discrimination against females carrying fertilized, shelled eggs by *Eubranchipus serratus* males, clasping was scored in two dishes. One contained five males and five receptive females with unfertilized eggs in their lateral pouches. The other contained five males and five females with fertilized, shelled eggs in their ovisacs. The 50x100 mm dishes were in a 15°C water bath. The experiment involved three observation periods of one hour each. We used different animals in each of the replicates. We replaced females that mated during the observation periods with new females. Total claspings of nonreceptive (fertilized) females were 16, 57, and 30. Those for receptive females were 31, 44, and 14. The results indicate no discrimination by the males of *E. serratus* ($X^2 = 1.02$, df = 1, $P > .25$) (103:89 against expectation of 1:1). During the course of these three experiments, several of the females accepted a male and copulated. Copulations by receptive females were 5, 6, and 2. The comparable results for nonreceptive females were 0, 1, and 0. This is the only time I ever observed a female carrying shelled eggs in her ovisac to accept intromission.

One observation of a male *Eubranchipus serratus* attempting to mate a conspecific female carrying shelled eggs seems especially noteworthy. This observation occurred during preliminary trials. The males and nonreceptive females were in a dish at 5°C. One of the males clasped one of the females and held onto her for five minutes, 11 secs. The female alternately struggled and relaxed before finally freeing herself. The male made 22 intromission attempts, all of which were unsuccessful. This observation seems to strengthen Wiman's (1981) speculation that females may block or allow intromission by controlling the gonopore flap. At the same time, it seems to weaken his speculation that a very vigorous male may eliminate the female's choice by rape.

Male anostracans not only fail to discriminate receptive from nonreceptive conspecific females; they also respond to congeneric females of other species. This enabled Wiman (1979a, b) and Clark and Bowen (1976) to study hybridization in *Streptocephalus* and *Artemia* species. Using the methods presented above, mating interactions were scored for five *Eubranchipus serratus* males in dishes with five *E. bundyi* females during three one-hour observation periods. The dishes were in a 5°C water bath. The *serratus* males made 39, 29, and 27 attempts to clasp *E. bundyi* females. There were 19, 1, and 6 successful claspings, respectively. The females struggled free in all these cases in from 1 to 30 secs. There were no copulations.

DISCUSSION

The reliance of anostracan males on cues that are less than precise for detection, orientation, station taking, and clasping most likely derives from the nature of the competitive scramble for the rare receptive female. Wiman (1981) observed that 80–95 percent of all females in his field-collected samples carried fertilized, shelled eggs. After sampling more than 900 anostracan populations, I agree with Dr. Wiman that receptive females are not common in collections from natural populations. Sex ratios are essentially 1:1 at the start of the reproductive period. While the ratio may shift late in the cycle, Moore (1955) concluded from a review of literature reporting sex ratios that there is no consistant bias in favor of either sex. Three studies of natural populations since Moore's review, Daborn (1976, 1977) and Hildrew (1985), report higher mortality for males. However, Hildrew (1985) noted a statistically significant sample size effect such that sample units of one or two animals contained mainly males while those with more than three contained relatively more females. In a study that avoided sampling effects by studying laboratory cultures, Browne and Sallee (1984) found that females lived longer than males in all the bisexual *Artemia* populations they examined. Thus it seems reasonable to assume, as Wiman (1981) did, that the male anostracan is in a competitive environment in which the few seconds required in detection, orientation, station taking, and clasping attempts are of little consequence as compared to the risk of missing a chance to clasp a receptive female.

Anostracan females do not store sperm. Thus a separate copulation is required to fertilize each clutch. Also, anostracan females typically accept intromission only when they have ripe eggs in their lateral pouches. Bowen (1962) demonstrated both these situations for *Artemia* in an elegant series of experiments using a recessive red eye mutant of *Artemia franciscana*. Prophet (1963) reported that the four freshwater anostracans he cultured to study egg production appeared to require a separate copulation for each clutch. He concluded this based on the reabsorption of unfertilized eggs by females held without males. Munuswamy and Subramoniam (1985) present evidence that *Streptocephalus dichotomus* does not store sperm. In studying *Eubranchipus serratus*, I have repeatedly isolated mated females and allowed them to mature and expel their eggs. Once the clutch is expelled, ripe eggs move from the ovaries into the lateral pouches. They remain there for weeks if the female is held in isolation from males. This demonstrates that sperm used for one clutch are not used for a second clutch. It may also indicate that stimulation of the ovisac by the apical penis induces the female to release her eggs into the ovisac and thus make use of the male's sperm. This is the "internal courtship" function proposed for the penis by Eberhard (1985:183). Of course, it is possible that some factor in the seminal fluid may trigger movement of eggs into the ovisac. Research into what stimulates egg movement from lateral pouches to ovisac could prove useful in understanding penis morphology and sexual selection.

Artemia differs from the other Anostraca in having an endogenous egg cycle. Even in the absence of insemination, ripe eggs will move from the lateral pouches into the ovisac after the usual 1 to 40 hr retention period. Then, following the usual three-to-five-day retention period in the ovisac, the unfertilized eggs will be expelled from the brood pouch (Bowen 1962). Criel (1980) discovered that vitelline membrane formation starts immediately after the eggs enter the ovisac, even in virgin females. The existence in *Artemia* of an endogenous egg cycle and the absence of such a cycle in the other Anostraca is the most likely reason parthenogenesis evolved in *Artemia* but not in any other anostracan.

In *Artemia*, precopulatory clasping lasts hours to days, and females accept amplexus even when they are carrying fertilized eggs. *Artemiopsis stefanssoni* seems to behave in this way also (Johansen 1922; Daborn 1977), and Mathias (1937) reported union for up to one hour in *Branchipus stagnalis*. Amplexus in the other species studied lasts only a few seconds to a few minutes, and females will not usually accept amplexus when they are carrying fertilized eggs. While it is not clearly understood why those differences exist, there is undoubtedly a high degree of uncertainty associated with continually moving from one female to the next in search of the rare receptive female. In *Artemia*, this uncertainty is compounded by the existence of an endogenous egg cycle that sets limits to the period during which the female may be successfully inseminated. In the other anostracans examined, ripe eggs remain available until copulation occurs. Also, in at least some nonartemians, the receptive females make themselves more available to males while the nonreceptive females avoid areas with high densities of cruising males (see below). The *Artemia* male probably overcomes this uncertainty and assures at least one mating every four to six days by clasping a female and staying with her.

There seems to be good reason to suspect that it is to the *Artemia* female's advantage to cooperate in this prolonged courtship. The limits set by an endogenous egg cycle on the effective period for insemination are likely to make it to her advantage to have a male at hand when ripe eggs enter the lateral pouches. Otherwise, she could lose a clutch. If the female of *Artemia* is able to free herself from clasping males and thus choose her mate, a prolonged courtship could give her additional time to assess potential males. Lent (1971) demonstrated for *Artemia* that precopulatory clasping has no adverse effect on the efficiency of swimming, thus respiration and feeding, or the escape response of the paired animals.

Kuenen (1939) pointed out that *Artemia* remain in amplexus for "some time" after copulation. he also observed cases in which, after one male released a female, another male clasped this same female and copulated with her. Alcock (1980) noted that scramble-competition polygyny in the piggyback bee included mate guarding behavior. Might this be the reason *Artemia* males continue in amplexus following copulation?

The uncertainties faced by the male in moving from one female to the next in search of the rare receptive female seem to be decreased, in the nonartemian Anostraca studied so far, by differences in the behavior of receptive and nonreceptive females. Pearse (1913) reported that copulation produces an immediate change in the behavior of females of *Eubranchipus serratus* (=*E. dadayi*). He observed that they would no longer swim or rest near the surface but would go to the bottom of the container and remain quiet. Building on this observation, Wiman (1978) hypothesized that a sexually receptive female anostracan may increase her chances of encountering a male by becoming more active. He tested this hypothesis by scoring for *Streptocephalus mackini* the distances moved in two directions during one-minute observation periods. The receptive females were more than twice as active as those carrying fertilized eggs, moving an average of 44 cm/min versus 18 cm/min. However, he considered these results less than clear, since males were present in the observation chambers and could have caused the increased activity of the receptive females by preferentially approaching them. The lack of preference shown by *Eubranchipus serratus* males in clasping receptive and nonreceptive females discussed above suggests that Wiman's concern is probably unfounded.

Table 7.1 presents results clearly demonstrating for *Eubranchipus serratus* that females spend more time swimming when they are receptive and more time resting at the bottom when they are carrying shelled eggs. These data also show that *E. serratus* males spend most of their time where the ratio of receptive to nonreceptive females is most favorable. The female brooding fertilized eggs

TABLE 7.1. Activity patterns of *Eubranchipus serratus*. Anostracans segregated according to sex and female reproductive condition, five animals per 1000 ml beaker in 800 ml of 15°C water. Each anostracan's activity was scored 14 times at five-minute intervals. Anostracans were placed in beakers 12–22 hrs before testing. Test 1 included three replicates with receptive females and three with males. Test 2 included three replicates with receptive females and three with bred females.

Test	Sex	Egg Location	Number Resting on Bottom	Number Swimming in Water Column
1	Female	Lateral pouches	54	156
	Male		0	210
2	Female	Lateral pouches	44	166
	Female	Ovisac	128	82

presumably decreases the cost associated with fending off unneeded suitors by staying in an area less frequented by males. She may also be in a situation that decreases her exposure to avian predators. At the same time, she will likely increase her exposure to invertebrate predators. However, for a fairy shrimp in clear water a brood pouch filled with dark-shelled eggs may make danger from birds greater than increased exposure to less visual predators.

CONCLUSIONS AND DIRECTIONS OF FUTURE RESEARCH

Courtship in the Anostraca relies on visual and tactile cues. Males seem to rely on vision and rather imprecise cues in choosing females. Females seem to rely on tactile cues in choosing males. Male antennal structures provide the primary tactile cues used by females in mate selection. The basal parts of the penes may provide additional cues for the females. Species-specific differences in the basal penes found in the family Chirocephalidae strongly suggest such a role in courtship. Sexual selection by female choice is most likely the dominant driving force in the evolution of the species-specific characteristics of these male reproductive structures. During the scramble for mates, a single anostracan female is exposed to contact with these structures on a considerable number of different males, thus meeting one of the major requirements of the sexual selection by female choice hypothesis (Eberhard 1985:184).

The function of the eversible penes, in addition to sperm delivery, may be to induce the female to move ripe eggs from the lateral pouches into the ovisac. This may or may not hold true for *Artemia* species that have an endogenous egg cycle. The morphology of the eversible penes appears to be relatively uniform within clearly defined genera. I suggest that evolution at the generic level in Anostraca results from infrequent bouts of runaway sexual selection by female choice in which morphological characters of the eversible penes become objects of female choice. This hypothesis offers a sound theoretical basis for using eversible penis morphology as a primary factor in defining anostracan genera.

Linder (1941) has already emphasized the importance of genital characters in establishing higher level taxonomic categories.

The bisexual *Artemia* species present some interesting differences in anostracan reproductive biology about which we lack information. The comparative value of understanding these differences, along with the economic importance of *Artemia*, suggests that it would be worthwhile to carefully study mating behavior in *Artemia*. The following are a few of the important questions needing answers: Is the female able to escape the male's clasp and thus choose the male that will fertilize each clutch of her eggs? Or is the male able to effectively eliminate sexual selection by female choice in *Artemia*? Does copulation override the endogenous egg cycle and stimulate movement of eggs from the lateral pouches into the ovisac? Or is it necessary for the male to continue in amplexus after copulation to guard against mating by another male and thus force the female to use his sperm? Does the female let the male know when it is time to mate? Or must the male make periodic attempts at intromission until he is successful?

Streptocephalus dorothae and *S. mackini* are so alike morphologically that only one feature, associated with the medial process of the second antenna, is available to distinguish between them. This is a spine on a swollen area near the tip of a structure called the finger (fig. 7.3). The spine occurs in *S. mackini* but is

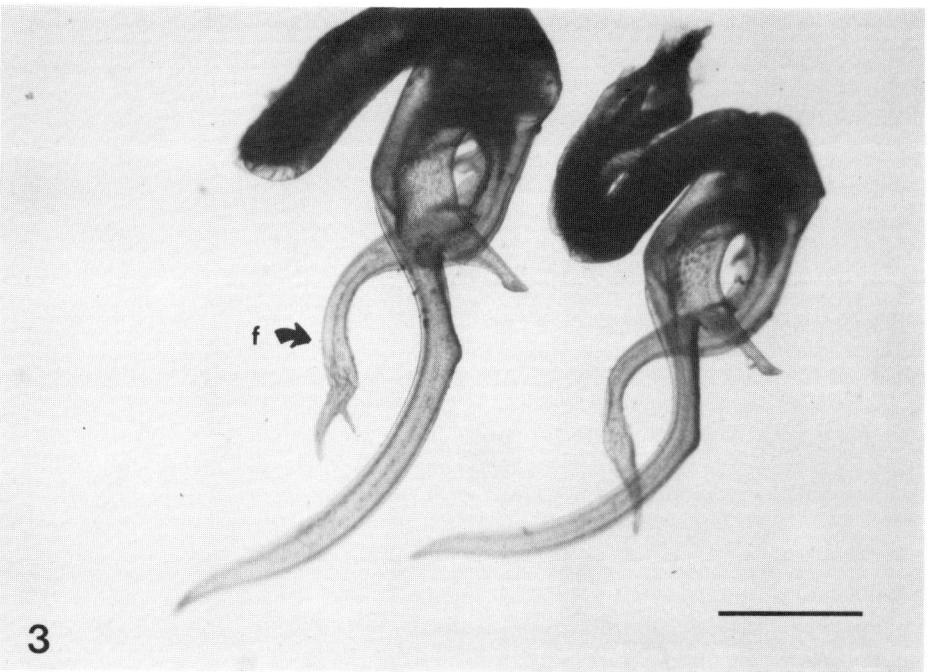

FIGURE 7.3. Lateral views of the left medial processes from *Streptocephalus mackini* Moore (left) and *S. dorothae* Mackin (right). The distal cheliform structure is called the hand. The ventral process is called the finger (f). The upward bend of the distal end of the finger from *S. mackini* is a result of handling and is not natural. The spine on the swollen region at the distal end of the finger in *Streptocephalus mackini* is the only morphological character separating these two species. Scale = 1.0 mm.

absent in *S. dorothae*. However, neither the presence nor the absence of this spine provides the females of these two species with an absolute character upon which they can recognize and accept only conspecific males. Wiman (1979a) demonstrated this when he obtained fertile hybrid offspring from both reciprocal crosses in no-choice mating tests and when he found evidence of hybridization in naturally occurring mixed-species populations (Wiman 1979b). After studying mating patterns in five species of *Streptocephalus*, Wiman (1979a) concluded that speciation in the genus appears limited to the slow accumulation of genetic differences in allopatric populations without selection for antihybridization mechanisms. Wiman's work (1979a,b, 1981) seems to indicate that morphological differences of the second antennae used to separate the North American species do not function as structures by which females recognize conspecific males. However, I am confident that the entire medial process of the antenna in these anostracans is important for male reproductive success just as the antennal appendage is in *Eubranchipus serratus*.

Antennal appendages, frontal processes, and medial processes are prime candidates for runaway sexual selection by female choice. As such, they are likely to develop excessive elaborate structural complexity (see Eberhard 1985). In the absence of selection for tactile cues useful in avoiding hybridization, much of the fine detail of these elaborate male structures may not be carefully evaluated by females. This is in contrast to the value of these structural complexities in showing the accumulation of genetic differences, a situation that forms the basis of their usefulness in species identification and taxonomy. Thinking of these evidences of genetic difference as reproductive isolating mechanisms is probably counterproductive. Instead, they are probably best viewed as arbitrary additions to the complexity of structures providing tactile courtship cues.

Acknowledgments

I thank my wife, Mary Schug Belk, for helping with the behavioral observations. I completed the behavior studies during an appointment as visiting assistant professor of zoology at Arizona State University. My appreciation is extended to the Smithsonian Institution for supporting two collecting trips to India with funds from their Special Foreign Currency Program. I thank Dr. Jenny Day for sending me anostracans from South Africa. I thank Dr. Sarane T. Bowen for comments on an earlier draft. I appreciate very much the valuable suggestions made by three annonymous reviewers.

Literature Cited

Alcock, J. 1980. Natural selection and mating systems of solitary bees. *Am. Sci.* 68:146–153.
Baird, W. 1850. *The Natural History of the British Entomostraca*. London: Ray Society, reprinted 1968 by Johnson Reprints, New York.
Baqai, I. U. 1963. Studies on the postembryonic development of the fairy shrimp *Streptocephalus seali* Ryder. *Tulane Stud. Zool.* 10:91–120.
Belk, D. 1973. *Streptocephalus moorei* n. sp., a new fairy shrimp (Anostraca) from Mexico. *Trans. Am. Microsc. Soc.* 92:507–512.

Belk, D. 1984. Antennal appendages and reproductive success in the Anostraca. *J. Crustacean Biol.* 4:66–71.
Borgia, G. 1979. Sexual selection and the evolution of mating systems. In M. S. Blum & N. A. Blum, eds., *Sexual Selection and Reproductive Competition in Insects.* New York: Academic Press.
Bowen, S. T. 1962. The genetics of *Artemia salina*. I. The reproductive cycle. *Biol. Bull. (Woods Hole)* 122:25–32.
Browne, R. A. & S. E. Sallee. 1984. Partitioning genetic and environmental components of reproduction and lifespan in *Artemia*. *Ecology* 65:949–960.
Brtek, J. 1966. Einige Notizen zur Taxonomie der Familie Chirocephalidae Daday 1910. *Annot. Zool. Bot.* 33:1–65.
Brtek, J. 1967. *Eubranchipus (Creaseria) moorei* n. sp. *Annot. Zool. Bot.* 36:1–7.
Clark, L. S. & S. T. Bowen. 1976. The genetics of *Artemia salina*. VII. Reproductive isolation. *J. Hered.* 67:385–388.
Criel, G. 1980. Morphology of the female genital apparatus of *Artemia*: A review. In G. Persoone, P. Sorgeloos, O. Roels & E. Jaspers, eds., *The Braine Shrimp Artemia,* 1:75–86. Wettern, Belgium: Universa Press.
Daborn, G. R. 1976. The life cycle of *Eubranchipus bundyi* (Forbes) (Crustacea: Anostraca) in a temporary vernal pond of Alberta. *Can. J. Zool.* 54:193–201.
Daborn, G. R. 1977. On the distribution and biology of an arctic fairy shrimp *Artemiopsis stefanssoni* Johansen, 1921 (Crustacea: Anostraca). *Can. J. Zool.* 55:280–287.
Eberhard, W. G. 1985. *Sexual Selection and Animal Genitalia.* Cambridge, Massachusetts: Harvard University Press.
Gissler, C. F. 1883. VI. Miscellaneous notes on the reproductive habits of Branchiopodidae. In A. S. Packard, Jr., A monograph of the Phyllopod Crustacea of North America with remarks on the order Phyllocarida. In F. V. Hayden, *Twelfth Annual Report of the U.S. Geological and Geographical Survey of the Territories: A Report of Progress of the Exploration in Wyoming and Idaho for the year 1878,* part 1, section 2, pp. 420–31. Washington: U.S. Geological Survey.
Hildrew, A. G. 1985. A quantitative study of the life history of a fairy shrimp (Branchiopoda: Anostraca) in relation to the temporary nature of its habitat, a Kenyan rainpool. *J. Anim. Ecol.* 54:99–110.
Johansen, F. 1922. *Report of the Canadian Arctic Expedition 1913–18.* Vol. 7, *Crustacea,* part G: Euphyllopoda. Ottawa: F. A. Acland.
Kuenen, D. J. 1939. Systematical and physiological notes on the brine shrimp *Artemia*. *Arch. Neerl. Zool.* 3:365–449.
Lent, C. M. 1971. Metachronal limb movements by *Artemia salina:* Synchrony of male and female during coupling. *Science (Washington, D.C.)* 173:1247–1248.
Linder, F. 1941. Contributions to the morphology and the taxonomy of the Branchiopoda Anostraca. *Zool. Bidr. Upps.* 20:103–303.
Lochhead, J. H. 1941. *Artemia*, the "brine shrimp." *Turtox News* 19(2):1–4.
Lynch, J. E. 1960. The fairy shrimp *Branchinecta campestris* from northwestern United States (Crustacea: Phyllopoda). *Proc. U.S. Natl. Mus.* 112:549–561.
McLaughlin, P. A. 1980. *Comparative Morphology of Recent Crustacea.* San Francisco: W. H. Freeman.
Mathias, P. 1937. *Biologie des Crustacés Phyllopodes.* Paris: Hermann & Cie.
Moore, W. G. 1955. The life history of the spiny-tailed fairy shrimp in Louisiana. *Ecology* 36:176–184.
Moore, W. G. & L. H. Ogren. 1962. Notes on the breeding behavior of *Eubranchipus holmani* (Ryder). *Tulane Stud. Zool.* 9:315–318.
Munuswamy, N. & T. Subramoniam. 1985. Influence of mating on ovarian and shell gland activity in a freshwater fairy shrimp *Streptocephalus dichotomus* (Anostraca). *Crustaceana (Leiden)* 49:225–232.
Pearse, A. S. 1913. Observations on the behavior of *Eubranchipus dadayi*. *Bull. Wisconsin Nat. Hist. Soc.* 10(3–4):109–117.
Pereira, G. & D. Belk. 1987. Three new species of *Dendrocephalus* (Anostraca: Thamnocephalidae) from Central and South America. *J. Crustacean Biol.* 7:572–580.
Prophet, C. W. 1963. Egg production by laboratory cultured Anostraca. *Southwest. Nat.* 8:32–37.
Schrehardt, A. 1987. A scanning electron-microscope study of the post-embryonic development of *Artemia.* In P. Sorgeloos, D. A. Bengtson, W. Declier, & E. Jaspers, eds., *Artemia Research and Its Applications,* 1:5–32. Wetteren, Belgium: Universa Press.
Šrámek-Hušek, R., M. Straškraba & J. Brtek. 1962. *Fauna ČSSR 16 Branchiopoda.* Praha: Československé AkademiŠe Věd.

Thornhill, R. & J. Alcock. 1983. *The Evolution of Insect Mating Systems.* Cambridge. Massachusetts: Harvard University Press.
Van Cleave, H. J. & S. M. Hogan. 1931. A comparative study of certain species of fairy shrimps belonging to the genus *Eubranchipus. Trans. Ill. State Acad. Sci.* 23:284–290.
Wiman, F. H. 1978. Speciation in the fairy shrimp genus *Streptocephalus.* Ph.D. dissertation, University of Wisconsin, Madison.
Wiman, F. H. 1979a. Mating patterns and speciation in the fairy shrimp genus *Streptocephalus. Evolution* 33:172–181.
Wiman, F. H. 1979b. Hybridization and the detection of hybrids in the fairy shrimp genus *Streptocephalus. Am. Midl. Nat.* 102:149–156.
Wiman, F. H. 1981. Mating behavior in the *Streptocephalus* fairy shrimps (Crustacea: Anostraca). *Southwest. Nat.* 25:541–546.

EIGHT
Mating and Insemination in the American Lobster, Homarus americanus

S. L. WADDY and
D. E. AIKEN

Department of Fisheries and Oceans, Aquaculture and Invertebrate Fisheries Division, Biological Station, St. Andrews, New Brunswick.

Abstract

Most female American lobsters mate when they are newly molted, but mating can occur at any molt stage. In this chapter we explore the differences between postmolt and intermolt mating and demonstrate the necessity for intermolt mating to the reproductive success of lobsters. Although large lobsters (≥ 120 mm carapace length) are able to fertilize two successive clutches of eggs from a single insemination, 20 percent store an insufficient amount of sperm and require reinsemination prior to their second spawning. For these lobsters, intermolt insemination is necessary to ensure their reproductive potential is realized. Intermolt mating is also necessary in smaller lobsters that for some reason did not mate at molt. Under laboratory conditions, all uninseminated preovigerous females mate prior to spawning. Intermolt mating has not been confirmed in feral populations, but circumstantial evidence is strong.

Male lobsters have an impressive capability to mate. One male successfully inseminated nine females over an 11 day period. Males are able to mate throughout the year regardless of water temperature. Male aggression is a major factor in the success of intermolt mating, and males can discriminate between immature and mature females and between inseminated and uninseminated females. Female receptivity is affected by both ovarian stage and the presence of stored sperm but not by molt stage. Females become unreceptive after insemination, but receptivity is restored when the supply of stored sperm is exhausted.

ALTHOUGH MANY crustaceans mate when the female is hard-shelled, American lobsters *(Homarus americanus)* usually mate just after the female molts. Until recently it was thought that female lobsters had a limited receptive period and that mating seldom occurred more than 24 hrs after the molt (Templeman 1934; Hughes & Matthiessen 1962; Aiken & Waddy 1980; Atema 1986). Earlier reports of female lobsters mating while in the hard-shell intermolt condition (Dunham & Skinner-Jacobs 1978; Dunham 1979) were thought to be the result of crowded laboratory conditions (Dunham 1988), and it was questioned whether mating and insemination could even occur once the cuticular flaps covering the seminal receptacle were no longer flexible (Nelson & Hedgecock 1971; Talbot, Thaler & Wilson 1986).

In apparent contrast to earlier reports (Aiken & Waddy 1980; Atema 1986, for reviews), we have found that American lobsters have a dual mating strategy: mating usually occurs at the female molt, but, when necessary, mating can occur anytime prior to spawning. In this chapter we reexamine previously held concepts of lobster sexual behavior and biology, explore the role of intermolt mating in the reproductive strategy of female lobsters, and summarize available information on artificial insemination and the control of mating in American lobsters.

MATING IN RECENTLY MOLTED FEMALE LOBSTERS

Mating behavior

Mating in lobsters is thought to be elicited by a suite of pheromonal, chemotactile and visual stimuli (McLeese et al. 1977; Atema et al. 1979; Atema & Cobb 1980). Atema (1986) described complex mating behavior involving pair formation, mate selection by the female, and male protection of the female by the male occurring over several days prior to and after the female molt.

When female lobsters reach late premolt they begin to show interest in males and indicate their choice by frequently approaching a male's shelter over the course of several days (Atema 1986). As the female approaches molt, she and the dominant male cohabit intermittently and then continuously. It has been suggested that intermolt females are rejected by the male, possibly because they lack the proper sex pheromone (Atema 1986; but cf. this section). It is not known how the female recognizes the dominant male nor what criteria she uses to select a mate, but there is evidence that males "advertise" (Cowan & Atema 1985; Atema & Cowan 1986). If the chosen male is already cohabiting with a female, another male may be selected or the female may wait for the preferred male to be available (Atema 1986). Mature females are said to delay molting if the male they have selected is unavailable (Cowan & Atema in press). Unfortunately, these females were not monitored for stage of the molt cycle (Aiken 1973), so the concept of molt delay requires confirmation.

Mating is initiated by the male and usually occurs about 30 mins after the female molt. The male mounts the female from the posteriodorsal aspect and moves slowly forward, turning the female over with his pereiopods. The female cooperates by stretching her claws forward, unfolding her abdomen, and remaining passive. The male inserts his gonopods into the seminal receptacle, and the two halves of the gonopods from a duct that conducts the spermatophores into the receptable (Atema 1986).

After mating, the male and female continue to cohabit for several days and the male guards the female, repels other males, and thereby reduces the opportunity for multiple insemination.

Factors Influencing Mating

Female Molt Stage. Templeman (1934) felt that almost all matings occurred within 24 hrs of the female molt and most occurred within the first 12 hrs. He observed matings as soon as 20 mins after the female molt and found a negative correlation between time following the female molt and the incidence of successful mating. He reported that 77 percent of pairings were successful if females had molted less than 3 hrs previously but that the percentage of successful matings dropped to 26 percent if females had molted 8–24 hrs earlier. In 20 pairings attempted after the first 24 hrs, none was successful. The time that passed before the male attempted to mate with the female also increased with time after molt.

Other studies appeared to confirm Templeman's (1934) observation that females rarely mate if they are not newly molted (Hughes & Mattheissen 1962; Aiken & Waddy 1980; Atema & Cobb 1980; Atema 1986) and, until recently, matings after molt stage A were considered unusual. We now know that is not the case; although there is a peak in female sexual receptivity immediately after molting, uninseminated females remain receptive until spawning. Receptivity peaks again during the weeks prior to spawning when most females are in molt stage C_4-D_0. However, the receptivity of unmated females is more dependent on ovarian stage than molt stage; apparent changes in receptivity during molt stages B, C, and D_0 are related more to ovary maturity and seasonal temperature than to molt stage. No limitations to mating and insemination have been found because of molt stage.

Female Reproductive Stage. Female reproductive stage is a major factor in the interest males show in females. Incidence of mating is strongly correlated with ovary development. In lobsters from the Canadian Maritimes it is unusual for preovigerous females (ovary stages 4b–6; Aiken & Waddy 1982) not to be carrying a sperm plug, and females with immature (stage 1, white) ovaries are rarely inseminated. We have examined preovigerous females from the southern Gulf of St. Lawrence (N=338), Northumberland Strait (N=272), southwest Nova Scotia (N=110), and the Bay of Fundy (N=252) and found that virtually all (>99.5%)

were carrying spermatophores in the seminal receptacle. Mature females without sperm have been reported from some stocks (Krouse 1973; Ennis 1980), but when those studies were done, criteria for assessing maturity were not as well defined as they are today and the samples may have included females that would molt again prior to spawning.

Sperm from a pubertal female's first insemination are rarely used in fertilization because females are usually inseminated when ovarian development is just beginning (ovary stages 2 or 3), two molts before spawning. The sperm are lost at the next molt, the female is reinseminated (ovary stage 4a), and spawning occurs later that same summer or the following year. If for some reason insemination does not occur at molt, the female remains receptive until spawning.

Male Potency. There is little information on the reproductive capacity of male lobsters. Hughes and Matthiessen (1962) reported that a male lobster mated with two females in one day but showed no interest in two different females the following day. Although males are often capable of mating with two females on the same day, their recovery time is variable. Some will mate again the following day while others may have no interest in mating for several days. In our laboratory, one male mated with nine newly molted females in 11 days. Three of the inseminations occurred within 21 hrs, and all females subsequently produced fertile eggs. This male mated with every female that was available to him, so his capability may have been even greater (Waddy, unpublished). Several males have demonstrated similar capability. However, there is considerable variation in potency within and between individuals.

Male Virility. Relative size of male and female influences mating success in lobsters. Small males that are functionally mature may still be incapable of mating if the female is significantly larger (Hughes and Matthiessen 1962), possibly because they lack the physical strength to subdue the female. Small males of only 40–50 mm carapace length (CL) may produce gametes, but they are often functionally incapable of mating (Aiken and Waddy 1980). In both the Gulf of St. Lawrence and the Bay of Fundy males become functionally mature at about the same size as females, several molts after they first produce sperm (Aiken & Waddy 1980, 1989, and unpublished).

It is difficult to predict whether or not a particular male will mate. Earlier studies suffered from lack of criteria for assessing functional maturity in males, and it is probable that immature males were incorrectly assumed to be mature. Size at maturity can vary from less than 60 mm to more them 110 mm CL depending on geographical origin and temperature history. Recent observations indicate that size of the crusher propodus is a reliable index of functional maturity in males. The volume of the crusher propodus changes to a higher level of allometric growth as the lobster matures and the cheliped propodite index (CPI) shows an inflection when plotted against carapace length (fig. 8.1). CPI values greater than 22–24 are indicative of maturity regardless of whether the male is 70 mm CL from the southern Gulf of St. Lawrence or 120 mm CL from the Bay of Fundy (Aiken & Waddy 1989, and unpublished).

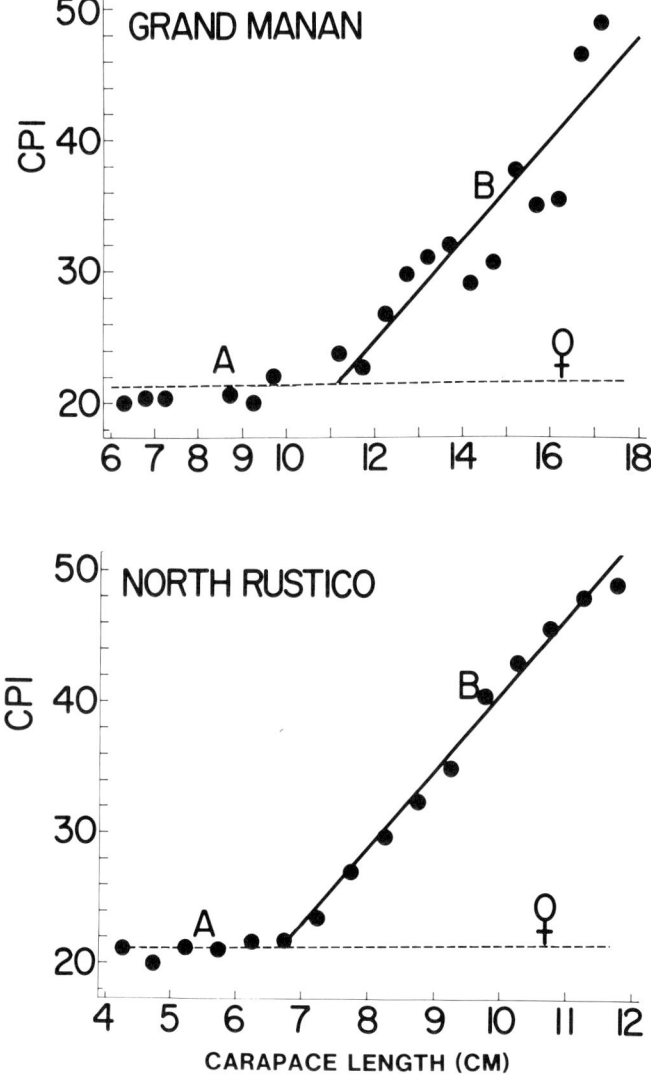

FIGURE 8.1. Allometric growth of the crusher cheliped of male American lobsters as indicated by the plot of crusher propodite index (CPI) against carapace length (CL). Intersect of mature male regression line (B) on female regression line (A) indicates size at onset of sexual maturity of male lobsters from the Bay of Fundy (Grand Manan) and the southern Gulf of St. Lawrence (North Rustico). Individual points are means of 5 mm CL size groups. From Aiken & Waddy 1989.

Male Molt Stage. Although male lobsters can mate in all molt stages from buckle-shell (early stage C) through intermolt and into premolt (Templeman 1934), there is a strong influence of molt stage on male potency. Mature males from the Bay of Fundy molt only once ever 2–3 years, and we have found that frequency of mating in individuals varies dramatically between years. When the mating

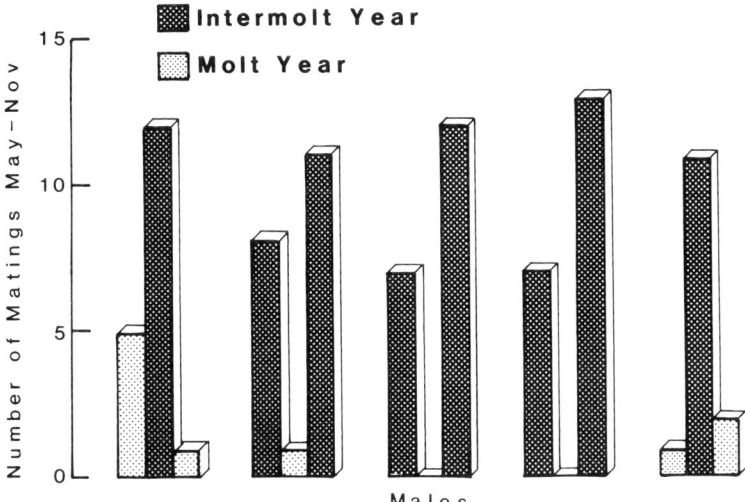

FIGURE 8.2. Variation in mating incidence in male lobsters with stage of the molt cycle. Mating activity of five males was monitored over three successive years (1987–1989).

activity of five males was monitored over three successive years (and phases of the molt cycle) it was found that almost all mating activity was confined to the years between molts and little, if any, mating occurred in the molt years (fig. 8.2). Until the effect of molt stage was recognized, loss of interest in mating was attributed to social and environmental conditions (Aiken et al. 1984). Long-term studies are needed to further define the relationship between molting and reproductive capability; small mature males in some localities molt each year and are obviously able to mate at some time in that year.

Environmental Effects. Male lobsters are capable of mating throughout the year, even in winter when water temperatures are 0–2°C (Waddy, unpublished). We have found mating incidence to be low (28%, N = 18) when water temperature is <5°C if shelters are available, but if lobsters are confined without shelter mating occurs readily (70%, N = 20). There are no indications of a strong seasonal influence on male mating capability. In support of this idea, an earlier study found no evidence of cyclic seasonal changes in the weight of the vasa deferentia (Aiken & Waddy 1986).

Despite the fact that temperature does not appear to be a major factor regulating the reproductive cycle of male lobsters, long-term temperature history during the juvenile phase determines size at maturity, and exposure to constant elevated temperature during maturation has prolonged negative effects on both sperm and spermatophore production. Males hatched and reared to maturity at constant 20°C are poor producers of sperm (small spermatophores with little or no sperm), but if they are conditioned for a year or longer at seawater temperatures typical of their natural habitat, more spermatophores and a greater quantity of sperm will be produced (determined by electroejaculation; Aiken et al.

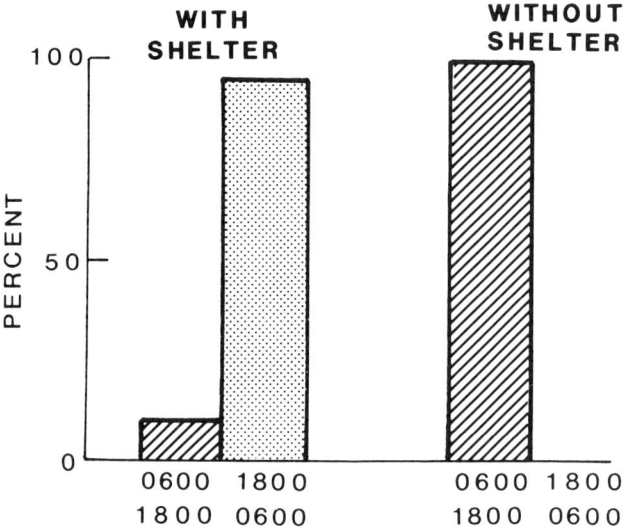

FIGURE 8.3. Effect of presence of shelter on daily timing of matings in hard-shelled females held communally with males. Mating occurs during the photophase if shelters are not available, but if there are shelters, mating occurs during the scotophase.

1984 and unpublished). The problem can be reduced by transferring the males to normal seasonal temperatures (0–14°C) when they are about 40–50 mm CL, well below the size at onset of functional maturity. Cultured males transferred at this size breed almost as well as wild males by the time they reach 65–75 mm CL.

The presence of shelter affects timing of mating; without shelters matings occurred during daylight hours (100%, N=46), but with shelters, most matings (90%, N=120) occurred between the hours of 1800 and 0600 (fig. 8.3).

Mechanisms to Ensure Insemination and Fertilization

Female Receptivity. Receptivity in female lobsters is related to both sperm supply and ovary condition. Male lobsters can distinguish between immature and mature females and between inseminated and uninseminated females (*pace* Dunham & Skinner-Jacobs 1978). Once inseminated, females are usually no longer receptive, so incidence of multiple insemination is relatively low. Males rarely attempt to mate with a female that is carrying stored spermatophores, even when they are held for long periods in confined quarters. In a recent study, only 10 percent of 203 females were inseminated twice, despite ample opportunity. Receptivity is restored once the sperm supply is depleted (Waddy unpublished).

Stage of ovarian maturation is also important. When males are held communally with both uninseminated preovigerous females (ovary stages 4b–6) and mature but nonpreovigerous females (ovary stage 4a), incidence of mating is significantly higher in the group that is approaching spawning (100 versus 61 percent) (Waddy & Aiken 1987). This indicates that lobsters are able to recognize uninseminated prespawning females. Similarly, females that spawn twice between molts become receptive again if they use all their stored spermatozoa to fertilize the first spawning, and in 20 females that were inseminated twice (within hours or days), the amount of material in the seminal receptacle was judged to be less than the normal amount and did not fill the receptacle. Possibly the second mating was induced by inadequate insemination, a phenomenon well documented in insects (Parker 1970; Walker 1980).

Behavior of Uninseminated Females. Mature uninseminated females broadcast behavioral and possibly chemical signals to alert male lobsters to their lack of insemination, and these signals become more pronounced as time of spawning approaches (Waddy, unpublished). Females that have not been inseminated often become more active as the time of spawning approaches, leaving their shelters more frequently and spending longer periods away from their shelters (several hours each night rather than less than an hour). Once insemination occurs, activity patterns become indistinguishable from those of inseminated females (Waddy, unpublished). Such a behavioral change has also been observed in preovigerous spiny lobsters *(Panulirus argus)* that had not copulated (Lipcius & Herrnkind 1985).

Multiple Insemination. According to Templeman (1934) a female may mate with many males (he noted one case in which a female mated with 10 males). In spite of this potential capability, he felt that multiple insemination would be unlikely in nature, as the dominant male would discourage suitors. When males are left with females after mating, the male seldom mates with the female a second time and seems preoccupied with repelling other males. Our data support this conclusion. Most females in our studies (90 percent) were inseminated only once, but if a spermatophore was not transferred during mating (or appeared small), the male persisted. In the occasional case where newly molted females mated with many males in rapid succession, we were reluctant to interfere and therefore were not able to determine how many of the males actually transferred a spermatophore.

Multiple insemination can also occur in lobsters that spawn twice within an intermolt period. In the laboratory, these females may mate a second time, 2–3 years after molting and the first mating. Since lobsters spawn whether or not they have been inseminated, reinsemination is necessary for these females to fulfill their reproductive potential. Reinsemination is difficult to confirm in nature, but all 84 Bay of Fundy female lobsters we examined at the appropriate size and reproductive condition were carrying sperm plugs, an unlikely event unless reinsemination occurred.

Sperm Viability. Lobster sperm retain their viability for long periods in the female seminal receptacle. Templeman (1934) concluded that sperm would remain viable for at least two years, and our studies have extended the viability to at least three years (Waddy & Aiken 1986).

In crabs there is thought to be progressive deterioration in the quality of stored sperm, since the quantity and quality of eggs produced decreases with each successive clutch (Morgan, Goy & Costlow 1983; Paul 1984). If given the opportunity, the crabs will mate prior to spawning and thus avoid using sperm that has been stored for long periods (Paul 1984). In the case of American lobsters, sperm is often stored for 2–3 years before use and we have been unable to detect any difference in fertilization or hatching success between successive broods (Waddy & Aiken 1986).

MATING IN INTERMOLT FEMALES

Methods

The following descriptions of intermolt mating are based on experiments in which more than 6000 hrs were recorded with a time-lapse video recorder over a three-year period (1987–1989). More than 120 matings of previously uninseminated females were recorded, and another 152 intermolt inseminations occurred during similar experiments that were not videotaped. Matings occurred among male and female lobsters held communally with shelters in a $2 \times 2 \times 1$ m tank illuminated at night with red lamps (light from 25 watt incandescent bulbs through Kodak No. 1A filters) so that lobster activity could be recorded continuously. The red lights used for videotaping at night had no detectable effect on lobster behavior.

Lobsters were identified by numbers etched on the dorsal carapace and highlighted with a white paint marker (Testors Corporation). Lobsters (75–90 mm CL) were from Miminegash, Prince Edward Island (Gulf of St. Lawrence), and were known to be mature; the females had spawned in captivity two years earlier, and the males had successfully inseminated other females. Females used in the studies had molted in isolation 6–12 months prior to the beginning of the studies and were not inseminated. Densities in the tank ranged from 2–5 lobsters/m^2, similar to the 3.2 lobsters/m^2 estimated in some wild populations (Cobb & Wang 1985). The male to female ratio varied from 3:1 to 1:1. Temperature varied seasonally (0–14°C) and natural photoperiod was provided through a west-facing window. A surplus of drainage tiles was provided as shelter for individuals. The tiles were too small to allow cohabitation, so matings occurred in the open where they could be recorded on videotape.

Videotapes were reviewed every 12–24 hrs so that observed matings could be confirmed by the presence of a spermatophore (the spermatophores are visible when the receptacle is opened with "barraquer" forceps; see fig. 8.8). Inseminated females were retained through spawning and egg incubation to determine the success of egg attachment and hatching from intermolt matings.

General Behavior

Lobsters provided with shelter in a confined area are active primarily at night. Only rarely do they leave the security of their shelters during daylight hours.

Activity begins at dusk and continues until dawn, with the peak of activity occurring between approximately 2200 and 0400 hrs (hours of activity vary somewhat with changing length of the scotophase). They spend the daylight hours in their shelters, usually returning early each morning (0400–0600 hrs) to the same shelter they occupied the night before. When a different shelter is selected it will often be adjacent to the one occupied the previous day. Their preference for a specific shelter may last several days, sometimes even weeks, and they are protective of this preferred space, returning quickly to the shelter if another lobster approaches the entrance. Similar behavior occurs under more natural conditions (Karnofsky & Price 1989; Karnofsky, Atema & Elgin 1989a).

Male lobsters are far more active than females under these conditions, leaving their shelters more frequently and remaining outside for longer periods of time. Between May and August, 95 percent of the males (N=5) were outside their shelters for an average of 5 hrs each night. In contrast, only 20 percent of the females left their shelters each night, and these remained outside, on average, only one hour (figs. 8.4, 8.5). The difference in activity between the sexes seems to be due to the fact that females leave their shelters primarily to feed while mature males spend considerable time exploring their surroundings and other shelters. In contrast, Karnofsky, Atema & Elgin (1989b) found no difference in the activity of immature male and female lobsters in the field.

Males also spent a great deal of time at the entrance of their shelters, with their abdomen inside and their thorax and chelipeds outside. From this position the male monitored events outside the shelter and often rushed out to accost

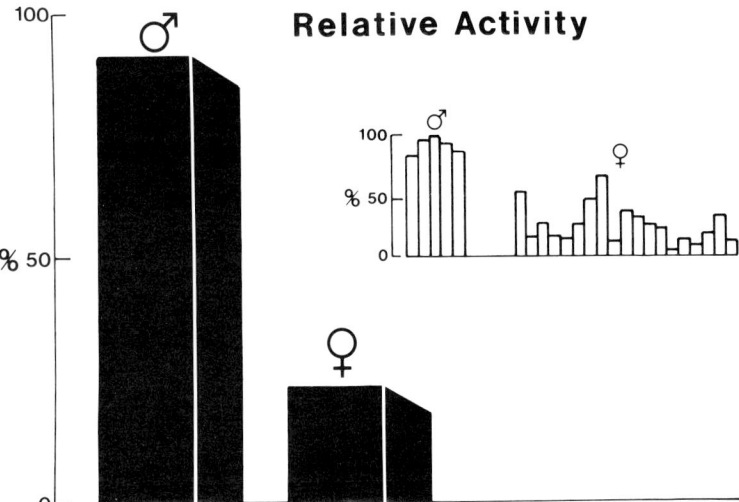

FIGURE 8.4. Relative activity of mature male (N=5) and female (N=19) lobsters held communally with shelter during June and July. Large histograms indicate the percentage of lobsters that left their shelters during each 24 hour period. Upper right: percentage of days during June and July that individual lobsters left their shelters. Lobsters were counted as having left their shelter if the entire lobster was out of the shelter.

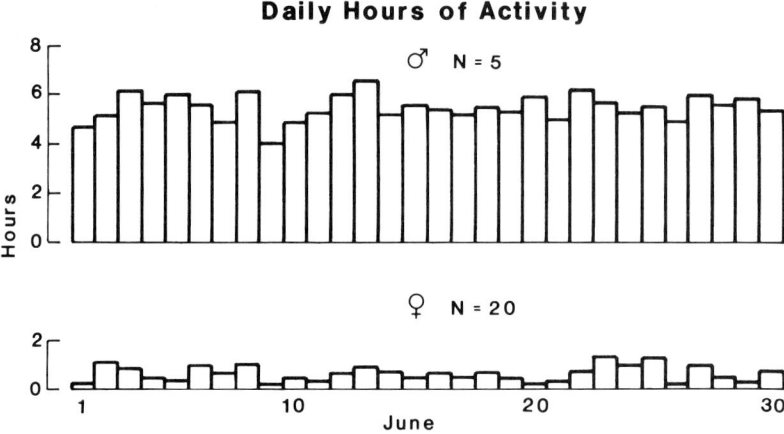

FIGURE 8.5. Daily hours of activity (time spent outside of shelters) of mature male (N = 5) and female (N = 19) lobsters during June.

passing lobsters, returning quickly if the intruder approached. If the visitor managed to enter the shelter the resident male would try to coax or drag him out, usually persisting until successful. Similar behavior was not seen in females.

Activity was greater in the warm summer months than in the winter, and male and female activity differed less in the winter than in the summer (fig 8.6). There also appeared to be differences in the activity levels of males that were in intermolt as opposed to early premolt. Limited data indicate that males in premolt behave more like females and spend less time outside their shelters than intermolt males (fig. 8.6).

Females were more reclusive than males. They spent little time socializing

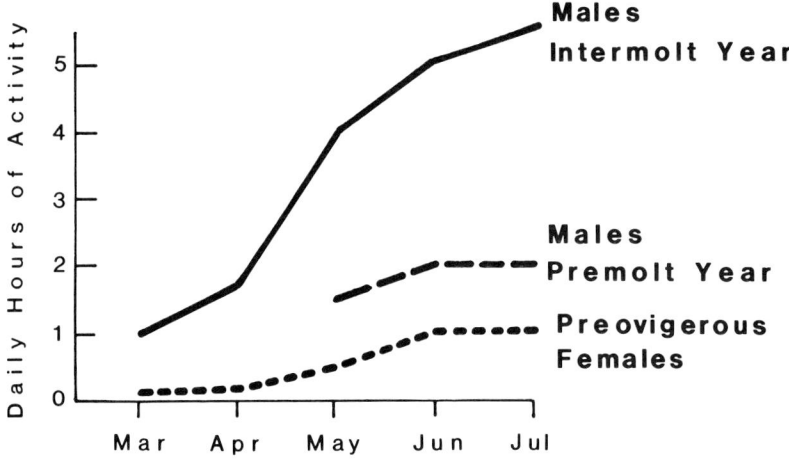

FIGURE 8.6. Variation in activity levels in male and female lobsters with molt stage and season.

and retreated when others approached, returning to their shelter if there was much activity in the area. Females often remained in their shelters for days and sometimes even weeks at a time, not even leaving to feed. In one study, a female spent two months in her shelter (May–June) without even appearing at the entrance. However, females that had not been inseminated by the time the spawning season approached became more active, leaving their shelters more frequently and remaining outside for longer periods of time.

Mating Behavior

Even though uninseminated preovigerous females became more active prior to spawning, mating activity was almost always initiated by a roving male and intermolt matings often appeared to result from random interactions of males and females.

Intermolt matings were of two different types, *consenting* and *forced*, and they occurred with equal frequency. In a "consenting" mating, the male and female lobsters appeared to meet by chance while outside their shelters. They would face one another while raised high on their walking legs and spar with chelae extended and antennae flicking rapidly. This head to head interaction often continued for as long as five minutes, after which the female turned in front of the male so that her abdomen was accessible to him. If she "presented" this way but got no response, she often turned back and touched the male with her chelipeds as if to attract his attention. This occurred most often in preovigerous females that were very close to spawning and had not yet been inseminated. Although this indicates definite interest on the part of the female, such interest has only been observed after the initial head to head interaction between the male and female.

In "forced" intermolt matings the female showed no interest in the male and tried to avoid his advances by retreating into her burrow. The male would pursue, grasp the female, and—if necessary—pull her out of her shelter. There was no interaction and no head to head contact between the sexes prior to the male grasping the abdomen of the female. In these cases the female would actively resist the male and was often able to escape. Even after the male had grasped the female and was attempting to turn her over, the female would continue trying to escape. On these occasions when the female did escape, the male often persisted (sometimes for hours).

In both the consenting and forced types of intermolt mating (which occur with equal frequency), insemination was usually completed in less than 2 mins, and as soon as the male released the female she would flap her abdomen, turn over, and retreat to the security of her shelter. Often the male would strike the departing female a forceful blow with his chelipeds in what appeared to be an aggressive act. Competition among males for uninseminated preovigerous females was rare. Despite the fact that mating in our tanks occurred in confined quarters in the presence of other males, there was seldom much interest expressed by the other males.

The aggressive nature of male lobsters is a major factor in the success of intermolt mating, and consummation of the act depends on the degree of aggression and persistence of the male. Males appear able to recognize females that have not been inseminated and rarely attempt to mate with a female that is already inseminated. Multiple matings do occur in females that are carrying small spermatophores, perhaps to ensure sufficient sperm for fertilization (see Walker 1980).

Intermolt mating is common in crustaceans (Hinsch 1968; Ingle & Thomas 1974; Paul 1984; Ameyaw-Akumfi 1987; Hamano 1988), and although the mating behavior described here for intermolt females contrasts with that observed in newly molted females (see Atema 1986), it is not unique among decapods. The interception and forced mating of reluctant females has been noted in crabs, as have male recognition of uninseminated females and the relationship of male aggression to female molt stage (Hartnoll 1969; Salmon 1983; Christy 1987; Diesel 1988; Donaldson & Adams 1989).

Natural Incidence

The natural incidence of intermolt mating is unknown. It has even been suggested that intermolt mating is an anomaly produced by confinement and is not something that occurs in nature (Dunham 1988). Several lines of evidence suggest that intermolt mating is a natural phenomenon: (1) eggs fertilized as the result of intermolt mating are as viable and attach as well as those resulting from postmolt mating; (2) lobsters of a size that spawn more than once during a molt cycle have not been found in the wild without stored spermatophores; (3) incidence of insemination in most wild populations is virtually 100 percent; and (4) under laboratory conditions, all uninseminated preovigerous females mate if given the opportunity.

Many decapods routinely mate when the female is hard-shelled, and some, such as the tanner crab, *Chionoecetes bairdi*, are able to mate irrespective of female shell condition. In that species, male and female mating behavior varies with the hardness of the female cuticle; males become more aggressive and precopulatory interaction is reduced when females are hard-shelled and more able to resist (Donaldson & Adams 1989). In many crustaceans, including crabs, shrimp, and spiny lobsters, males are known to recognize preovigerous females and the criterion for mate selection is the female's proximity to spawning (see Berry 1970; Diesel 1988; Yano et al. 1988).

Female lobsters spawn whether or not they have been inseminated. Intermolt mating is a mechanism that ensures fertilization of spawned eggs. Intermolt mating mitigates the effect of unsuccessful matings (cf. Saila & Flowers 1965) and appears to be a mechanism that allows lobsters to mate opportunistically at any time during the molt cycle. This is advantageous in areas where it may be difficult to locate a mate due to low population density or biased sex ratios. It is essential in about 20 percent of females that spawn consecutively, and females of this size are common in the Bay of Fundy, Gulf of Maine, and offshore areas where population densities can be low and fishery regulations may have caused a bias in sex ratios in these size classes (see Campbell & Pezzack 1986; Daniel, Bayer & Waltz 1989).

ARTIFICIAL INSEMINATION

It is not possible to obtain successful natural matings with an unresponsive male, and artificial insemination is one solution (figs. 8.7, 8.8). Ejaculation of the spermatophore can be induced by stimulating the male gonopore with 10–12 milliamps obtained with 7–8 metered RMS volts AC (Aiken et al. 1984). An AC ammeter should be connected in series to the transformer so that amperage can be controlled and tissue damage minimized.

However, a subthreshold current can also be problematic. If insufficient stimulation is used the spermatophore will only partially extrude and a portion will remain and harden in the distal portion of the vas deferens. Unless this plug is removed, further ejaculations will be impossible. Hardened plugs from partial ejaculations have never been observed in males allowed to mate normally (see also Talbot 1984).

To facilitate spermatophore placement, the seminal receptacle is held open with "barraquer" forceps (fig. 8.8). The extruded spermatophore is dipped in seawater for lubrication and then inserted into the receptacle, distal end first, just as it would enter during copulation. Talbot, Thaler, and Wilson (1986) used a similar technique, except that a common screwdriver was used to open the receptacle. They felt that inseminations had to occur within 48 hrs of the female molt because after this time the receptacle might be damaged. However, with the proper instrument the receptacle can be opened at any time during the reproductive cycle, either for insemination or to check for the presence of a sperm plug.

The number and quality of spermatophores obtained with electroejaculation vary with environmental conditions, social conditions, and long-term temperature history. Newly captured feral male lobsters that have recently molted will consistently ejaculate spermatophores bilaterally, but their potency declines over time, apparently due to progression through the molt cycle and time spent in isolation. Males hatched and reared in a culture system at year-round temperatures of 20°C are poor producers of sperm and are rarely interested in mating (Aiken et al. 1984; Waddy & Aiken in press; cf. Talbot et al. 1983). Cultured males often ejaculate small spermatophores devoid of a sperm tube. It is not known whether such a male would mate naturally and transfer an empty spermatophore, but this apparently does happen in insects (Ridley 1988). When cultured males are transferred to seasonally varying seawater temperatures before maturity, they produce more spermatophores with larger sperm tubes and are usually able to mate. This difference is associated with temperature history and has implications for selective breeding programs using cultured broodstock. We have better success when cultured males are transferred to seasonally varying temperature regimes (0–14°C) well before maturity than when transfer occurs after maturity.

Molt stage of the female at the time of insemination has no effect on fertilization success or egg retention (Waddy & Aiken 1986). This finding is important, as the value of artificial insemination would be compromised unless it gave the culturist considerable latitude concerning time of insemination. The only prob-

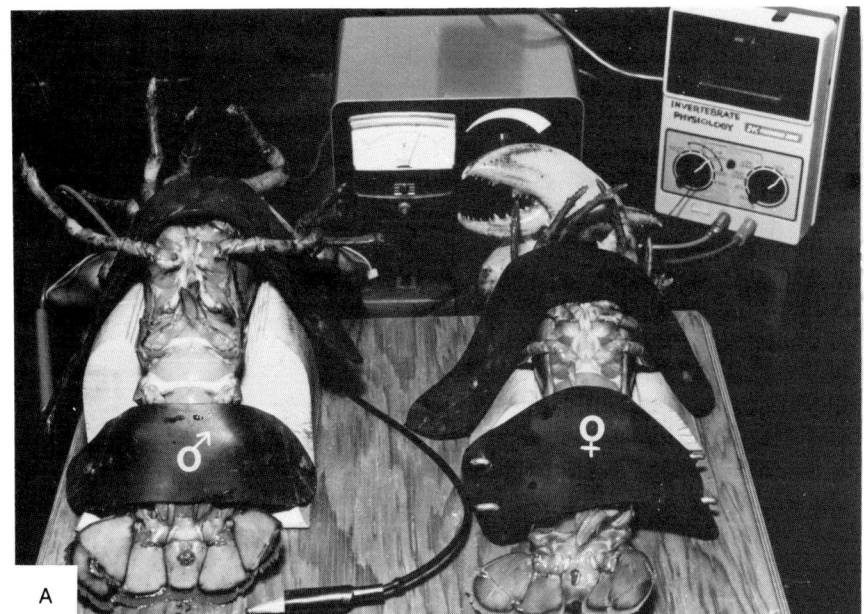

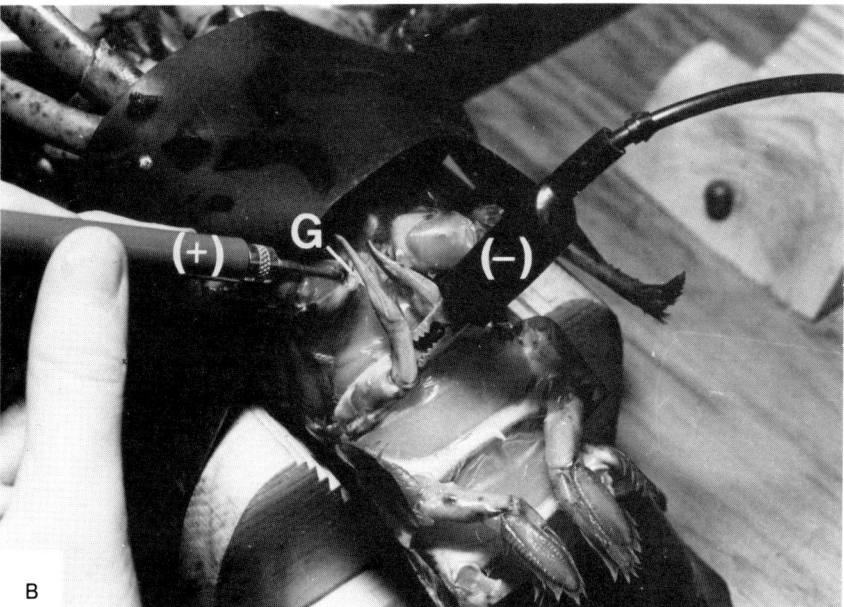

FIGURE 8.7. Apparatus for electroejaculation and artificial insemination of *Homarus*. A. Platform used to secure male and female lobsters. Electrodes, multimeter and Hammond VAC transformer are shown. B. Position of anode (+) adjacent to the gonopore (G) and the cathode (−) on the sternal cuticle prior to electrical stimulation. From Aiken et al. 1984.

Mating in the American Lobster 141

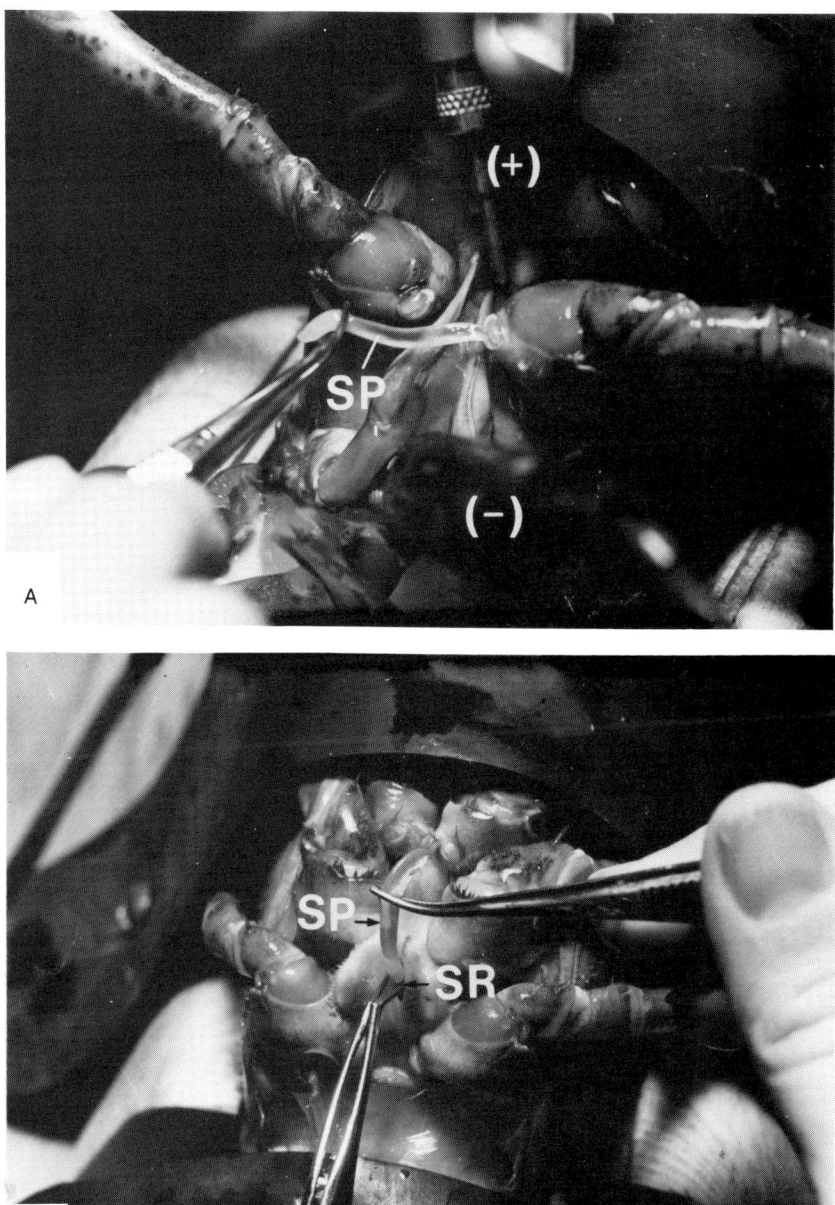

FIGURE 8.8. Electroejaculation and insemination of *Homarus*. A. Spermatophore (SP) extruded from the gonopore following electrical stimulation. B. Insertion of the spermatophore into the seminal receptacle (SR), held open with spring-loaded blunt "barraquer" forceps. From Aiken et al. 1984.

lem currently associated with artificial insemination is a reduced level of egg attachment and retention. In our studies only 25 percent of artificially inseminated females of feral stock hatched a normal complement of eggs. Most females incubated and hatched only 50 percent of their spawned brood. Results in cultured females were far worse; only 2 percent of females hatched a normal number of eggs (Waddy & Aiken 1985). Unfortunately, this work was done before it was recognized that temperature history of the maternal female had an impact on egg attachment. It is likely that the poor results obtained with cultured females were due as much to environmental factors experienced by the maternal females during early ovary development as to artificial insemination. Only limited data are available on females from feral stocks, so this work should be repeated now that the effect of temperature on egg attachment is recognized.

Histological sections of seminal receptacles from artificially inseminated females indicate that placement of the spermatophore in the receptacle may also be a factor; location of the sperm in the receptacle appears more variable in artificially inseminated females than in females mated naturally (T. P. Foyle, personal communication). In addition, there are noticeably fewer spermatozoa in the receptacles of artificially inseminated females than in those mated naturally (only one spermatophore was transferred in artificial insemination). Obviously, artificial insemination should not be dismissed as a technique for controlling insemination until further work is done.

SUMMARY

The mating habits of American lobsters are more flexible than originally thought. While most matings occur at the female molt ("postmolt"), lobsters can mate at any time between molt ("intermolt"). The receptivity of females appears to be dependent on ovarian development and the presence or absence of sperm in their seminal receptacle. Mating during the intermolt period seems to be a normal component of the reproductive strategy of American lobsters; it occurs in females that were not mated at molt, in those that were inseminated with an inadequate amount of sperm, and in those that will spawn twice between molts and that have insufficient sperm to fertilize the second clutch of eggs.

Our recognition of intermolt mating has altered traditional concepts of mating, fertility, and egg production in American lobsters and has answered previously puzzling questions. Without intermolt mating, the significant reproductive potential of the very large lobsters ($\geq$ 120 mm CL) common to the Bay of Fundy, Gulf of Maine, and offshore areas would be wasted, because a significant proportion of their second spawnings would be infertile. Intermolt mating also mitigates the effect of unsuccessful postmolt matings, as it allows lobsters to mate opportunistically at any time between molts. This is advantageous in areas where lobsters exist in low density or the sex ratios are biased because of selective fishing.

Similar mating alternatives are well known in other crustaceans. Male recognition of uninseminated and preovigerous females, the forced mating of reluctant females, and the relationship between degree of male aggression and female molt stage are found in other decapods as well.

Literature Cited

Aiken, D. E. 1973. Proecdysis, setal development and molt prediction in the American lobster *(Homarus americanus). J. Fish. Res. Board Can.* 30:1337–1344.
Aiken, D. E. & S. L. Waddy. 1980. Reproductive biology. In J. S. Cobb & B. F. Phillips, eds., *The Biology and Management of Lobsters*, 1:215–276. New York: Academic Press.
Aiken, D. E. & S. L. Waddy. 1982. Cement gland development, ovary maturation and reproductive cycles in American lobsters. *J. Crustacean Biol.* 2:315–327.
Aiken, D. E. & S. L. Waddy. 1986. Growth of the vasa deferentia of mature *Homarus americanus:* Conflicting results from field and laboratory studies. *Can. J. Fish. Aquat. Sci.* 43:1453–1457.
Aiken, D. E. & S. L. Waddy. 1989. Allometric growth and onset of maturity in male American lobsters *(Homarus americanus):* The crusher propodite index. *J. Shellfish Res.* 8:7–11.
Aiken, D. E. & S. L. Waddy, K. Moreland & S. M. Polar. 1984. Electrically induced ejaculation and artificial insemination of the American lobster *Homarus americanus. J. Crustacean Biol.* 4:519–527.
Ameyaw-Akumfi, C. 1987. Mating in the lagoon crab *Cardisoma armatum* Herklots. *J. Crustacean Biol.* 7:433–436.
Atema, J. 1986. Review of sexual selection and chemical communication in the lobster, *Homarus americanus. Can. J. Fish. Aquat. Sci.* 43:2283–2390.
Atema, J. & J. S. Cobb. 1980. Social behavior. In J. S. Cobb & B. F. Phillips, eds. *The Biology and Management of Lobsters*, 1:409–450. New York: Academic Press.
Atema, J. & D. F. Cowan. 1986. Sex-identifying urine and molt signals in lobster *(Homarus americanus). J. Chem. Ecol.* 12:2065–2080.
Atema, J., S. Jacobson, E. Karnofsky, S. Oleszko-Szuts & L. Stein. 1979. Pair formation in the lobster, *Homarus americanus:* Behavioral development, pheromones and mating. *Mar. Behav. Physiol.* 6:277–296.
Berry, P. F. 1970. Mating behavior, oviposition and fertilization in the spiny lobster *Panulirus homarus* (Linnaeus). *Oceanogr. Res. Inst. (Durban) Invest. Rep.* 24:1–16.
Campbell, A. & D. S. Pezzack. 1986. Relative egg production and abundance of berried lobsters, *Homarus americanus*, in the Bay of Fundy and off southwestern Nova Scotia. *Can. J. Fish Aquat. Sci.* 43:2190–2196.
Christy, J. H. 1987. Competitive mating, mate choice and mating associations of brachyuran crabs. *Bull. Mar. Sci.* 42:177–191.
Cobb, J. S. & D. Wang. 1985. Fisheries biology of lobsters and crayfishes. In A. J. Provenzano, Jr., ed., *The Biology of Crustacea.* 10:167–247. New York: Academic Press.
Cowan, D. & J. Atema. 1985. Serial monogamy, mate choice and pre- and post-copulatory guarding in lobsters. *Biol. Bull. (Woods Hole)* 169:550.
Cowan, D. F. & J. Atema. In press. Molt staging and serial monogamy in American lobsters. *Anim. Behav.*
Daniel, P. C., R. C. Bayer & C. Waltz. 1989. Egg production of V-notched American lobsters *(Homarus americanus)* along coastal Maine. *J. Crustacean Biol.* 9:77–82.
Diesel, R. 1988. Male-female association in the spider crab *Inachus phalangium:* The influence of female reproductive stage and size. *J. Crustacean Biol.* 8:63–69.
Donaldson, W. E. & A. E. Adams. 1989. Ethogram of behavior with emphasis on mating in the tanner crab *Chionecetes bairdi* Rathbun. *J. Crustacean Biol.* 9:37–53.
Dunham, P. J. 1979. Mating in the American lobster: Stage of molt cycle and sex pheromone. *Mar. Behav. Physiol.* 6:1–11.
Dunham, P. J. 1988. Pheromones and behavior in Crustacea. In H. Laufer & R. G. H. Downer, eds., *Invertebrate Endocrinology*, 2:375–392. New York: Alan R. Liss.
Dunham, P. J. & D. Skinner-Jacobs. 1978. Intermolt mating in the lobster *(Homarus americanus). Mar. Behav. Physiol.* 5:209–214.
Ennis, G. P. 1980. Size-maturity relationships and related observations in Newfoundland populations of the lobster *(Homarus americanus). Can. J. Fish. Aquat. Sci.* 37:945–956.
Hamano, T. 1988. Mating behavior of *Oratosquilla oratoria* (De Haan. 1844) (Crustacea: Stomatopoda). *J. Crustacean Biol.* 8:239–244.
Hartnoll, R. G. 1969. Mating in the Brachyura. *Crustaceana (Leiden)* 16:161–181.
Hinsch, G. W. 1968. Reproductive behavior in the spider crab, *Libinia emarginata* (L.). *Biol. Bull. (Woods Hole)* 135:272–278.
Hughes, J. T. & G. C. Matthiessen. 1962. Observations on the biology of the American lobster, *Homarus americanus. Limnol. Oceanogr.* 7:414–421.

Ingle, R. W. & W. Thomas. 1974. Mating and spawning of the crayfish *Austropotamobius pallipes* (Crustacea: Astacidae). *J. Zool. (Lond.)* 173:525–538.

Karnofsky, E. B., J. Atema & R. H. Elgin. 1989a. Field observations of social behavior, shelter use, and foraging in the lobster, *Homarus americanus*. *Biol. Bull. (Woods Hole)* 176:239–246.

Karnofsky, E. B., J. Atema & R. H. Elgin. 1989b. Natural dynamics of population structure and habitat use of the lobster, *Homarus americanus*, in a shallow cove. *Biol. Bull. (Woods Hole)* 176:247–256.

Karnofsky, E. B. & H. J. Price. 1989. Dominance, territoriality and mating in the lobster, *Homarus americanus:* A mesocosm study. *Mar. Behav. Physiol.* 15:101–121.

Krouse, J. S. 1973. Maturity, sex ratio and size composition of a natural population of American lobster, *Homarus americanus*, along the Maine coast. *U.S. Natl. Mar. Fish. Serv. Fish. Bull.* 71:165–173.

Lipcius, R. N. & W. F. Herrnkind. 1985. Photoperiodic regulation and daily timing of spiny lobster mating behavior. *J. Exp. Mar. Biol. Ecol.* 89:191–204.

McLeese, D. W., R. L. Spraggins, A. K. Bose & B. N. Pramanik. 1977. Chemical and behavioral studies of the sex attractant of the lobster *(Homarus americanus)*. *Mar. Behav. Physiol.* 4:219–232.

Morgan, S. G., J. W. Goy & J. D. Costlow. 1983. Multiple ovipositions from single matings in the mud crab *Rhithropanopeus harrisii*. *J. Crustacean Biol.* 3:542–547.

Nelson, K. & D. Hedgecock. 1971. Electrophoretic evidence of multiple paternity in the lobster *Homarus americanus* (Milne-Edwards). *Am. Nat.* 3:361–365.

Parker, G. A. 1970. Sperm competition and its evolutionary consequences in the insects. *Biol. Rev. Camb. Philos. Soc.* 45:525–567.

Paul, A. J. 1984. Mating frequency and viability of stored sperm in the tanner crab *Chionoecetes bairdi* (Decapoda, Majidae). *J. Crustacean Biol.* 4:375–381.

Ridley, M. 1988. Mating frequency and fecundity in insects. *Biol. Rev. Camb. Philos. Soc.* 63:509–549.

Saila, S. B. & J. M. Flowers. 1965. A simulation study of sex ratios and regulation effects with the American lobster, *Homarus americanus*. *Proc. Gulf Caribb. Fish. Inst.* 18:66–78.

Salmon, M. 1983. Courtship, mating systems, and sexual selection in decapods. In S. Rebach & D. W. Dunham, eds., *Studies in Adaptation*, ch. 6, pp. 143–169. New York: Wiley.

Talbot, P. 1984. Problems and progress in controlling reproductive biology in the American lobster *(Homarus)*. In W. Engels, W. H. Clark, Jr., A. Fischer, P. J. W. Olive & D. F. Went, eds., *Advances in Invertebrate Reproduction*, 3:473–480. Amsterdam: Elsevier.

Talbot, P., D. Hedgecock, W. Borgeson, P. Wilson & C. Thaler. 1983. Examination of spermatophore production by laboratory-maintained lobsters *(Homarus)*. *J. World Maricult. Soc.* 12:271–278.

Talbot, P., C. Thaler & P. Wilson. 1986. Artificial insemination of the American lobster *(Homarus)*. *Gamete Res.* 14:25–31.

Templeman, W. 1934. Mating in the American lobster. *Contrib. Can. Biol. Fish. (NS)* 8:423–432.

Waddy, S. L. & D. E. Aiken. 1985. Fertilization and egg retention in artificially inseminated female American lobsters, *Homarus americanus*. *Can. J. Fish. Aquat. Sci.* 42:1954–1956.

Waddy, S. L. & D. E. Aiken. 1986. Multiple fertilization and consecutive spawning in large American lobsters, *Homarus americanus*. *Can. J. Fish. Aquat. Sci.* 43:2291–2294.

Waddy, S. L. & D. E. Aiken. 1987. Potential of intermolt mating in broodstock management for lobster culture. *Bull. Aquacult. Assoc. Canada* 87–2:28–29.

Waddy, S. L. & D. E. Aiken. 1988. Intermolt mating and insemination in preovigerous American lobsters. *Am. Zool.* 28:110A.

Waddy, S. L. & D. E. Aiken. In press. Egg production in American lobsters. In A. M. Wenner & A. Kuris, eds., *Egg Production in Crustacea*. Amsterdam: Balkema Press.

Walker, W. F. 1980. Sperm utilization strategies in nonsocial insects. *Am. Nat.* 115:780–799.

Yano, I., R. A. Kanna, R. N. Oyama & J. A. Wyban. 1988. Mating behavior in the penaeid shrimp *Penaeus vannamei*. *Mar. Biol. (Berl.)* 97:171–175.

NINE
Sperm Competition and the Evolution of Mating Behavior in Brachyura, with Special Reference to Spider Crabs (Decapoda, Majidae)

RUDOLF DIESEL

Laboratoire Arago,
Université Pierre et Marie Curie Paris,
Banyuls-sur-Mer, France.
Max-Planck-Institut für Verhaltensphysiologie
Abt. Wickler,
Seewiesen, F.R.G.

Abstract

This paper deals with sperm competition in the Brachyura and its implications for male mating behavior. Females of all brachyuran species possess sperm storage organs, of which, according to current knowledge, there are three types: (1) the thelyca of Podotremata and (2) the dorsal- and (3) ventral-type seminal receptacles of Eubrachyura. In seminal receptacles of many eubrachyuran crabs a male's sperm is likely to encounter viable sperm from previous mates, either retained from the previous breeding cycle or from earlier insemination during the same cycle. Hence sperm from different males compete for fertilization of ova. The form of male adaptations that improve the chances of fertilization largely depends on the morphology of the sperm storage organs. Male adaptations have been best studied in spider crabs (Majidae), which possess receptacles in which insemination and fertilization occur ventrally (ventral-type). Male spider crabs displace rivals' sperm dorsally and seal it with hardening seminal plasma, leading to last male sperm precedence. After copulation males guard their mate until spawning to assure paternity. To reduce guarding time, males prefer to mate with females whose hatching and spawning is imminent. Eubrachyuran species with receptacles where gonopod access is ventral and fertilization takes place dorsally (dorsal-type), as in several cancrid and portunid crabs, have considerably different reproductive behavior from Majidae. Possible sperm-priority and paternity-assurance patterns are detailed, but more studies are needed to shed light on the sperm competition pattern in these and many other brachyuran crabs.

THROUGHOUT THE animal kingdom selection has shaped conspicuous male courtship patterns, subtle male abilities for locating a mate, and large male body size for successful mate acquisition. For crustaceans, especially decapods, a large amount of literature has dealt with courtship behavior (Salmon 1983), aggressive intrasexual competition for mates (Schöne 1968), precopulatory mate guarding (Ridley 1983), mate location and recognition by sex pheromones (Dunham 1978), and mating behavior itself (Hartnoll 1969; Hazlett 1975).

For a long time it was believed that competition for fertilization terminated with the transfer of sperm, until Parker (1970) showed in his classic review on sperm competition in insects that selection continues between copulation and fertilization. Parker (1970) defined sperm competition as the competition between ejaculates of at least two males for fertilization of a female's eggs. He argued that there is a great potential for sperm competition in species where females store viable sperm for extended periods in specialized organs and mate with more than one male before eggs are fertilized. These conditions favor selection of specialized male organs or behaviors that allow a male to give its sperm an advantage in fertilization over rival's sperm. These conditions simultaneously select for a male's ability to prevent its own sperm from becoming replaced or displaced, i.e., to assure paternity.

Sperm competition has now been shown to have significant implications for the mating behavior of different animal taxa (examples in Smith 1984), but the Crustacea, one of the largest animal groups, have not been reviewed. In this paper I will demonstrate that sperm competition is also important in Crustacea and I review its possible mechanisms in Brachyura and its consequences for crab mating behavior. Primarily because of space limitations, but also because of the fragmentary nature of the relevant literature, I will focus on the spider crabs (Majidae).

PREADAPTATIONS FOR SPERM COMPETITION IN BRACHYURA

General Remarks

A review of the large variability in brachyuran reproductive features is beyond the scope of this paper, but I will briefly point out some basic patterns. Mating in Brachyura is either (1) restricted to a brief period during which females are anatomically receptive to the female molt (e.g., Cancridae, Portunidae) or to temporary operculum decalcification in species with a normally immobile vulval operculum (e.g., some Grapsidae and Ocypodidae) or (2) not restricted, as in species with mobile opercula (e.g., some Grapsidae and Ocypodidae) and those with permanent soft vulval openings (e.g., Majidae). Females that anatomically can mate at any time will be referred to as *continuously receptive*, although they may vary in their "attractiveness" to males, depending on their breeding status.

In species with continuously receptive females, male precopulatory and postcopulatory behavior varies depending on whether males mate with multiparous females or with primiparous females, which after the molt of maturity spawn (1) immediately or (2) after a delay of several weeks or months. Female spider crabs have a final molt of maturity (Hartnoll 1963; Conan et al. 1988), after which they usually breed continuously throughout their lives. Spawning of a new egg clutch follows within hours or a few days of hatching, and therefore spawning of the next clutch can be predicted from the developmental stage of the eggs carried on the pleopods. Females that carry well-developed eggs will fertilize a new clutch soon. I will refer to them as *prespawning* females.

Sperm Transfer, Storage, and Utilization

Male Reproductive Anatomy. There is an extensive literature on brachyuran male reproductive anatomy, histology, and biochemistry (for details see Ryan 1967a; Hinsch & Walker 1974; Pochon-Masson 1983; Adiyodi & Anilkumar 1988; Diesel 1989). Here I will concentrate on structures and products of the reproductive system that may be relevant to sperm competition.

Basically, the male reproductive system consists of paired testes, vas deferentia, and ejaculatory ducts that open in a small genital papilla. Ejaculate consists of nonmotile sperm and a varying amount of seminal plasma. Sperm are enclosed in numerous small spermatophores containing a dozen to several hundred spermatozoa. Spermatophores are formed and stored in the proximal (anterior) vas deferens. The distal (posterior) vas deferens produces and stores pure seminal plasma and is particularly enlarged in spider crabs and some other brachyurans (see Mouchet 1931; plates V–VI); it is obviously capable of producing and sorting a large amount of seminal plasma. Complex and diverse injection devices have evolved from first and second pleopods for ejaculate transfer to female sperm storage organs (fig. 9.1; see below).

Female Reproductive Anatomy. The Brachyura possess sperm storage organs that can be divided into two types by their morphology, their position in relation to the ovary, and their functioning during spawning: (1) the *thelycum* of the Podotremata Guinot 1977 (comprising the Dromiacea, Homoloidea, Tymoloidea, and Raninoidea) and (2) the *seminal receptacle* of the Eubrachyura de Saint Laurent 1988 (Heterotremata Guinot and Thoracotremata Guinot). The thelyca are paired or unpaired sternal invaginations, without connection to the ovary, that open on the coxa of the 3rd pereiopod (see Gordon 1963; fig. 9.2a). In Podotremata fertilization is external.

The seminal receptacles of Eubrachyura are enlargements of the paired female genital tracts (Ryan 1967b; Hartnoll 1968; Johnson 1980; Hinsch 1986; Adiyodi & Anilkumar 1988; Beninger et al. 1988; Diesel 1989). Typically they

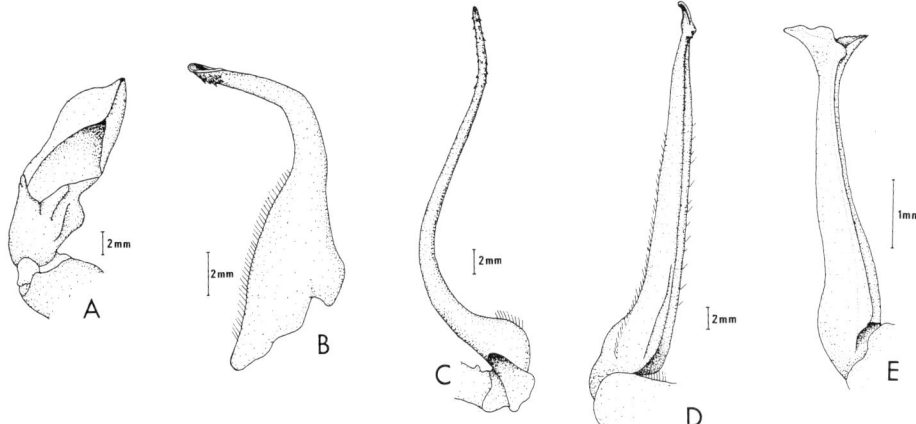

FIGURE 9.1. First pleopods of brachyuran crabs. A. *Dromia personata*. B. *Liocarcinus depurator*. C. *Portunus sanguinolentus*. D. *Maja squinado*. E. *Pisa tetraodon*. A redrawn from Hartnoll 1975; B redrawn from George 1963.

have a pouchlike form, with a mesodermal dorsal part that is an enlargement of the ovary and an ectodermal ventral part that opens via the vagina and vulva on the 6th thoracic sternite (figs. 9.2B–E). The vagina is typically a deflated tube with lateral muscles that run to the sternite and when contracted open the vagina and vulva for spawning and for pleopod access during mating. The ovary is connected to the seminal receptacle by an oviduct, which usually forms an open passage shortly before spawning. The dorsal part of the seminal receptacle consists of an outer flexible connective tissue and an inner glandular epithelium. The ventral part is lined with a cuticle layer, which is shed with the exuvium during the molt (e.g., Ryan 1967b). The dorsal part of the receptacle is highly elastic and can contain large amounts of ejaculate. In *Inachus phalangium* and *Pisa tetraodon*, an iris-diaphragm–like muscular membrane, the "velum," is situated at the border of dorsal and ventral chamber (figs. 9.2E, 9.3A,B) (Ryan 1967b and the "epithelial sheet," fig. 278 in Johnson 1980). During fertilization, the velum contracts to a small central opening, separating the dorsal sperm-storage chamber from the ventral chamber, where fertilization takes place (fig. 9.4). This separation may regulate sperm expenditure during spawning.

There is some variation in the way the seminal receptacles of eubrachyuran

FIGURE 9.2. Schematic diagrams of brachyuran sperm storage organs. A. Thelycum (th) of *Dromia personata*. Diagrams B–C seminal receptacles (sr) of B. *Liocarcinus depurator*; C. *Portunus sanguinolentus*; D. *Maja squinado*; E. *Pisa tetraodon*. All females had recently been inseminated except *L. depurator*. The female of the latter molted in captivity but did not mate; the small receptacle contains old sperm residue (so); m, muscle; o, ovary; od, oviduct; sp, seminal plasma, forming the external (esp) and the internal (isp) sperm plug; spp, aggregated sperm or spermatophores forming a sperm packet; st, sternite; va, vagina; ve, velum; vu, vulva. The arrow indicates the area where the oviduct opens into the receptacle. A. redrawn and modified after data from Hartnoll 1975; C. after data from George 1963 and Ryan 1967b.

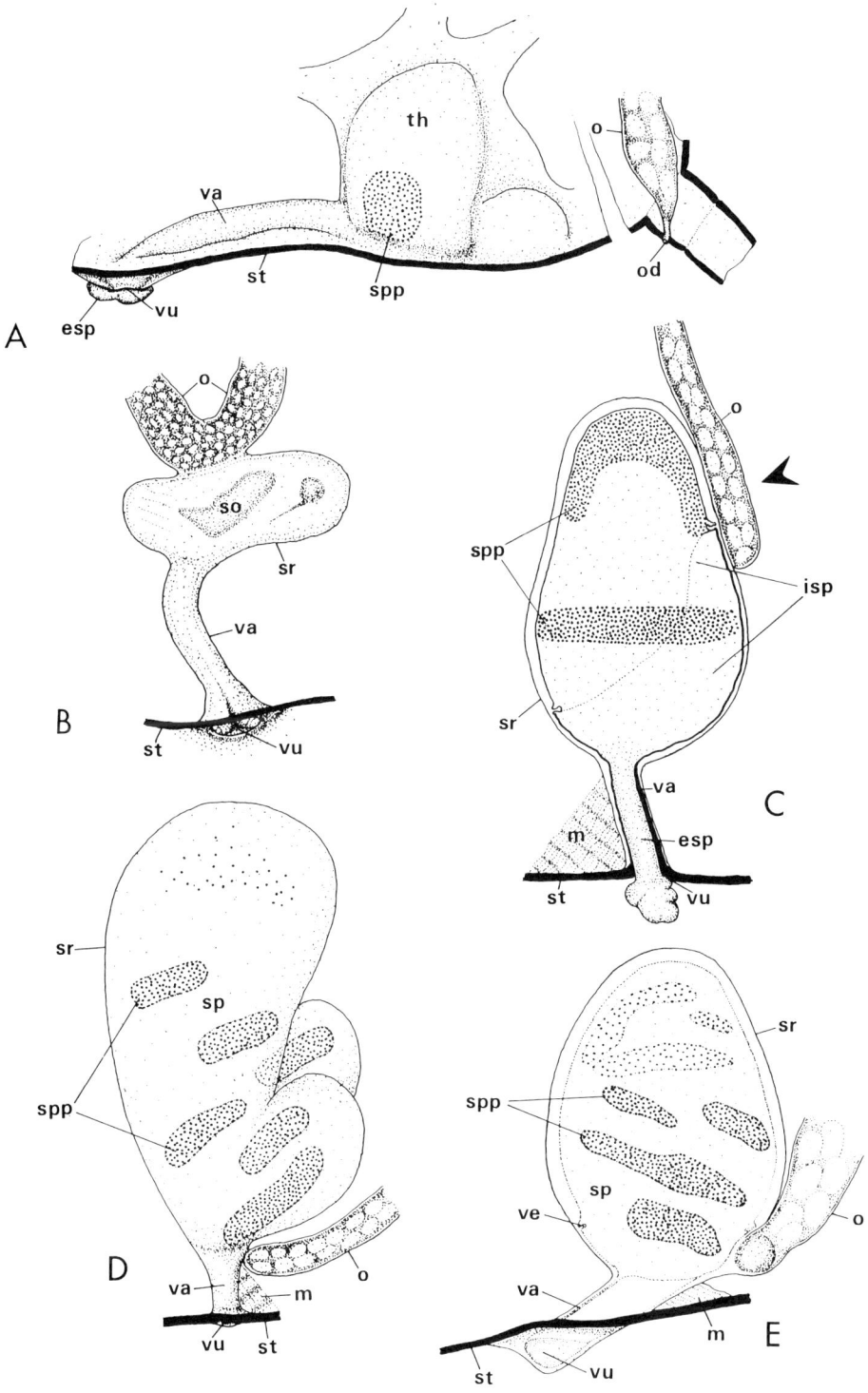

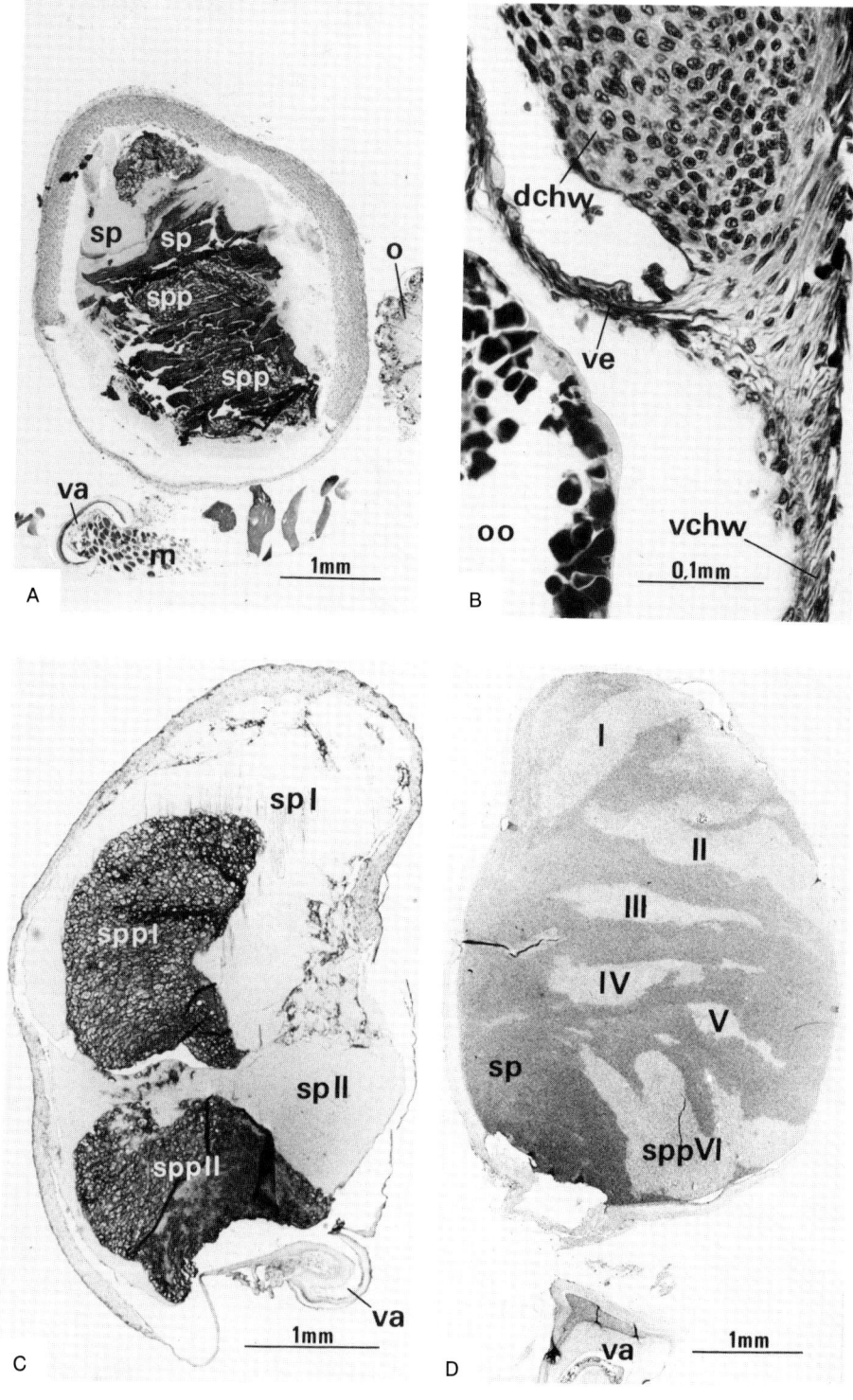

crabs are integrated into the reproductive tract. Basically, there are two types, which I will refer to as *dorsal-type* and *ventral-type* seminal receptacles. The dorsal-type resembles an enlarged tube in which the oviduct opening is more or less dorsal and the vagina, at some distance, lies ventrally. Dorsal-type receptacles occur in Portunidae *(Portunus sanguinolentus* [George 1963: 297; fig. 9.2c], *P. pelagicus* [Bawab & El-Sherief 1988: fig. 4], *Carcinus maenas,* [Hartnoll 1968: fig. 3], *Callinectes sapidus* [Johnson 1980:354], *Liocarcinus depurator* [fig. 9.2B], and Xanthidae *[Pilumnus hirtellus;* personal observation]). The ventral-type is saclike, with oviduct and vagina opening ventrally, close to each other (figs. 9.2D, E). This type occurs in Calappidae *(Calappa granulata),* Geryonidae *(Paragalene longicrura),* Leucosiidae *(Ilia nucleus),* Parthenopidae *(Parthenope angulifrons,* [see Cano 1891: plate 17]), Parathelphusidae *(Parathelphusa hydrodromous* [Anilkumar & Adiyodi 1977: fig. 1]), Corystidae *(Corystes cassivelanus* [personal observation]), Ocypodidae *(Uca lactea* [Murai, Goshima et al. 1987: 1340]), and in all spider crab species of which descriptions and observations are available, i.e., *Pisa armada, Inachus dorsettensis* (Cano 1891: plate 17), *Hyas araneus* and *H. coarctatus* (Hartnoll 1968: fig. 7), *Chionoecetes opilio* (Beninger et al. 1988: 324), *Inachus phalangium* (Diesel 1989a: fig. 8), *Pisa tetraodon, Maja squinado* (figs. 9.2D, E), *I. communissimus,* and *Macropodia rostrata* (personal observation).

The dichotomy in receptacle morphology may have major implications for the fertilization process. Oocytes are fertilized when passing from the oviduct to the vagina, and spermatozoa placed close to the oviduct opening are the most likely to fertilize ova. Hence males would be expected to show adaptations that guarantee that spermatozoa are placed close to the oviduct opening.

Ejaculate Transfer. Male products are transferred sequentially during insemination in order of their position in the vas deferens (Ryan 1967a, b; Diesel 1989): *seminal plasma* first, followed by numerous small *spermatophores.* This sequential-release pattern from the vas deferens is probably typical for the brachyura. Ejaculate transfer is achieved by the gonopods. The large diversity in brachyuran gonopod shape may be in many cases an adaptation to deliver sperm in the most favorable position into the receptacle, i.e., close to the oviduct, in order to increase fertilization chances.

Ejaculate Storage. Few studies describe in detail the overall storage pattern of ejaculates in seminal receptacles, in particular the location of sperm in relation to the seminal plasma and the oviduct and vagina openings. In the nonmajids

FIGURE 9.3. Seminal receptacle of *Inachus phalangium.* A. Light micrographs of a sagittal paraffin section of a multiple mated female. B. Detail from a receptacle during spawning, with an oocyte (oo) passing through the ventral chamber. dchw, dorsal chamber wall; m, muscle; o, ovary; oo, oocyte; sp, seminal plasma (differently staining Mallory); spp, sperm packet; va, vagina; ve, velum; vchw, ventral chamber wall. A from Diesel 1989. C–D. Sagittal paraffin sections of seminal receptacles of multiply mated C, *Inachus phalangium,* D, *Pisa tetraodon.* Roman numerals indicate the seminal plasma (sp) and the sperm packet (spp) from first (I), second (II), etc., matings.

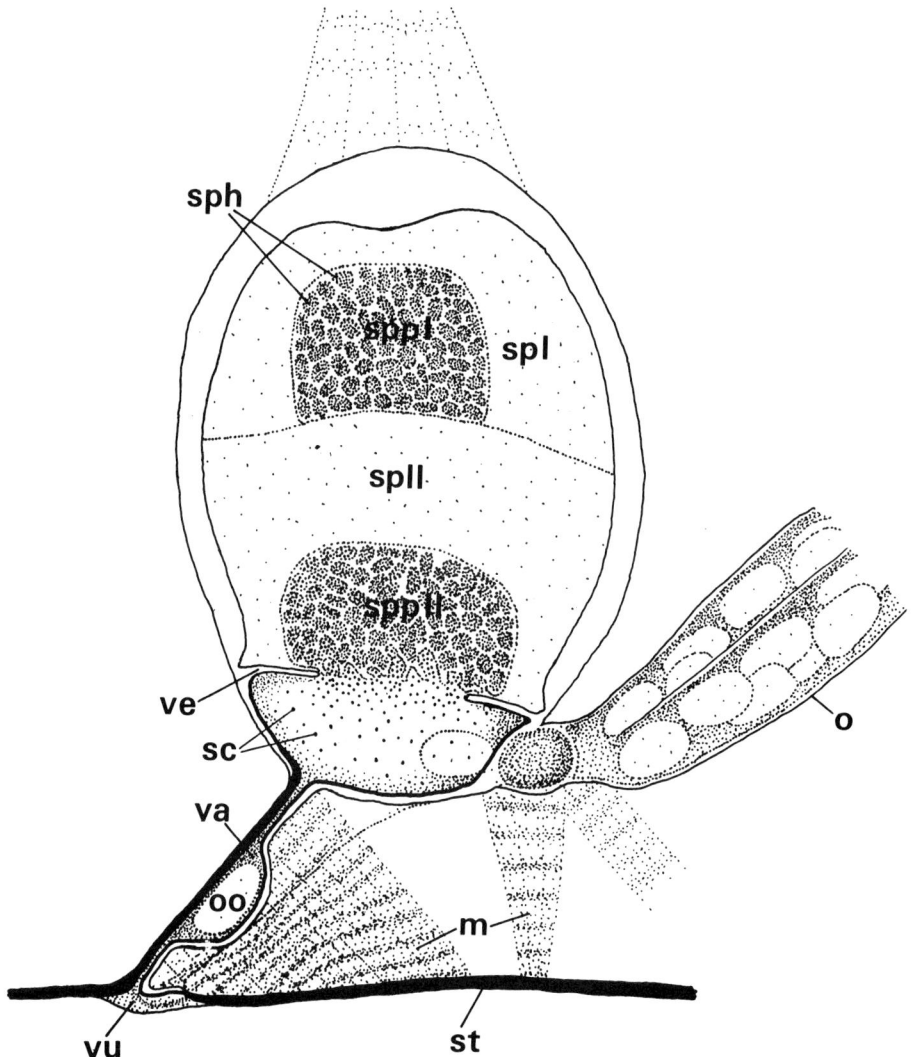

FIGURE 9.4. Schematic diagram of a seminal receptacle of *Inachus phalangium* during spawning. m, muscle; o, ovary; oo, oocyte; sc, sperm cells; sp, seminal plasma; spp, sperm packet; st, sternite; va, vagina; ve, velum; vu, vulva. Roman numerals I and II indicate the sperm and seminal plasma of the first and second mate. Redrawn from Diesel 1990.

Portunus sanguinolentus (Ryan 1967b), *Callinectes sapidus* (Johnson 1980), *Ovalipes ocellatus* (Hinsch 1986), and *Geryon fenneri* (Hinsch 1988) spermatophores and seminal plasma are stored separately until spawning. This is similar to the spider crab *Chionoecetes opilio*, whose spermatophores are densely aggregated and remain intact after transfer (Beninger et al. 1988). Histological sections of receptacles of *Inachus phalangium*, *I. communissimus*, *Pisa tetraodon*, *Macropodia rostrata*, and *Maja verrucosa* (Diesel 1989) and the dissections of receptacles of *Maja squinado* (personal observation) revealed that numerous small spermatophores form a dense *sperm packet* that is surrounded by the initially transferred and hardened seminal plasma *(sperm gel)* (fig. 9.4).

For *Inachus phalangium* it was demonstrated using unilaterally gonopod-amputated males and autoradiography of ^{3}H-thymidine labeled sperm that (1) during *one copulation* only *one sperm packet* is transferred into each receptacle and (2) each packet is deposited into the lowermost position in the receptacle, mantled by sperm gel except on its ventral side (Diesel 1989, 1990); figs 9.3c, 9.4).

Spawning. Little is known about the processes that take place in the seminal receptacle during spawning. However, it is obvious that for fertilization spermatozoa must be liberated from the spermatophore envelopes. It has been suggested that the spermatophore envelopes are broken down by secretions (enzymes) released from the glandular epithelium of the seminal receptacle (Ryan 1967b; Adiyodi & Anilkumar 1988; Diesel 1989). Beninger et al. (1988) suggested an alternative spermatophore dehiscence: seawater influx through the vagina may lead to a dehiscense of spermatophores and liberation of spermatozoa. However, whether enzymes of seawater account for sperm liberation, only sperm from spermatophores exposed to either one can be liberated; those embedded in sperm gel would be hindered. In *Inachus phalangium*, spermatophores from sealed-off sperm packets remain unchanged throughout spawning, and only sperm from the ventral, sperm gel-free side of the lowermost sperm packet are released into the ventral receptacle compartment and participate in fertilization (Diesel 1989).

Sperm Viability Potential

Sperm retention following spawning and multiple fertilizations using stored sperm have been reported in numerous eubrachyurans, e.g., *Portunus sanguinolentus* (Ryan 1967a), *Menippe mercenaria* (Cheung 1968), *Uca thayeri* (Salmon 1987), *Parathelphusa hydrodromous* (Adiyodi & Anilkumar 1988), and for the spider crabs *Pisa tetraodon* (Vernet-Cornubert 1958), *Microphrys bicornutus* (Hartnoll 1965), *Chionoecetes opilio* (Watson 1970), *Chionoecetes bairdi* (Adams & Paul 1983; Paul 1984), *Libinia emarginata* (Hinsch 1968), And *Inachus phalangium* (Diesel 1989). Spider crabs are able to store viable sperm for long periods. In *C. bairdi*, for example, 71 percent of the females produced viable clutches although they were isolated from males for 2 years (Paul 1984), and *P. tetraodon* may fertilize up to 8 clutches using stored sperm only (Vernet-Cornubert 1958).

Multiple Mating

Sperm from different males co-occur in a female's seminal receptacles if she (1) copulates with more than one male during a breeding cycle or (2) mates only once per cycle, but with different males in each cycle, and viable sperm is retained from the previous cycle. Mating with more than one male during a breeding cycle has been observed in Portunidae *(Portunus sanguinolentus* [Ryan 1967b]), Gecarcinidae *(Cardisoma guanhumi,* [Taissoun 1974]), Ocypodidae *(Uca vocans, U. thayeri* [Salmon 1984, 1987 respectively]), Cancridae *(Cancer borealis,* [Elner, Gass & Campbell 1985]), and Majidae *(Libinia emarginata* [Hinsch 1968], *Chionoecetes bairdi* [Paul 1984; Donaldson & Adams 1989], *Inachus phalangium, Inachus communissimus, Maja verrucosa, Macropodia* spp., and *Pisa tetraodon* [Diesel 1985]).

On the other hand, multiparous females of many eubrachyuran taxa retain viable sperm and remate prior to spawning. Hence they are likely to store sperm of more than one male in their seminal receptacles. Multiple ejaculates (from the same or different males) have been found in dissected receptacles of *Chionoecetes opilio* collected in the field (Taylor, Hooper & Ennis 1985), in histological gross sections of seminal receptacles of wild-caught *Inachus phalangium* (Diesel 1988a), *Inachus communissimus, Pisa tetraodon, Maja verrucosa,* and *Macropodia* spp. (Diesel 1985), and in the dissected receptacles of *Maja squinado* (personal observation; fig. 9.2D) and *C. bairdi* (Paul 1984) females were kept with males in a tank.

SPERM COMPETITION IN MAJIDAE

Sperm Displacement

Detailed knowledge of the mechanism of sperm competition in Brachyura is currently only available for *Inachus phalangium* (Diesel 1990). Male ghost spider crabs transfer the ejaculate by inserting only the tips of their gonopods into the vaginae (as in *Chionoecetes bairdi* [Adams 1982] and possibly in *Uca* spp. [Christy personal communication]). Seminal plasma fills the ventral part of a receptacle and displaces stored ejaculate(s) dorsally into the apex of the flexible dorsal chamber; a sperm packet is then placed at the lowermost position in the seminal plasma (fig. 9.4). After transfer from the vasa deferens the seminal plasma hardens, forming the sperm gel which seals predecessors' sperm packets in a dorsal position, displaced from the ventral oviduct and vagina openings and totally engulfed by sperm gel. During spawning, the sperm gel prevents liberation of rivals' sperm. Only sperm from the last male's sperm packet are released to fertilize oocytes passing from the oviduct to the vagina. Using the sterile-male method (Parker 1970) it was shown that the last male's sperm is used for fertilization (last male sperm precedence). The mechanism has only been ascertained in the ghost spider crab; all spider crab species for which histological cross sections of the seminal receptacle are available store sperm packets and sperm gel in a similar way to *I. phalangium*. Furthermore, Taylor, Hooper & Ennis (1985) found in dissected seminal receptacles of *Chionoecetes opilio* new (white) ejaculate in the ventral part of the receptacle and the displaced old (brown) ejaculate in the receptacle apex. Old and new ejaculate have been found in receptacles of *C. bairdi* (e.g., Paul 1984), but their location was not mentioned. Hence current evidence indicates that spider crab males displace rivals' ejaculate from the ventral oviduct and vagina.

Paternity Assurance

Whenever last male sperm precedence arises in a species, a male risks losing its entire mating investment in a female if she remates with another male. Parker (1970) pointed out that selection would thus favor physiological, morphological, and behavioral adaptations that allow males to secure their paternity. These include *sperm plugs* (male secretions that block the female genital tract to prevent further access of male copulatory appendages), *prolonged copulation* (remaining in copula without further sperm transfer), and *postcopulatory mate guarding* (males remain in physical contact or in the near vicinity guarding the female).

In brachyuran species with sperm plugs, males can leave their mate shortly after insemination (when the plug has hardened) and search for further prospective mates. However, in species where males remain in physical contact with their mates, a male has to stay until the female becomes unreceptive (e.g., when the carapace or the operculum has hardened) or until she spawns (e.g., in species with continuously receptive females). This implies that there is a trade-off between time spent securing paternity and time searching for additional mates (see Parker 1974). One would therefore expect males of species with continuously receptive females and lacking sperm plugs to recognize female breeding status and preferentially mate with and guard *prespawning females*, but they may mate with, and subsequently leave, females that spawn weeks or months later.

Sperm plugs and prolonged copulation are not known for spider crabs (although the hard receptacle content is frequently termed a sperm plug in the literature, it is not a plug in the above sense, since it does not prevent pleopod access), but there are numerous reports of male behaviors such as "grasping," "mating embrace," or "guarding." Most of these descriptions include precopulatory and/or postcopulatory mate guarding. *Precopulatory mate guarding* ensures a prospective mate and is common in species with restricted female receptivity (Ridley 1983). This behavior is common in spider crabs when males monopolize females until the molt of maturity, in species where primiparous females spawn right after the terminal molt (e.g., *Chionoecetes opilio*, [Watson 1972]), or else where prespawning multiparous females are highly synchronized within a short period of time (e.g., *C. opilio*, [Hooper 1986]).

Postcopulatory mate guarding may be shown by males after mating with primiparous and multiparous females. Only a few studies present sufficient detail on male mating behavior and female status, but in most cases male postcopulatory behavior can be inferred from descriptions given in the literature. Males that are attracted to, and mate with, prespawning females have been reported for *Pugettia producta* (Knudsen 1964), which "copulates prior to the hatching of the current brood of eggs" (p. 64); a similar report for *Maja squinado* is found in Števčić (1971; see below). In *Chionoecetes opilio*, males pair with prespawning females and guard them after mating until eggs hatch (Taylor, Hooper & Ennis 1985; Hooper 1986). Similarly, in *C. bairdi* "males . . . are attracted to multiparous females at the time the females' eggs hatch" (Adams 1982:266), and "having completed intromission . . . (pairs) remained together until the female extruded eggs" (p. 258). In *Maja verrucosa* males mate with prespawning females and then aggressively defend them until spawning (films E895 and E1053, cited in Schöne 1968; personal observation). *Libinia emarginata* males were attracted to "females about to release zoeae," and guard them ("obstetrical behavior") until the female's eggs have hatched (Hinsch 1968). Most females spawned within 12 hours of hatching. Males were not attracted to "females with eggs in the early stages of development." Hartnoll (1969:169) reported that a *Hyas coarctatus* male, after mating with a female that has molted, "resumes his . . . post standing over the female, for several days." He states that males mated with multiparous females and did not guard them, but gives no information on the females' breeding condition. Hooper (1986) observed male *H. araneus* in the sea being strongly attracted to prespawning *Chionoecetes opilio* females but not to those that had spawned. One observed mating in *Pleistacantha mosely* involved a multiparous female that had recently spawned; she was not guarded after mating (Berry & Hartnoll 1970). Vernet-Cornubert (1958) noted that *Pisa tetraodon* males mate with with nonprespawning females and did not mention any guarding behavior. *Stenorhynchus lanceolatus* males were found guarding prespawning females in the sea (Wirtz personal communication). *Inachus phalangium* (Diesel 1986a), *Maja verrucosa*, *Macropodia* spp., *Pisa tetraodon*, *Acanthonyx lunulatus*, and *Achaeus cranchii* males (Diesel 1988b) were also observed guarding prespawning females in the sea and, along with *I. communissimus*, in captivity (personal observation).

All the spider crabs listed above meet the predictions: males are generally attracted to and mate with prespawning females, which they guard until spawning. Males may also mate with females that have recently spawned or those that are not yet ready to spawn, but then leave this mate unguarded. Hence the morphology of the seminal receptacles, the pattern of sperm storage, and the mating and postmating behavior in the spider crabs provide independent evidence for last male sperm precedence by sperm displacement.

Last Male Sperm Precedence and Implications for the Majid Mating System

Field observations on spider crab mating behavior are limited to a few species. However, for *Chionoecetes opilio*, *C. bairdi*, *Maja squinado*, and *Inachus phalangium*, sufficient information on mating behavior is available to attempt a synthesis of their mating systems and a discussion of the consequences of sperm competition for male mating behavior.

The Snow Crab, Chionoecetes opilio. An important fisheries industry has developed around the large snow crab, *Chionoecetes opilio*, in the Northwestern Atlantic, North Pacific, and Sea of Japan. In Bonne Bay, Newfoundland (Canada), multiparous female snow crabs migrate annually in spring (April, May) from deeper to shallow water to eclose larvae (Taylor, Hooper & Ennis 1985; Hooper 1986; Ennis, Hooper & Taylor 1988). Taylor, Hooper, and Ennis (1985) suggested that females migrate to higher temperatures in shallower water for faster embryonic development and to increase larval survival. (Števčić [1971] made a similar suggestion for *Maja squinado*; see below.) During the rest of the year adult snow crabs live in deeper water, with large males segregated (deeper) from females, which are found with small males (shallower) (Miller & Keefe 1981). Hooper's (1986) observations indicate that at the onset of the breeding migration large males (89–140 mm carapace width [CW]; Taylor, Hooper & Ennis 1985) select females and then arrive in pairs in shallow water. The paired snow crabs remain there for about one month, until females have released their larvae. The pairs then separate and most females turn to deeper water to spawn, although some spawn in shallow water. Male snow crabs and even males of another spider crab species *(Hyas araneus)* are highly attracted to females carrying eggs ready to hatch. Intense competition for prespawning females results in the largest males being paired and smaller males unpaired. Competition also leads to heavy injuries in males and females (26 and 44 percent with one or more missing legs, respectively). Males monopolize females for an extended time prior to copulation and after copulation guard them until the eggs hatch. Taylor, Hooper & Ennis (1985) examined the seminal receptacles of 77 paired females and found that 97 percent held old sperm in store; most had sufficient sperm to fertilize at least one additional brood (Elner & Gass 1984). From the end of April to the end of May the proportion of paired females storing old brownish ejaculate in the dorsal parts of the receptacles and new whitish ejaculate in the ventral increased from 48.5 to 92.9 percent (Taylor, Hooper & Ennis 1985). If one assumes that the new ejaculate stems from the guarding male, then he has displaced the "old" sperm and defends paternity.

Females benefit from being guarded in the form of reduced risk of injury if their mate is large and able to prevent takeover attempts by smaller rivals. Hence females may be reluctant to mate with small males. In captivity, mating attempts of small males with hard females are usually unsuccessful because females "forcefully escaped from smaller males" (*Chionoecetes bairdi* [Adams 1982:240; Donaldson & Adams 1989]). There has even been controversy over whether in nature small males with functional testes are able to mate successfully (see Conan & Comeau 1986; smallest male observed to mate in captivity was 47 mm CW [Conan et al. 1988]). However, small *C. opilio* males invest relatively more in testis size than do large males (size range: 62–144 mm CW; Conan et al. 1988: fig. 4), and fertilization capacity of small males' (62–92 mm CW) ejaculate is no different from that of large males (>95–126 mm CW [Ennis et al. 1988]). Where are the small males that were found with the females for the rest of the year? There are no field observations available on the mating success of small males. In captivity, small males copulated successfully only with soft newly molted primiparous females (Conan et al. 1988), which are less fecund than multiparous females (e.g., 30 percent fewer eggs in *C. bairdi* [Somerton & Meyers 1983]) and thus may be less attractive to large males. In nature, small males may take their chance to mate in deeper water with soft primiparous females when they pass their synchronized terminal ecdysis in February and provide little or no resistance. They could also mate with some of the multiparous females that return from breeding migration but have not yet spawned. Adams (1982) stated that *C. bairdi* females, released by the guarding male hours or days before the new egg mass was spawned, remated with other males prior to egg extrusion.

The Spiny Spider Crab, Maja squinado. The large spiny spider crab, *Maja squinado*, occurs in the European North Atlantic and the Northern Mediterranean Sea. Annual deep- to shallow-water breeding migrations have been reported for different locations (Števčić 1971). In the Northern Adriatic Sea only mature crabs migrate during May and early June, into the warmer shallow water (5 to 10 m depth) where females release larvae and subsequently return to deeper water to spawn. Females form large aggregations ("heaps"), probably as a protection against heavy octopus predation (Carlisle 1957). Males are attracted to these female aggregations. Števčić (1971) conducted experiments on heap formation in captivity and found females and small males in the center of heaps, whereas large males were either on the heap surface or in its vicinity. Similar distributions are reported for *M. squinado* aggregations on the English coast, where heap formation occurs from July to September (Carlisle 1957). The females pass their final molt of maturity in the heap and subsequently mate with the larger males on the heap surface before they spawn. Although there are no observations, it is possible that the small males mate within the heap. Probably because females are highly aggregated, large males cannot monopolize them prior to copulation. Instead, they compete for females departing from the heap to deeper water. If large males mate with the emerging females they will displace the sperm of any small male that may have mated in the heap.

The Ghost Spider Crab, Inachus phalangium. The ghost spider crab, *Inachus phalangium*, is a relatively small species (21 mm carapace length) from the Northeastern Atlantic and the Mediterranean Sea, usually associated with the sea anemone *Anemonia sulcata* (Diesel 1986b). Hidden among the stinging tentacles of the anemones, *I. phalangium* is protected from predators (Wirtz & Diesel 1983). In the Mediterranean Sea, near Banyuls-sur-Mer (southern France), females breed continuously from December onward, with up to 6 clutches, until August of the following year, when they die (Diesel 1986a). The breeding cycles of females are unsynchronized. Individual females are sedentary, usually associated with a particular sea anemone. The distribution of anemones and females is patchy. Males also live on anemones but move among them to search for prespawning females.

Some days before the eggs hatch and the new clutch is spawned, up to 4 males may visit, mate, and guard a prespawning female. Frequently a guarding male is replaced by a larger one on a subsequent day until finally the largest (dominant) male in an area guards her after copulation until spawning. A dominant male patrols an area containing up to 8 females, each of which is visited in turn one or two days before they spawn. Males can maximize mating success if they arrive at a female when she is in prespawning condition. This would require males to keep a record of each female *locality* and her *breeding cycle*. Although male spider crabs possibly recognize the prehatching condition of eggs carried on the pleopods (Hinsch 1968, Diesel 1986a), it is unlikely that males can allocate distant, prespawning females downcurrent. However, the hypothesis is supported by dominant males appearing at the expected time at a female's anemone when she has died in the meantime (Diesel 1986a).

EVOLUTIONARY ASPECTS OF SPERM COMPETITION IN BRACHYURA

The Brachyura is a large group with considerable variation in the morphology of the female reproductive system and sperm storage organs. Hence sperm competition may have different forms. Austad (1984) related sperm priority patterns in spiders to the type of female sperm storage organs. He distinguished a *"cul-de-sac"*-type, in which only one duct is connected to the sperm storage organ and through which insemination occurs and sperm are released for fertilization, and a *"conduit"*-type, which has two connections: an external for insemination and an internal for fertilization, the latter delivering sperm to a separate reproductive duct. Austad suggested that in species with cul-de-sac organs, last male sperm precedence of mixing of sperm should be most common, whereas in species with conduit organs first male sperm precedence should dominate.

In Brachyura the thelyca of Podotremata is clearly of the cul-de-sac-type (see Hartnoll 1975; fig 9.2A) and thus may favor last male sperm precedence or sperm mixing. For the dorsal- and ventral-type seminal receptacles of Eubrachyura the nearest approach to Austad's categories would be that the ventral-type receptacles resemble a cul-de-sac organ, because insemination and fertilization occur close to each other. The dorsal-type receptacle, on the other hand, functions like a conduit organ, with insemination duct and fertilization site far apart. Ventral-type receptacles are found in species from other crab families besides the Majidae. For nonmajids there is only one report of sperm priority pattern for the continuously receptive fiddler crab, *Uca lactea*. Murai, Goshima, and Henmi (1987) suggested that last male sperm precedence is achieved in this species by sperm displacement. *Uca lactea* males defend paternity by guarding their pre-spawning mate in the burrow until she spawns.

If dorsal-type receptacles in several portunid and cancrid species are conduit organs, the first male should place its sperm dorsally, close to the fertilization site, and plug the ventral part of the receptacle with hardening seminal plasma (sperm plug). This is only possible if the male pleopods are long enough to reach into the receptacle apex. Seminal plasma transferred initially in copula will then be pushed ventrally by the spermatophores, which take the dorsal position. A subsequent male would have to deliver its ejaculate ventral to the plug. In *Portunus sanguinolentus*, in which the male gonopods reach into the apex of the seminal receptacles, sperm is actually positioned dorsally and seminal plasma ventrally (Ryan 1967a,b; fig. 9.2c). Similar dorsal sperm storage is reported for *P. pelagicus* (Bawab & El-Sherief 1988) and possibly occurs in *Carcinus maenas* (see Spalding 1942). What happens when a second male copulates? Ryan (1967b) studied the receptacle contents of a multiple mated female. He removed the first male at the end of the copulation and added another male to the female. After the second insemination the receptacle held two discrete ejaculates, one dorsally and one ventrally. Ryan attributed the ventral ejaculate to the second male, a situation that led to first male sperm precedence. But it is conceivable that last male sperm precedence can also occur in *P. sanguinolentus*, depending on time between inseminations. If a subsequent male would mate before the seminal plasma had hardened, its pointed gonopods (fig. 9.1c) could penetrate to the receptacle apex, displace the previous mate's ejaculate ventrally, and place its own dorsally. Observations by Elner, Gass, and Campbell (1985) on *Cancer borealis* indicate that 3 males could mate in short succession with a female before the sperm plugs hardened and prevented a fourth male from copulation. Until experiments with individually marked ejaculates have been conducted, it is not possible to decide whether the situation in *Portunus sanguinolentus* appears to be in accordance with Austad's (1984) predictions for conduit sperm storage organs.

Sperm plugs are usually defined as secretions applied by males to block the females' genital tracts (Parker 1970). Sperm plugs are typical for Cancridae and Portunidae (Hartnoll 1969). They may be "internal," i.e., a hard substance within the receptacles, and/or "external," i.e., filling the vaginal duct and protruding from the vulva (Elner et al. 1985: fig. 1; figs. 9.2a–c). Both structures are regarded as being produced by the males. However, no description is available of how they are produced, nor are there reports of the specialized male behavior expected if males fill the vagina and seal the vulval opening with seminal plasma. It is indeed unlikely that males form an "external" sperm plug, since they terminate copulation with spermatophore transfer. An alternative possibility is that seminal plasma infiltrates the vagina from the receptacle as the male's pleopods are withdrawn. This might be supported by a contraction of the receptacle and must be supported by contraction of the vaginal musculature to prevent the vagina from collapsing before the seminal plasma has hardened. It seems very unlikely that a male could block a female's vagina (external plug) without her involvement. Males produce internal plugs to prevent their own sperm from being displaced, and females produce the external plug by forcing seminal plasma into the vagina. This external plug may be in the females' "interest," as it prevents further copulation and thus reduces competition and disturbances during a postmolt period when she is extremely vulnerable.

At present nothing is known about sperm competition in Cancridae and Portunidae or many other brachyuran crabs. Sperm competition patterns as described here for spider crabs could occur in many other taxa with ventral-type receptacles, e.g., in fiddler crabs. Sperm displacement may be widespread among Brachyura. Further studies on the content of sperm storage organs of multiply mated females are needed. Experimental manipulations, e.g., with unilaterally gonopod-amputated males, would provide a powerful tool to analyze the mechanisms by which males compete for fertilizations.

Acknowledgments

My grateful thanks are due to J. H. Christy, R. G. Hartnoll, and P. I. Ward for comments on an earlier draft of the manuscript and especially to H. Hofer for his criticism of the final draft, as well as to L. Gardiner for improving the English. During the preparation of the manuscript I was supported by the Max-Planck-Gesellschaft (FRG) and the Centre National de la Recherche Scientifique (France).

Literature Cited

Adams, A. E. 1982. The mating behavior of *Chionoecetes biardi*. In B. Melteff, ed., *Proceedings of the International Symposium of the Genus Chionoecetes*. Sea Grant Report 82–10, pp. 235–271. University of Alaska, Fairbanks, Lowell Wakefield Fisheries Symposia Series.

Adams, A. E. & A. J. Paul. 1983. Male parents size, sperm storage and egg-production in the crab *Chionoecetes bairdi* (Decapoda, Majidae). *Int. J. Invertebr. Reprod.* 6:181–187.

Adiyodi, K. G. & G. Anilkumar. 1988. Arthropoda-Crustacea. In K. G. Adiyodi & R. G. Adiyodi, eds., *Reproductive Biology of Invertebrates*. Vol. 3 *Accessory Sex Glands*, pp. 261–318. New York: Wiley.

Anilkumar, G. & K. G. Adiyodi. 1977. Spermatheca of the freshwater crab, *Parathelphusa hydrodromus* (Herbst) in relation to the ovarian cycle. In K. G. Adiyodi & R. G. Adiyodi, eds., *Advances in Invertebrate Reproduction*, 1:269–274. Karivellur, Kerala, India: Peralam-Kenoth.

Austad, S. N. 1984. Evolution of sperm priority patterns in spiders. In R. L. Smith, ed., *Sperm Competition and the Evolution of Animal Mating Systems*, pp. 223–249. New York: Academic Press.

Bawab, F. M. & S. S. El-Sherief. 1988. Stages of the reproductive cycle of the female crab *Portunus pelagicus* (L., 1758) based on the anatomical changes of the spermatheca (Decapoda Brachyura, Portunidae). *Crustaceana (Leiden)* 54:139–148.

Beninger, P. G., R. W. Elner, T. P. Foyle & P. H. Odense. 1988. Functional anatomy of the male reproductive system and the female spermatheca in the snow crab *Chionoecets opilio* (O. Fabricius) (Decapoda: Majidae) and a hypothesis for fertilization. *J. Crustacean Biol.* 8:322–332.

Berry, P. F. & R. G. Hartnoll. 1970. Mating in captivity of the spider crab *Pleistacantha moseleyi* (Miers) (Decapoda, Majidae). *Crustaceana (Leiden)* 19:214–215.

Cano, G. 1891. Morfologia dell apparecchio femminile, glandole del cemento e fecondazione nei crostacei decapodi. *Mitt. zool. Stn. Neapel* 9:503–532.

Carlisle, D. B. 1957. On the hormonal inhibition of moulting in decapod Crustacea. 2. The terminal anecdysis in crabs. *J. Mar. Biol. Assoc. U.K.* 36:291–307.

Cheung, T. S. 1968. Transmolt retention of sperm in the female stone crab, *Menippe mercenaria* (Say). *Crustaceana (Leiden)* 15:117–120.

Conan, B., M. Moriyasu, M. Comeau, P. Mallet, R. Cormier, Y. Chiasson & H. Chiasson. 1988. Growth and maturation of snow crab *Chionoecetes opilio*. In G. S. Jamieson & W. D. McKone, eds., *Proceedings of the International Workshop on Snow Crab Biology*, December 8–10, 1987, pp. 45–66. Montreal, Quebec.: Can. MS Rep. Fish. Aquat. Sci. 2005.

Conan, G. Y. & M. Comeau. 1986. Functional maturity and terminal molt of male snow crab, *Chionoecetes opilio. Can. Fish. Aquat. Sci.* 43:1710–1719.

Diesel, R. 1985. Fortpflanzungsstrategie und Spermienkonkurrenz der Seespinne Inachus phalangium *(Decapoda, Majidae)*. Ph.D. dissertation, Albert-Ludwig-Universität, Freiburg.

Diesel, R. 1986a. Optimal mate searching strategy in the symbiotic spider crab *Inachus phalangium* (Decapoda). *Ethology* 72:311–328.

Diesel, R. 1986b. Populations dynamics of the commensal spider crab *Inachus phalangium* (Decapoda: Majidae). *Mar. Biol. (Berl.)* 91:481–489.

Diesel, R. 1988a. Discrete storage of multiple-mating sperm in the spider crab *Inachus phalangium*. *Naturwissenschaften* 75:148–149.

Diesel, R. 1988b. Male-female association in the spider crab *Inachus phalangium:* The influence of female reproductive stage and size. *J. Crustacean Biol.* 8:63–68.

Diesel, R. 1989. Structure and function of the reproductive system of the symbiotic spider crab *Inachus phalangium* (Decapoda, Majidae): Observations on sperm transfer, sperm storage and spawning. *J. Crustacean Biol.* 9:266–277.

Diesel, R. 1990. Sperm competition and reproductive success in a decapod *Inachus phalangium* (Majidae): A male ghost spider crab that seals off rivals' sperm. *J. Zool. (Lond.)* 220:213–224.

Donaldson, W. E. & A. E. Adams. 1989. Ethogram of behavior with emphasis on mating for the Tanner crab *Chionoecetes bairdi* Rathbun. *J. Crustacean Biol.* 9:37–53.

Dunham, P. J. 1978. Sex pheromones in Crustacea. *Biol. Rev. Camb. Philos. Soc.* 53:555–583.

Elner, R. W. & C. A. Gass. 1984. Observations on the reproductive condition of female snow crabs from N. W. Cape Breton Islands, November 1983. *Can. Atl. Fish. Sci. Adv. Comm. Res. Doc.* 84:1–20.

Elner, R. W., C. A. Gass & A. Campbell. 1985. Mating behavior of the jonah crab, *Cancer borealis* Stimpson (Decapoda, Brachyura). *Crustaceana (Leiden)* 48:34–39.

Ennis, G. P., R. G. Hooper & D. M. Taylor. 1988. Functional maturity in small male snow crab *(Chionoecetes opilio). Can. J. Fish. Aquat. Sci.* 45:2106–2109.

George, M. J. 1963. The anatomy of the crab *Neptunus sanguinolentus* (Herbst). Part V. Reproductive system and embryological studies. *J. Madras Univ. Sect. B* 33:289–304.

Gordon, I. 1963. On the relationship of Dromiacea, Tymolinae and Raninidae to the Brachyura. In H. B. Whittington & W. D. Rolfe, eds., *Phylogeny and Evolution of Crustacea*, pp. 51–57. Cambridge, Massachusetts: Museum of Comparative Zoology.

Hartnoll, R. G. 1963. The biology of Manx spider crabs. *Proc. Zool. Soc. Lond.* 141:423–496.

Hartnoll, R. G. 1965. The biology of spider crabs: A comparison of British and Jamaican species. *Crustaceana (Leiden)* 9:1–16.

Hartnoll, R. G. 1968. Morphology of the genital ducts in female crabs. *J. Linn. Soc. Lond. Zool.* 47:279–300.

Hartnoll, R. G. 1969. Mating in the Brachyura. *Crustaceana (Leiden)* 16:161–181.

Hartnoll, R. G. 1975. Copulatory structure and function in the Dromiacea, and their bearing on the evolution of the Brachyura. *Pubbl. Stn. Zool. Napoli* 39:657–676.

Hazlett, B. A. 1975. Ethological analyses of reproductive behavior in marine Crustacea. *Pubbl. Stn. Zool. Napoli* 39:677–695.

Hinsch, G. W. 1968. Reproductive behavior in the spider crab, *Libinia emarginata* (L.). *Biol. Bull. (Woods Hole)* 135:273–278.

Hinsch, G. W. 1986. A comparison of sperm morphologies, transfer and sperm mass storage between two species of crab, *Ovalipes ocellatus* and *Libinia emarginata*. *Int. J. Intervebr. Reprod.* 10:79–87.

Hinsch, G. W. 1988. Morphology of the reproductive tract and seasonality of reproduction in the golden crab *Geryon fenneri* from the eastern Gulf of Mexico. *J. Crustacean Biol.* 8:254–261.

Hinsch. G. W. & M. H. Walker. 1974. The vas deference of the spider crab *Libinia emarginata*. *J. Morphol.* 143:1–19.

Hooper, R. G. 1986. A spring breeding migration of the snow crab, *Chionoectes opilio* (O. Fabr.), into shallow water in Newfoundland. *Crustaceana (Leiden)* 50:257–264.

Johnson, P. T. 1980. *Histology of the Blue Crab*, Callinectes sapidus: *A Model for the Decapoda*. New York: Praeger Publishers.

Knudsen, J. W. 1964. Observations of the mating process of the spider crab *Pugettia producta* (Majidae, Crustacea). *Bull. South. Calif. Acad. Sci.* 63:38–41.

Miller, R. J. O. & P. G. Keefe. 1981. Seasonal and depth distribution, size, and molt cycle of the spider crabs, *Chionoecetes opilio*, *Hyas araneus* and *H. coarctatus* in a Newfoundland Bay. *Can. Tech. Rep. Fish. Aquat. Sci.* 1003:1–18.

Mouchet, S. 1931. Spermatophores des Crustaces, Decapodes, Anomoures et Brachyoures et castration parasitaire chez quelques Pagures. *Ann. Stn. Oceanogr. Salambo* 6:1–204.

Murai, M., S. Goshima & Y. Henmi. 1987. Analysis of the mating system of the fiddler crab, *Uca lactea*. *Anim. Behav.* 35:1334–1342.

Parker, G. A. 1970. Sperm competition and its evolutionary consequences in the insects. *Biol. Rev. Camb. Philos. Soc.* 45:525–568.

Parker, G. A. 1974. Courtship persistence and female-guarding as male time investment strategies. *Behaviour* 48:157–184.

Paul, A. J. 1984. Mating frequency and viability of stored sperm in the tanner crab *Chionoecetes bairdi* (Decapoda, Majidae). *J. Crustacean Biol.* 4:374–381.

Pochon-Masson, J. 1983. Arthropoda-Crustecea. In K. G. Adiyodi & R. G. Adiyodi, eds., *Reproductive Biology of Invertebrates*. Vol. 2, *Spermatogenesis and Sperm Function*, pp. 407–450. New York: Wiley.

Ridley, M. 1983. *The Explanation of Organic Diversity*. Oxford: Clarendon Press.

Ryan, E. P. 1967a. Structure and function of the reproductive system of the crab *Portunus sanguinolentus* (Herbst) (Brachyura: Portunidae). 1. The male system. *Mar. Biol. Assoc. India Symp. Ser.* 2:506–521.

Ryan, E. P. 1967b. Structure and function of the reproductive system of the crab *Portunus sanguinolentus* (Herbst) (Brachyura: Portunidae). 2. The female system. *Mar. Biol. Assoc. India Symp. Ser.* 2:522–524.

Salmon, M. 1983. Courtship, mating systems, and sexual selection in decapods. In S. Rebach & D. Dunham, eds., *The Behavior of Higher Crustacea*, pp. 143–169. New York: Wiley.

Salmon, M. 1984. The courtship, aggression and mating system of a "primitive" fiddler crab (*Uca vocans*: Ocypodidae). *Trans. Zool. Soc. Lond.* 37:1–50.

Salmon, M. 1987. On the reproductive behavior of the fiddler crab *Uca thayeri*, with comparison to *U. pugilator* and *U. vocans*: Evidence for behavioral convergence. *J. Crustacean Biol.* 7:25–44.

Schöne, H. 1968. Agonistic and sexual display in aquatic and semi-terrestrial brachyuran crabs. *Am. Zool.* 8:641–654.

Smith, R. L. 1984. Human sperm competition. In R. L. Smith, ed., *Sperm Competition and the Evolution of Animal Mating Systems,* pp. 601–660. New York: Academic Press.

Somerton, D. A. & W. S. Meyers. 1983. Fecundity differences between primiparous and multiparous female Alaskan tanner crab *(Chionoecetes bairdi). J. Crustacean Biol.* 3:183–186.

Spalding, J. F. 1942. The nature and formation of the spermatophore and sperm plug in *Carcinus maenas. Q. J. Microsc. Sci.* 83:399–422.

Števčić, Z. 1971. Laboratory observations on the aggregations of the spiny spider crab *(Maja squinado* Herbst). *Anim. Behav.* 19:18–25.

Taissoun, E. 1974. El cangrelo de terra *Cardisoma guanhumi* (Latreile) en Venezuela: Distriucion, ecologia, biologia y evaluacion poblaional. *Bol. Cent. Invest. Biol. Univ. Zulia* 10:8–50.

Taylor, D. M., R. G. Hooper & G. P. Ennis. 1985. Biological aspects of the spring breeding of snow crabs, *Chionoecetes opilio,* in Bonne Bay, Newfoundland (Canada). *U.S. Natl. Mar. Fish. Serv. Fish. Bull.* 83:707–711.

Vernet-Cornubert, G. 1958. Biologie générale de *Pisa tetraodon* (Pennant). *Bull. Inst. Oceanogr. Monaco* 1113:1–52.

Watson, J. 1970. Maturity, mating and egg laying in the spider crab, *Chionoecetes opilio. J. Fish. Res. Board Can.* 27:1607–1616.

Watson, J. 1972. Mating behavior of the spider crab, *Chionoecetes opilio. J. Fish Res. Board Can.* 29:447–449.

Wirtz, P. & R. Diesel. 1983. The social structure of *Inachus phalangium,* a spider crab associated with the sea anemone *Anemonia sulcata. Z. Tierpsychol.* 62:209–234.

TEN
Sex Change, Mating, and Sperm Transfer in *Crangon crangon* (L.)

R. BODDEKE, J. R. BOSSCHIETER, and P. C. GOUDSWAARD

Netherlands Institute for Fishery Investigations (RIVO), IJmuiden, the Netherlands.

Abstract

Crangon crangon (Caridea, Crangonidae) is the dominant shrimp species in coastal areas with a soft bottom in the Eastern Atlantic. Males and primary females of *C. crangon* develop from larvae: secondary females develop from spent males. Males exhibit protandric hermaphroditism. They copulate several times during a spawning period and then change sex. Evidence is given for internal fertilization, contradicting the general observation of external fertilization in caridean shrimps. A model of insemination is presented. It seems likely that only one oviduct is inseminated during a copulation, and the appendix masculina is not used. The spermatophore disintegrates during copulation. The covering stays behind in the ejaculatory duct of the male while the sperm mass is deposited in one of the oviducts of the female. In inseminated females large numbers of sperm cells can be observed along the inner wall of the oviduct. The internal sperm storage enables *C. crangon* to spawn fertile eggs at some interval after copulation. Consequently, during a two-month period without fertile males, only females that previously carried eggs at the beginning of this period spawn. The mode of insemination (internal versus external) may be an important character in reexamination of the phylogeny of the Caridea.

CRANGON CRANGON (Linné 1758) (Caridea, Crangonidae) inhabits shallow coastal areas, mainly with a sandy or silty bottom, in the Eastern Atlantic. The distribution ranges from the White Sea (USSR) to the Atlantic coast of Morocco and includes the Baltic, Mediterranean, and Black Sea as well. Along the coast of the Netherlands four populations can be distinguished (Boddeke 1982).

Crangon crangon is one of the smallest shrimp species fished for human consumption. Commercial size shrimps are usually females, 55–85 mm long and 1–4 g in weight. Fisheries with an average annual yield of more than 5000 tons exist in the Federal Republic of Germany and the Netherlands. Smaller and more local fisheries exist in Denmark, the United Kingdom, Belgium, France, Spain, Morocco, and Italy.

Because of its commercial importance, this species has been the subject of a large number of publications. A synopsis of biological data was published by Tiews (1970). An indexed bibliography by Redant (1984) covers no fewer than 560 papers dealing with biology, exploitation, management, and culture. However, the sexual cycle of *Crangon crangon* has been the subject of few studies and has been insufficiently described.

A sex change from male to female in *Crangon crangon* has been demonstrated on the basis of extensive research on the structure of the gonads, although the results were published only in a preliminary form (Boddeke 1966). Previous studies of caridean shrimps have shown that external spermatophores are used in fertilization (Höglund 1943; Chow, Ogasawara & Taki 1982; Bauer 1976). However, observations by Boddeke (1975, 1982) suggested internal sperm storage in *Crangon crangon*. These observations stimulated the study on mating and sperm transfer of *C. crangon* presented here.

METHODS

All animals used in histological research and for observations on copulation and sperm transfer were caught in different areas along the Dutch coast with a beach seine or by vessels operating at distances of not more than 3 km offshore in summer and up to 40 km in winter. Shrimp were transported live to the laboratory and used as soon as possible (maximum period approximately 4–5 weeks) to avoid possible aberrations caused by laboratory conditions.

To study the structure of the gonads, paraffin sections were prepared of the entire cephalothorax of the shrimp after removing the exoskeleton and gills. Selection of a fixative was done on the basis of criteria discussed by Baker (1958). Very satisfying results were obtained with fixation for 24 hrs in San Felice. Because this fixative is not stable, equal amounts of 1.5 percent chromiumtrioxide and 30 percent formaldehyde + 9.6 percent acetic acid were mixed directly before use. The low pH (1.6) of this fixative favors use of gallocyanin-eosin staining.

The long time needed for gallocyanin staining (48 hrs at room temperature) is more than compensated for by the high selectivity of this dye for sperm, nuclei, and, in particular, chromosomes. Sections were fixed and preserved on glass plates of 10×15 cm. For squashes, aceto-carmine (45 ml acetic acid, 55 ml H_2O, 1 g carmine) was used. This dye stains the cytoplasm of oocytes bright red and nucleoli in the nucleus dark red. Spermatophores were stained in a 0.5 percent aqueous solution of methylene blue for 1–2 min, after removal from the ejaculatory duct and a rinse in water. In this way sperm cells are stained bright blue. After removal from the female, oviducts were fixed in 96 percent alcohol for 1–5 mins and the presence of sperm cells determined by staining with polychrome methylene blue or new methylene blue for 5–60 mins.

SEX DETERMINATION AND SEX CHANGE

External Characteristics

Ehrenbaum (1890) described a method for distinguishing externally between males and females based on the difference in length of the endopods of the first pair of pleopods in *Crangon crangon*. The difference in endopod length was used by Havinga (1930), Lloyd and Yonge (1947), Meredith (1952), and Tiews (1954) for *C. crangon* and by Price (1962) for *C. septemspinosa*. The endopod of the male first pleopod differs rather clearly in form from that of the female. In males the endopod is round and pointed, has the form of a stretched S, lies against the exopod, and is bare (fig. 10.1A). The length (L_e) increases very little with the body length (L_b), following the equation: $L_e = 0.03 L_b - 0.56$. In females, the endopod is flat and spatulate, has a rounded top, is connected to the pleopod at an angle <30°, bears setae, and is modified to carry the eggs (fig. 10.1D). In primary females > 30 mm the size of the endopod increases regularly with body length following the equation $L_e = 0.11 L_b - 2.67$. Approximately 1,200 nonberried females with endopods fitting this equation were examined by us for the presence of oocytes by squashing the anterior part of the gonad. The results were 100 percent positive. Secondary females, starting as spent males of 37–60 mm, need several molts to bring the form and size of the endopod to the level normal for primary females of the same length. Therefore, an endopod measure falling between these two equations is a useful characteristic for identifying secondary females (fig. 10.1 B–C).

Nouvel (1939) described the *Crangon crangon* appendix masculina, a small squamiform protuberance of the second pleopod positioned closely against the endopod in males. Females lack the appendix. In *C. crangon* the appendix is very small, most likely has no function, and seems to disappear in one molt during the process of sex change, while in *Pandalus borealis* and *P. montagui* the comparatively larger appendix masculina decreases in size in one or more molts before disappearing totally (Leopoldseder 1934; Mistakidis 1957).

Lloyd and Yonge (1947) reported a difference between males and females in

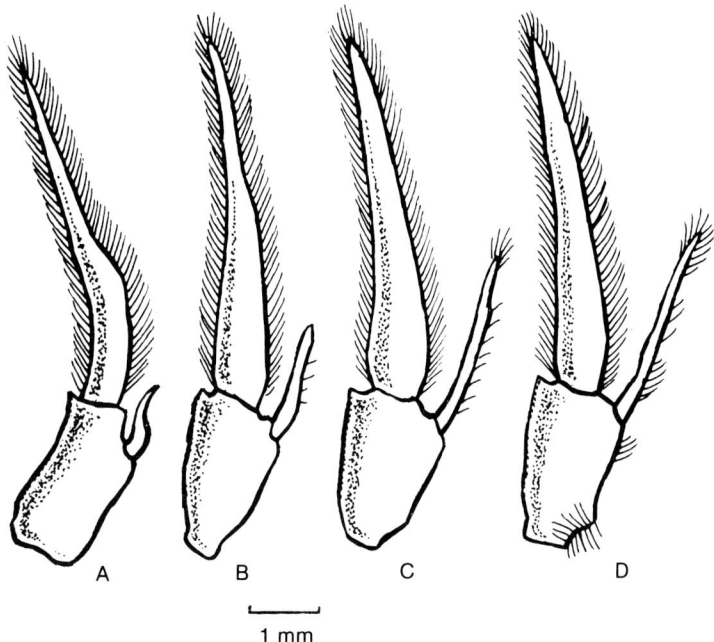

FIGURE 10.1. First pleopod of *Crangon crangon*. A. Male, 51 mm total body length (=TL). B. Secondary female, 46 mm TL. C. Secondary female, 50 mm TL. D. Primary female, 50 mm TL.

the width and the number of segments of the outer flagellum of the antennule but gave only a single example. Dornheim (1969) demonstrated that this difference could be shown only by calculating average values. Tiews (1954) reported a larger number of hairs on all segments of the outer flagellum of the antennule of males, but Dornheim (1969) found gradual differences between male and female shrimps < 40 mm and included no shrimps > 44 mm.

The structure of the gonads of *Crangon crangon* was investigated by Boddeke (1966). A total of 700 specimens was selected for sectioning on the basis of a relatively short endopod of the first pleopod. Martens and Redant (1986) investigated the sex change of *C. crangon* along the Belgian coast. They selected 630 specimens with male external characteristics from samples collected in August–October and examined the sexual organs by squashing. Oocytes and sperm cells were simultaneously present in 3.0, 9.2, and 2.5 percent of the specimens in August, September, and October, respectively.

Sex Change

Sex change is not uncommon in caridean shrimps. Of 32 decapod species reported as sex changers, 26 are caridean shrimps (Bauer 1986a). The phenomenon is well known and extensively described in pandalids such as *Pandalus borealis* (Jägersten 1936; Allen 1959), *P. montagui* (Mistakidis 1957), and for *P. jordani*, *P.*

platyceros, *P. hypsinotus*, *P. danae*, and *Pandalopsis dispar* occurring off the coast of British Columbia (Butler 1964). Sex change has also been observed in the crangonid shrimp *Argis dentata* (Fréchette, Corivault & Couture 1970). *Crangon crangon* exhibits protandric hermaphroditism, generally following the pattern of pandalids. Males and primary females develop from the larval stage, while secondary females develop after sex reversal of spent males.

Stages observed in the development of the gonads of *Crangon crangon* are largely similar to those described and depicted by Mistakidis (1957). The clear presence of degenerated vasa deferentia in secondary females is highly similar to that described in *Pandalus borealis* by Jägersten (1936).

Differences between *Crangon crangon* and most pandalids are probably explained by (1) the much faster sexual cycle of *C. crangon*, which occurs in months instead of years as in northerly stocks of *P. borealis*, and (2) the prolonged spawning season (10.5 months) of *C. crangon* in comparison to the restricted spawning season (approximately 2 months) of the above mentioned pandalids (Butler 1964; Sigurdson & Hallgrimsson 1965; Howard 1983). Because of its long spawning season, *C. crangon* has a practically continuous recruitment to the adult stock. Males recruit at a smaller size than primary females. During most of the year, males in all stages of sexual development are present simultaneously, making quantification of the effect of sex change very difficult. This certainly contributes significantly to the rebuilding of the stock of mature females in May–June after the annual minimum in March–April. The migratory behavior of mature males is similar to that of the (much larger) mature females (Boddeke 1976). Consequently, the impact of sex change on the mature female population is especially relevant in deep water/open sea, while most earlier work on this species has been carried out in shallow inshore nursery areas where juveniles prevail.

PRIMARY SEXUAL CHARACTERISTICS

Gonads

The external features of the female ovary and the male testis were described by Ehrenbaum (1890). Both gonads are paired organs, consisting of two longitudinal tubes with closed ends, situated dorsally in the caudal part of the cephalothorax, ventral to the heart but dorsal to the stomach. A connection between the two lobes at one fifth of the length from the anterior end gives the gonad an H-shape. The extension of the gonad in longitudinal and lateral directions depends on sex and stage of development. The lobes of the testis are intricately curled and are approximately 10 mm long and 0.2 mm wide and they extend within the first abdominal segment.

In females that are ready to spawn, the ovary is rather flat and white and covers a considerable part of the dorsal side of the cephalothorax and the first two abdominal segments. During the development of the ovary, both lobes grow closer together. In the tubal halves of the male and female gonad are strings of finely granulated cells completely or partially surrounded by gametocytes. These

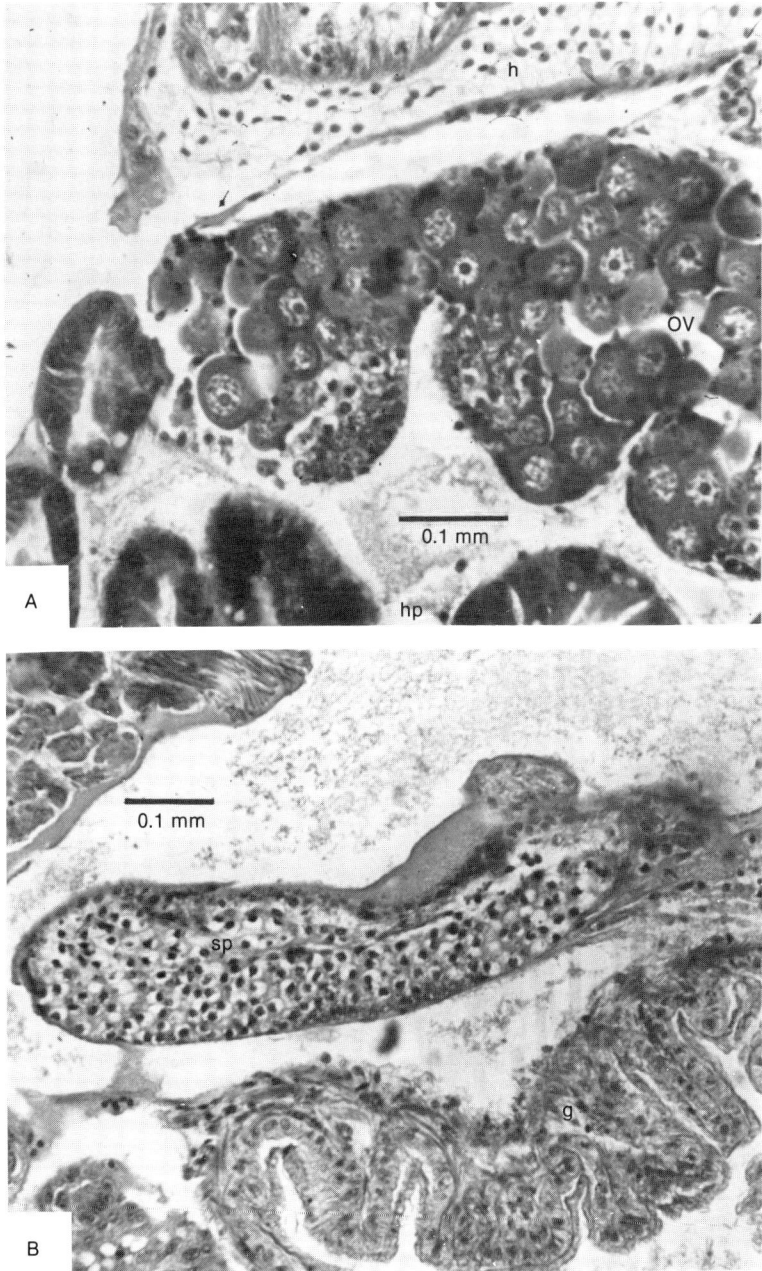

FIGURE 10.2. Stages in the development of the gonad of *Crangon crangon*. A. Transverse section, primary female 27 mm TL. h, heart; hp, hepatopancreas; ov, ovary. B. Longitudinal section, male 33 mm TL, at the beginning of sperm formation. g, gut; sp, spermatocytes.

gametocytes develop into oocytes (fig. 10.2A) and spermatocytes (fig. 10.2B). Oocytes are rapidly removed from the gonad by spawning. It is likely that gametocyte tissue is restored from the finely granulated cells. Ripening of the gonad normally starts at the front end, rapidly transforming the gametocytes over the entire length of the gonad. When ripe, the ovary is filled with oocytes of 0.2–0.3 mm diameter. A fully ripe testis has a spongy appearance (fig. 10.3A–B). In paraffin section the tissue is not stained by eosin, and sperm cells stained black by gallocyanin are regularly dispersed in the tissue.

Careful examination of serial sections of gonads has made clear that the discharge of sperm to the vas deferens is a gradual process. Different stages in a gonad changing from testis to ovary are: spongy tissue containing sperm, a structureless eosinophilic mass (Fig. 10.4A–B) and the formation of oocytes (Fig. 10.4B–D). The last two stages can be observed in the gonad, while in the vasa deferentia sperm is still present (fig. 10.4A). Gradual reconstruction of the gonad occurs in winter, but in spring and summer change is more radical. During development of the voluminous oocytes the formerly curled testis straightens out and increases in length and width (fig. 10.4D).

Vas Deferens and Oviduct

The vas deferens originates from the middle of both testicular lobes. It is very short (± 1 mm) and leads into the conspicuous ejaculatory duct, with a diameter of 0.5–1.0 mm and a length of 5–10 mm, ending in a pore on the coxa of the fifth pereiopod (Ehrenbaum 1890; fig. 25). The large size of the ejaculatory duct in *Crangon crangon* contrasts with *Macrobrachium rosenbergii*, in which a long, convoluted vas deferens only slightly increases in diameter distally (Chow, Ogasawara & Taki 1982; fig. 1). The very large size of the ejaculatory duct in *Crangon crangon* is mainly caused by its thick wall. This suggests that the spermatophore is squeezed out during copulation. In the duct lumen spermatophoric sacs are formed that cover the sperm masses. Sperm may present in vasa deferentia of specimens with developing oocytes, especially in August/September, when copulation is very unlikely to occur. In fresh secondary females the degenerated vasa deferentia are still clearly visible (fig. 10.5A). This is more difficult to observe in specimens routinely preserved in ethanol or formaldehyde because of shrinkage and discoloration of tissue.

The oviducts arise from the ovarian lobes at about one third of the length from its anterior end. The oviduct (4–6 mm long) is straight and has a wide lumen with a thin epithelial wall. The diameter, 0.5 mm at the ovary, narrows to 0.3 mm near the genital pore. In mature inseminated females, the oviducts have a distinct bluish color, making them easy to distinguish from the surrounding musculature. The bluish color is caused by a substance, probably excreted by the wall of the oviduct, present in the lumen as drops and smears. Lloyd and Yonge (1947) noted that the oviduct secretes the inner chitinous membrane around the eggs. In functional males, undeveloped oviducts could not be distinguished with certainty because of the intricate curled form of the testis.

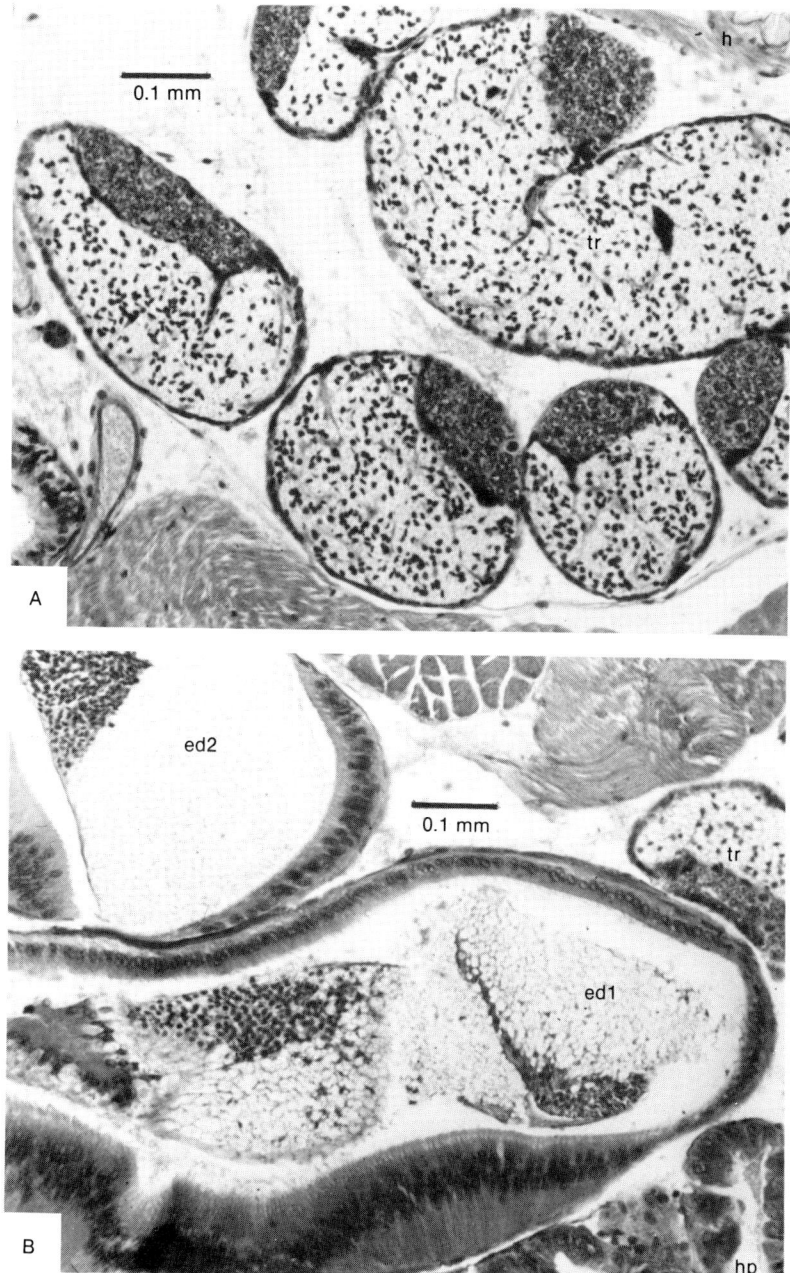

FIGURE 10.3. Longitudinal sections of the testis of a *Crangon crangon* male 53 mm TL. A. Maximum sperm formation. h, heart; tr, ripe testis. B. Curls of ejaculatory duct (edl, ed2) showing the gradual loss of structure in the sperm mass from the ripe testis (tr).

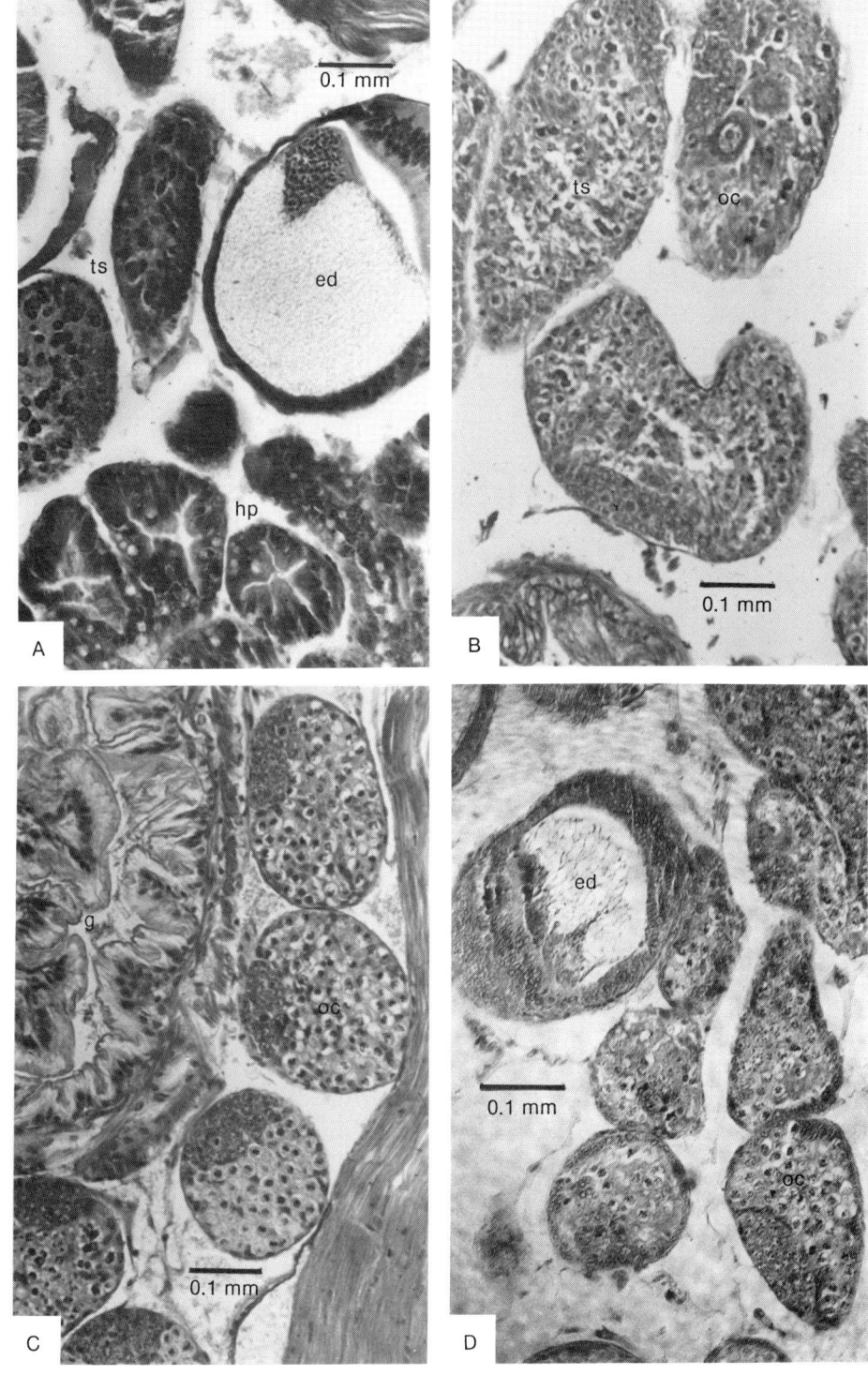

Spermatophores and Sperm Cells

Spermatophores found in the ejaculatory ducts are oblong, elastic bodies 1–2 mm long. In paraffin sections spermatophores seem to have a comparable structure to the sperm mass in a ripe testis, but the concentration of sperm in these sections may be an artifact caused by fixation and dehydration (fig. 10.3B). In fresh spermatophores the sperm is regularly dispersed throughout the sperm mass. Structures beside the sperm mass in the spermatophore, such as adhesive eosinophilic matrix used for attaching the spermatophore to the sternite of the female and a basophilic protective matrix as decribed by Chow, Ogasawara, and Taki (1982) for *Macrobrachium rosenbergii* have not been observed by us in the spermatophores of *Crangon crangon*.

Sperm cells in the *Crangon crangon* spermatophore before copulation are the usual crustacean thumbtack type (fig. 10.6A). The sperm cell is a round disc with a diameter of 10–12μ and a width of 4–5μ. In the center of one side the disc bears a pin, a small cap with a spike described and depicted by Felgenhauer, Abele, and Kim (1988). The spike is 4–5μ long. During copulation most of the pins appear to stay behind in the spermatophoric sac (fig. 10.6B). Consequently, most sperm cells brought into the oviduct during copulation lack these pins or lose them during the stay in the oviduct (fig. 10.5B). The place where the caps had been attached is marked by a round spot (fig. 10.5c).

COPULATION AND SPERM TRANSFER

Copulation of *Crangon crangon* has been described by several authors, but there is no consensus about the position of the male and female during copulation. Nouvel (1939) described the male as sliding its body under that of the female. Lloyd and Yonge (1947) added that after preliminary behavior the male turned the passive female on its back and then bent its body in a *U*-shape transversely across the female at about the junction of the thorax and abdomen so that the ventral regions of the two animals were in contact. These observations by Lloyd and Yonge, as well as their opinion that the appendices masculinae are not used for seizing the female, have been confirmed by Tiews (1954).

During our work, 75 copulations were observed and partially recorded on 8 mm film. This enabled us to analyze the course of copulation in detail.

About sperm transfer in *Crangon crangon*, Ehrenbaum (1890) stated: "Unfortunately I never succeeded in observing a spermatophore excreted naturally neither on a male nor on a female." During preparations, Ehrenbaum often

FIGURE 10.4. Further developmental stages in the gonad of *Crangon crangon* (longitudinal sections). A. Male (44 mm TL) with spent testis (ts) and ejaculatory duct (ed) still containing sperm. hp, hepatopancreas. B. Spent male (39 mm TL) in the process of sex change. oc, oocytes; ts, spent testis. C. Male (same as A) with young oocytes in the anterior part of the gonad. g, gut; oc, oocytes. D. Secondary female (52 mm TL) with well-developed oocytes (oc) in gonad and with sperm remaining in an ejaculatory duct (ed).

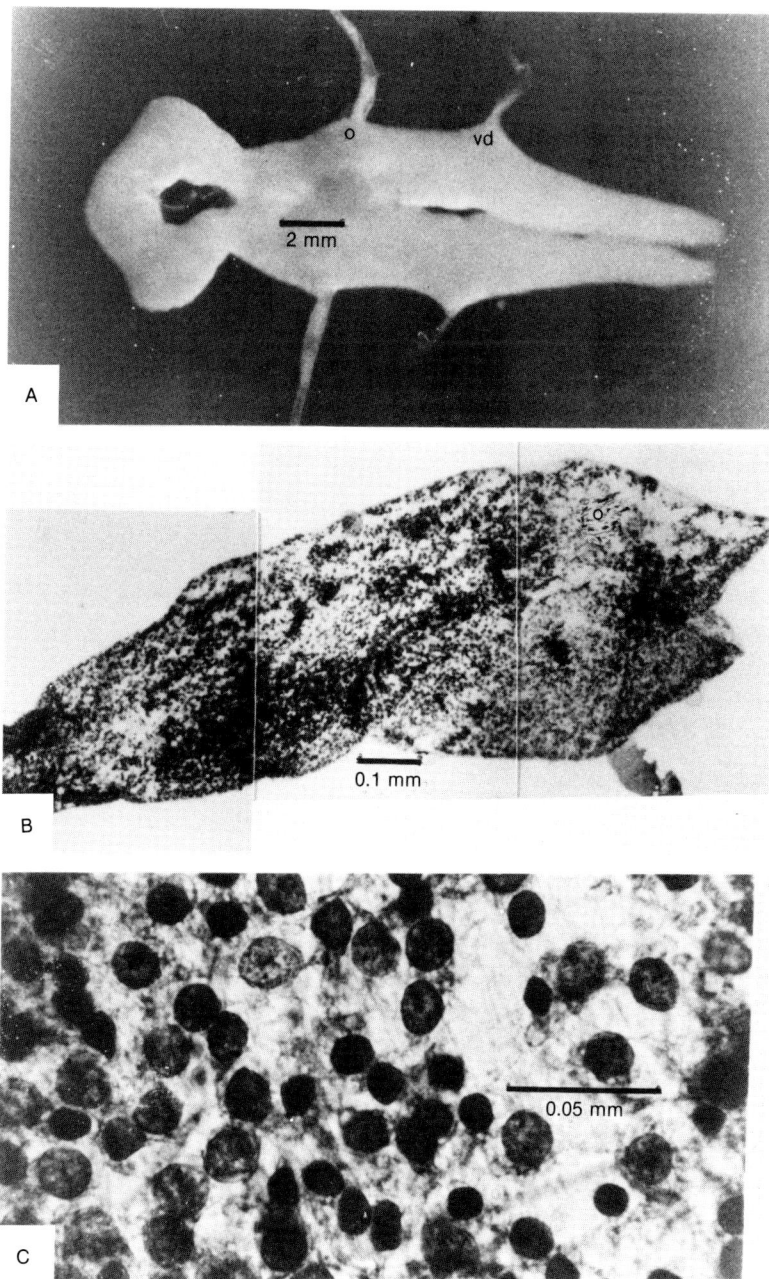

FIGURE 10.5. A. Ovary of secondary female showing oviducts (o) and degenerated vasa deferentia (vd). B. Part of oviduct of inseminated female, with wall covered with sperm cells. Staining: methylene blue. C. Detail of oviduct wall of inseminated female showing sperm cells. Staining: methylene blue.

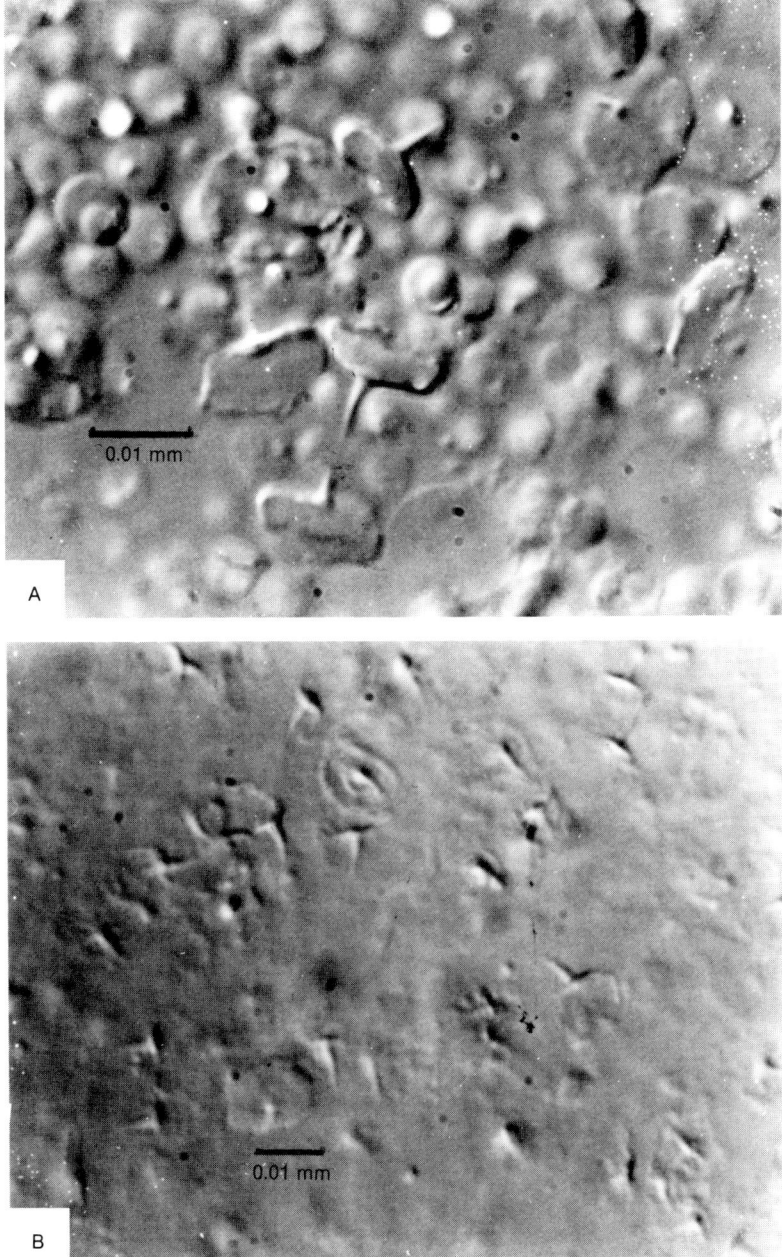

FIGURE 10.6. A. Detail of spermatophore before copulation showing sperm cells with "pins." B. Detail of spermatophore after copulation showing lose "pins" and a few complete sperm cells.

observed spermatophores emerging from the ejaculatory duct, but this gave "no indication about the way sexual unification takes place." Not much has changed since then. Tiews (1954) suggested that the two first pairs of pleopods should play a role in the transport of the spermatophore from the male sexual pore to the female. This was based on the finding of "remainings of spermatophores" on the inner rami of the two first pairs of pleopods of the male in 35 cases out of 89 observed copulations. Tiews added that the possibility of direct placement of a spermatophore onto the female during insemination could not be excluded. This statement, although vague, suggests external fertilization. In the descriptions of copulation of *C. crangon* by Nouvel and Tiews, the presence of sperm or spermatophore on the females after copulation was not mentioned. Lloyd and Yonge (1947) stated that "spermatophores being applied to the ventral side of the female usually more or less adjacent to the genital opening." No evidence was given to support this statement.

Organization of Experiments

Research on copulation and sperm transfer was carried out from November 1978–June 1979. Hundreds of shrimps were caught periodically in different areas along the Dutch coast and transported quickly to the laboratory. Earlier observations had shown that insemination only can take place in females with a ripe ovary, directly after molting (Tiews 1954). Female shrimps (n = 120) with externally obvious ripe ovaries were divided among 20 compartments built from transparent persplex in large basins with a regular water exchange and a sand-covered bottom. Molting occurred mainly at night. The presence in the morning of shed exoskeletons in these cages made it possible to quickly pick the molted soft-shelled specimens. Observations on copulation were made on 162 females; 75 successful copulations were observed. Histological observations had shown that from November–June a large percentage of individuals 30–40 mm long had male sexual characteristics. Thirty such males were placed in each of two glass basins of $35 \times 25 \times 20$ cm with a sand-covered bottom, for use in copulation observations and experiments.

Copulation

To observe copulation, we released a soft-shelled female in a basin with 30 male specimens during the morning hours. The following sequence of events was typical of mating and copulation. Before the release of the female, all males were usually buried in the sediment as a response to high incident illumination. Upon release the female swims slowly to the bottom and remains still. Within a few minutes the males start to react by moving their antennae in an upward direction. Subsequently several males come out of the sand and start to swim around, regularly passing the female at short distance. Suddenly a male approaches the female laterally, lays its third pair of maxillipeds and its first pair of pereiopods on the female dorsum, and pushes it with pereiopods and pleopods on its back (fig. 10.7A,B). The male then bends its body, touching its cephalothorac-abdominal junction to the ventrolateral edge of the female's cephalothorax (fig. 10.7c).

This fixed position is maintained for 0.5–2.0 secs. Then the male moves away from the female, sinks to the bottom, stays immobile for approximately 10 secs, and starts swimming around again. Approximately 5 secs after copulation the female swims away and starts to bury herself in the sediment, constantly contacted by other males. The female resists another copulation by fleeing.

After copulation, the female and the male were immediately caught for further inspection. In practically all cases in which a copulation was observed, an empty spermatophore was still present in an ejaculatory duct, often hanging partially out of the genital pore. No spermatophore could be observed elsewhere. In the postcopulatory spermatophore only small numbers of intact sperm cells could be observed, but large numbers of loose "pins" (caps bearing spikes) were present (fig. 10.6B). In 11 of 75 females, large numbers of sperm cells were present in the oviduct, mainly without pins. Intact sperm cells and loose pins could be observed. Sperm cells without pins could be identified by their size, form, pin mark, and reaction with stains for nuclei (fig. 10.5c). The rather low number of successful copulations was perhaps caused by the abnormally high light intensity necessary for filming. The presence of sperm in the oviducts of unberried females as the result of an earlier impregnation is very unlikely in November–March. In berried females captured during August and September, large numbers of sperm cells can be found regularly in the oviducts (fig. 10.5 B).

Insemination Model

On the basis of these observations and the inspection of 8 mm films of copulation, the process of insemination is hypothesized as follows (fig. 10.8 A–D): when the male is touching his ventral side against the ventrolateral female cephalothorax, the male fifth pereiopod crosses the female third pereiopod. The male presses one of his genital pores against one female genital pore (Fig. 10.8B). The male maintains this position for 1–2 seconds. In this short period, the spermatophore is pushed, by peristaltic contractions of the muscular wall of the ejaculatory duct, partially out of the genital pore and emptied in the oviduct by a forceful contraction (fig. 10.8c). Relaxation of the ejaculatory duct brings the now empty spermatophore largely back into its original position (fig. 10.8D). It is later removed if a new spermatophore is formed. If this is not the case, remains of the sperm mass stay in the ejaculatory duct (fig. 10.4B).

The organization of our experiments did not allow us to demonstrate whether the same male inseminates the other oviduct of the female as well. Under natural conditions, densities of sexually mature *Crangon crangon* are normally low (1–0.01/m^2), and the sex ratio among mature specimens is in favor of females most of the year (Boddeke 1982; Goudswaard 1981). This implies that the first spawning of a female is usually based on sperm of one male. In later spawnings, several males will contribute to the fertilization of the eggs if a new insemination precedes the spawning. In August–September, when fertile males are virtually absent, this is not the case. Spawning in these months is restricted to females carrying eggs (and thus being inseminated) in July (fig. 10.9).

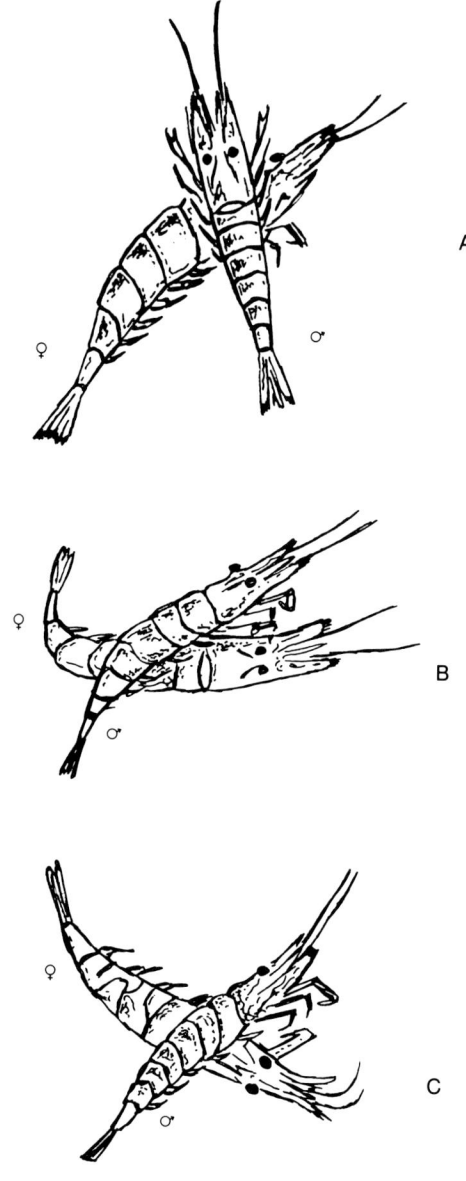

FIGURE 10.7. Copulation of *Crangon crangon* (drawn from individual film frames). A. Male starts to push female with pereiopods and first pair of pleopods. B. Same as A, but further advanced. C. Male has reached copulatory position.

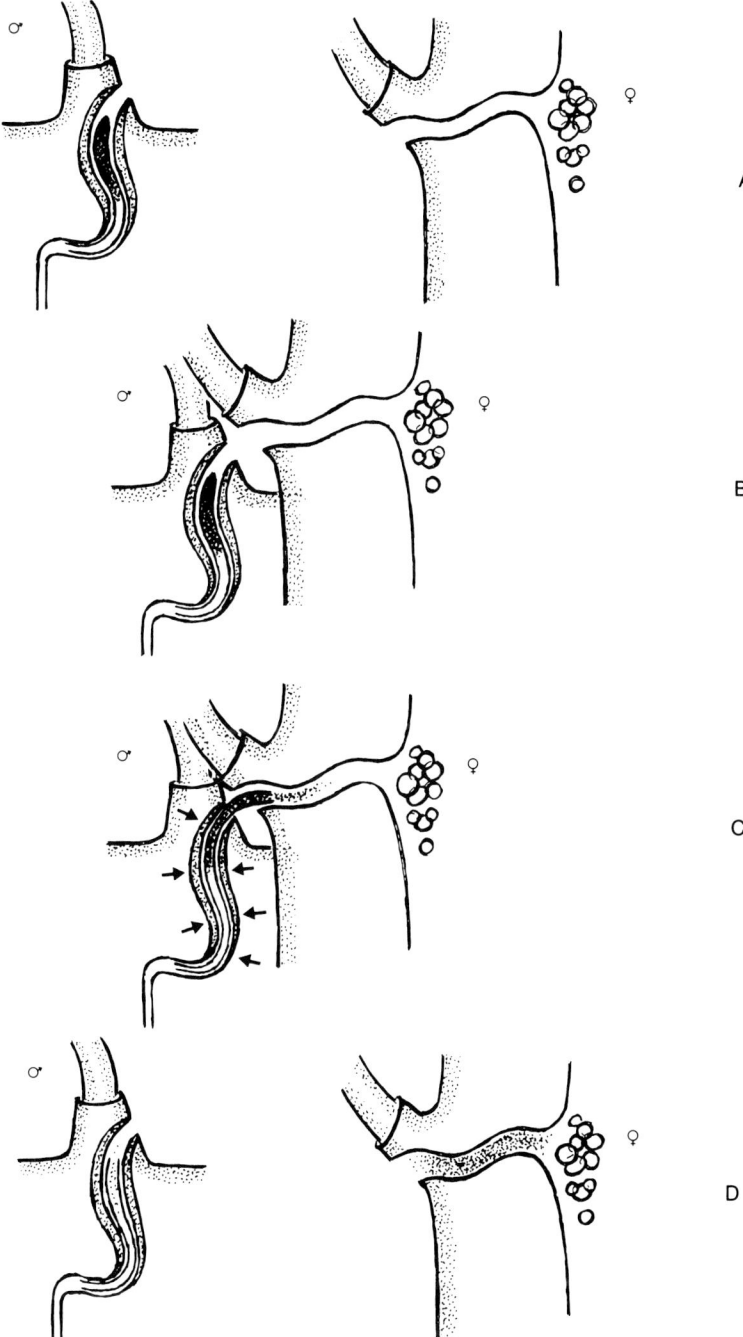

FIGURE 10.8. Hypothesized series of events in the insemination of *Crangon crangon*. A. Beginning positions of male ejaculatory duct and gonopore and female oviduct and gonopore. B. Male presses one gonopore against the opening of one female gonopore. C. Male empties spermatophoric sac into the oviduct. D. Empty spermatophoric sac stays in ejaculatory duct. Inner side of oviduct is covered with sperm cells.

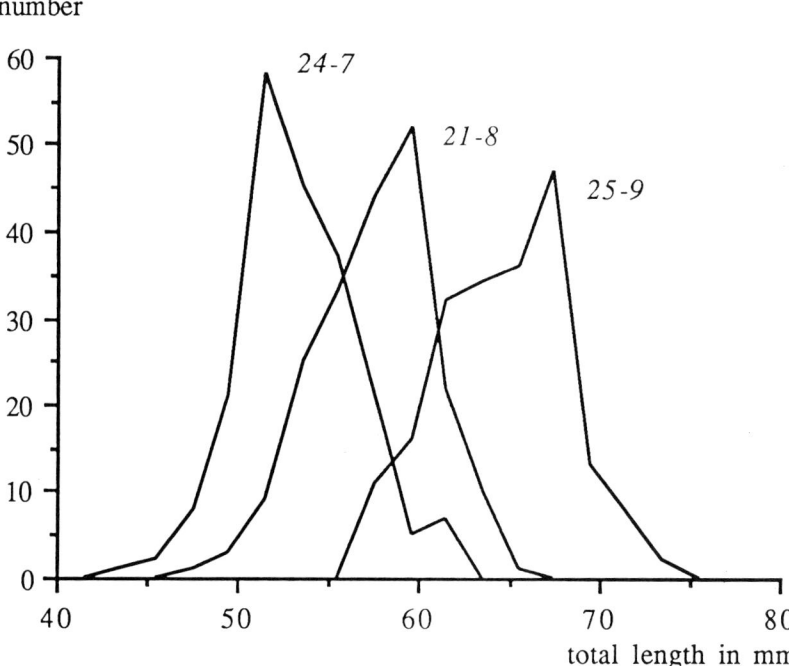

FIGURE 10.9. Length frequency distribution of berried females in July and following months, showing growth during molts. Spawnings after July are fertilized with stored sperm.

DISCUSSION AND FUTURE RESEARCH

Evidence presented here for internal fertilization in *Crangon crangon* sheds new light on reproduction in caridean shrimps. These findings contradict the general observation that in caridean shrimps oocytes are fertilized externally by a spermatophoric mass attached to the ventral side of the female (e.g., Bauer 1976; Chow, Ogasawara & Taki 1982) *Crangon crangon* is at present the only decapod known besides brachyuran crabs in which sperm is deposited internally in the female so that fertilization is internal. In this respect it must be noted that caridean shrimp species in which external fertilization has been described with certainty (belonging to the families Palaemonidae and Hippolytidae) live in more restricted environments (weed beds, tidal pools, riverbanks) and presumably much higher densities than is normally the case with adult *C. crangon*. Future research can reveal whether the internal fertilization of *C. crangon* is unique or more widespread among caridean shrimps, in particular among related species with a comparable spatial distribution. By checking oviducts of berried females for sperm cells, internal sperm storage and fertilization can be demonstrated.

The concept of internal fertilization seems to be linked to anatomical and behavioral differences between *Crangon crangon* and caridean shrimps with external fertilization as mentioned in preceding sections. These differences include the structure of vas deferens–ejaculatory duct, the structure of the spermatophore, the lack of a brood chamber, and the mating position. In particular the absence of an adhesive matrix in the spermatophore seems to be a functional consequence of internal insemination. The spermatophore of *C. crangon*, in which the covering stays behind in the ejaculatory duct, disintegrates during copulation and not during oviposition as in caridean shrimps with external fertilization. The lateral mating position of *C. crangon*, in which the male touches the ventrolateral edge of the female's cephalothorax, differs clearly from those observed in species with external fertilization. In the latter species the ventral sides of male and female are brought closely together in the final mating position (Höglund 1943; Bauer 1976; Chow, Ogasawara & Taki 1982). The hypothesis of internal sperm transfer in *C. crangon* is supported by this difference in mating position.

In *Crangon crangon* the sperm cell spikes seem to have no function because they mainly stay behind in the spermatophoric sac during copulation. Burkenroad (1947) doubted a function for these spikes. He stated that in *Palaemonetes* a naked oocyte can capture and engulf a passive sperm cell at the moment of extrusion of the oocyte from the oviduct. For oocytes of *C. crangon*, surrounded in the oviduct by sperm cells, this is even more plausible.

About the actual sperm transfer in *Crangon crangon* some questions remain. Although from our observations the use of the appendix masculina or any other part of a pleopod is not very likely, it cannot be totally excluded, because of difficulties in observing and establishing the exact moment of sperm transfer. Experiments on copulation with only one male per female will be necessary to reveal whether the same male inseminates the other oviduct as well.

Spawning by females without an observed preceding copulation has been reported in *Leander squilla* (Höglund 1943). The unfertilized eggs, however, did not become attached. The internal storage of sperm enables *Crangon crangon* to spawn fertile eggs some time after a copulation. This considerably favors the chances of successful spawning of individual females and extends the spawning period in a period of sexual inactivity.

Our observations on mating and sperm transfer in this crangonid species support the view of Felgenhauer and Abele (1983) that the Caridea is a heterogeneous group whose taxonomic relationships should be reexamined. The mode of insemination (external versus internal) thus may be an important character in the systematics of the Caridea.

Literature Cited

Allen, J. A. 1959. On the biology of Pandalus borealis Krøyer, with reference to a population off the Northumberland coast. *J. Mar. Biol. Ass. U.K.* 38:189–220.

Baker, J. R. 1958. *Principles of Biological Microtechnique. A Study of Fixation and Dyeing.* New York: Wiley.

Bauer, R. T. 1976. Mating behaviour and spermatophore transfer in the shrimp *Heptacarpus pictus* (Stimpson) (Decapoda: Caridea: Hippolytidae). *J. Nat. Hist.* 10:415–440.

Bauer, R. T. 1986a. Sex change and life history pattern in the shrimp *Thor manningi* (Decapoda: Caridea): A novel case of partial protandric hermaphroditism. *Biol. Bull. Woods Hole* 170:11–31.

Bauer, R. T. 1986b. Phylogenetic trends in sperm transfer and storage complexity in decapod crustaceans. *J. Crustacean Biol* 6:313–325.

Boddeke, R. 1966. Sexual cycle and growth of brown shrimp *(Crangon crangon)*. *ICES Shellfish Comm. C.M. M:6.* Mimeo.

Boddeke, R. 1975. The use of biological tags in shrimp research. *ICES, Shellfish and Benthos Comm. C.M. K:45.* Mimeo.

Boddeke, R. 1976. The seasonal migration of the brown shrimp *Crangon crangon*. *Neth J. Sea Res.* 10:103–130.

Boddeke, R. 1982. The occurence of winter and summer eggs in the brown shrimp *(Crangon crangon)* and the pattern of recruitment. *Neth. J. Sea Res.* 16:151–162.

Burkenroad, M. D. 1947. Reproductive activities of decapod Crustacea. *Am. Nat.* 81:392–398.

Butler, T. H. 1964. Growth, reproduction and distribution of pandalid shrimps in British Columbia. *J. Fish. Res. Board Can.* 21:1403–1452.

Chow, S., Y. Ogasawara & Y. Taki. 1982. Male reproductive systems and fertilization of the palaemonid shrimp *Macrobrachium rosenbergii*. *Bull. Jpn. Soc. Sci. Fish.* 48:177–183.

Dornheim, H. 1969. Beiträge zur Biologie der Garnale *Crangon crangon* (L.) in der Kieler Bucht. *Ber. Dtsch. Wiss. Kommn. Meeresforsch.* 20:179–215.

Ehrenbaum, E. 1890. Zur Naturgeschichte von *Crangon vulgaris* Fabr. *Mitteilungen der Sektion für Küsten-und Hochseefischerei (Sonderbeilage)*, Berlin, p. 124.

Felgenhauer, B. E. & L. G. Abele. 1983. Phylogenetic relationships among shrimp-like decapods (Penaeoidea, Caridea, Stenopodidea). In E. R. Schram, ed., *Crustacean Phylogeny*, pp. 291–311. Rotterdam: Balkema.

Felgenhauer, B. E., L. G. Abele & Won Kim. 1988. Reproductive morphology of the anchialine shrimp *Procaris ascensionis* (Decapoda, Procarididae). *J. Crustacean Biol.* 8:333–339.

Fréchette, J., G. W. Corivault & R. Couture. 1970. Hermaphrodisme protérandrique chez une crevette de la familie des Crangonidés, *Argis dentata* Rathbun. *Naturaliste Can. (Que-)* 97: 805–822.

Goudswaard, P. C. 1981. *Onderzoek naar de ♂ - ♀ verhouding bij de garnaal* Crangon crangon *(L.)*. ZE 81-01. IJmuiden:RIVO

Havinga, B. 1930. Der Granat (*Crangon vulgaris* Fabr) in den holländischen Gewässern. *J. Cons. Perm. Int. Explor. Mer* 5:57–87.

Höglund, H. 1943. On the biology and development of *Leander squilla* (L.) forma typica de Man. *Svensk Hydrogr. Biol. Komm. Skr. Ny Serie Biol.* 2 (6):44.

Howard, F. G. 1983. The United Kingdom fishery for the deep water shrimp *Pandalus borealis* in the North Sea. *Scott. Fish. Res. Rep.* 30:16.

Jägersten, G. 1936. Über die Geslechtsverhältnisse und das Wachstum bei *Pandalus*. *Ark. Zool.* 28 A (20):26.

Leopoldseder, F. 1934. Geslechtsverhältnisse und Heterochromosomen bei *Pandalus borealis*. *Z. Wiss. Zool.* 145:337–350.

Lloyd, A. J. & F. R. S. Yonge. 1947. The biology of *Crangon vulgaris* L. In the British Channel and Severn estuary. *J. Mar. Biol. Ass. U.K.* 26:626–661.

Martens, E. & F. Redant. 1986. Protandric hermaphroditism in the brown shrimp *Crangon crangon* (L.), and its effect on recruitment and reproductive potential. *ICES Shellfish Comm.C.M.* K: 37. Mimeo.

Meredith, S. S. 1952. A study of Crangon vulgaris in the Liverpool Bay area. *Proc. Trans. Liverpool Biol. Soc.* 58:75–109.

Mistakidis, M. N. 1957. The biology of *Pandalus montagui* Leach. *Fish. Invest. Lond. Ser. 2*, 21(4):52.

Nouvel, L. 1939. Observation de l'accouplement chez une espèce de crevette *Crangon crangon*. *C.R. Acad. Sci. Paris.* 209:639–641.

Price, K. S. 1962. Biology of the sand shrimp, *Crangon septemspinosa*, in the shore zone of the Delaware Bay region. *Chesapeake Sci.* 3:244–255.

Redant, F. 1984. An indexed bibliography on the common shrimp, *Crangon crangon* (L.) complementary to Tiews, 1967. *ICES Shellfish Comm. C.M. K:2.* Mimeo.

Sigurdson, A. & I. Hallgrimsson. 1965. The deep-sea prawn *(Pandalus borealis)* in Icelandic waters. *Rapp. P.-v. Réun. Cons. Perm. Int. Explor. Mer* 156:105–108.

Tiews, K. 1954. Die biologischen Grundlagen der Büsumer Garnelenfischerei. *Ber Dtsch. Wiss. Kommn. Meeresforsch.* 13:235–269.

Tiews, K. 1970. Synopsis of biological data on the common shrimp, *Crangon crangon* (Linnaeus, 1758). *FAO Fish. Biol. Synop.* 91:58.

ELEVEN
Sperm Transfer and Storage Structures in Penaeoid Shrimps: A Functional and Phylogenetic Perspective

RAYMOND T. BAUER

*Center for Crustacean Research,
University of Southwestern Louisiana,
Lafayette.*

Abstract

Basic functional and evolutionary aspects of penaeoid shrimp insemination morphology remain poorly known or understood in spite of the ecological, commercial, and phylogenetic importance of these animals. Sperm is transferred from male to female in spermatophores of varying structural complexity. Externally attached spermatophores are most complex, while those stored in paired seminal receptacles inside the female cephalothorax are reduced to simple sperm masses. The petasma, a complex structure on the male's first pair of abdominal appendages, is often assumed to function mechanically in transferring spermatophores. Petasmata vary among penaeoid males in their degree of ventral openness and presence of distal horns or spouts. Associated morphoclines in male petasmata, female spermatophore attachment/storage genitalia, and spermatophore complexity are described. Correlation of these morphoclines with a reduction in serial homology (branchial characters) suggests that external attachment of a complex spermatophore to an open thelycum is primitive, while storage of simple sperm masses in internal seminal receptacles is derived. Alternative hypotheses about petasma function are proposed (spermatophore transfer structures; female-stimulating devices). Results of manipulative mating experiments with *Sicyonia* demonstrate the pattern of seminal receptacle filling. An experimental approach is necessary to resolve questions about petasma function and related theoretical questions about mating systems in penaeoid shrimps.

PENAEOID SHRIMPS (order Decapoda, suborder Dendrobranchiata, superfamily Penaeoidea) are important crustaceans in several respects. They are commercially important, making up the bulk of the catch in shrimp fisheries around the world. In recent years, they have become a focus of crustacean mariculture. Penaeoids are often a diverse and abundant element of bottom faunas in tropical and subtropical marine communities. In the study of decapod phylogeny they are a key group, unique among the Decapoda in their primitive method of reproduction in which fertilized embryos are not brooded but spawned free into the water, hatching as nauplii. Yet in spite of their commercial, ecological, and phylogenetic significance, many aspects of penaeoid reproductive biology are poorly known. In this chapter I will focus on penaeoid morphology that is concerned with insemination, i.e., the transfer of sperm from males to attachment or storage structures of females, illustrated with examples from several penaeoid species.

Penaeoid shrimp have morphologically complex external male and female genitalia. External features of the genitalia, so important in penaeoid taxonomy, have been well described and figured for many species by authors such as Burkenroad (1934, 1936), Heldt (1938a), Kubo (1949), and especially Pérez Farfante (1969, 1971a, b, 1975, 1977, 1982, 1985, 1988). Male genitalia consist of two structures: the *petasma*, or modified endopods of the male's first abdominal appendages (pleopods), and the *appendix masculina* and (in some) the appendix interna, processes on the endopod of the second male pleopod. Female genitalia consist of the *thelycum* and/or *seminal receptacles*, thoracic sternal structures associated with spermatophore attachment or storage.

Although the modified anterior pleopods of other decapod males have been shown to function in insemination (crayfish [Andrews 1911]; caridean shrimp [Bauer 1976; Berg & Sandifer 1984], brachyuran crabs [Hartnoll 1969]), it is not known if and how the male genitalia function in copulation and insemination in penaeoid shrimp. In this chapter, I will address the hypothesis that the petasma transfers spermatophores in penaeoid shrimp.

Another focus of this chapter will be to examine the hypothesis, originally proposed by Burkenroad (1934, 1936) and especially by Heldt (1938b), that there are correlated morphoclines in spermatophore, male petasma, and female thelycum/seminal receptacle structure in penaeoid genera. Bauer's (1986) restatement of these apparent morphoclines indicated that they had both functional and phylogenetic significance. I will describe and illustrate in greater detail these correlated morphoclines that are associated with increased internalization of sperm storage in penaeoids.

SPERMATOPHORES

The term *spermatophore* has been variably used by different authors. The definition of spermatophore used here is: the structure emitted from the male ejaculatory duct (= terminal ampoule) during copulation. Spermatophore structure and complexity will be described from some original observations on spermatophore structure as seen in histological sections of the male ejaculatory duct and from information in the literature.

Penaeoid spermatophores exhibit considerable variation. At one extreme are structurally complex external spermatophores, characterized by various wings, flanges, plates, and adhesive materials (families Aristeidae, Solenoceridae, and the subgenus *Litopenaeus*, genus *Penaeus*, in the Penaeidae). At the other extreme are the highly internalized simple spermatophoric masses found in the Sicyoniidae.

The spermatophore of the white shrimp *Penaeus (Litopenaeus) setiferus* typifies the complex type deposited externally on the female (Pérez Farfante 1975). A *P. setiferus* female receives a compound structure that is composed of two single spermatophores emitted from the left and right ejaculatory ducts of the male. Figure 11.1 illustrates the general features of the male reproductive tract, includ-

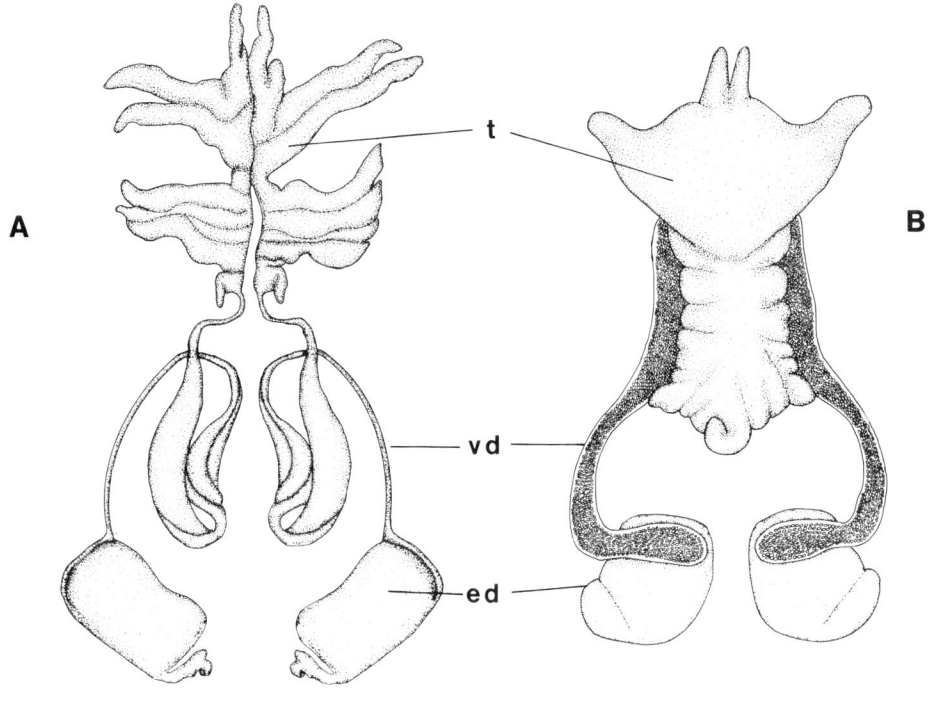

FIGURE 11.1. Male reproductive tracts, dorsal view of A. *Penaeus setiferus* (modified from King 1948) and B. *Trachypenaeus similis*. ed, ejaculatory duct; t, testes; vd, vas deferens. Scale bar = 10 mm in A, 6 mm in B.

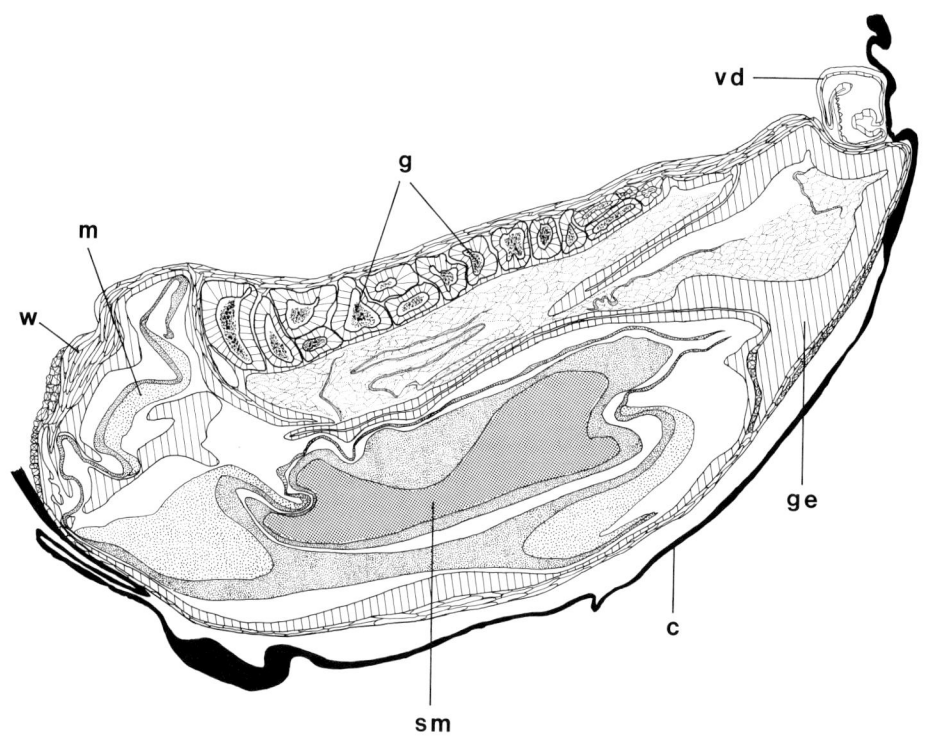

FIGURE 11.2. Transverse section, right ejaculatory duct of *Penaeus setiferus*. Vas deferens (vd) marks the duct's proximal end and wing (w) the distal end. The sperm mass (sm) is surrounded by various sperm-free materials (stippled). c, cuticle; g, glands; ge, glandular epithelium (hatched); m, muscle. Scale bar = 1.0 mm.

ing the ejaculatory duct (= the "terminal ampoule" of penaeoid specialists). In *P. setiferus*, the interior of the ejaculatory duct is a complex, "busy" space (figs. 11.2, 11.3A–E). Posteriorly, the glandular epithelium is produced into a series of folds or incomplete septa that partition the duct's lumen. King (1948) and Chow et al. (in press) have shown that the sperm (fig. 11.3E) is enclosed by various substances in the vas deferens, whereas other spermatophore parts, e.g., the

FIGURE 11.3. A–D. Transverse section through right ejaculatory duct, *Penaeus setiferus* (see fig. 11.2 for orientation). A. Glandular area. B. Gland with contained secretion. C. Distal end of duct showing wing and sperm mass surrounded by sperm-free materials. D. Magnification of sperm mass in C and surrounding materials. E. Sperm cells in matrix (magnified from sperm mass in D). F. Spermatophore emitted from one ejaculatory duct, *P. aztecus*, showing the main body and appendage (fixed immediately after emission). G. *P. aztecus* spermatophore, showing reaction of appendage in seawater. H. Transverse section through left ejaculatory duct, *P. duorarum*, showing spermatophore sperm mass, sperm-free layers, and part of the appendage in its chamber. apd, appendage; g, gland; ma, matrix; mb, main body; ms, mesial; s, sperm cell; sf, sperm-free spermatophore material; sm, sperm mass; v, ventral; w, wing. Scale bar in A = 540 μ in A, 115 μ in B, 770 μ in C, 310 μ in D, 3 μ in E, 1.1 mm in F, 1.2 mm in G, 950 μ in H.

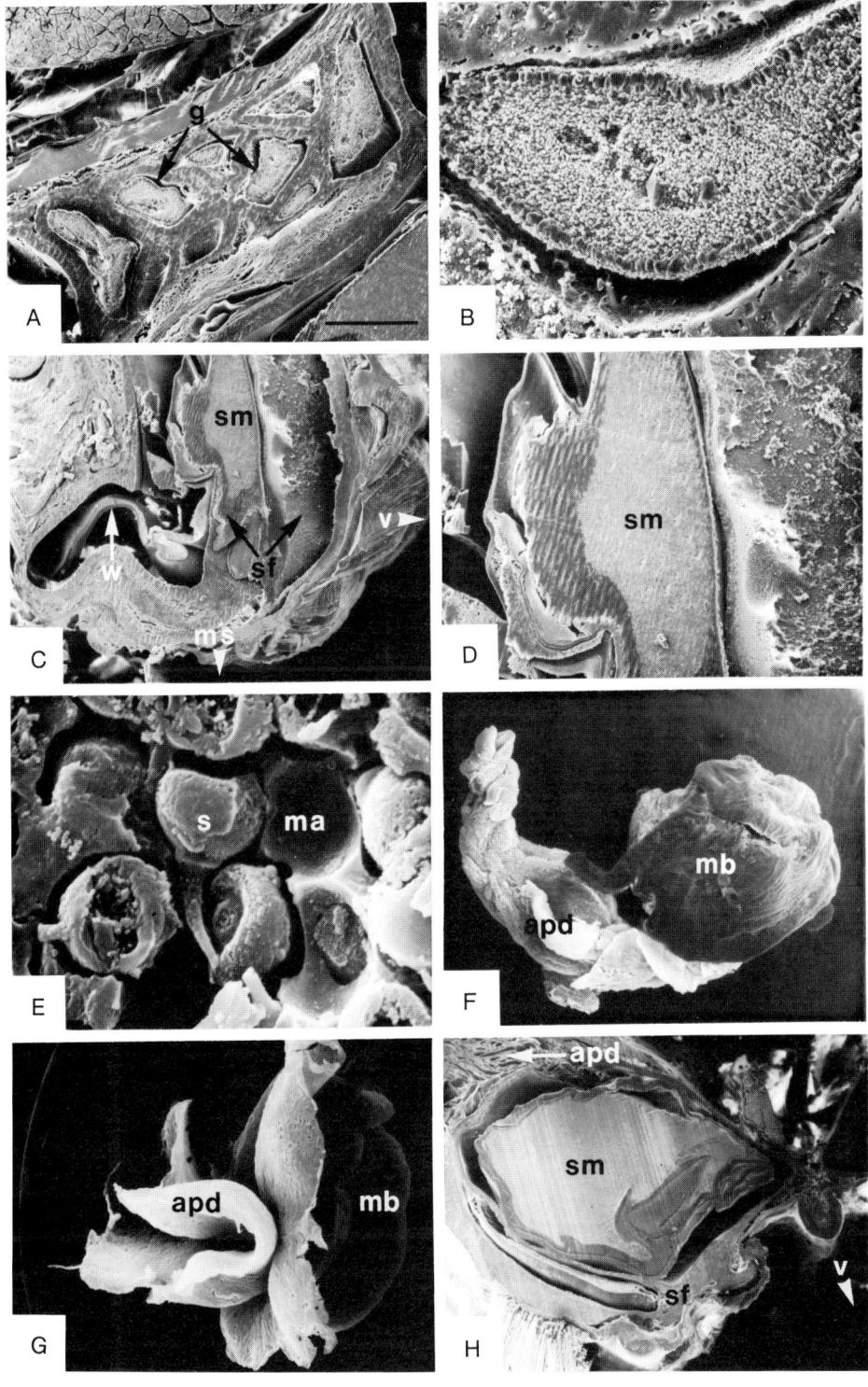

wing (figs. 11.2, 11.3A, C) are secreted entirely in the ejaculatory duct. The septa of glandular epithelium surround and are intermingled among the spermatophore parts (fig. 11.2), and layers formed from their secretion droplets and particles are added onto the spermatophore. Anteriorly, the duct is filled with tubular particle- or droplet-filled glands (figs. 11.2, 11.3A, B) that extend over the spermatophore and appear to empty their secretion around the sperm capsule, apparently secreting the adhesive material that causes the spermatophore to adhere to the external surface of the female. (See Chow et al. in press for their interpretation of spermatophore formation and ejaculatory duct structure.)

Orsi Relini and Tunesi (1987) and Tunesi (1987) have shown in the aristeids *Aristeus antennatus* and *Aristaeomorpha foliacea* (respectively) that the external spermatophores are complex. Emitted spermatophores of solenocerid penaeoids have been well described only in *Pleoticus* and *Mesopenaeus* (Pérez Farfante 1977). A section through the ejaculatory duct of *Solenocera vioscai* is illustrated in fig. 11.4. Although further study is warranted, it is obvious from the above-mentioned studies that the spermatophore is complex, composed of several structures and substances in addition to sperm.

Spermatophores deposited within some sort of seminal receptacle are morphologically less complex than those deposited externally on the female. Except for the subgenus *Litopenaeus*, all *Penaeus* species have spermatophores stored in a median pocket behind the female thelycum. In *P. aztecus* and *P. duorarum*, the

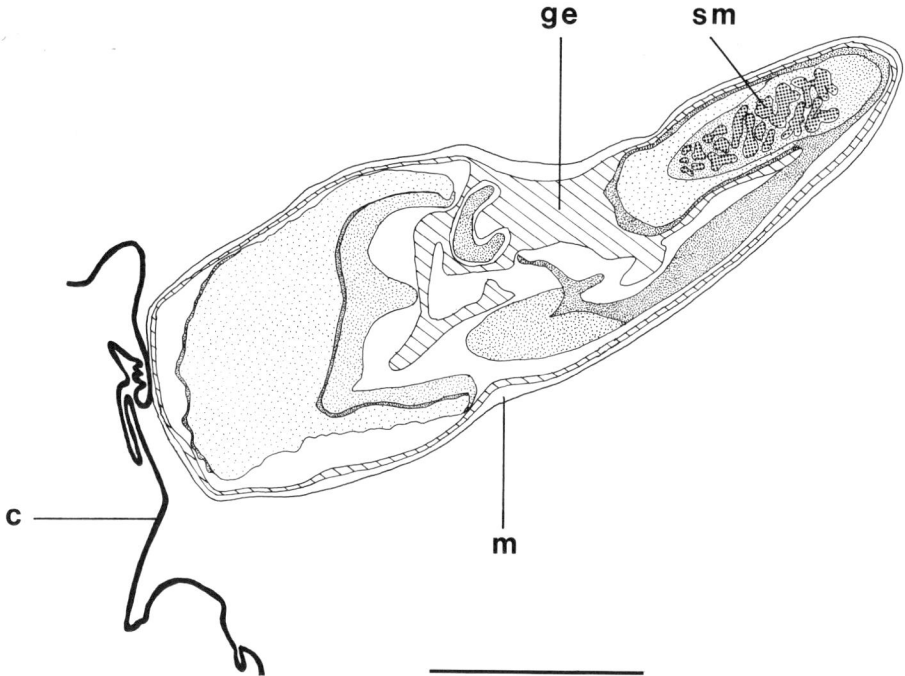

FIGURE 11.4. Transverse section, right ejaculatory duct of *Solenocera vioscai*. The top right of the figure is proximal; cuticular side is distal. Note sperm masses (sm) in sperm chambers and various sperm-free spermatophore substances (stippling). c, cuticle; ge, glandular epithelium (hatched); m, muscle. Scale bar = 1.0 mm.

spermatophore ejaculated from one duct consists of a large main body, containing sperm surrounded by layers of sperm-free material, and a tail-like appendage (this structure is referred to as a "wing" by Malek and Bawab 1974a, but it is clearly neither homologous nor analogous to the wing of *Penaeus setiferus*) (Fig. 11.3c, F). When the spermatophores of each side are emitted from the male, the main bodies of each adhere to form a compound spermatophore (personal observation on *P. duorarum*). The appendage undergoes a swelling and delamination after several minutes' exposure to seawater (fig. 11.3F, G) (illustrated in Eldred 1958 for *P. duorarum*). Sections through the male duct show that, posteriorly, the main body is produced into two folds separated by a septum of glandular epithelium with surrounding secreted layers of material (fig. 11.5). More ante-

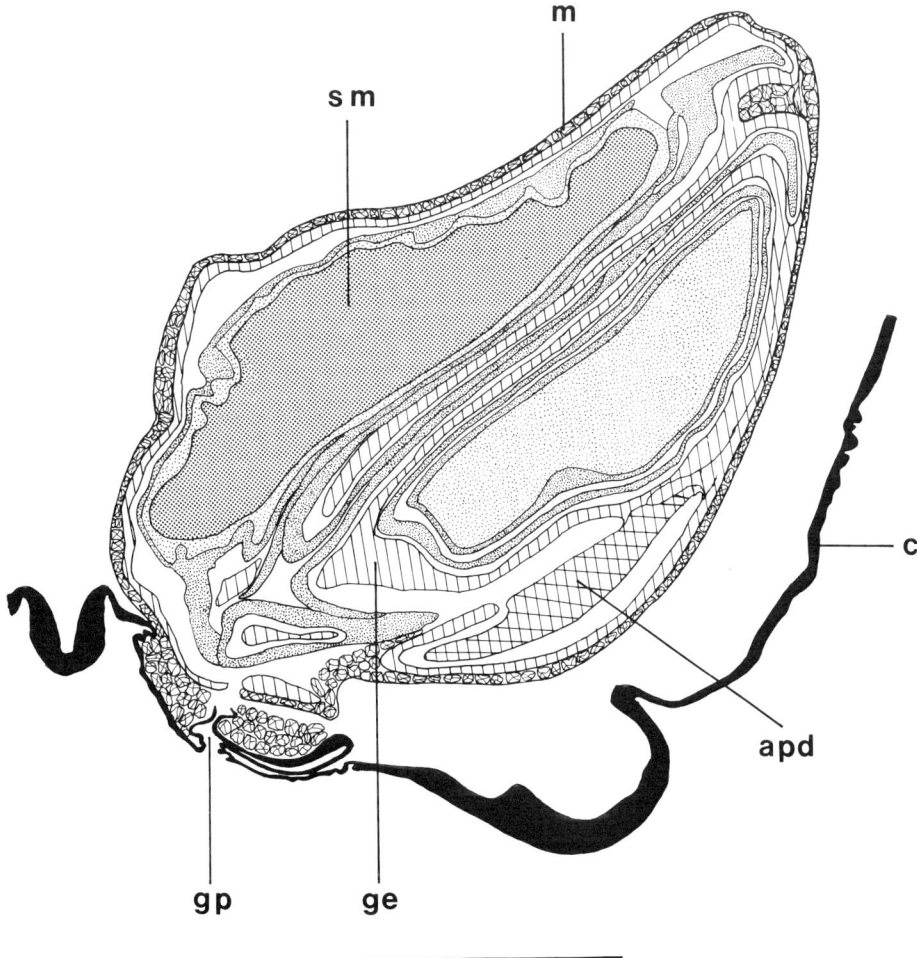

FIGURE 11.5. Transverse section, right ejaculatory duct, *Penaeus aztecus*. The gonopore (gp) marks the distal end of the duct. See text for explanation. apd, appendage material (crosshatched); c, cuticle; ge, glandular epithelium (hatched); m, muscle; sm, sperm mass; stippling indicates various layers of sperm-free material of the spermatophore main body in this region. Scale bar = 2.0 mm.

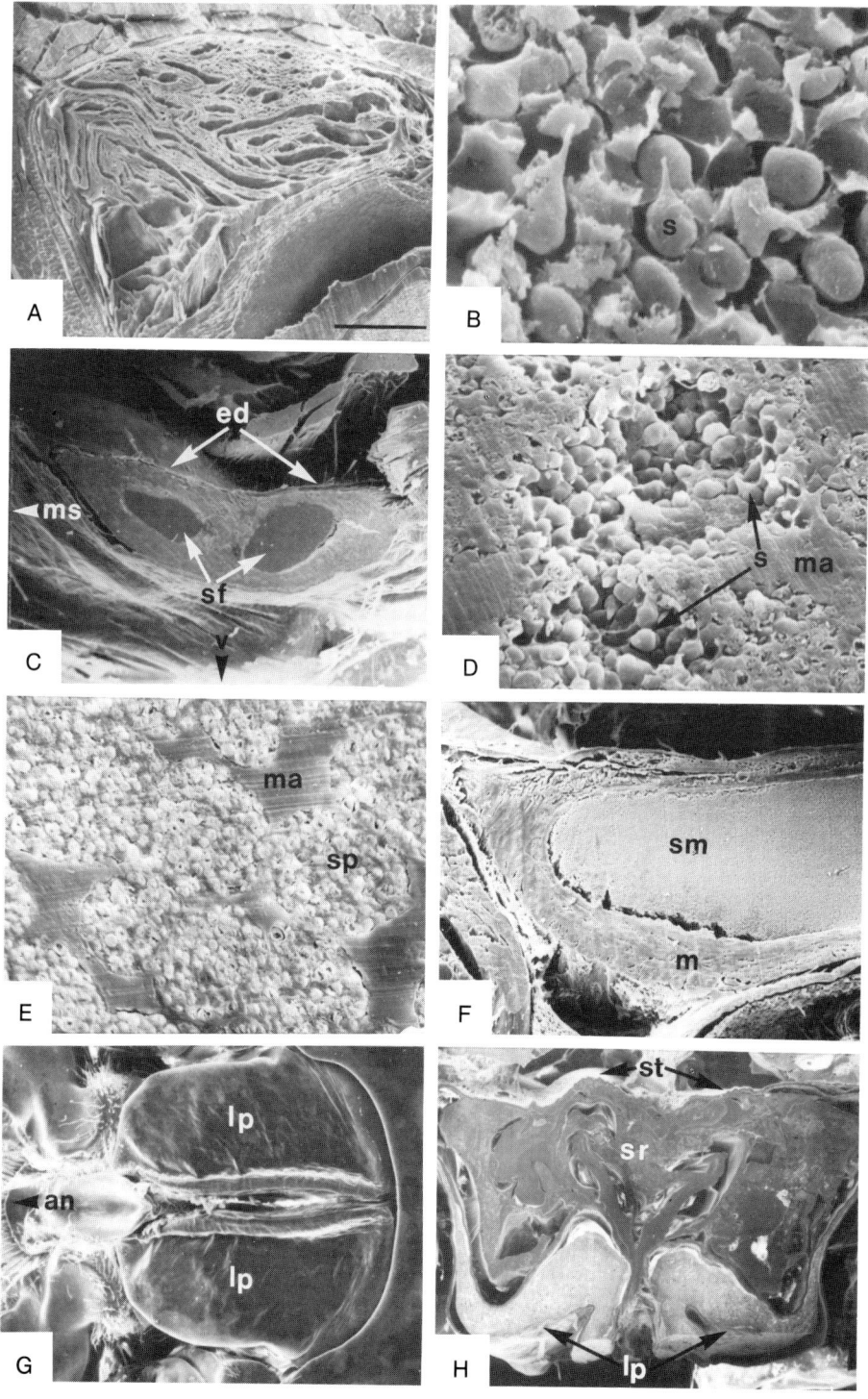

riorly, sections show that the posterior folds coalesce into a single sperm-filled structure (figs. 11.3H, 11.6B) surrounded by layers of sperm-free material. The appendage is found mainly in a chamber that occupies most of the anterior ejaculatory duct. In *P. aztecus*, the posteriormost part of the appendage material connects to the main body proximally at the level of the gonopore (fig. 11.5). A section through the appendage material (fig. 11.6A) shows it to be composed of a multitude of anastomosing layers that apparently undergo the delamination reaction described above. Chow et al. (in press) regard the appendage chamber to be a "glue duct," comparable to the adhesive glands in white shrimp (fig. 11.2). A comparison of the gland secretions of *P. setiferus* (fig. 11.3A, B) with the material making up the appendage in *P. aztecus* and *P. duorarum* (fig. 11.6A) shows that, at the very least, their mode of formation is quite different in the two types (open versus closed thelycum) of *Penaeus* spp.

The above observations, as well as those of Tirmizi (1958) on *Penaeus japonicus*, Malek and Bawab (1974a,b) on *P. kerathurus*, and Champion (1987) on *P. indicus*, indicate a reduced level of complexity in the insemination morphology of non–*Litopenaeus Penaeus* species. There are fewer structures and materials composing the spermatophores, and there is obviously less secretory activity and morphological complexity in the ejaculatory duct of species such as *P. aztecus* and *P. duorarum* than in species with externally attached spermatophores.

The genus *Trachypenaeus* shows reduction of spermatophore complexity in a different manner. The emitted spermatophores consist of a short cord of sperm and surrounding matrix plus a short block of sperm-free material. The ejaculatory duct of *T. similis* (figs. 11.1B, 11.7) consists, posteriorly, of a large glandular pocket in which the sperm-free material of the spermatophore is secreted. Anteriorly, the expanded distal end of the vas deferens lies above the glandular pocket (fig. 11.1B), but their lumens are confluent, forming the anterior portion of the ejaculatory duct. Thus a posterior section of the duct shows glandular epithelium (under a muscular coat) surrounding a large mass of sperm-free material (fig. 11.6C). More anteriorly, the packets of sperm, characteristic of *Trachypenaeus* (Figs. 11.6D, E), surround the anterior part of the sperm-free mass (fig. 11.7). At the gonopore, the ejaculatory duct is filled with sperm packets. Thus, when the musculature of the duct contracts, the sperm packets leave the gonopore first, followed by a cord of sperm-free material (personal observation).

FIGURE 11.6. A. Section through part of appendage chamber, *Penaeus duorarum* ejaculatory duct, from fig. 11.3H. B. Sperm cells in sperm mass in fig. 11.3H. C. Transverse section, posterior right ejaculatory duct, *Trachypenaeus similis*. D. Sperm cells in surrounding matrix from anterior part of *T. similis* ejaculatory duct. E. Sperm in packets surrounded by matrix from vas deferens of *T. similis*. F. Proximal half, left ejaculatory duct, *Sicyonia brevirostris*, showing sperm mass surrounded by duct's muscular sheath. G. Thelycum of *P. aztecus*, ventral view. H. Transverse section through posterior part of *P. aztecus* seminal receptacle, posterior view, showing seminal receptacle filled with the compound (twin) spermatophore. an, anterior; ed, ejaculatory duct; lp, lateral plate; m, muscle; ma, matrix; ms, mesial; s, sperm cell; sf, sperm-free spermatophore material; sm, sperm mass; sp, sperm packet; sr, seminal receptacle; st, sternum; v, ventral. Scale bar in A = 330 μ in A, 5 μ in B, 290 μ in C, 12 μ in D, 26 μ in E, 330 μ in F, 1.2 mm in G, 1.3 mm in H.

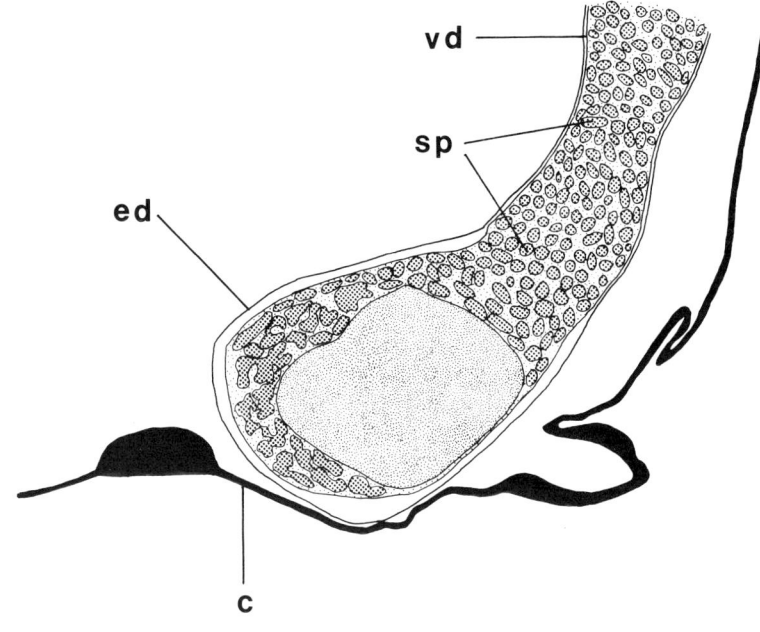

FIGURE 11.7. Transverse section, right ejaculatory duct of *Trachypenaeus similis*. Vas deferens (vd) marks proximal end of the duct. c, cuticle; ed, ejaculatory duct; sp, sperm packet; stippled area shows sperm-free material of the spermatophore. Scale bar = 1.0 mm.

The extreme in reduced spermatopore complexity is seen in the genus *Sicyonia*. The vas deferens and terminal ejaculatory duct are filled with the same material, i.e., sperm mixed in seminal matrix (figs. 11.6F, 11.8). There are no sperm-free substances, wings, or appendages like those found in so many penaeoid spermatophores.

THELYCA, SEMINAL RECEPTACLES, AND SPERMATOPHORE DISPOSITION

The term *thelycum*, which has been variously used in the literature, has been defined by Bauer (1986) as any external modifications of the female's posterior (somites 12–14) thoracic sternites and/or coxae that are related to sperm transfer and storage. Penaeoid females with externally deposited spermatophores are said to have *open* thelyca, and modifications of the posterior coxae and sternites to which the spermatophores attach compose the thelycum. Open thelyca are characteristic of the families Aristeidae, Solenoceridae, and Benthesicymidae and the penaeid subgenus *Litopenaeus*. Because of their utility in taxonomy, thelycal features have been described and illustrated for several species by authors such as Kubo (1949), Burkenroad (1934, 1936), and Pérez Farfante (1969, 1975, 1977, 1988). A *closed* thelycum refers to sternal plates that may enclose a noninvaginated seminal or sperm receptacle (spermatheca), cover a space that

Penaeoid Shrimp Sperm Transfer and Storage Structures 193

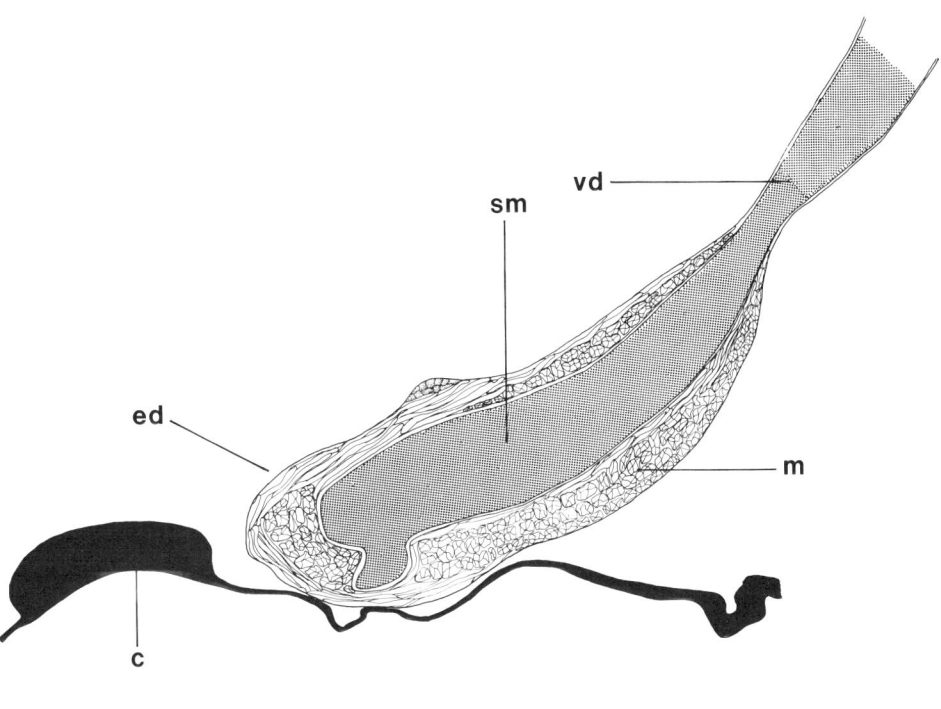

FIGURE 11.8. Transverse section, right ejaculatory duct of *Sicyonia brevirostris*. Vas deferens marks proximal end of duct. c, cuticle; ed, ejaculatory duct; m, muscle; sm, sperm mass; vd, vas deferens. Scale bar = 1.0 mm.

leads to spermathecal openings, or form an external shield guarding the spermathecal openings (Bauer 1986). *Seminal receptacle* will be used for any enclosed space in which spermatophores with sperm or spermatophoric masses are stored (slightly different from Bauer 1986). In the Penaeoidea, seminal receptacles can be uninvaginated and unpaired, i.e., a median pocket behind thelycal plates, or paired sacs or tubes invaginated into the cephalothoracic cavity.

In most members of the Penaeidae and in the Sicyoniidae, insemination is internalized in the sense that sperm are stored after transfer in some sort of seminal receptacle, associated in various ways with a closed thelycum. In *Penaeus aztecus*, the twin spermatophores emitted by the male are stored in an unpaired space below the sternum and above the thelycum plates, doorlike evaginations of the lateral sternal wall that meet at the midline (fig. 11.6G). The transferred pair of spermatophores completely fills the unpaired seminal receptacle (fig. 11.6H). The same spermatophore elements found in the male ejaculatory duct can be seen in the transferred spermatophores, e.g., the sperm mass and the sperm-free layers surrounding it (figs. 11.6H, 11.9, 11.10A). It is unclear if the spermatophore appendage material is completely transferred (also see Champion 1987; Chow et al. in press). The function of the appendage delamination reaction is unknown, although various hypotheses that might be suggested include (1) formation of an impermeable protective barrier around the sperm

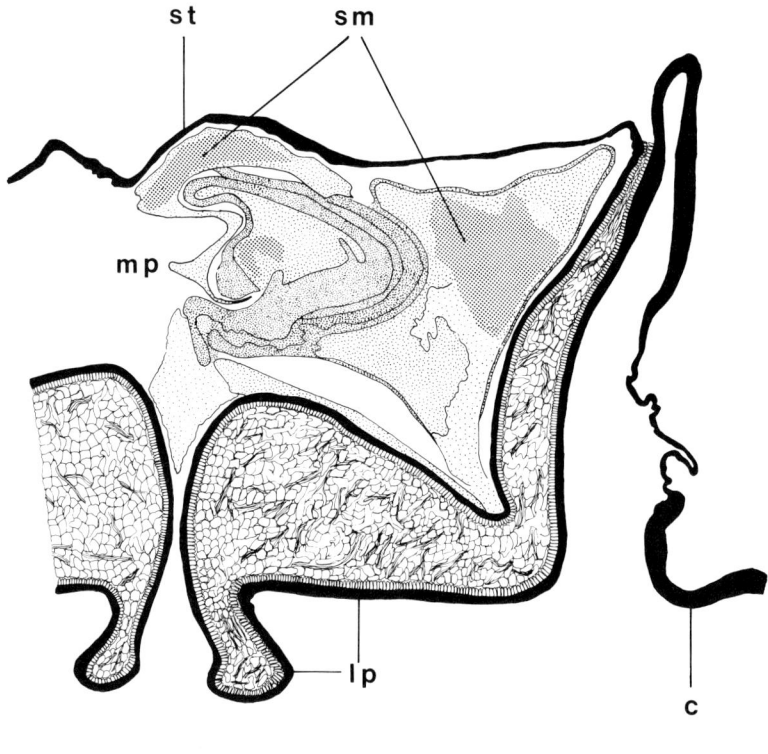

FIGURE 11.9. Transverse section, middle and right side, posterior part of the seminal receptacle, *Penaeus aztecus*. c, cuticle; lp, thelycum lateral plate; mp, median pocket; sm, sperm mass; st, sternum; various sperm-free spermatophore materials are shown stippled. Scale bar = 2.0 mm.

FIGURE 11.10. A. Part of one spermatophore in seminal receptacle of *Penaeus aztecus* (see fig. 11.6H for orientation). B. Thelycum of *Trachypenaeus similis*, ventral view, showing spermatophore material protruding from the median pocket. C. Dorsal view of *T. similis* seminal receptacle inside the cephalothorax. D. External apertures to the seminal receptacles, thelycum of *Sicyonia brevirostris*. E. Transverse section through sternite XIV, posterior view, showing sperm masses in the seminal receptacles of *S. brevirostris*. F. Magnification of sperm mass in seminal receptacle from E. G. Cincinnuli from the mesial borders of *P. setiferus* petasmal endopods (see also fig. 11.13B). H. Flexible cuticle in unstretched condition from medial lobule of *Solenocera vioscai* petasma. an, anterior; ap, aperture; cn, cincinnulum; lp, lateral plate; mpl, median plate; pe, petasmal endopod; sf, sperm-free spermatophore material; sm, sperm mass; sr, seminal receptacle; st, sternum; tp, thelycum plate. Scale bar in A = 590 μ in A, 370 μ in B, 1.2 mm in C, 830 μ in D, 1.0 mm in E, 490 μ in F, 22 μ in G, 39 μ in H.

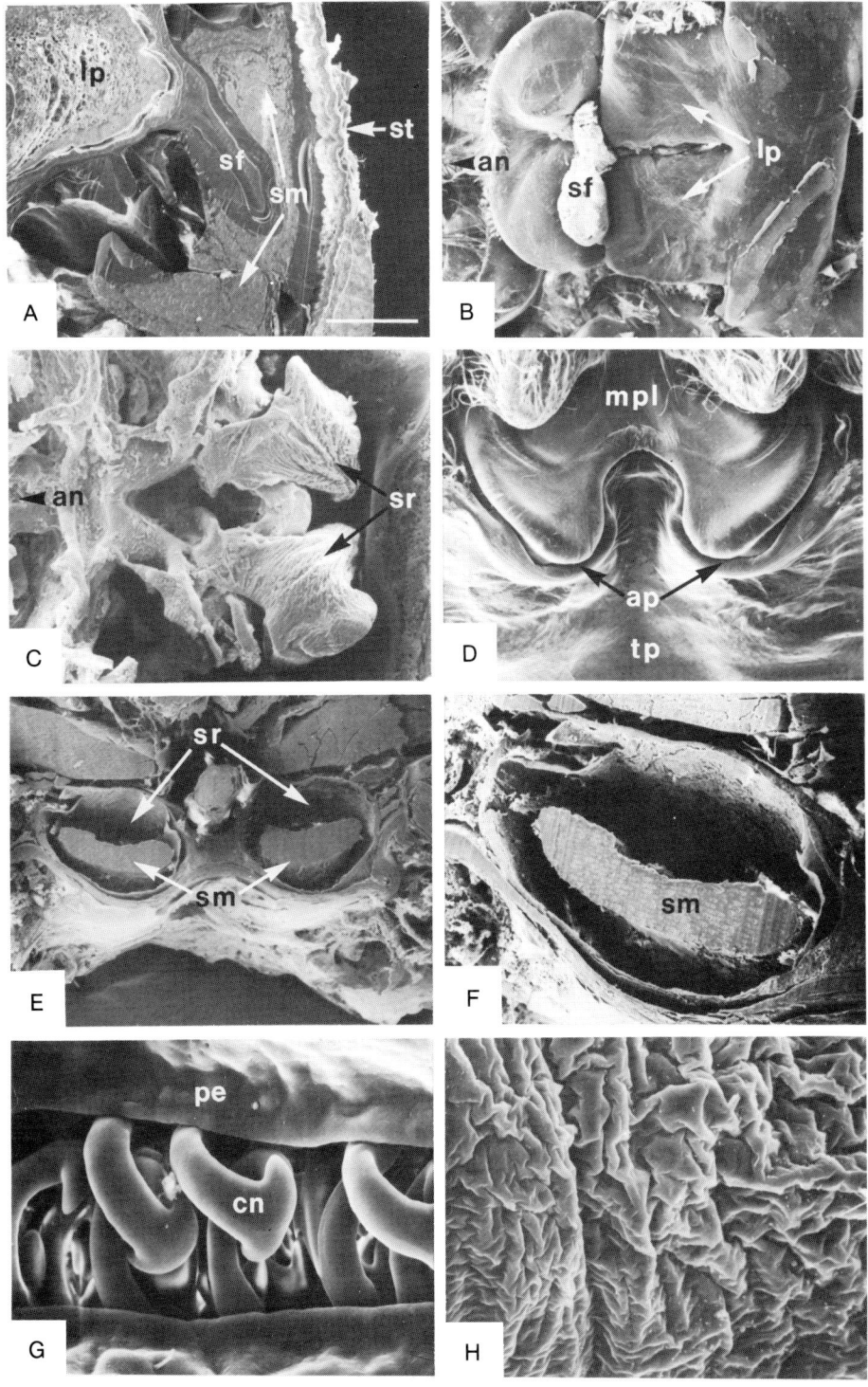

during storage, (2) a thelycum seal that prevents replacement of the spermatophores and insemination by another male, or (3) involvement with positioning or firmly lodging the spermatophores within the receptacle. Further investigation is clearly required on this point.

Further internalization in sperm storage is found in *Trachypenaeus similis*. The closed thelycum (figs. 11.10B, 11.11) has twin lateral plates that enclose a median space under the sternum. However, unlike *Penaeus* closed thelycum species, the sperm (and surrounding matrix) is stored in seminal receptacles (figs. 11.10c, 11.11; see Pérez Farfante 1971a for *T. fuscina*) that are paired invaginations of the exoskeleton into the cephalothoracic cavity. The median space behind the thelycal plates, homologous with that of *P. aztecus*, contains no sperm, only the sperm-free material that originated in the posterior ejaculatory duct of the male. An excess of this material frequently can be seen externally on the thelycal plates (fig. 11.10B).

Complete internalization of sperm storage occurs in *Sicyonia*. As in *Trachypenaeus*, the sperm is stored in seminal receptacles that are paired cuticular invaginations (see Bauer 1986, fig. 1G,H; Pérez Farfante 1985). The posterior portion of the thelycum consists of the single plate (sternite 14) with a median depression, i.e., there are no lateral plates that enclose a median pocket of any kind (figs. 11.10D, 11.12). Rather, on each side there is a slitlike aperture (between the posterolateral median plate and thelycal plate) into a small canal that leads to a smaller anterior and larger posterior seminal receptacle (figs. 11.10 D–F, 11.12; Bauer 1986: fig. 1G, H; Pérez Farfante 1985). The "spermatophore," without sperm-free materials of any kind, is attached internally within the

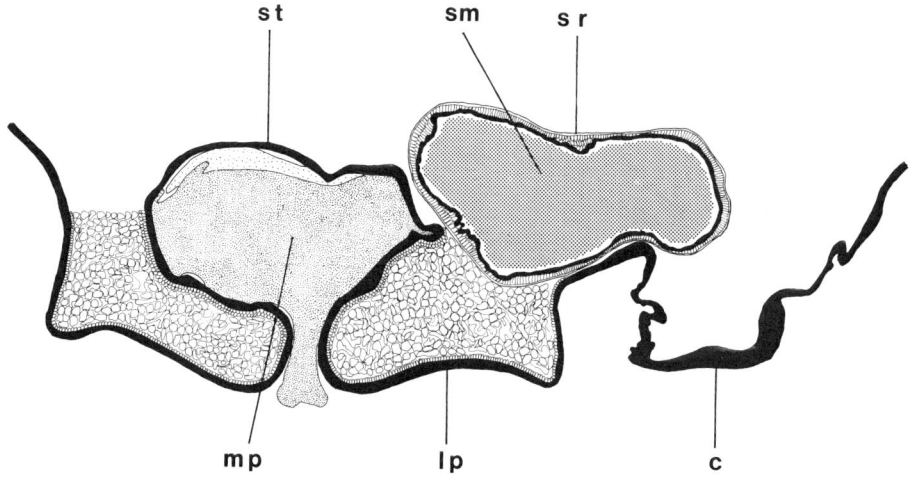

FIGURE 11.11. Transverse section through thelycum and right posterior seminal receptacle in *Trachypenaeus similis*. c, cuticle; lp, thelycum lateral plate; mp, median pocket (filled with sperm-free spermatophore material, stippled); sm, sperm mass; sr, seminal receptacle; st, sternum. Scale bar = 1.0 mm.

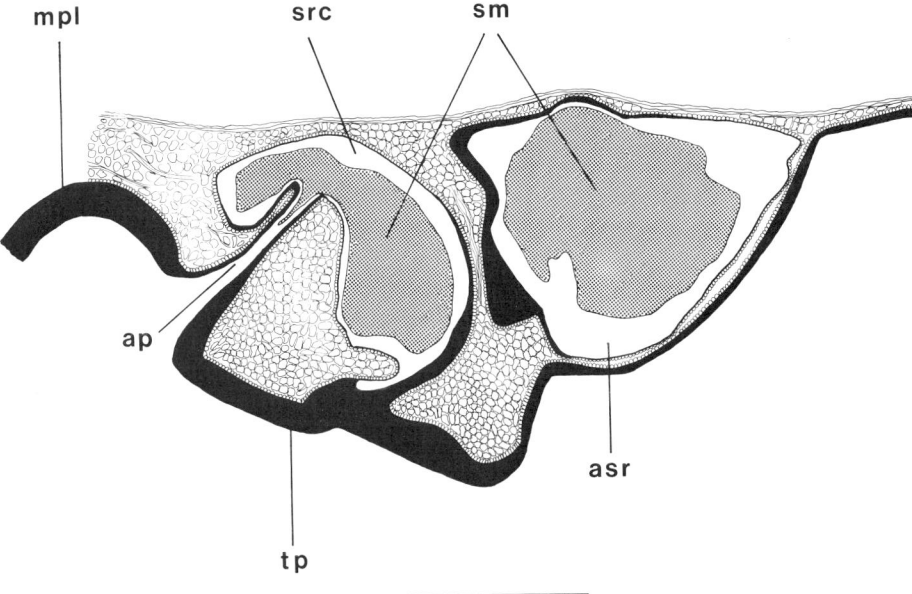

FIGURE 11.12. Transverse (oblique) section through right side of thelycum area in *Sicyonia brevirostris*. ap, aperture; asr, anterior seminal receptacle; mpl, median plate; sm, sperm mass; src, seminal receptacle canal; tp, thelycum plate. Scale bar = 1.0 mm.

seminal receptacles around the complex topography of the aperture (fig. 11.12) so that the spermatophoric mass (sperm and matrix) appears suspended within the seminal receptacles (fig. 11.10E, F).

Thus it can be seen that, with increased internalization of sperm storage, transferred spermatophores become less complex, with a reduction in the number and amount of sperm-free materials retained in the female.

PETASMATA

The petasma is a complex structure formed from the joined endopods of the first pleopods in male penaeoid shrimps. As with the structure of female genitalia, male ejaculatory ducts, and spermatophores, there is considerable variation in petasma structure in penaeoids. (The reader is referred to authors such as Burkenroad, Kubo, and especially Pérez-Farfante in "Literature Cited" for detailed descriptions and the complex terminology of the petasma.) In this section, some characteristics of penaeoid petasmata that may have both functional and phylogenetic significance will be discussed.

Two features that may have special importance in understanding the evolution of the petasma are its degree of openness and flexibility. Each half (modified endopod) of the petasma is composed of a median lobe and a lateral lobe. The median lobes of the complete petasma are dorsomesially joined, in adult males, by small hooks (cincinnuli) (fig. 11.10G). The lateral lobes are turned ventrally (posteriorly) to some degree in many species. Thus in some species the free edges of the lateral lobes (= ventral costae) may nearly meet midventrally (fig. 11.13 C, D). The "openness" of the petasma might be defined both by the degree to which the ventral costae approach each other along the midline and the degree to which they can be stretched apart when the appendage basipods (to which the lateral lobes articulate, fig. 11.13B,C,F) move laterally. Openness of the petasma depends, in part, on the flexibility of its various lobes. In most species, at least some part of the petasma is composed of thin, uncornified, relatively uncalcified, and thus flexible cuticle.

In the families Aristeidae and Solenoceridae, the petasma can be considered open. The petasmata of *Aristeus* (personal observation), *Aristaeomorpha*, and *Plesiopenaeus* (Pérez Farfante 1988) are completely open, appearing as flattened plates, with the venral costae not turned ventrally. In *Solenocera vioscai*, the ventral costae are somewhat turned ventrally and the median lobes are quite thin and flexible throughout most or all of their length (figs. 11.13A, 11.10H). The distance between the ventral costae can be increased and decreased (opened), but even when the ventral costae are nearest each other, the petasma is still very open ventrally (posteriorly). The petasma of *Penaeus (Litopenaeus) setiferus* (fig. 11.13B) can be closed in the sense that the ventral costae nearly meet at the midline, but the petasma is very flexible and can be opened or stretched apart greatly (see Pérez Farfante 1969). In *P. aztecus*, the petasma is similar to that of *P. setiferus* in general form and can be opened widely due to its flexibility. However, the *P. aztecus* petasma is more closed in that the internal space of the unstretched petasma is more compressed and less spacious than that of the podlike petasma of *P. setiferus*. A further stage of petasma "closure" is found in *Trachypenaeus*, in which the ventrolateral lobules are reflexed back dorsally against the rest of the petasma (fig. 11.13c) and are hardened (cornified, calcified) structures. Even though these lobules can be stretched apart somewhat, the petasma can never be opened as it can, for example, in *P. setiferus*. The extreme in petasma "closure" is in *Sicyonia* (fig. 11.13D), in which the petasma has virtually no flexibility. The rigid ventrolateral lobules almost meet under (posterior to) the rest of the petasma (fig. 11.13D), nearly enclosing a space or cavity within the petasma.

Another important petasma feature is presence, absence, or degree of development of distal horns or spouts. The following examples illustrate some of the variations found in the group. In *Solenocera* and *Penaeus* examined, the petasma

FIGURE 11.13. A. Petasma, *Solenocera vioscai*, dorsal (anterior) view. B. Petasma, *Penaeus setiferus*, dorsal view. Arrow on median lobe indicates line of petasmal endopod attachment (by cincinnuli, see fig. 11.10G). C. Petasma, *Trachypenaeus similis*, ventral (posterior) view *in situ* (pleopodal exopods removed). D. Petasma, *Sicyonia brevirostris*, ventral (posterior) view. E. Dorsal (anterior) view of left lateral horn, *T. similis* petasma. F. Dorsal view of articulation between left petasmal endopod and basipod of first pleopod, *P. setiferus*. bp, basipod; ch, channel; dp, distal projection of dorsolateral lobule; lh, lateral horn; ll, lateral lobe; ml, median lobe; pe, petasmal endopod; vl, ventrolateral lobule. Scale bar in A = 1.3 mm in A and B, 1.0 mm in C, 950 μ in D, 300 μ in E, 540 μ in F.

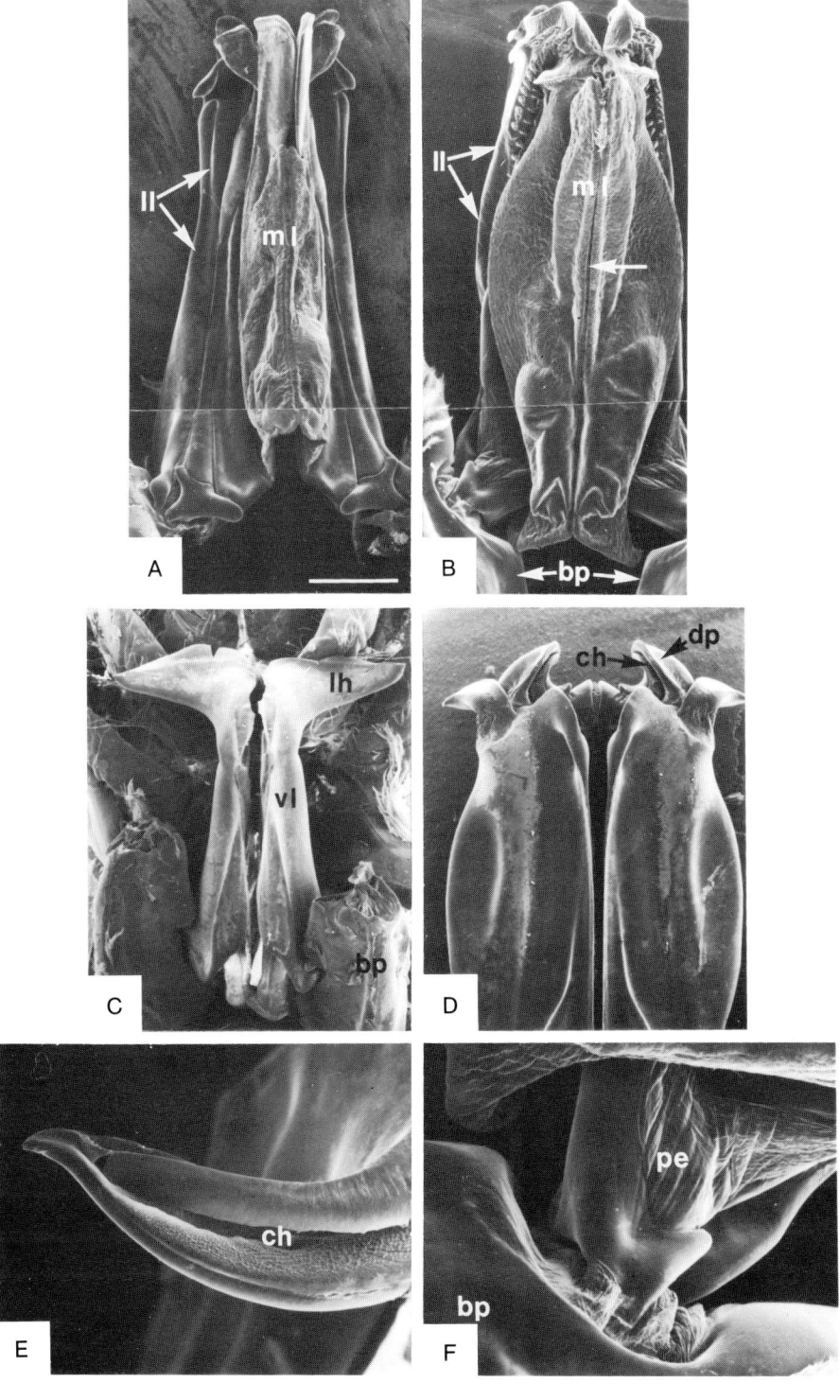

may show some complex folding distally (fig. 11.13A, B), but there are no horns or spoutlike projections. In *Trachypenaeus*, the ventrolateral lobule is produced distally into long horns with dorsal channels and ending in cornified points (fig. 11.13C,E; Pérez Farfante 1971a). In *Sicyonia*, both the dorsolateral and ventrolateral lobules terminate in projections, with the dorsolateral ones having grooves or channels (fig. 11.13D; Bauer 1986: fig. 1B–C), resulting in spoutlike structures.

HYPOTHESES ON PETASMA FUNCTION

The most reasonable assumption about the petasmata of male penaeoid shrimps is that they function in insemination, either directly or indirectly. These complex structures, so variable among the penaeoids, are located near the male gonopores where spermatophores are emitted. In those penaeoids in which mating has been observed, the anterior pleopods of the male are in close proximity to the female thelycum. The appendix masculina (and, in solenocerids, the appendix interna), a smaller, also morphologically complex structure on the endopod of the second pleopod, may also act with the petasma in sperm transfer and copulation. However, at this time there is so little evidence on its role that speculations on its function are even more premature than those on petasmata.

Burkenroad (1934) was a strong proponent of the view that the petasma is directly involved in handling and transferring emitted spermatophores to the female. He called attention to the fact that twin spermatophores of *Penaeus setiferus* fit into the ventral space of the petasma quite well. Presumably, the spermatophores would be pressed onto the female thelycum by the petasma during copulation. How the spermatophores might be fitted into the petasma is not known. The ventral (posterior) side of the petasma, which can be spread open, does not face the gonopore. However, the petasma is well articulated at its base (fig. 11.13F) so that its proximal end might swing anteriorly to permit loading from the gonopores. Mating partners would have to copulate ventral surface to ventral surface, facing the same direction. This has been observed in *P. vannamei* (Yano et al. 1988), a species with insemination morphology similar to that of *P. setiferus*. However, as Pérez Farfante (1975) has pointed out, the compound spermatophore is emitted with the same orientation to the male body as that found on the inseminated female, i.e., at some point during copulation, the spermatophore complex has to make a 180° rotation given the mating position observed in *P. vannamei*. The paradox of the compound spermatophore rotation and petasma loading (if it occurs) in *Penaeus* species in which spermatophores are deposited externally on the female (*Litopenaeus* spp.) can only be resolved by experimental work on copulation.

It should be noted briefly here that in all species with a closed thelycum and seminal receptacle(s) mating is known to occur or is thought to occur soon after molting. Any spermatophore material stored in the seminal receptacles is cast off since the receptacle is formed or lined with exoskeleton, which is molted (e.g., Pérez Farfante 1969; personal observation on *Trachypenaeus*, *Xiphopenaeus*, and *Sicyonia* spp.). Just after the molt, the thelycal plates and slits are soft, a presumed advantage in opening and subsequently filling of the seminal receptacles during copulation. However, mating has been observed to take place several days after molting in *Sicyonia ingentis* (W. H. Clark, Jr., personal communication). In *S. parri* and *S. laevigata*, mating takes place within one day after molting (personal observation).

Burkenroad (1934) proposed that the petasma of *Penaeus aztecus* was modified for accepting emitted spermatophores and for entering the seminal receptacle of the female through the median slit of the thelycum. Burkenroad also proposed that the petasmata of *Trachypenaeus* species served as syringelike devices in which spermatophoric material flowed from the gonopores into the female seminal receptacles via the channeled lateral horns (fig. 11.13c,E). The rigid, semitubular petasma of *Sicyonia*, equipped with terminal spoutlike projections (fig. 11.13D), might similarly function as an injection device during sperm transfer.

Hypotheses on petasma function, proposed on the basis of morphology (Burkenroad 1934), must be tested with experiments. In lieu of experiments, observations on male and female positions during copulation can help to support or reject a given hypothesis. Mating was observed and figured in *Sicyonia carinata* by Palombi (1939), and I have made observations (including video recording) of mating in *S. parri* and *S. laevigata*. In these latter two species, seminal receptacles of intermolt females invariably contain sperm masses (fig. 11.14A). After molting, the receptacles of captive, isolated females are empty (fig. 11.14B), and molted receptacles contain sperm masses, confirming that females must be reinseminated after a molt. When such newly molted females are exposed to males in laboratory aquaria, mating and insemination take place. Laboratory inseminations are apparently "normal," since both seminal receptacles are usually filled (fig. 11.14c,D). During mating, the male reacts strongly to the female upon contact. The male will push the female abdomen upward with his rostrum and anterior cephalothorax (figured in Palombi 1939). When the female abdomen is exposed, the male dips below her, and the copulatory position is at right angles, with the ventral cephalothoracic/abdominal junction of the male opposing that of the female. In *S. parri* and *S. laevigata*, there are always two copulations per mating, one from each side. After the second copulation, the female usually rejects further mating attempts by the male.

The mating position observed indicated that a symmetrically bilateral injection of sperm mass via the petasma spouts into the seminal receptacles was unlikely. In order to determine the pattern of seminal receptacle filling, interruption of mating was done with several mating pairs after copulation from one side of the female was completed. Only one seminal receptacle was filled in such matings, and the receptacle filled was on that side of the female on which the male was copulating (fig. 11.14E,F). These results do not support a hypothesis of simultaneous injection of sperm via the petasma. Although unlikely, injection could occur through only one of the distal spouts. Alternative hypotheses that are more in accord with these results are that (1) the petasma projections pry open the female receptacle apertures, with sperm masses flowing directly from one male gonopore into one female receptacle, or that (2) the petasma is not directly (mechanically) involved in sperm transfer.

Brinton (1978) proposed that the petasmata of euphausiids might not be involved in the mechanics of spermatophore transfer, a common assumption

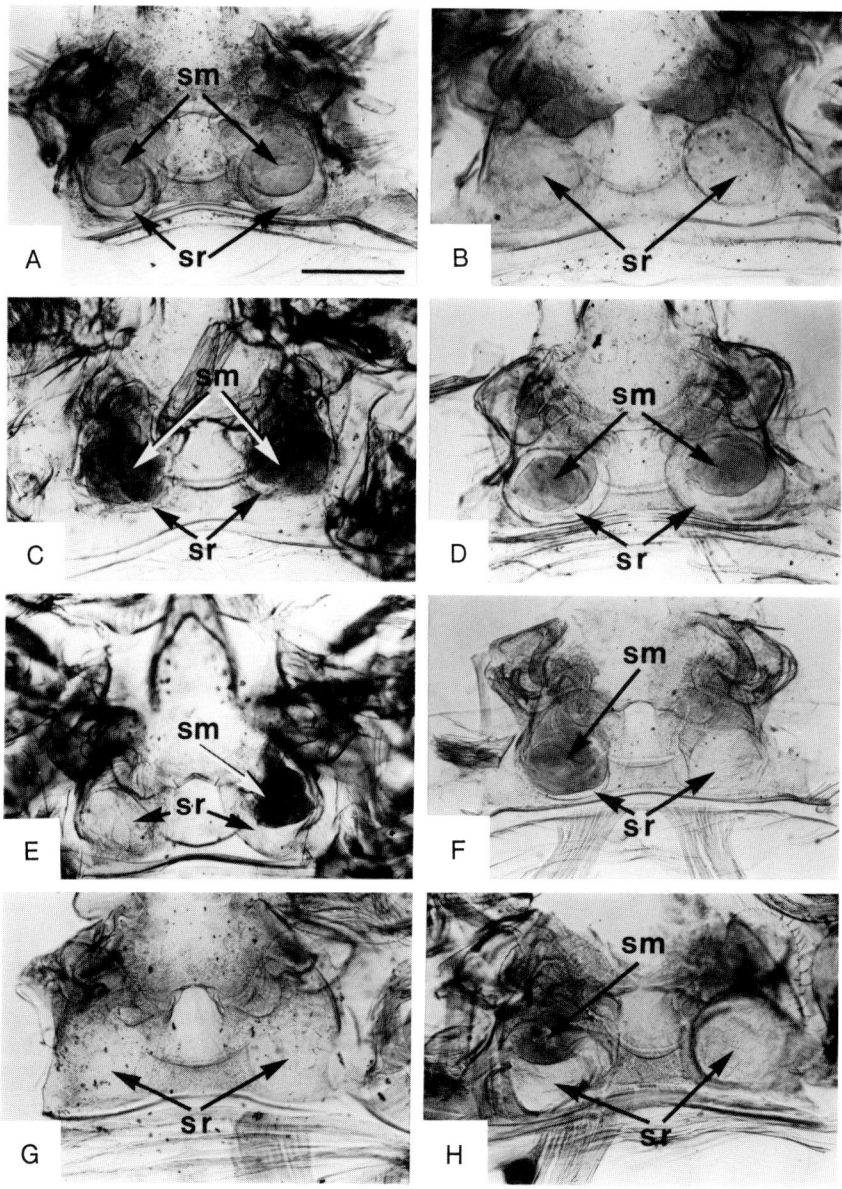

FIGURE 11.14. Pattern of female seminal receptacle filling in matings of *Sicyonia parri*. See text for explanation of mating observations and experiments. Shown are seminal receptacles dissected from: A. Female collected from wild population. B. Newly molted captive female. C–D. Females normally (copulation from each side) mated in the laboratory. E–F. Females in which mating was terminated after male copulated from one side only. G. A female after copulation from one side only (no sperm transferred). H. A female normally mated (copulation from each side), sperm transferred (partially) to only one receptacle. sr, seminal receptacle; sm, sperm mass. Scale bar in A = 760 μ in A–H.

that Pérez Farfante (1982) also questioned for penaeoids. Brinton speculated that the petasma might be sensory or stimulatory in function. Although the euphausiid petasma is not strictly homologous to the penaeoid petasma (Burkenroad 1963), the idea that the petasma might be a male sensory appendage or one used to stimulate the female in copulation is a reasonable alternative hypothesis to investigate.

Eberhard (1985) documented that males of many animal species have genitalia that appear much more complex than mechanically necessary to carry out insemination. He favored the hypothesis that complex genitalia are often courtship devices that come into play during actual copulation. Females may choose and discriminate among males on the basis of such genitalic courtship devices, leading to the evolution of complex genitalia that may have no special mechanical advantage during copulation or that are not indicative of overall male fitness. The evolution of highly developed, often inadaptive (in terms of survival) courtship plumage of many male birds can be considered analogous to the evolution of complex genitalia as courtship devices. One piece of supporting evidence for this hypothesis in penaeoids comes from mating observations in *Sicyonia* described above. In some matings, males appeared to copulate normally with females, but examination of the female's seminal receptacles revealed that no spermatophoric material had been deposited in a particular copulation. In nature, other males could mate and fill uninseminated receptacles. If acceptance or rejection of a particular male's spermatophore is mediated by female choice during copulation, based on some characteristic of the male petasma, then Eberhard's hypothesis could account for the complex petasmata of *Sicyonia* and other penaeoid genera.

MORPHOCLINES IN INSEMINATION MORPHOLOGY AND PHYLOGENETIC SIGNIFICANCE

As with any study in functional morphology, it is important to discuss insemination morphology of penaeoids in an evolutionary context. One goal of functional morphology is to hypothesize the evolutionary polarities (primitive to advanced) of variable characters. Character analysis is the most critical step in determining the phylogeny, or evolutionary history, of any group of organisms. In character analysis, one must try to determine whether similarities among taxa reflect homology (common ancestry) or homoplasy (e.g., parallelism, character reversal). In this section, clines in penaeoid insemination morphology will be outlined and hypotheses on the direction of evolution in these clines will be proposed.

One extreme in insemination morphology is represented by species characterized by an open thelycum (no seminal receptacle), complex external spermatophores, and open, flexible petasmata. Among those examples within this grouping discussed in previous sections, the aristeids *Aristeus* and *Aristaeomorpha* show perhaps the most open petasma, followed by the solenocerid *Solenocera* and, in the Penaeidae, the subgenus *Litopenaeus*. The first step towards internalization of sperm storage is found in *Penaeus* spp. other than *Litopenaeus*. In these species, the spermatophores are placed in a pocket behind a closed thelycum, but this seminal receptacle is not invaginated into the cephalothoracic cavity. The petasma is more compressed, less "open" than in *Litopenaeus,* and the spermatophores are much less complex, composed of fewer materials and with fewer parts. *Trachypenaeus* is given as an example of a further increase in internalization of spermatophores, with only part of the spermatophore in a median pocket and the sperm in truly invaginated paired seminal receptacles. The spermatophores are composed of only two substances, the sperm (and seminal material) and the sperm-free "plug" material. The petasma is more closed and less flexible than in *Penaeus.* The petasma also shows elaborate distal projections or horns. Members of the Sicyoniidae show the extreme in the trends toward spermatophore internalization and petasma closure and distal elaboration. Are these different steps in spermatophore internalization (with associated changed in thelycum, spermatophore, and petasma structure) morphological grades or do they represent a morphocline with phylogenetic information? The general consensus among systematists has been that the open thelycum and open, flexible petasma end of the insemination morphocline is primitive (Burkenroad 1934, 1936; Kubo 1949). However, objective evidence for this view has not been clearly stated. Internalization of sperm storage and associated characters can be shown to be an advanced state by means of the correlated characters method (Maslin 1952; Bauer 1984). A morphocline in one group of characters is considered to have the same evolutionary polarity (primitive to advanced) as another morphocline of known polarity if the two sets of characters are highly correlated. The number of branchial characters (fewer = advanced) in penaeoid genera can be used as a morphocline of known direction with which the insemination morphocline can be compared. Thus, using the gill formulas given in Kubo (1949), the trend toward internal sperm storage can be correlated with a reduction in the total number of branchial elements (podobranchs, arthrobranchs, pleurobranchs, epipods): *Aristaeomorpha* (31), *Aristeus* (29), *Solenocera* (28), *Penaeus* (25), *Trachypenaeus* (20–22), and *Sicyonia* (20). This brief, tentative analysis supports the conclusion that external complex spermatophores attached to an open thelycum and male with open, flexible petasma is the primitive condition and that simple sperm masses in paired internally invaginated seminal receptacles behind closed thelycum and male with "closed," less flexible petasma is the derived state.

The family Benthesicymidae was not included in the above discussion because Burkenroad (1936) pointed out that the genera in this group show a parallel but nonhomologous trend in spermatophore internalization. Thus some species have open thelyca with completely external spermatophores while in others there are true invaginated seminal receptacles that hold the sperm masses. However, these receptacles are not homologous to those of the Penaeidae and Sicyoniidae because in the latter the receptacles are invaginations between somites 13 and 14, while those of the Benthesicymidae are found between somites 12 and 13. In the Benthesicymidae, males of those species with internalized sperm storage do not show the same trends in petasma structure as in the Penaeidae and Sicyoniidae. The spermatophores are always complex in the Benthesicymidae, according to Burkenroad, but the major part of the spermatophores not entering the sperm receptacles (in those species with them) are apparently soon lost after sperm transfer.

DIRECTIONS FOR FUTURE RESEARCH

This brief treatment of penaeoid shrimp insemination morphology is intended as a framework or outline in which further research can be expanded. Although much descriptive work has been done on external genitalia, detailed research must be performed on the microscopic internal anatomy and histology of the male reproductive system, with emphasis on spermatophore formation and chemical composition. The final structure and composition of the transferred, stored spermatophores in female seminal receptacles must be described and analyzed. Above all, work with live animals on spermatophore emission, mating behavior, and copulation must be performed. Experimental work along the lines of Bauer (1976) and Berg and Sandifer (1984) must be conducted in order to test hypotheses on mode of spermatophore transfer and petasma function.

Insight into more theoretical aspects of penaeoid reproductive biology, such as the evolution of sperm storage, sexual selection, male investment in reproduction, and mating strategy, await this descriptive and experimental work. For example, male time and energy investment in spermatophore formation certainly varies in penaeoids. Species of genera such as *Penaeus* must produce a massive, complex spermatophore while in *Sicyonia* species the spermatophore is simply an emitted portion of a continuous supply of material in the male tract. *Sicyonia* males can mate several times successively (Palombi 1939, personal observation) while *Penaeus* males may have to wait days before mating again (7–11 days for spermatophore regeneration in *P. monodon* [Lin & Ting 1986]). Differences in spermatophore investment must affect mating strategies in penaeoids. Much of the variation in penaeoid insemination morphology might be explained by selection pressures related to mating strategies. In order to infer what the selection pressures are or have been, the functional morphology of insemination has to be analyzed adequately by descriptive and experimental work on a variety of penaeoid species.

Acknowledgments

I gratefully acknowledge Cora Cash for her help with the manuscript, specifically for assistance in inking the figures, drawing fig. 11.1, and careful darkroom work. My thanks to Frank Truesdale for his valuable advice and assistance in collecting penaeid shrimps. My thanks to Seinen Chow for sending me the Chow et al. (in press) paper. Comments on the manuscript by Austin B. Williams and anonymous reviewers, as well as discussions with Isabel Pérez Farfante, were quite helpful and were greatly appreciated by the author. This work on penaeid insemination morphology would not have been possible without the financial support of the NOAA Sea Grant Program (Louisiana Sea Grant NA85AA-D-SG141) and Louisiana Educational Quality Grant LEQSF-1988-ENH-BS-10. This is Contribution No. 19 of the Center for Crustacean Research.

Literature Cited

Andrews, E. A. 1911. Sperm transfer in certain decapods. *Proc. U.S. Natl. Mus.* 39:419–434.
Bauer, R. T. 1976. Mating behaviour and spermatophore transfer in the shrimp *Heptacarpus pictus* (Stimpson) (Decapoda: Caridea: Hippolytidae). *J. Nat. Hist.* 10:415–440.
Bauer, R. T. 1984. Morphological trends in the genus *Heptacarpus* (Decapoda: Caridea) and their phylogenetic significance. *J. Crustacean Biol.* 4:201–225.
Bauer, R. T. 1986. Phylogenetic trends in sperm transfer and storage complexity in decapod crustaceans. *J. Crustacean Biol.* 6:313–325.
Berg, A. B. & P. A. Sandifer. 1984. Mating behavior of the grass shrimp *Palaemonetes pugio* Holthuis (Decapoda, Caridea). *J. Crustacean Biol.* 4:417–424.
Brinton, E. 1978. Observation of spermatophores attached to pleopods of preserved male euphausiids. *Crustaceana (Leiden)* 35:241–248.
Burkenroad, M. D. 1934. The Penaeidea of Louisiana, with a discussion of their world relationships. *Bull. Am. Mus. Nat. Hist.* 68:61–143.
Burkenroad, M. D. 1936. The Aristaeinae, Solenocerinae, and pelagic Penaeinae of the Bingham Oceanographic Collection. Materials for a revision of the oceanic Penaeidae. *Bull. Bingham Oceanogr. Collect. Yale Univ.* 5:1–151.
Burkenroad, M. D. 1963. The evoulation of the Eucarida (Crustacea: Eumalacostraca) in relation to the fossil record. *Tulane Stud. Geol.* 2:3–16.
Champion, H. F. B. 1987. The functional anatomy of the male reproductive system in *Penaeus indicus*. *S. Afr. J. Zool.* 22:297–307.
Chow, S., P. A. Sandifer, M. M. Dougherty & W. J. Dougherty. In press. Spermatophore formation in penaeid shrimps. *Proc. 2nd Asian Fish. Forum*.
Eberhard, W. G. 1985. *Sexual Selection and Animal Genitalia*. Cambridge, Massachusetts: Harvard University Press.
Eldred, B. 1958. Observations on the structural development and the impregnation of the pink shrimp *Penaeus duorarum* Burkenroad. *Fla. Board Conserv. Mar. Res. Lab. Tech. Ser.*, no. 23.
Hartnoll, R. G. 1969. Mating in the Brachyura. *Crustaceana (Leiden)* 16:161–181.
Heldt, J. H. 1938a. La reproduction chez les crustacés décapodes de la famille des peneides. *Ann. Inst. Oceangr. Monaco* 18:31–206.
Heldt, J. H. 1938b. De l'appareil genital des Penaeidae. Relations morphologique entre spermatophore, thelycum, et petasma. *Trav. Stat. Zool. Wimereux* 13:349–358.
King, J. E. 1948. A study of the reproductive organs of *Penaeus setiferus* (Linnaeus). *Biol. Bull. (Woods Hole)* 94:244–262.
Kubo, I. 1949. Studies on penaeids of Japanese and adjacent waters. *J. Tokyo Coll. Fish.* 36:1–467.
Lin, M. N. & Y. Y. Ting. 1986. Spermatophore transplantation and artificial fertilization in grass shrimp. *Bull. Jpn. Soc. Sci. Fish.* 52:585–589.
Malek, S. R. A. & F. M. Bawab. 1974a. The formation of the spermatophore in *Penaeus kerathurus* (Förskal, 1775) (Decapoda, Penaeidae). I. The initial formation of a sperm mass. *Crustaceana (Leiden)* 26:273–285.
Malek, S. R. A. & F. M. Bawab. 1974b. The formation of the spermatophore in *Penaeus kerathurus* (Förskal, 1775) (Decapoda, Penaeidae). II. The deposition of the main layers of the body and of the wing. *Crustaceana (Leiden)* 27:73–83.
Maslin, T. P. 1952. Morphological criteria of phyletic relationships. *Syst. Zool.* 1:49–70.
Orsi Relini, L. & L. Tunesi. 1987. The structure of the spermatophore in *Aristeus antennatus* (Risso, 1816). *Invest. Pesq.* 51 (Supl. 1):461–470.
Palombi, A. 1939. Note biologiche sui Peneidi. La fecondazione e la deposizione delle uova in *Eusicyonia carinata* (Olivi). *Boll. Zool.* 10:223–227.
Pérez Farfante, I. 1969. Western Atlantic shrimps of the genus *Penaeus*. *U.S. Fish Wild. Serv. Fish. Bull.* 67:461–591.
Pérez Farfante, I. 1971a. A key to the American Pacific shrimps of the genus *Trachypenaeus* (Decapoda: Penaeidae), with the description of a new species. *U.S. Natl. Mar. Fish. Serv. Fish. Bull.* 69:635–646.
Pérez Farfante, I. 1971b. Western Atlantic shrimps of the genus Metapenaeopsis (Crustacea, Decapoda, Penaeidae), with descriptions of three new species. *Smithson. Contrib. Zool.*, no. 79.
Pérez Farfante, I. 1975. Spermatophores and thelyca of the America white shrimps, genus *Penaeus*, subgenus *Litopenaeus*. *U.S. Nat. Mar. Fish Serv. Fish. Bull.* 73:463–486.

Pérez Farfante, I. 1977. American solenocerid shrimps of the genera *Hymenopenaeus, Haliporoides, Pleoticus, Hadropenaeus* new genus, and *Mesopenaeus* new genus. *U.S. Nat. Mar. Fish. Serv. Fish. Bull.* 75:261–346.

Pérez Farfante, I. 1982. The geminate shrimp species *Parapenaeus longirostris* and *P. politus* (Crustacea: Decapoda: Penaeoidea). *Quad. Lab. Tecn. Pesca Ancona.* 3:187–205.

Pérez Farante, I. 1985. The rock shrimp genus *Sicyonia* (Crustacea: Decapoda: Penaeoidea) in the eastern Pacific. *U.S. Natl. Mar. Fish. Serv. Fish. Bull.* 83:1–78.

Pérez Farfante, I. 1988. Illustrated key to the penaeoid shrimps of commerce in the Americas. *NOAA Tech. Rep. NMFS* no. 64.

Tirmizi, N. M. 1958. A study of some developmental stages of the thelycum and its relation to the spermatophores in the prawn *Penaeus japonicus* Bate. *Proc. Zool. Soc. Lond.* 131:231–244.

Tunesi, L. 1987. La formation du spermatophre chez *Aristaeomopha foliacea* (Risso, 1827). *Invest. Pesq.* 51 (Supl. 1):471–476.

Yano, I., R. A. Kanna, R. N. Oyama & J. A. Wyban. 1988. Mating behavior in the penaeid shrimp *Penaeus vannamei. Mar. Biol. (Berl.)* 97:171–175.

TWELVE
Functional and Evolutionary Aspects of the Sexual System in the Rhizocephala (Thecostraca: Cirripedia)

JENS T. HØEG

*Institute of Cell Biology and Anatomy,
the Zoological Institutes,
University of Copenhagen,
Denmark.*

Abstract

The sexual systems in all described Rhizocephala are reviewed and concluded to be gonochoristic. The Peltogastridae, Lernaeodiscidae, and Sacculinidae have morphologically dissimilar male and female cyprids. The male cyprids settle on a juvenile female parasite and metamorphose into trichogons, which are implanted into a pair of receptacles in the female. Failing to receive at least one male, the female parasite does not develop beyond the juvenile stage. The implanted male is retained for the entire life span of the female parasite.

In *Clistosaccus* and *Sylon*, "male" and "female" cyprids are morphologically similar and "functional" males do not metamorphose into trichogons. Instead the male cyprid uses antennular penetration to implant spermatogonia into a juvenile female, where spermatogenesis takes place in a single receptacle *(Clistosaccus)* or within the ovary *(Sylon)*.

In most remaining rhizocephalans spermatogenesis takes place in so-called spermatogenic islands, which are hollow epithelium-lined capsules that originate peripherally in the mantle but are later released into the mantle cavity.

A number of species combine characters from two or all three of these sexual systems. By comparison with the Cirripedia *sensu stricto* and with the Ascothoracida it is argued that the trichogon-receptacle system is the most plesiomorphic, while the antennular penetration system is advanced. It is furthermore argued that the male cells in species with spermatogenic islands also originate from cypris males.

The scant available information on the mechanism of sex attraction in the Rhizocephala is also reviewed.

THE IMPRESSION left by the most modern texts on the Rhizocephala is that the problems of their sexual system have been solved (e.g., Barnes 1986). However, the life cycles presented are only valid for the Sacculinidae, Peltogastridae, and Lernaeodiscidae (Yanagimachi 1961; Ritchie & Høeg 1981; Lützen 1984). These three families, which encompass the majority of species, have a gonochoristic life cycle similar to that illustrated in fig. 12.1. They are also the only rhizocephalans where the presence of an infective kentrogon stage has been demonstrated, thus constituting the Kentrogonida *sensu stricto* (Høeg & Lützen 1985).

There remains, however, a number of rhizocephalans whose sexual systems are still much debated, viz. the Clistosaccidae, Sylonidae, Chthamalophilidae, and the provisional family Akentrogonidae (Spivey 1982). The morphology of these species differs extensively among themselves and from the three above mentioned kentrogonidan families, and they have variously been classified as being hermaphrodites or parthenogenetic. Recently, however, some of them, notably the Clistosaccidae, were also shown to be gonochorists (Høeg 1982, 1985). The present paper reviews for the first time our current knowledge of the sexual biology of all Rhizocephala and presents a scenario for its evolution.

THE TRICHOGON-RECEPTACLE SYSTEM

The Juvenile Female Parasite

Morphology and Growth of the Virgin Externa. In the three kentrogonid families (the Peltogastridae, the Lernaeodiscidae, and the Sacculinidae) the juvenile parasite emerging on the abdomen of the host is a virgin female (Ichikawa & Yanagimachi 1957, 1958, 1960; Ritchie & Høeg 1981; Lützen 1984). The general shape of this recently emerged externa resembles the adult parasite, and the externa has a distinct stalk (fig. 12.1). At emergence the lumen of the mantle cavity is already formed, but it lacks an aperture or passage to the exterior (fig. 12.2). In lernaeodiscids and peltogastrids the externa passes through two preparatory molts, the last of which results in the development of an open mantle aperture (Høeg & Ritchie 1985; Lützen 1987). But the actual entrance to the mantle cavity is reduced to narrow crevices by pieces of the old exuvium remaining within or above the mantle aperture (Reinhard 1942; Høeg 1987a).

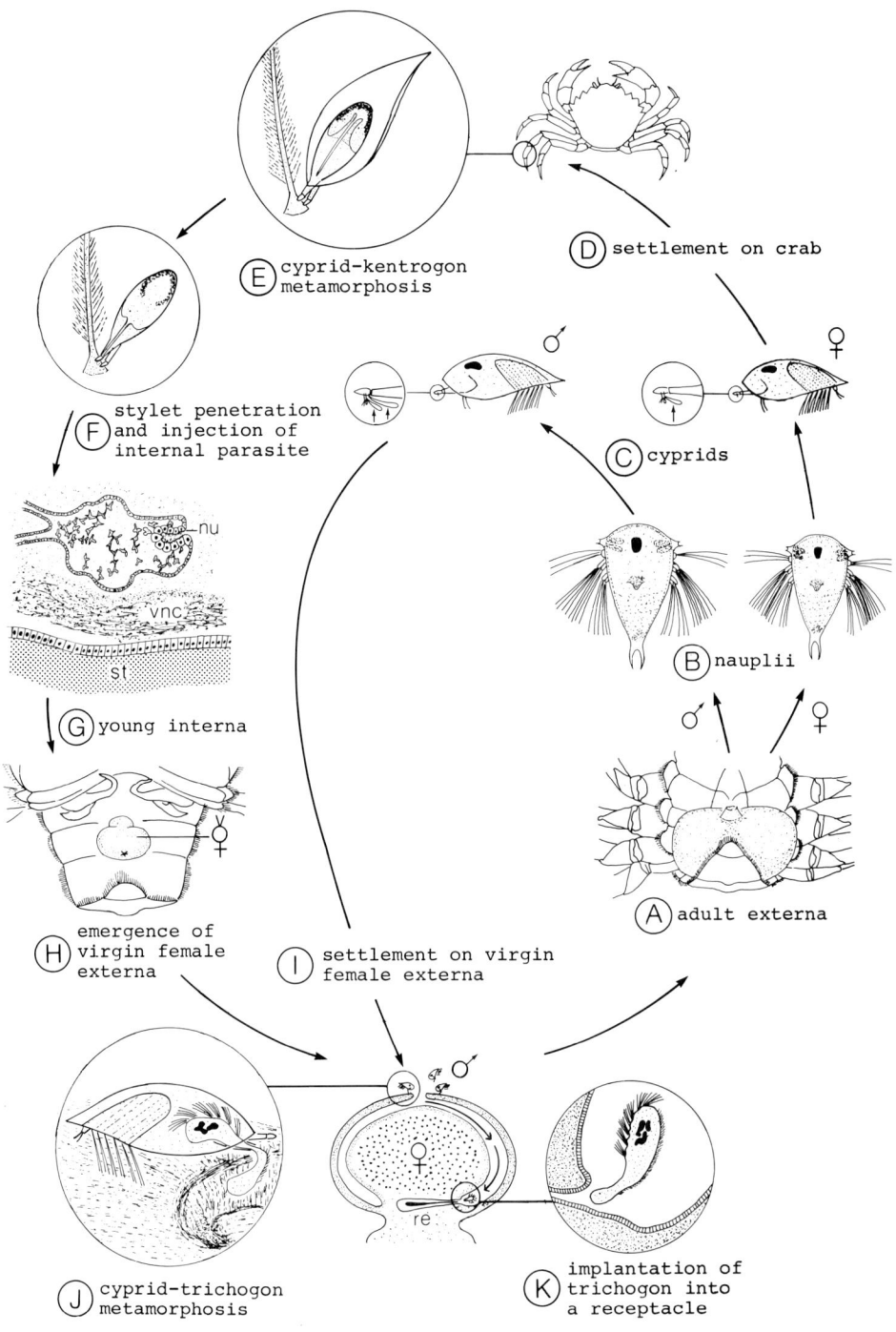

The Receptacles. All peltogastrids, lernaeodiscids, and sacculinids have two flask-shaped "seminal" receptacles located in the basal part of the visceral sac near the stalk, and these receptacles communicate with the mantle cavity through narrow canals (fig. 12.3.) The single receptacle reported in *Cyphosaccus* is actually two receptacles with independent canals ensheathed by a common layer of connective tissue (Reinhard 1958).

The receptacles are quite elaborate structures and are well known only from the Sacculinidae (Høeg 1987a). Basically they are simply tube-shaped invaginations of the general cuticle and epithelium of the mantle cavity (fig. 12.3). In the more distal part of the receptacle the narrow cuticular canal is surrounded by layers of connective tissue and circular muscles. The most proximal part of the canal, however, lacks cuticle even at the TEM level. In this so-called terminal canal the central lumen is surrounded by a low epithelium and thick layers of very large cells rich in rough endoplasmic reticulum (fig. 12.3). The receptacles in lernaeodiscids and peltogastrids are known only at the light microscope level, but there seems to have been considerable evolution in the detailed morphology of these organs (see Klepal 1987).

Settlement and Metamorphosis of Male Cyprids

Settlement on the Virgin Externa. Settlement of male cyprids on the female externa usually occurs soon after breakthrough of the mantle aperture and was first studied in detail by Ichikawa and Yanagimachi (1957, 1958, 1960). The cyprids attach within the mantle aperture (Peltogastridae and Lernaeodiscidae) or around its circumference (Sacculinidae) (fig. 12.1, 12.2, 12.4). The passage to the mantle cavity is too narrow in all three families to allow entrance of the cyprid. In fact, this was the main reason most early authors regarded the cyprids, long observed to attach around the aperture of juvenile externae, as nonfunctional males.

FIGURE 12.1. The gonochoristic life cycle of a typical kentrogonid rhizocephalan, *Sacculina carcini* Thompson (Sacculinidae) parasitizing the crab *Carcinus maenas* (L.). A. The female *S. carcini* externa is attached on the ventral side of the host's abdomen. B. Parasite larvae are released as nauplii. C. At the cyprid stage, male and female cyprids differ in size and in the number of antennular aesthetascs (circles, and see fig. 12.6). D. Female cyprids settle on crabs at the basis of plumose hairs, usually on the appendages. E. They metamorphose into a kentrogon, which penetrates into the host with a hollow stylet (F) and injects the internal parasite into the crab's hemocoel. G. After some time the young internal parasite is found in the abdomen dorsal to the ventral nerve cord (vnc); the nucleus (nc) is the anlage to the ovary and the receptacles; st, sternum of crab. H. The female parasite emerges from the crab's abdomen as a virgin externa, far removed from the original site of kentrogon penetration. I. Male cyprids settle around the mantle aperture of the virgin externa. J. They metamorphose into an unsegmented, limbless, and spine-covered larva—the trichogon—which escapes into the mantle aperture through one of the cyprid antennules. K. The trichogon migrates through the mantle cavity into one of the two receptacles (re). An implanted male produces sperm that fertilize the eggs spawned by the adult externa, and it is retained for the entire life span of the female parasite. Compiled from information in Delage 1884; Rubiliani, Turquier and Payen 1982; Lützen 1984; Walker 1985; Høeg 1987a,b.

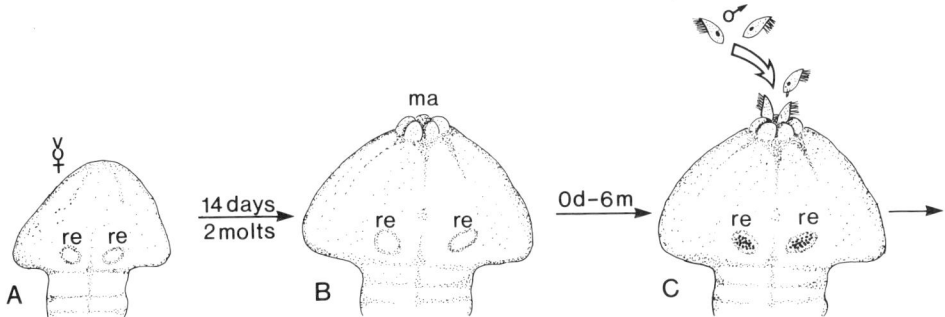

FIGURE 12.2. Sex attraction and juvenile development in the kentrogonid rhizocephalan, *Lernaeodiscus porcellanae*. A. The recently emerged virgin female externa has no mantle aperture and does not attract male cyprids, receptacles (re) empty. B. The mantle aperture is acquired at the second molt, whereafter the externa is attractive to male cyprids. C. The externa's molt cycle is now arrested until settlement of male cyprids. Settlement may occur immediately after the molt, but can be delayed for up to 6 months after acquisition of the mantle aperture; implantation of males into receptacles (re) induces resumption of the molt cycle, and the externa ceases to attract additional males within a few days.

Cyprid-Trichogon Metamorphosis. Within a few minutes after settlement, the male cyprid metamorphoses into the trichogon instar, which is homologous to the infective female kentrogon instar (see Høeg 1987a).

The trichogon (fig. 12.5) is undoubtedly the most reduced larval stage in the Crustacea, being entirely without external or internal segmentation and lacking even rudiments of appendages and internal organs. In comparison, the most highly reduced males of the Acrothoracica and Thoracica, e.g., those in *Ibla idiotica*, retain both rudimentary antennules and highly reduced testes (Batham 1945; Klepal 1987). In *Sacculina carcini* the trichogon is vermiform and covered with a cuticle only 15 um thick. This morphology allows the drastic changes in shape required during the trichogon's escape from the cyprid, passage through the mantle aperture, and implantation into the receptacle. The cuticle on the posterior half of the body is ornamented with a row of long cuticular spines (fig. 12.5). The trichogons in *Peltogaster paguri* (Peltogastridae) and *Lernaeodiscus procellanae* (Lernaeodiscidae), whose metamorphosis has not yet been closely studied, seem to resemble that in *Sacculina carcini*. The trichogon in *Peltogasterella*, however, lacks spines altogether (Veillet 1985; Høeg 1987a, b).

The Implantation of the Trichogon

Migration through the Mantle Cavity. In all species the trichogon escapes through one of the cyprid's antennules (fig. 12.4B). The distal end of the second antennular segment is encircled by a preformed abscission zone, which ruptures after cyprid settlement and leaves the antennule open-ended for the escape of the trichogon (fig. 12.6; Glenner et al. 1989).

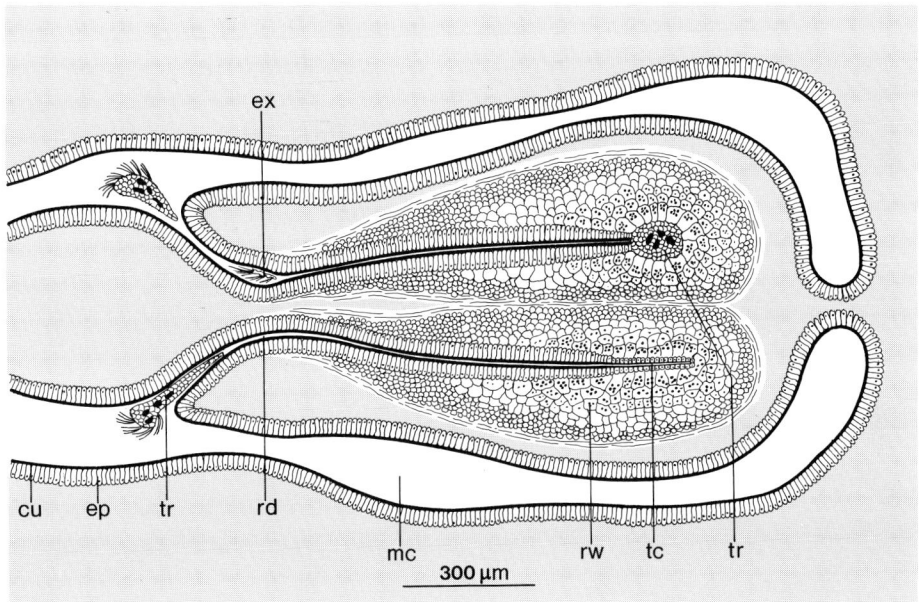

FIGURE 12.3. Diagrammatical section through the paired receptacles in the kentrogonid rhizocephalan *Sacculina carcini*. One still virgin receptacle is about to receive a trichogon (bottom). The other receptacle (top) has already received a trichogon, whose exuvium lies in the distal receptacle canal and prevents implantation of later-arriving trichogons (redrawn from Høeg 1987a). cu, cuticle; ep, epithelium; ex, exuvium of implanted trichogon; mc, mantle cavity; rd, receptacle duct; rw, receptacle wall; tc, terminal canal; tr, trichogon.

Unlike the cyprid, the trichogon is slim enough to pass through the narrow apertural crevices leading into the mantle cavity. Since the trichogon lacks muscles, it is most likely transported to the entrance of one of the two receptacles by the peristalsis of the mantle musculature, which starts with the formation of the mantle aperture. This may also explain why the mantle aperture is partially blocked, since otherwise the trichogon could quite easily be expelled.

Penetration into a Receptacle. Externae have been observed both in laboratory experiments and in field collections to carry many cyprids (up to 50), and numerous trichogons may accordingly be found within the same mantle cavity, apparently all trying to reach a receptacle. But when a trichogon migrates up through the narrow receptacle canal it sheds its cuticle, which is left behind and effectively blocks the entrance for any subsequently arriving males (fig. 12.3). This effectively limits each receptacle to receiving a single trichogon (Høeg 1987a).

An externa will develop normally even if only one receptacle receives a male, but the empty receptacle will atrophy before the parasite spawns its first brood (personal observation).

214 *Jens T. Høeg*

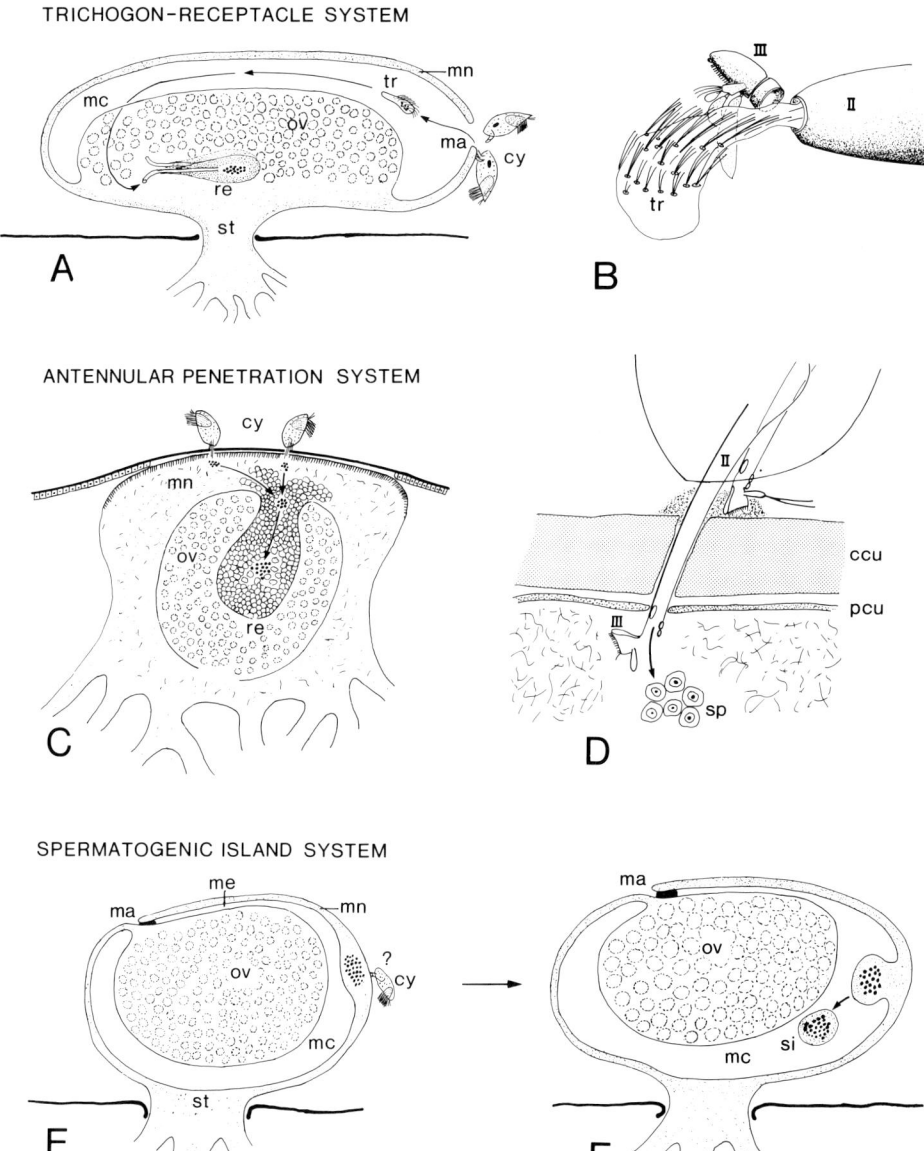

FIGURE 12.4. The three sexual systems found in extant Rhizocephala. A. In kentrogonid rhizocephalans trichogons are implanted through the open mantle aperture and into a pair of receptacles. Male cyprids (cy) settle in the mantle aperture (ma) of a virgin female externa; they metamorphose into trichogons (tr), which are implanted into the paired receptacles (re). B. Detail of A showing the escape of the trichogon (tr) through a cypris antennule; the second segment (II) has broken in a preformed abscission zone (see fig. 12.5), leaving the antennule as an open ended tube. C. In *Clistosaccus paguri* spermatogonia (small black cells) are implanted by cyprids (cy) before emergence of the female parasite from the host; the injected spermatogonia (black cells) migrate through the mantle tissue (mn) and into the single receptacle (re). D. Detail of C showing the penetration of the cypris antennule

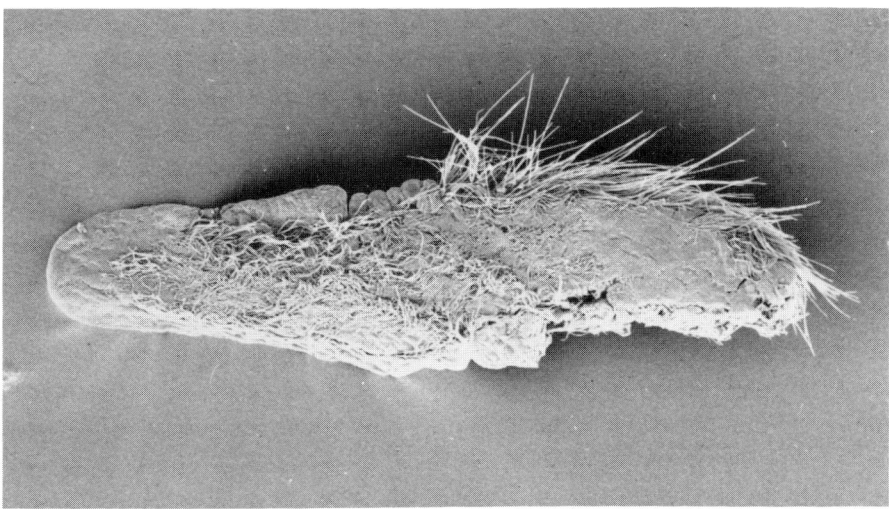

FIGURE 12.5. SEM micrograph of a *Sacculina carcini* trichogon; anterior end is left; note the large posteriorly directed spines along the posterior body part (x 560).

The Implanted Male

Spermatogenesis. Following implantation, the male begins to produce spermatozoa. The externa itself, being wholly female, cannot produce sperm, so the one or two originally implanted males remain in the parasite for its entire life span. This is in contrast to almost all Cirripedia Thoracica and Acrothoracica, in which the females or hermaphrodites generally outlive the males (Klepal 1987). Spermatogenesis in the rhizocephalan receptacle is cyclic and synchronized to oogenesis, so mature sperm is available whenever a new batch of eggs is released into the mantle cavity.

Morphological Relations Between the Externa and the Male. After implantation the male and female cells are in direct contact with no intervening cuticle, and they cannot readily be distinguished when spermiogenesis is well advanced without TEM techniques and a detailed knowledge of prior events (Høeg 1987a). This highly advanced dwarf male system prevents the male from being expelled

through crab cuticle (ccu) and underlying parasite cuticle (pcu); the third antennular segment (III) is dislodged after penetration. E. In chthamalophilid rhizocephalans spermatogonia (black cells) are first seen in the mantle (mn) of the juvenile female parasite. The mantle aperture (ma) is blocked, so males cannot enter the mantle cavity (mc) through the mesenteric canal (me). Instead it is theorized that male cyprids (cy) implant spermatogonia through the integument as in *Clistosaccus*. F. Later stage of E; evagination and liberation of the spermatogonia as spermatogenic island (si) into the mantle cavity (mc). For further explanation, see text. ccu, crab cuticle; cy, cyprid; ma, mantle aperture; mc, mantle cavity; me, mesenteric canal; mn, mantle; ov, ovary; pcu, parasite cuticle; st, stalk; tr, trichogon; re, receptacle; si, spermatogenic island; sp, spermatogonia injected by cyprid.

when the cuticle of the mantle cavity and the more distal part of the receptacle canal are molted following release of a brood of larvae. It furthermore facilitates the nourishment of the male by the female, which must occur, since some rhizocephalans may live for more than two years while retaining the originally implanted males in a functional state (Ritchie & Høeg 1981; Lützen 1987).

The Nature of the Rhizocephalan Receptacle. It follows from the description above that rhizocephalan receptacles are not ordinary seminal receptacles. Rather than merely receiving spermatozoa or their precursors, the receptacles in the Rhizocephala accommodate an entire male organism, which is effectively reduced to a "gonad" nourished and controlled by the female. Such a condition was aptly termed cryptogonochorism by Bresciani & Lützen (1972). But since the receptacles are basically a pair of cuticular infoldings, they probably evolved from much simpler structures.

Cyphosaccus and Boschmaia. Two peltogastrid genera, *Cyphosaccus* and *Boschmaia*, differ from all other kentrogonids in that the mantle aperture does not break through until after the eggs are released and fertilized in the mantle cavity (Reinhard 1958). It is unknown how males are implanted into the receptacles under these circumstances, but implantation might take place by direct impregnation of male cells through the externa integument as described in the next section.

THE ANTENNULAR PENETRATION SYSTEM

Rhizocephalans other than the Peltogastridae, Lernaeodiscidae, and Sacculinidae are known or strongly suspected to lack both a kentrogon and a trichogon (Høeg 1985b, 1990). In these species (viz. the Clistosaccidae, the Sylonidae, the Chthamalophilidae, and the provisional family Akentrogonidae *sensu* Spivey 1982) the mantle aperture forms only long after emergence of the parasite or there is no such aperture at all, and only the Clistosaccidae possess receptacles. In the absence of any open access for the entrance of external males into the juvenile parasite, these species were previously considered to be either parthenogenetic or self-fertilizing hermaphrodites (Bocquet-Védrine 1961). But recent studies indicate that most if not all of them are cryptogonochorists, in which females receive males in a highly specialized manner.

Clistosaccidae

In the sole species of this family, *Clistosaccus paguri*, the cyprid is the terminal larval instar, since neither a kentrogon nor a trichogon is formed after settlement (Høeg 1985b, 1990). It furthermore seems that male and female cyprids are at least morphologically similar (table 12.1).

Morphology of the Juvenile Female. The morphology of the recently emerged female *Clistosaccus paguri* parasite (fig. 12.4c) differs from those of the kentro-

TABLE 12.1. The distribution in the Rhizocephala of characters associated with the sexual system.

Taxon	Trichogon	Antennular Penetration	Male and Female Cyprids Dissimilar	Mantle Aperture in Juvenile Externa	No. of Receptacles	Spermatogenic Islands
Peltogastridae	+	−	+	+	2	−
Lernaeodiscidae	+	−	+	+	2	−
Sacculinidae	+	−	+	+	2	−
Clistosaccus	−	+	−	−	1	−
Sylon	−	+(1)	−	−	0	−
Mycetomorpha	−	+(1)	+(2)	−	0	+
Thompsonia	−	+(1)	−	−	0	+(3)
Pirusaccus	−	?	?	−	0	+
Cryptogaster	−	?	?	−	0	+
Duplorbis	−	?	?	−	0	+
Chthamalophilidae	−	?	?	−	0	+

Note: Trichogons and antennular penetration by the cyprid are mutually exclusive means of implanting males into female parasites. (1) Indirect evidence for antennular penetration is very strong in *Sylon, Mycetomorpha*, and *Thompsonia*. (2) Only male cyprids have been studied in *Mycetomorpha*. (3) The spermatogenic islands in *Thompsonia* are reduced.

gonid families in being disc-shaped and lacking a stalk, so that it is almost level with the surface of the host (Høeg 1982). The single receptacle lies in the ovary and is a solid mass of cells (fig. 12.4c). The two receptacle ducts are also solid strings of cells, which only develop a lumen much later, when sperm is released at oviposition. At this event the mantle cavity forms by separation of two hitherto tightly adjoined epithelia and the mantle aperture breaks through (Høeg 1982).

Settlement and Penetration of Male Cyprids. In the absence of any open access from the exterior to the receptacle, functional male cyprids settle either on the general surface of recently emerged externae or on the host cuticle directly above late stage, but still internal, primordia (Høeg 1985). The settled cyprids use one of the antennules to penetrate into the female parasite and implant spermatogonia (fig. 12.1c–d). When settled above still internal primordia, the cyprids must also penetrate through the abdominal integument of the hermit crab before they reach the female parasite (fig. 12.4d). The cyprid's antennules are extremely long and slender and thus well adapted to penetration, which may be facilitated by lytic secretions from antennular glands (Høeg 1985). The third antennular segment is dislodged after penetration, leaving the antennule as an open syringe for the passage of spermatogonia (fig. 12.4d). The cyprid disintegrates after penetration, but a scar is left in the cuticle of the juvenile externa (Høeg 1982).

Clistosaccus paguri is the only known rhizocephalan species in which males are implanted into the female before emergence from the host. Whether males settle before or after emergence seems to be a matter of chance, and an internal primordium will develop through emergence regardless of whether males have been received (Høeg 1982).

The Number and Development of the Males. The injected spermatogonia, which are not enveloped by cuticle or epidermis, migrate through the mantle connective tissue and into the receptacle (fig. 12.4c). The receptacle protrudes into the mantle with a mushroom-shaped group of cells, which withdraw when they have received the spermatogonia (fig. 12.4c). A lumen now develops in the center of the receptacle, and spermatogenesis starts.

In *Clistosaccus paguri* numerous male cyprids may settle and implant spermatogonia into the mantle of a single female parasite (Høeg 1982, 1985). It is furthermore possible that spermatogonia from several males may successfully invade the single receptacle, since a mechanism to prevent this is apparently lacking. This would of course generate an even more complicated genetic situation than would the two males normally found in kentrogonid rhizocephalans.

Sylonidae

Much evidence suggests that life history events in *Sylon hippolytes*, sole species of the Sylonidae (see Høeg & Lützen 1985), are very similar to those in *Clistosaccus paguri*, strongly indicating a sister-group relationship. The morphology and development of the recently emerged *S. hippolytes* externa resembles that of *C. paguri*, but in *S. hippolytes* there is no receptacle at all, and the species was previously believed to be parthenogenetic. Lützen (1981a), however, concluded that groups of spermatogonia found among the ovarian tubules of recently emerged externae were implanted by cyprid males, since no precursors for these cells could be found in the immediately preceding internal stages. Glenner et al. (1989) furthermore found a close similarity between the cyprids of *C. paguri* and *S. hippolytes*, especially in the almost identically shaped antennules. On this evidence little doubt remains that *S. hippolytes* also employs antennular penetration for the injection of male cells. In the absence of a receptacle in *S. hippolytes*, the implanted cells instead migrate from the mantle to the place of the would-be receptacle, i.e., among the ovarian tubules in the visceral sac, where they seem to grow at the expense of some oocytes (Lützen 1981a).

Thompsonia

Species of the poorly known genus *Thompsonia* are usually gregarious and occur in great numbers on the hosts (Høeg & Bruce 1988). True receptacles are absent at any stage in the life cycle, and Potts (1915) believed them to be parthenogenetic. In *Thompsonia cubensis*, however, Reinhard and Steward (1956) found what they believed to be two testes embedded in the ovarian mass. These "testes" consist of a few large nurse cells surrounded by numerous cells undergoing spermatogenesis. In *T. japonica*, Yanagimachi and Fujimaki (1967) found similar organs situated in the mantle beneath small canal-shaped scars in the external cuticle reminiscent of those left by penetrating male cyprids in *Clistosaccus*

(Høeg 1982). This occurrence and the presence of extremely long and slender antennules in *Thompsonia* cyprids provides evidence that injection of male cells also takes place in this genus (Glenner et al. 1989). Reinhard and Steward (1956) demonstrated the anlage of both a mantle aperture and cavity in *T. cubensis*. But the eggs are apparently fertilized in the ovary, where they remain and develop until hatching, which seems to take place by fatal rupture of the entire externa. Such lost externae are replaced by regeneration from the interna (Potts 1915). Production of externae by asexual budding is also known in the peltogastrid *Peltogasterella* and is suspected for *Cyphosaccus*, but is otherwise rare in the Rhizocephala (Lützen 1981b; Høeg 1982).

Mycetomorpha

The mantle tissue of this monotypic genus contains several bodies consisting of a few large cells surrounding a cavity filled with numerous small cells in spermatogenesis. Reinhard and Evans (1951) suggested that these "mantle bodies" are implanted by cyprids, which were found attached to the external cuticle. Once again this interpretation agrees with the cyprids having very long and slender antennules seemingly adapted for penetration (Glenner et al. 1989). Otherwise the *Mycetomorpha* cyprids resemble those of kentrogonids in that the number of antennular aesthetascs seems to differ between males and females. By comparison with, e.g., *Clistosaccus*, where the implanted spermatogonia are received by large nurse cells from the receptacle, it seems most likely that the large cells forming the spermatogenic bodies in *Mycetomorpha* originate in the female, while only the spermatogonia proper derive from cyprid males. Fertilization probably takes place in the spacious mantle cavity into which the eggs are released by way of paired oviducts.

THE SPERMATOGENIC ISLAND SYSTEM

Development of the spermatogenic island

This sexual system is the most enigmatic in the Rhizocephala and prevails in the alleged akentrogonid family Chthamalophilidae and in the two closely related genera *Duplorbis* and *Cryptogaster* (Bocquet-Védrine 1961; Bocquet-Védrine & Bourdon 1984). In all these species the male organ originates in the mantle as a local thickening of large cells intercalated between the inner and outer mantle epithelia (fig. 12.4E). As it increases in size, this anlage begins to bulge into the mantle cavity and is finally liberated altogether as a free floating cell group (fig. 12.4F). It is initially solid throughout, but later develops into an epithelium-delimited bladder where spermiogenesis takes place in the central lumen, whence it is called a spermatogenic island *(ilot spermatogenique)*. According to Bocquet-Védrine (1961), it is the large cells of the localized mantle thickening that develop into the cells surrounding the central island lumen. But TEM investigations (Høeg et al. 1990) demonstrate that the wall cells of the island are derived from the inner mantle epithelium, which comes to surround the island after its liberation.

Number and Fate of the Spermatogenic Islands

In *Chthamalophilus* and *Boschmaella* the original single and solid spermatogenic island disintegrates and forms multiple islands, each of which forms its own lumen, while in *Cryptogaster* and *Duplorbis* the island seems to remain single (Bocquet-Védrine & Bourdon 1984). At least for the Chthamalophilidae it has been shown that the parasite produces a large number of broods. But unlike kentrogonids, the very thin cuticle lining the mantle cavity is not shed in between the broods, and this explains why the original spermatogenic islands are not expelled but persist in a functional state in the mantle cavity. Furthermore, both in the Chthamalophilidae and in *Cryptogaster* and *Duplorbis*, the mantle aperture takes the form of a very long mesenteric canal, which is so narrow that it prevents accidental expulsion of the spermatogenic islands when the cypris larvae are liberated (fig. 12.4E).

Are the Spermatogonia Implanted by Male Cyprids?

Bocquet-Védrine (1961, 1972) considered the male organ of the Chthamalophilidae to be a true testis in a self-fertilizing hermaphrodite. This description, however, was published well before the cryptogonochoristic system was accepted for almost all other Rhizocephala (see Høeg 1987a), and it now seems that the original arguments in favor of self-fertilizing hermaphroditism in chthamalophilids can be refuted. First, self-fertilization is very rare in the Cirripedia, and in species with low population densities and a patchy habitat, such as parasites, evolution seems to favor a sexual system with dwarf males over simple hermaphroditism (Crisp 1983). Absence of an anlage to an ovotestis also speaks against hermaphroditism. Instead, the male organ first appears peripherally in the mantle and only after the externa has emerged. It was exactly this kind of observation that led Reinhard and Evans (1951), Lützen (1981a), and Høeg (1982) to suspect hypodermic impregnation of male cyprid cells in the rhizocephalans they studied. Finally, Bocquet-Védrine and Bourdon (1984) argued convincingly for a homology between the mantle bodies in *Mycetomorpha*, where the evidence for antennular penetration is strong, and the true spermatogenic islands in *Cryptogaster*, *Duplorbis*, and chthamalophilids. I therefore suggest that the male cells in all rhizocephalans with spermatogenic islands are derived from cyprid males (fig. 12.4 E–F).

Pirusaccus, a "Missing Link" to Thompsonia?

The monotypic genus *Pirusaccus* was recently described by Lützen (1985) and has a general resemblance to *Thompsonia* in the shape of the externa and in being gregarious on the host. However, unlike *Thompsonia*, the eggs in *Pirusaccus* develop in a true mantle cavity containing spermatogenic islands like those present in the Chthamalophilidae. Therefore, just like *Mycetomorpha*, *Pirusaccus* seems to "bridge" the morphological gap between species with antennular penetration and those with spermatogenic islands. Unfortunately, however, the larval development in *Pirusaccus* remains entirely unknown.

SEX ATTRACTION IN THE RHIZOCEPHALA

Male Cyprid Settlement and Externa Growth

In *Lernaeodiscus porcellanae*, Høeg and Ritchie (1985) showed that male cyprids are neither attracted to nor do they settle on the female externa before the breakthrough of the mantle aperture (fig. 12.2). At this stage the molt cycle of the still very small female externa is arrested; it ceases to grow and there is no maturation of the ovary. But the externa will stay alive and be capable of successfully receiving males for a period of up to six months. The implantation of males induces the externa to resume growth and development into the adult stage, and within a few days after receiving the first male it ceases to attract additional male cyprids (Høeg & Ritchie 1985). The course of events seems to be quite comparable in sacculinids (Lützen 1984; Høeg 1987a) and in peltogastrids (Reinhard 1942; Ichikawa & Yanagimachi 1958). In *Clistosaccus*, development is also arrested after emergence until males are acquired (Høeg 1982). In *Sylon*, however, the parasite may at least occasionally develop through spawning of an unfertilized and hence abortive brood of eggs (Lützen 1981a).

The Location of Virgin Parasites by Male Cyprids

The problem faced by the male cyprids in locating a host animal carrying a virgin externa must necessarily be much more difficult than the female's location of an uninfested host. It is, therefore, noteworthy that male cyprids carry two large olfactory aesthetascs on the antennules in the Kentrogonida and in *Mycetomorpha*, while female cyprids carry only one much smaller aesthetasc (fig. 12.6).

It is entirely unknown by which stimuli male cyprids recognize the female. Many cirriped cyprids recognize settled conspecifics by so-called arthropodin proteins in the cuticle (Crisp 1985). Lewis (1978) suggested that male cyprids reacted to ecdysones released by the virgin female externae at the molt, whereby the mantle aperture is acquired. However, this particular molt is both preceded and followed by other molts of the externa, where males do not settle, so it is difficult to see why the mantle aperture-acquiring molt would differ. It seems more likely that the receptacles, which are the terminal targets for the male, produce a pheromone that is released into the exterior, when the mantle aperture breaks through. The occasional observation of juvenile externae, in which the mantle cavity was filled with male cyprids that had gained entrance through wounds in the mantle, also supports the notion that a sex pheromone originates in the mantle cavity (Reinhard 1942; Høeg personal observation).

In species lacking a mantle aperture in the virgin female, events are even more enigmatic. In *Clistosaccus paguri* any chemical compound responsible for

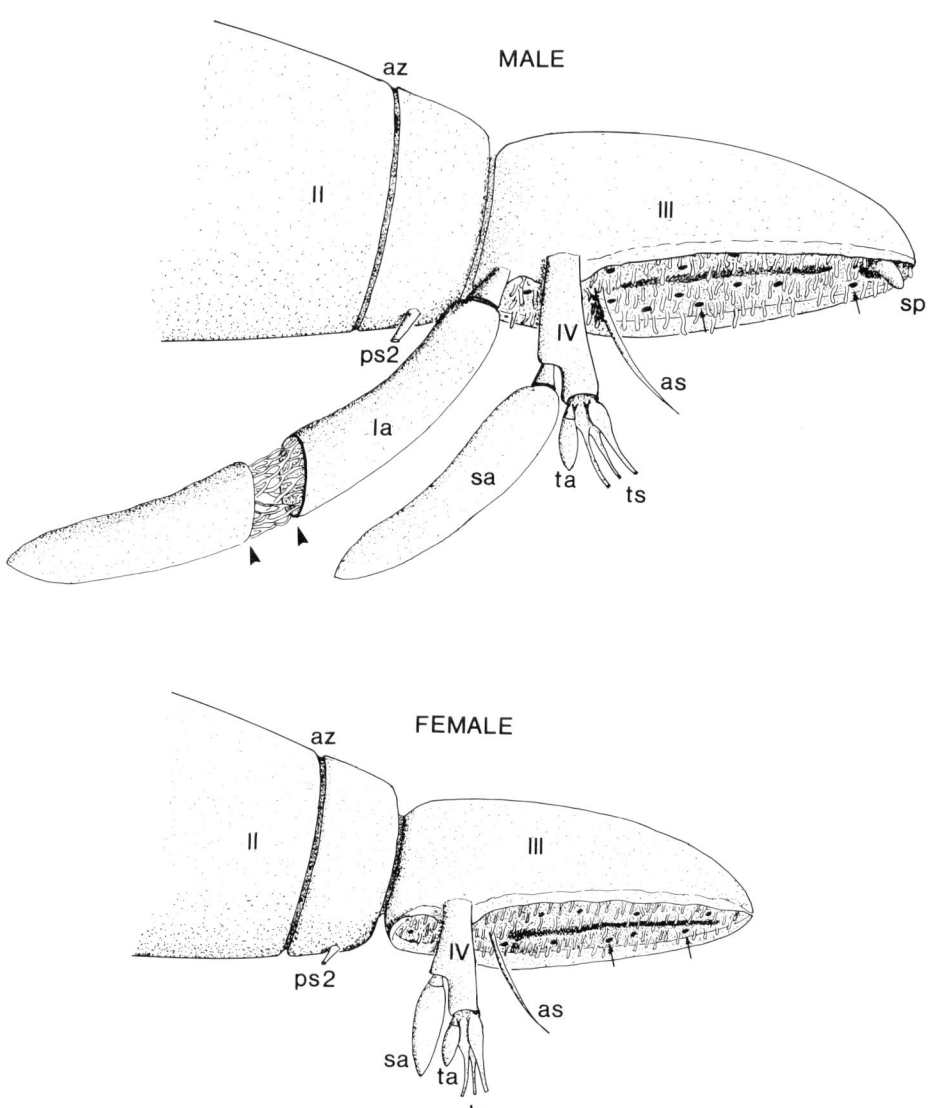

FIGURE 12.6. Antennular sense organs in cyprid larvae of a typical kentrogonid rhizocephalan. In the female the third segment lacks the large aesthetasc (la) and the spinous process (sp), and the subterminal, fourth segmental aesthetasc (sa) is relatively much smaller than in males. In the male the large aesthetasc (la) is cut open to reveal the branching outer dendritic segments (between arrowheads). Small arrows point to the exit pores of the antennulary glands. The cement pore is located in the large longitudinal groove in the attachment disc. Roman numerals designate segment numbers. as, axial seta; az, preformed abscission zone encircling distal end of second segment for the release of the trichogon (males) and the kentrogon (females); la, large third segmental aesthetasc unique to males; ps2, postaxial seta on second segment; sa, subterminal aesthetasc; sp, spinous process; ta, terminal aesthetasc; ts, terminal sensory setae.

sex attraction would have to be released from an internal parasite and pass through the overlying host tissue and cuticle before it could be detected by the cyprid. Nevertheless, *Clistosaccus paguri* cyprids have a reduced chemosensory armature without antennular aesthetascs (Glenner et al. 1989).

EVOLUTIONARY IMPLICATIONS OF THE SEXUAL SYSTEMS

The Basic Homology of the Sexual Systems

Although the three sexual systems found in the Rhizocephala are superficially very different, the preceding sections have demonstrated that several species combine characters from two or from all three systems. It follows that they are probably all specializations of a single ancestral cryptogonochoristic system. This was also the conclusion of Mourey (1974). *Clistosaccus*, while employing antennular penetration, also retains a single receptacle with a function basically similar to the homologous paired receptacles in kentrogonids. The observation that receptacle cells in *Clistosaccus* bulge into the mantle to receive implanted spermatogonia could indicate that the "mantle bodies" in *Mycetomorpha* evolved from an ancestral receptacle. In *Mycetomorpha*, male cyprids can be distinguished morphologically as in kentrogonids. But rather than metamorphosing into a trichogon, the male cyprids also seem to employ antennular penetration as in *Clistosaccus paguri*, and the injected cells develop into spermatozoa in spermatogenic islandlike mantle bodies as in the Chthamalophilidae. Finally, *Pirusaccus* has true spermatogenic islands in the mantle cavity and resembles *Thompsonia*, which employs antennular penetration for the injection of male cells. I will now discuss which of the three sexual systems is likely to be most plesiomorphic.

Males in Other Thecostraca

The Thecostraca *sensu* Grygier (1987) comprise the Facetotecta (y-larvae), the Ascothoracida, and the Cirripedia. Dwarf or complemental males are found in most Ascothoracida, all Cirripedia Acrothoracica, and some Cirripedia Thoracica. These males vary from being almost free-living in plesiomorphic Ascothoracida to highly reduced, lecitotrophic males without segmentation in some Cirripedia Thoracica such as *Ibla idiotica* (reviews by Klepal 1985, 1987). Usually the dwarf males reside near the mantle aperture, where they may be attached within preformed pocket-shaped receptacles such as in *Scalpellum scalpellum*. But the males may also be located within the mantle cavity of the hermaphrodite/female barnacle. In this case the male's stalk may be embedded in the tissue of the hermaphrodite, so it is retained when the cuticle of the mantle cavity is molted.

The Ancestral Sexual System of the Rhizocephala

The Trichogon-Receptacle System Is Plesiomorphic. In the traditional view, the Rhizocephala evolved from a true thoracican barnacle. Newman (1987), however, recently argued that the Rhizocephala should be removed from the Cirripedia *sensu stricto* and placed as the sister group to the Thoracica + Acrothoracica. In this interpretation, the Thoracica + Acrothoracica, and more distantly the Ascothoracida, become the appropriate outgroups for the Rhizocephala.

Whichever of these phylogenies is accepted, it is most reasonable to assume that the trichogon-receptacle system is plesiomorphic. In this sexual system an entire male instar is implanted into the female, so it provides a morphological bridge to the less reduced dwarf males found both in the Thoracica, Acrothoracica, and Ascothoracida.

It is, on the other hand, almost impossible to argue that the male metamorphosis in rhizocephalans employing antennular penetration is plesiomorphic. The direct impregnation of naked spermatogonia into the female parasite using the antennule in a penislike manner and the late development of the mantle cavity and aperture are mutually linked characters, to which nothing comparable is known from the possible outgroups within the Thecostraca. In the latter there are either true testes in a hermaphrodite or whole dwarf males rather than mere spermatogonia, and the mantle cavity and aperture form early in ontogeny. It is interesting, however, that the males of the more distantly related Tantulocarida seem to use their penis to fertilize the females by hypodermic impregnation (Boxshall & Lincoln 1987).

The Evolution of the Sexual System of the Rhizocephala. The scenario in Figure 12.7 is constructed from comparison with the sexual systems found in other Thecostraca. The ancestral rhizocephalan male probably lived in close association with the female on the host animal, as in some Ascothoracida. Eventually the males became attached within the female's mantle cavity. The paired sites of attachment, the future receptacles, gradually became specialized and invaginated for further protection and to allow the males to remain with the female when the mantle cavity is molted. Similar trends can also be seen in several lines of the Cirripedia Thoracica (Klepal 1987). The attachment sites finally evolved into the advanced receptacles of extant Rhizocephala, where the proximal part lacks cuticle. This allowed the males to be nourished by the female and thus to be retained permanently, obviating the risk involved in repeatedly reacquiring males. The male itself became gradually reduced to the vermiform trichogon, which facilitates penetration through the narrow duct leading into the receptacles. Implantation by means of the trichogon also allowed the mantle aperture to stay very constricted throughout implantation, thus preventing other organisms, such as egg predators, from entering the mantle cavity. The evolutionary advantage of having a late developing mantle aperture, as in all species without trichogons, may also be protection against such foreign elements in the brood chamber.

In summary, the paired deployment of receptacles in the otherwise very specialized externa of the Kentrogonida (table 12.1) evolved from homologous but simpler structures in the nonparasitic ancestor. A few advanced rhizocephalans evolved an abbreviated ontogeny, where the trichogon instar is lost and male cells are injected directly into the female. In those lineages the receptacle became reduced or lost altogether. The phylogenetic implications of this conclusion will be published elsewhere.

HYPOTHETICAL ANCESTRAL SEXUAL SYSTEM

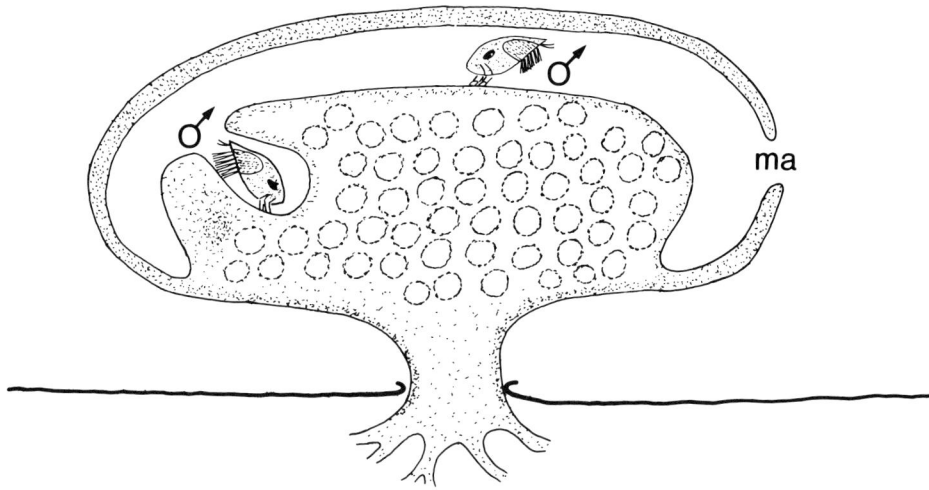

FIGURE 12.7. The hypothetical ancestral sexual system in the Rhizocephala. The males were initially free or attached to rather unspecialized sites within the mantle cavity of the female. For further protection and control of the males by the female, the attachment sites subsequently evolved into the deeply invaginated and highly specialized receptacles of extant Rhizocephala.

DIRECTIONS FOR FUTURE RESEARCH

The implantation of male cells and the structure of the male organ are still very poorly known in many rhizocephalans, notably the Chthamalophilidae, and these are therefore high priority research topics.

Information on the mechanism behind sex determination is very scarce (see Walker 1985, 1987). In *Clistosaccus, Sylon*, and the chthamalophilidlike genera, the absence of structurally different male and female larvae could indicate that the sex is determined by the substrate upon which they settle, as is the case in some pendunculate barnacles (Svane 1986).

The chemical basis for sex attraction in the Rhizocephala is still based on anecdotal evidence but should be a rewarding field for future studies. Finally, spermatogenesis, sperm release, oviposition, and fertilization need closer attention.

Acknowledgments

The financial support for this study from the Carlsberg Foundation (Denmark) (Grant No. 1986/87, 185/II) and the National Science Foundation (U.S.A.) is gratefully acknowledged. I am also indebted to three anonymous referees for constructive criticism. Finally, I wish to thank my wife for enduring an even less normal family life while this paper was written.

Literature Cited

Barnes, R. D. 1986. *Invertebrate Zoology*, 5th ed. Philadelphia: Saunders College.
Batham, E. J. 1945. Description of female, male and larval forms of a tiny stalked barnacle, *Ibla idiotica* n.sp.. *Trans. R. Soc. N.Z.* 75:347–356.
Bocquet-Védrine, J. 1961. Monographie de *Chthamalophilus delagei* J. Bocquet-Védrine, Rhizocéphale parasite de *Chthamalus stellatus* (Poli). *Cah. Biol. Mar.* 2:455–593.
Bocquet-Védrine, J. 1972. Les Rhizocéphales. *Cah. Biol. Mar.* 13:615–626.
Bocquet-Védrine, J. & R. Bourdon. 1984. *Cryptogaster cumacei* n. gen., n. spec.; premier rhizocéphale parasite d'un cumacé. *Crustaceana (Leiden)* 41:261–270.
Boxshall, G. A. & R. J. Lincoln. 1987. The life cycle of the Tantulocarida (Crustacea). *Philos. Trans. R. Soc. Lond. B Biol. Sci.* 315:267–303.
Bresciani, J & J. Lützen. 1972. The sexuality of *Aphanodomus* (Parasitic copepod) and the phenomenon of cryptogonochorism. *Vidensk. Medd. Dan. Naturhist. Foren.* 135:7–20.
Crisp, D. J. 1983. *Chelonobia patula* (Ranzani), a pointer to the evolution of the complemental male. *Mar. Biol. Lett.* 4:281–294.
Crisp, D. J. 1985. Recruitment of barnacle larvae from the plankton. *Bull. Mar. Sci.* 37:478–486.
Delage, Y. 1884. Evolution de la Sacculine (*Sacculina carcini* Thomps) crustacé endoparasite de l'ordre nouveau des kentrogonides. *Arch. Zool. Exp. Gén. ser.* 2(2):417–736.
Glenner, H., J. T. Høeg, A. Klysner & B. Brodin Larsen. 1989. Cypris ultrastructure, metamorphosis and sex in seven families of rhizocephalan barnacles (Crustacea: Cirripedia: Rhizocephala). *Acta Zool. (Stockh.),* 70:229–242.
Grygier, M. J. 1987. New records, external and internal anatomy, and systematic position of Hansen's Y-larvae (Crustacea: Maxillopoda: Facetotecta). *Sarsia* 72:261–278.
Høeg, J. T. 1982. The anatomy and development of the rhizocephalan barnacle *Clistosaccus paguri* Lilljeborg and relation to its host Pagurus bernhardus (L.). *J. Exp. Mar. Biol. Ecol.* 58:87–125.
Høeg, J. T. 1985. Male cypris settlement in *Clistosaccus paguri* Lilljeborg (Crustacea: Cirripedia: Rhizocephala). *J. Exp. Mar. Biol. Ecol.* 89:221–235.
Høeg, J. T. 1987a. Male cyprid metamorphosis, and a new male larval form, the trichogon, in the parasitic barnacle *Sacculina carcini* (Crustacea: Cirripedia: Rhizocephala). *Philos. Trans. R. Soc. Lond. B Biol. Sci.* 317:47–63.
Høeg, J. T. 1987b. The relation between cypris ultrastructure and metamorphosis in male and female *Sacculina carcini* (Crustacea, Cirripedia). *Zoomorphology* 107:299–311.
Høeg, J. T. 1990. "Akentrogonid" host invasion and an entirely new life cycle in the rhizocephalan parasite *Clistosaccus paguri* (Crustacea Thecostraca Cirripedia). *J. Crustacean Biol.*
Høeg, J. T. & A. J. Bruce. 1988. *Thompsonia lützeni*, new species (Cirripedia: Rhizocephala). A solitary parasite from the Alpheid shrimp *Alpheus parvirostris*. *Bull. Mar. Sci.* 42(2):246–252.
Hoeg, J. T., C. M. Kapel, P. Thor and P. Webster. In press. The anatomy and sexual biology of Boschmaella japonica, an akentrogonid rhizocephalan parasite on barnacles from Japan (Crustacea Cirripedia Rhizocephala). *Acta Zool. (Stockh.).*
Høeg, J. & J. Lützen. 1985. *Crustacea Rhizocephala*. *Marine Invertebrates of Scandinavia*, no. 6. Oslo: Norwegian University Press.
Høeg, J. T. & L. G. Ritchie. 1985. Male cypris settlement and its effects on juvenile development in *Lernaeodiscus porcellanae* Müller (Crustacea: Cirripedia: Rhizocephala). *J. Exp. Mar. Biol. Ecol.* 87:1–11.
Ichikawa, A. & Yanagimachi, R. 1957. The sexual nature of a rhizocephalan, *Peltogasterella socialis*. *J. Fac. Sci. Hokkaido Univ.* series 6, Zool. 13:384–389.
Ichikawa, A. & R. Yanagimachi. 1958. Studies on the sexual organization of the Rhizocephala. I. The nature of the "testes" of *Peltogasterella socialis* Krüger. *Annot. Zool. Jpn.* 31:82–96.
Ichikawa, A. & R. Yanagimachi. 1960. Studies on the sexual nature of the rhizocephala. II. The reproductive function of the larval (cypris) males of *Peltogaster* and *Sacculina*. *Annot. Zool. Jpn.* 33:42–56.
Klepal, W. 1985. *Ibla cumingi* (Crustacea, Cirripedia)—a gonochoristic species (anatomy, dwarfing and systematic implications). *Mar. Ecol.* 6:47–119.
Klepal, W. 1987. A review of the comparative anatomy of the males in cirripedes. *Oceanogr. Mar. Biol. Annu. Rev.* 25:285–351.
Lewis, C. A. 1978. A review of substratum selection in free-living and symbiotic cirripedes. In F.-S. Chia & M. E. Rice, eds., *Settlement and Metamorphosis of Marine Invertebrate Larvae*, pp. 207–218. New York: Elsevier.
Lützen, J. 1981a. Observations on the rhizocephalan barnacle *Sylon hippolytes* M. Sars. parasitic on the prawn *Spirontocaris lilljeborgi* (Danielssen). *J. Exp. Mar. Biol. Ecol.* 50:231–254.

Lützen, J. 1981b. Field studies on regeneration in *Sacculina carcini* Thompson (Crustacea: Rhizocephala) in the Isefjord, Denmark. *J. Exp. Mar. Biol. Ecol.* 53:241–249.

Lützen, J. 1984. Growth, reproduction, and life span in *Sacculina carcini* Thompson (Cirripedia: Rhizocephala) in the Isefjord, Denmark. *Sarsia* 69:91–106.

Lützen, J. 1985. Rhizocephala (Crustacea: Cirripedia) from the deep sea. *Galathea Report* 16: 99–112.

Lützen, J. 1987. Life history parameters calculated from growth rings in parasitic barnacles of the family Peltogastridae (Crustacea: Cirripedia: Rhizocephala). *J. Crustacean Biol.* 7: 493–506.

Mourey, M. 1974. *Duplorbis, Cirripède parasite*. Doctoral dissertation, L'Université de Nancy. (Unpubl.)

Newman, W. A. 1987. Evolution of cirripedes and their major groups. In A. J. Southward, ed., *Barnacle Biology, Crustacean Issues* 5:3–42. Rotterdam: Balkema.

Potts, F. A. 1915. On the rhizocephalan genus *Thompsonia* and its relation to the evolution of the group. *Publ. Carnegie Inst. Wash.* 212(8):1–32.

Reinhard, E. G. 1942. The reproductive role of the complemental males of Peltogaster. *J. Morphol.* 70:389–402.

Reinhard, E. G. 1958. Rhizocephala of the family Peltogastridae parasitic on West Indian species of Galatheidae. *Proc. U.S. Natl. Mus.* 108:295–307.

Reinhard, E. G. & J. T. Evans. 1951. The spermiogenic nature of the "mantle bodies" in the aberrant rhizocephalid *Mycetomorpha*. *J. Morphol.* 89:59–69.

Reinhard, E. G. & T. Stewart. 1956. The hermaphroditic nature of *Thompsonia* (Crustacea Rhizocephala) with the description of *Thompsonia cubensis*, n. sp.. *Proc. Helminthol. Soc. Wash.* 23:162–168..

Ritchie, L. E. & J. T. Høeg. 1981. The life history of *Lernaeodiscus porcellanae* (Cirripedia Rhizocephala) and co-evolution with its porcellanid host. *J. Crustacean Biol.* 1:334–347.

Rubiliani, C., Y. Turquier & G. G. Payen. 1982. Recherche sur l'ontogenese des rhizocephales. I. Les stades precoces de la phase endoparasitaire chez *Sacculina carcini* Thompson. *Cah. Biol. Mar.* 23:287–297.

Spivey, H. 1982. Rhizocephala. In S. P. Parker, ed., *Synopsis and Classification of Living Organisms*, 2:229–232. New York: McGraw-Hill.

Svane, I. 1986. Sex determination in *Scalpellum scalpellum* (Cirripedia Thoracica Lepadomorpha), a hermaphroditic goose barnacle with dwarf males. *Mar. Biol (Berl.)* 90:249–253.

Yanagimachi, R. 1961. Studies on the sexual organization of the Rhizocephala. III. The mode of sex-determination in *Peltogasterella*. *Biol. Bull. (Woods Hole)* 120:272–283.

Yanagimachi, R. & N. Fujimaki 1967. Studies on the sexual organization of the Rhizocephala. IV. On the nature of the "testis" of *Thompsonia*. *Annot. Zool. Jpn.* 40:98–104.

Veillet, A. 1985. Note preliminaire sur la phase larvaire mâle qui suit la fixation des cypris mâles chez les Rhizocéphales. *Bull. Soc. Lorraine Sci.* 24:141–146.

Walker, G. 1985. The cypris larvae of Sacculina carcini Thompson (Crustacea, Cirripedia, Rhizocephala). *J. Exp. Mar. Biol. Ecol.* 93:131–145.

Walker, G. 1987. Further studies concerning the sex ratio of the larvae of the parasitic barnacle *Sacculina carcini*. *J. Exp. Mar. Biol. Ecol.* 106:151–163.

THIRTEEN
Functional Morphology and Evolution of Isopod Genitalia

GEORGE D. F. WILSON

Marine Biology Research Division,
Scripps Institution of Oceanography,
La Jolla, California.

Abstract

Isopod Crustacea are ubiquitous and have a long evolutionary history. Their success may have been aided by their mating systems, of which an important feature is internal insemination. This work presents an overview of the function of isopod genitalia, with some discussion of their evolutionary aspects. Isopods have varying degrees and kinds of precopula, and the actual copulation has been described in detail only a few times. Mating generally takes place during the isopodan biphasic molt, although some groups have extended receptive periods. The nonmotile, "pennantlike" sperm are grouped into "spermatophores" that may be necessary for sperm transfer during internal insemination. Primitively, the male genital papillae are on the coxae of the last walking legs, but in most isopods they have moved onto the last thoracic sternite or, in some taxa, onto the pleotelson. Sperm transfer may be mediated by the appendix masculina of the male second pleopod, although this appendage's morphology shows a great deal of diversity. In the simple and primitive form, the second pleopod has a rod on the medial ramus. The copulatory function of this rodlike appendix masculina is not clear. In at least three independent lineages, the first two pleopods form a "funnel" system that acts as an extension of the penile papillae. The Asellota also have a unique copulatory apparatus on the second pleopod, the "arm and hammer" form, that in the successful superfamily Janiroidea has a highly constrained function owing to interlinked parts and a continuous sperm conduit through both anterior pairs of pleopods. The female genitalia primitively open on the coxae of the sixth thoracic limbs, although the opening is on the sternum in most groups. Some taxa, however, have coxal plates that replace the sternum, so the derivation of the oopore may not be apparent. Most isopods are suspected of practicing sperm holding, and instances of well-developed spermathecae in the oviduct are known. In the Asellota, the female genitalia are separated into two separate functions: an oopore to release the ova and a spermathecal duct that receives sperm from the male. Although the pore for the asellotan spermathecal duct is directly associated with the tissues of the ventral oviduct, this duct opens on the dorsal surface in the evolutionarily derived Janiroidea. The janiroidean spermathecal duct receives the male appendix masculina directly, and the accessory structures seen in other Asellota are not present. While most isopods can copulate only during the brief period during the molt to the brooding stage, the highly evolved copulatory system of the Janiroidea allows the female to mate over much longer periods. The mating system of these asellotes may be a preadaptation to low population densities that permitted their astounding evolutionary radiation in the deep sea.

ISOPOD CRUSTACEA have interesting mating systems that may have had a profound impact on their adaptations, evolutionary longevity, and success. The Isopoda are diverse and have undergone an extensive morphological radiation. The current classification (Bowman & Abele 1982; Wilson 1987; but see Wägele 1989) recognizes 9 diverse suborders including approximately 102 family-level taxa. Isopods can be defined by several reasonably constant, derived traits. Unlike many malacostracan Crustacea, the Isopoda lack a carapace fold over the segments (pereonites) bearing walking legs (pereopods). If eyes are present, they are sessile on the cephalon, not on movable stalks. Isopods lack outer rami (exites) on the thoracic limbs, and their abdominal limbs (pleopods) are generally broad and lamellar. The broad pleopods are used for respiration, and a dorsal heart is located in or near the pleotelson to supply these branchial appendages. As in most other peracarid Crustacea, most isopods retain developing embryos in ventral brood pouches formed from cuticular plates attached to the first segment (coxa) of the pereopods. The young are released from the brood pouch before the last pereopod develops (the manca instar). Finally, most isopods also practice internal insemination (Ridley 1983).

The isopods are an ancient group, with the first fossil, a phreatoicid, appearing in the Middle Pennsylvanian (Schram 1970) and with ancestral isopods perhaps living in the Devonian (Schram 1974). The long evolutionary history of the Isopoda might explain the great diversity of form observed among all the suborders. On the other hand, a secure method for transferring sperm between the male and female may be an important factor in their ability to colonize many different environments and to diversify morphologically. This ecological broadness could have ensured that isopods were not greatly affected by major biotic extinction events known to have taken place since the Paleozoic, when they first appeared (Raup 1986).

Although sexual adaptations are central to many aspects of isopod biology, the structure and the function of their copulatory organs are poorly understood, especially in the female. As a response to this need, this paper provides an overview of copulatory morphology in the Isopoda and touches on some interesting evolutionary aspects. I will consider only the general details from the larger, better known groups, because the genitalia have been studied in detail in only a few groups. Discussion of problematical taxa such as the Microcerberidae (Wägele 1983) and the Calabozoidea (Van Lieshout 1983) will be left for future study.

MATING BEHAVIOR

Prior to mating, many isopods may practice some sort of precopula or mate guarding to allow the male to be present when the female is receptive during the isopodan biphasic molt (best reviewed in Ridley 1983; W. Johnson & P. Stevens, personal communication). Precopula occurs over various lengths of time and can involve either passive attachment to the female or active carrying by the male. Serolid isopods, which are common in the Antarctic benthos, may have precopula periods of months (Luxmore 1982; Michel 1986) with the males clasped to the female's back. An odd form of passive precopula was observed by the author at Friday Harbor Laboratories in 1981: a male of the asellote *Munnogonium waldronense* clasped a female by its posterior pereopods on one side. Despite the female's attempts to dislodge the male, the male hung on passively. The dwarf males of parasitic epicaridean isopods, which are attached to much larger females, could be considered as having an extended form of passive precopula. Simply remaining near a precopulatory female may be an effective strategy if competition for mates is not great, such as in deep-sea isopods where population densities are very low. Tubiculuous or burrowing forms, such as the wood borer *Limnoria*, may extend the male-female association for long periods of time inside the burrow (Menzies 1954). Competition for mates must be significant in several species such as *Paragnathia formica, Paracerceis sculpta,* and *Gnathia calva,* where single males keep harems of females inside the cavity of sponges (Monod 1926; Shuster 1987; Wägele 1988). In *Paracerceis,* the reproductive rewards of mating with a concentrated group of females has selected for the evolution of smaller, genetically different femalelike and juvenilelike mature males that can elude the alpha male guarding the opening of the sponge and mate with the females (Shuster 1987). Males of several species, such as in the genera *Jaera* and *Munna,* are known to carry around potential mates (Veuille 1980, Hessler & Strömberg 1989). Active isopods that are in constant motion, like *Excirolana* spp., may not practice precopula (J. Weinberg, personal communication).

In the unspecialized mating process, copulation takes place during the female biphasic molt to the brooding stage (when the oostegites are fully deployed and the ova are released into the brood pouch). The gravid female (having fully developed ova in the ovaries) first molts the exoskeleton posterior to the fifth thoracic segment. While the female's exoskeleton is still soft, the male mounts the female, either dorsally or ventrally, and accomplishes insemination. After insemination (in the unspecialized taxa), the female will then molt the anterior half of the body and deploy the oostegites that form the brood pouch. The eggs, fertilized internally during their transit of the oviduct in the sixth thoracic segment, are released ventrally into a brood pouch shortly after molting is completed. In those species that have well-developed spermathecae (a female sperm reservoir), mating may take place at any time after the female is receptive and not necessarily only during the molt to the brooding stage. This type of behavior is best known in the Oniscidea (Mead 1976), but also occurs in the janiroidean Asellota (Veuille 1978b, personal observations). The sperm are held internally in the spermatheca until the ova are mature and ready for fertilization. Sperm holding is discussed further under female genitalia.

The details of copulation are generally unclear because it occurs so quickly (Ridley 1983). The Asellota have the best known copulatory behavior (e.g., Maercks 1931; Veuille 1978b). In the asellote *Jaera*, the details of copulation were discovered by the extreme procedure of pouring liquid nitrogen on coupling pairs (Veuille 1978a). When the female is receptive, the male mounts her back and quickly transfers the sperm using an appendix masculina on the second pleopods sequentially to each side of the female. The dwarf males of parasitic Epicaridea may inseminate the ova directly when they are released into the brood pouch (Hiraiwa 1935). The Asellota, the Oniscidea, and some groups of Valvifera use their complex male pleopods to transfer sperm from the penile papillae to the female. In other suborders, it is not known whether the simple rodlike appendices masculinae on the second male pleopods are involved in sperm transfer or if sperm move directly from the penile papillae to the female, with the appendix masculina performing some subsidiary function. The rodlike appendix masculina possibly could be used for pushing aside the oostegite of pereonite 5 and exposing the oopore for insemination. In the groups mentioned above, the first pleopods are also involved in passing the sperm to the female. The male pleopods are discussed more fully below.

SPERM AND SPERMATOPHORES

The nonmotile isopodan sperm, like the sperm of other peracarids (Cotelli et al. 1976; Reger, Itaya & Fitzgerald 1979; Wirth 1984), has an unusual elongate acrosome-associated process that is tail-like (but is not a true flagellum) and can be up to 300 microns long. Because peracarid sperm have elongate nuclear bodies projecting at acute angles from the acrosomal process, they have been called "pennant sperm" by Wirth (1984). The sperm are held together in elongate bundles (often referred to as spermatophores) by these long acrosomal processes. The processes helically coil around each other and the nuclear bodies of the sperm, projecting radially. Each bundle, containing many individual sperm, also may be held together further by a cap of noncellular macrotubules (Cotelli et al. 1976). The mechanics of spermatophore transfer are puzzling (e.g., Cotelli et al.

1976), although *Porcellio scaber* has two mitochondria-rich sphincter muscles in the lower vas deferens of the male that may push spermatophores along the intromittent organ into the female (Radu & Craciun 1972). Nevertheless, after the sperm leave the vas deferens and pass through cuticular canals of the male and female copulatory organs, the source of their movement is unknown. The parasitic epicaridean isopods, which may inseminate eggs directly in the brood pouch, have lost the unusual "pennant" sperm form (Wirth 1984). This suggests that the long acrosomal process is somehow necessary for passing the sperm from the male to the female and when internal insemination was lost the process was lost, too. Because the acrosomal processes hold the sperm together into spermatophores, the bundles themselves may be necessary for sperm transmission during copulation.

MALE GENITAL PAPILLAE

The penile papillae or "penes" (projections bearing the external openings of the vas deferens; some authors use *appendix masculina* for these structures, although this paper will apply *appendix masculina* only to structures of the endopod of the second pleopod) display important variability despite their general treatment as a simple monomorphic feature within the Isopoda (e.g., Pires 1987). In most groups, the penes emerge on the sternum of pereonite 7 and sometimes are fused into a cone. This common condition is not the ancestral condition for the Isopoda and may have been derived independently several times within the order. The Phreatoicidea have the penile papillae on the coxae (Nicholls 1943; see fig. 13.1A). This is the most likely primitive condition for the Peracarida in general because it occurs in potential outgroups such as Syncarida, Mysidacea, Spelaeogriphacea, and Mictacea. The penes of the phreatoicids are more complex than a simple extension of the vas deferens: they have a well-developed semicircular sphincter and groups of fine cuticular combs on the distal tip (fig. 13.1B). No other isopod group is known to have coxal penes, thereby establishing a synapomorphy for the remainder of the isopods: penes on the sternite of pereonite 7.

The Asellota show an intermediate condition, primitively at least (Wilson 1987). In the Aselloidea, Gnathostenetroidoidea, and the Stenetrioidea, the penes are long, thin, and well separated. Their position between the coxae of the last walking leg and the midline suggests a probable evolutionary scenario for the migration from the coxae to the sternum. During the evolution of the stem isopods, the penes migrated to the articular membrane of the coxa and then moved medially to become partially incorporated into the sternum. Sternal penes are therefore derivatives of the coxa that sometimes sit upon a small plate separate from the sternite of pereonite 7. Within the Janiroidea, the medial migration of the penes is complete: they are a pair of adjacent, short, conelike structures at the midline between the last pereonal segment and the first pleo-

FIGURE 13.1. Penile papillae of a male phreatoicid isopod (*Colubotelson*). A. Ventromedial view of coxa of thoracopod 8 and part of pleopod 1. Scale bar = 0.2 mm. B. Penile papilla medial view. Scale bar = 0.05 mm.

Morphology and Evolution of Isopod Genitalia 233

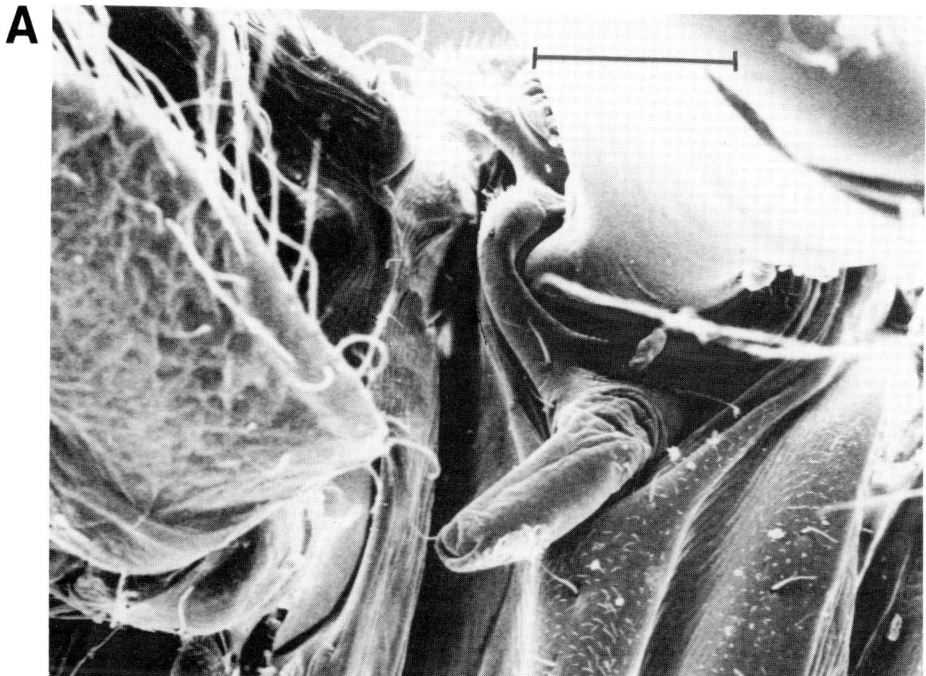

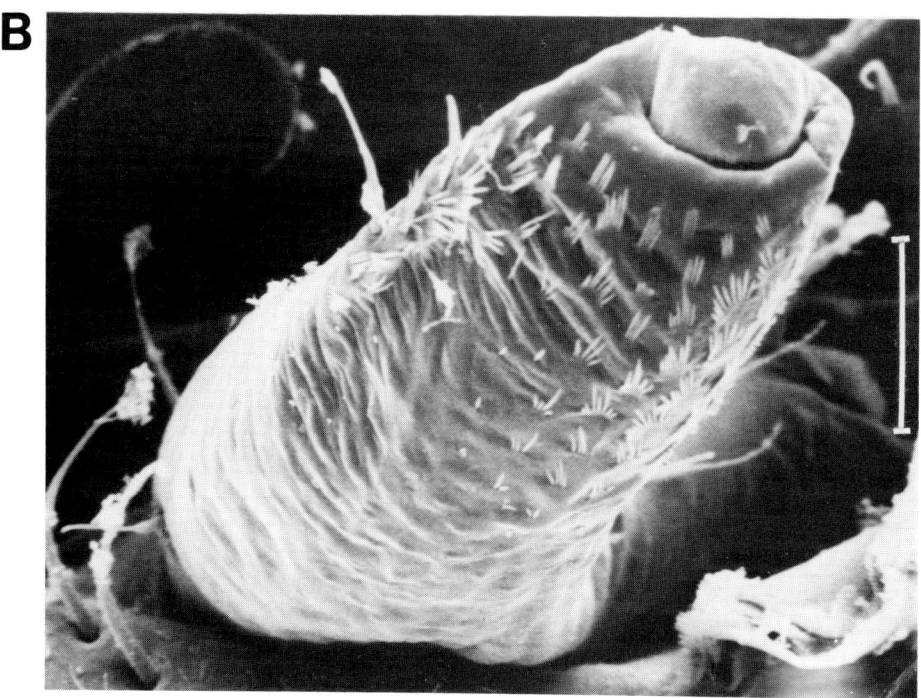

nite. Because the Janiroidea are the latest derived taxon in the Asellota (Wilson 1987), their condition must be derived convergently from the other isopods that have fused or adjacent penes.

The penile papillae vary in position, degree of fusion, and elongation throughout the remaining isopod groups (fig. 13.2). Most isopods, however, do not show the primitive penile states found in the Phreatoicidea or the long, well-separated penes of the asellids and the stenetriids (Asellota). In the Valvifera, the vas deferens and penes have moved from the last thoracic segment onto the pleotelson, perhaps to allow better protection within the uropod-covered pleopodal chamber. In the valviferan family Arcturidae (and related families), the penes additionally become elongate and fused into a single flattened penis (Sheppard 1957; Kussakin 1979; Brusca 1984).

APPENDIX MASCULINA AND ASSOCIATED STRUCTURES

Besides the penes, male isopods have modified pleopods that may function during copulation. Although their function is sometimes unclear, the male copulatory pleopods can be grouped into three nonphylogenetic morphological types: (1) the "rod," a rodlike appendix masculina on pleopod II, (2) the "funnel," a functional extension of the genital papillae by folds or tubes in the anterior pleopods, and (3) the "arm and hammer," the highly modified second pleopod of the Asellota. These types are not necessarily independent: a "funnel" also functions with the "arm and hammer" forms.

The basic rodlike form of the appendix masculina is simple and seemingly without obvious means of transferring the elongate spermatophores to the female. The appendix masculina consists of a curved rod either on the base or on the tip of the endopod of the male second pleopod (fig. 13.2). From its position, the rod seems to be homologous with the distal article of the endopod, so that a distal position of the rod (e.g., *Plakarthrium;* see Wilson, Thistle & Hessler 1976) may be the primitive form. This simple appendix masculina is found in the diverse "flabelliferan" forms, most Valvifera, the Anthuridea, and, significantly, the ancient Phreatoicidea and may be the primitive form of the isopodan appendix masculina for several reasons. (1) many Malacostraca have a modification of the male second pleopod for transferring sperm, so an appendix masculina of some form is likely to have occurred in the ancestral isopodan stock; (2) the second pleopod is found in a simple unmodified state in the Phreatoicidea (the sister group to the remaining isopod taxa, based on the evidence from the coxal penes); (3) it is more parsimonious to presuppose that the rod form is the primitive form and developed only once than to hypothesize that it arose separately in each of the suborders in which it appears; and (4) more derived forms of the appendix masculina (as seen in the Asellota) could be derived from the simple rod form. Some isopods have elaborate tips on the rod that may have a species recognition function. In the anthurid *Cyathura*, for example, several closely related species can be distinguished by the form of knobs and spines on the distal tips of the male second pleopod endopod (Wägele 1982).

The "funnel" is a functional grouping that has been attained independently at

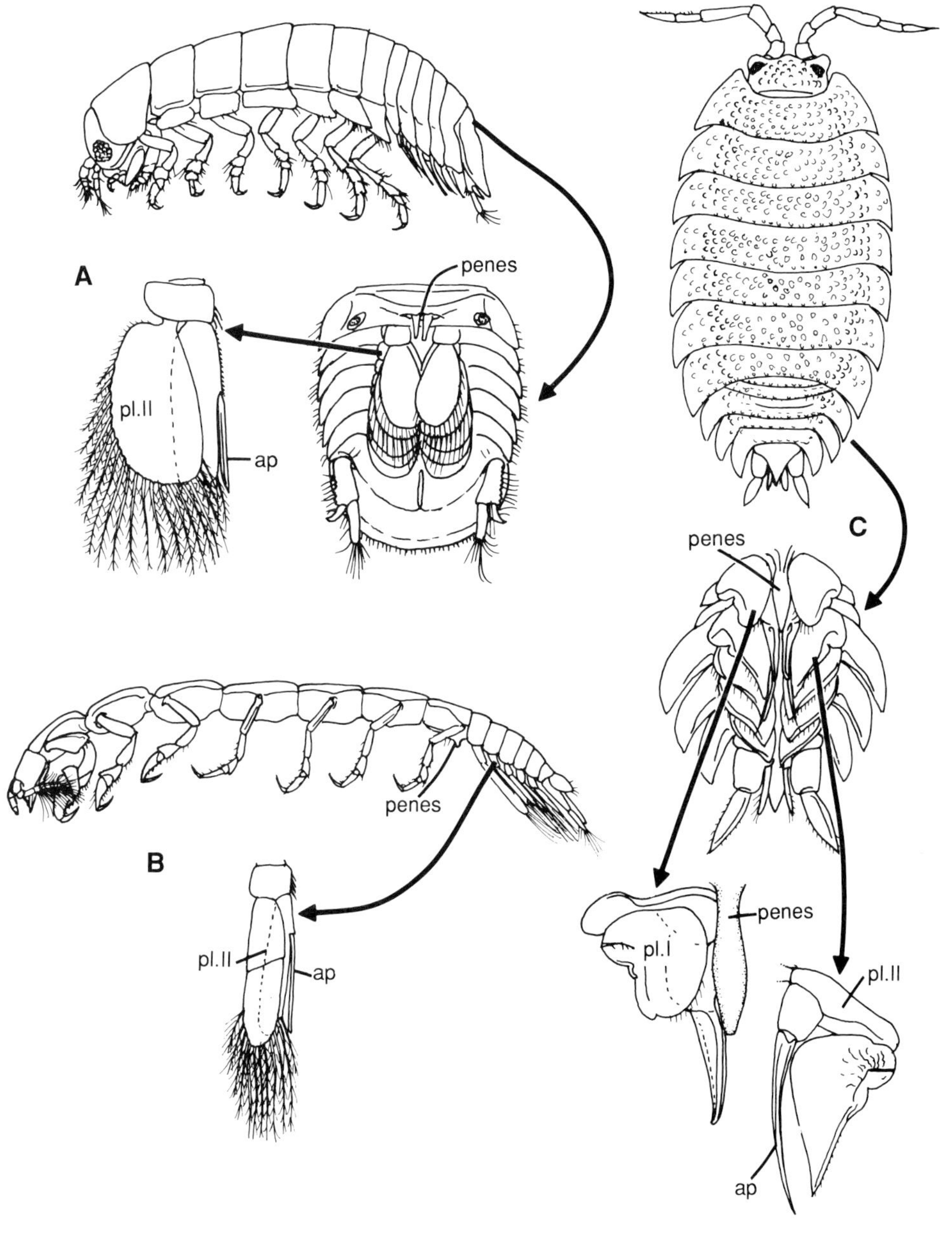

FIGURE 13.2. Male copulatory pleopods of several different isopods. A. *Limnoria* (Limnoriidae). B. *Leptanthura* (Paranthuridae). C. *Porcellio* (Oniscidae). All three taxa show a rodlike appendix masculina (ap) on the second pleopod (pl. II). In addition, the first pleopod (pl. I) of *Porcellio* illustrates one type of funnel-like extension of the penile papillae (penes). Illustrations after Sars 1899.

least three times: in the terrestrial Oniscidea (fig. 13.2C), in the derived Arcturidae and their allies in the Valvifera, and in the Asellota (fig. 13.3). The independent derivation of the funnel is demonstrated by its absence in forms derived earlier in the same subordinal clade (Valvifera [Sheppard 1957; Brusca 1984]; Asellota [Wägele 1983; Wilson 1987]) and by the dissimilar forms among the three groups. The funnel consists primarily of the medial parts of the first and sometimes the second pleopods that form a channel or tube from the penes to the female. The appendix masculina on the second pleopods may act as an extension of the penes in some instances. The morphological details differ considerably among the three groups. In the Arcturidae, the exopod of pleopod I has a ventral groove that is nearly or completely closed into an open-ended tube. The penes in this family have become fused and elongate, extending to the groove of the exopod. The Oniscidea show a great deal of variability in the male first and second pleopods, and much of their classification is based on these features. The funnel is found in the synochete and the crinochete oniscideans in two separate forms: the penes being extended primarily by the second pleopod in the former and by the first pleopod in the latter (fig. 13.2c). Interestingly, the Ligiidae (diplochete oniscideans) do not have a distinct funnel, and the second male pleopod is similar to the standard rod form of most other isopods. In the janiroidean Asellota, the penes are extended by a single long tube along the line of medial fusion of both first pleopods (fig. 13.3). The janiroidean form therefore provides only a single penile extension instead of a pair as in the Oniscidea or the Valvifera.

The "arm and hammer" appendix masculina, a synapomorphy of the Asellota, has a peculiar functional morphology (fig. 13.3). In this group, the protopod of the second pleopod is enlarged and houses elongate opposing pairs of muscles attached to the distal rami. Magniez (1974) shows a variety of forms that Asellotan male pleopods can take. The rami are no longer simple lamellae, generally having elliptical cross sections and dicondylic pivot points allowing movement in the plane of the pleopod. The exopod is the "arm" and usually has the largest protractor and retractor muscles. Its distal segment has a hook or curved projection that allows the exopod to grasp the proximal segment of the endopod. The endopod, then, is the "hammer." The distal segment of the endopod has a variety of forms, but always has structures for conveying sperm packets. In species where copulation has been observed (*Asellus* [Maercks 1931] *Jaera* [Veuille 1978a, b]), the exopod grasps the endopod and forces it into the female, a seemingly bizarre and indirect means of insemination.

Coadaptation between the "funnel" and "arm and hammer" in the janiroidean Asellota creates a highly sophisticated conduit for the sperm once they leave the vas deferens and the penes (fig. 13.3). The "hammer" of the Janiroidea might be described as a "pick hammer" or more accurately a stylet that is long and pointed at the distal end, with an internal tube opening at its midpoint on the dorsal side and at its tip. The first and second pleopods fit together as a functional unit. The fused first pleopods have small tabs that hold the protopod of the second pleopod in place. The stylet of the endopod fits into a dorsal groove on the distal part of the first pleopod. The medially fused penes lie inside a funnel in the proximal part of the fused first pleopods. These parts fit together to

Morphology and Evolution of Isopod Genitalia 237

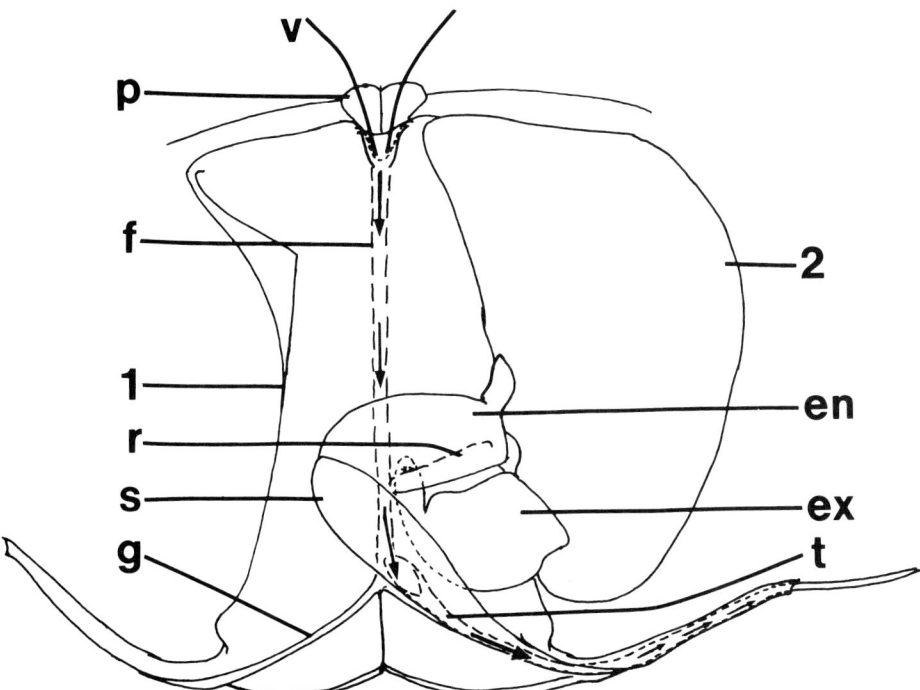

FIGURE 13.3. Copulatory pleopods of a male janiroidean isopod (Asellota), showing the interlinking of first and second pleopods (one and two) and the sperm channels. Small arrows show the sperm flow through the following structures: the vas deferens (v), the penile papillae or penes (p), the funnel and sperm tube (f) in the medial line of fusion in the first pleopods, and finally the sperm tube (t) in the stylet (s) of the second pleopod. In copulatory position, the endopodal stylet slides in a distal groove (g) of the first pleopods and is thrusted by the short exopod (ex) of the second pleopod, which is hooked to a ridge (r) on the basal part of the endopod (en). Simplified drawing of *Jaera* male pleopods modified from Veuille (1978a). These structures are similar in all Janiroidea, except for the extensions of the stylet groove on the first pleopod ("copulatory horns" that act as a sheath for the stylet), which are present only in some species of *Jaera*.

make a complete sperm channel from the penes to the tip of the stylet. In addition, the stylet of the second pleopod is mechanically restrained by the stylet guide in the first pleopod, allowing only a thrusting motion. During copulation, the stylet is inserted directly into the spermathecal duct of the female (described below).

This complex system did not arise all at once, but developed in evolutionary steps with each stage being fully functional (Wilson 1987). However, once the fully contained male sperm conduit appeared, it appears to be correlated with an enormous adaptive radiation in which the Janiroidea invaded all parts of the ocean and became one of the most common Crustacea in the deep-sea benthos. This is discussed further in "Mating Systems and Deep-Sea Isopods."

FEMALE GENITALIA

The female reproductive system of isopods consists of elongate paired ovaries in the dorsal body cavity. The ovaries are connected to the outside via oviducts that descend to oopores near the medial base of the fifth pereopods (in the sixth thoracic segment; (fig. 13.4). Recent work indicates that the female system is not as simple as was once thought (Wilson 1986b). Not only are oopore position homologies in doubt throughout the order, but the oviduct may not be a simple one-tube structure (e.g., fig. 13.4A). Most evidence indicates that isopods are inseminated internally, although some doubts remain. Despite the possibility that internal fertilization may be a synapomorphy of the isopod suborders, a review of mating in the isopods (Ridley 1983) shows that, in most cases, information is limited.

In many Malacostraca, the oopore exits on the coxa of the sixth thoracopod, probably the primitive condition. This is also the oopore position in the Phreatoicidea (fig. 13.5). In all other Isopoda (fig. 13.4), the oopore seems to have moved to the ventral surface, where it exits separately from the base of thoracopod 6 (pereopod V). The Asellota are clear examples because they have retained full flexibility of the coxa (Hessler 1982) and have the oopore on the ventral surface of pereonite 5. In other groups, the position of the oopore with respect to the coxa is less clear. In some groups, such as the Valvifera, the coxa generally becomes strongly fused to the main part of the body (Hessler 1982) and the coxa expands to cover the ventral surface, thereby replacing the sternite (Sheppard 1957). The valviferan oopore exits separately from the origin of the pereopod, but because the coxa has expanded to cover the ventral surface, the oopore exits through the coxa. Parsimony would suggest that the oopore simply migrated medially with the coxa, although an alternative explanation is that the oopore migrated medially before the coxa and then was surrounded by the latter's migration. This question currently remains unresolved.

The oviduct has a cuticular lining that assumes a variety of shapes and positions in the Isopoda. This cuticular lining may be associated with sperm holding and the spermatheca, the sperm holding organ. The spermatheca itself may vary considerably within the Isopoda. A survey of female reproductive organs (Menzies 1954; Ridley 1983; Wilson 1986b) might lead one to the conclusion that the Isopoda have two generalized types of reproductive systems: with and without distinct spermathecae. The isopods that have a well-defined spermatheca associated with the oviduct (e.g., Oniscidea [fig. 13.4B], Asellota [fig. 13.4 E–F]) are certain to have some sort of internal insemination, although the manner in which the sperm are received and stored is unknown in many taxa (fig. 13.4A,C).

The isopods thought to lack a spermatheca could be suspected of not practicing internal insemination. A clear example is the crustacean parasites Epicaridea (fig. 13.4G), where a dwarf male is attached to a large parasitic female. The epicaridean male generally lives in or near the brood pouch and lacks an appendix masculina. Fertilization may take place directly in the brood pouch as the ova are released by the female. Thus no sperm holding is necessary in the Epicaridea.

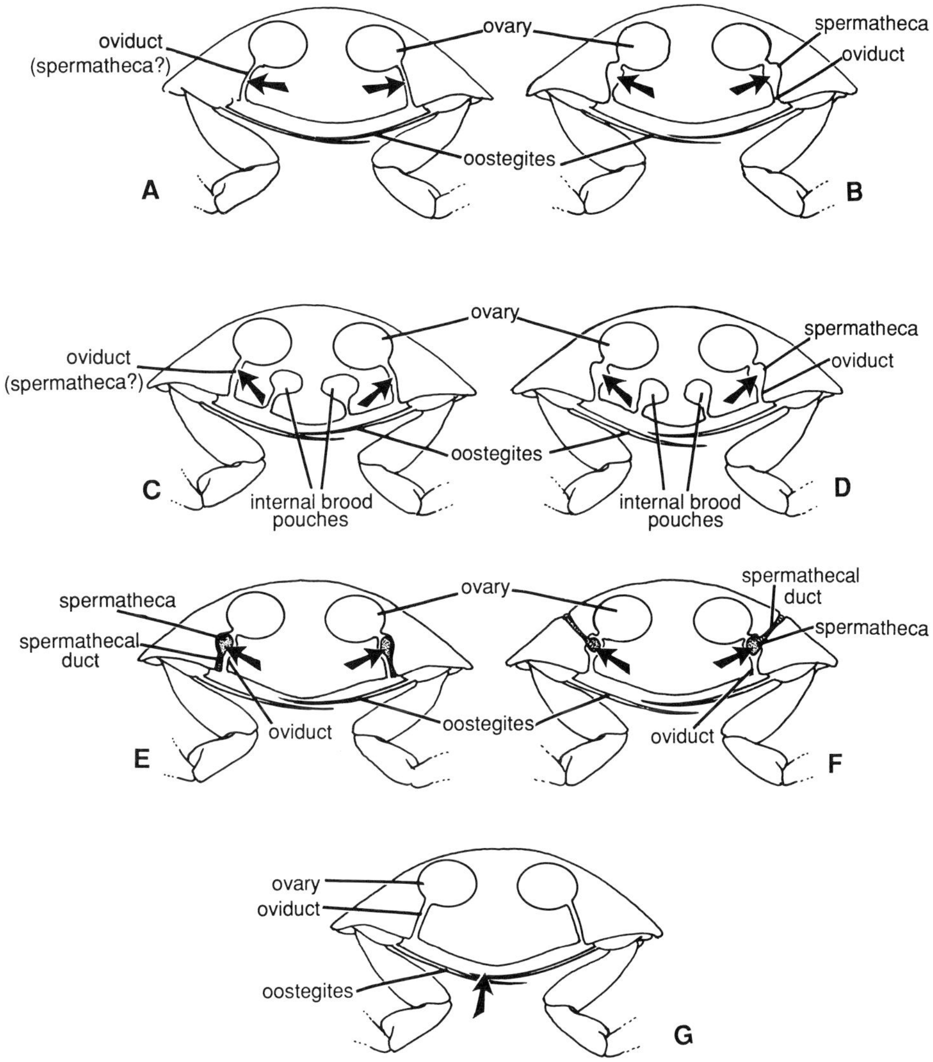

FIGURE 13.4. Isopod female genital anatomy; diagrammatic cross sections of pereonite 5. Dark arrows mark presumed or known position of sperm holding after insemination. A. *Limnoria* (Limnoriidae). B. Oniscidea. C. *Sphaeroma* (Sphaeromatidae). D. *Dynamenella*. E. *Asellus* (Asellidae). F. Janiroidea. G. *Epipenaeon* (Bopyridae). Derived from Ridley 1983, Wilson 1986 a,b; S. Shuster personal communication.

Nevertheless, much circumstantial evidence exists for copulation and sperm holding in many isopod taxa. Most groups have an appendix masculina, suggesting that internal insemination is common to all Isopoda. For example, development is internal in the ovoviviparous *Excirolana* (Klapow 1970), therefore requiring internal insemination. In isopods that have precopula, males may hold the females until the latter enter their molt to oostegite-bearing stage. The fertilized eggs may be deposited in the brood pouch well after the male has left the female. Copulation apparently takes place during the molt, and sperm are retained internally in some manner. These isopods, although not known to have a spermatheca, are nevertheless holding the sperm at least for a short time. A spermathecal structure of some sort may be present after all.

Insemination through the oopore is thought to take place in most groups (Ridley 1983), although this is not the case in the Asellota (Wilson 1986b). The Asellota have been known to have spermathecae for many years (e.g., Mac-Murrich 1895; Maercks 1931), but only recently have details of its internal structure been elaborated. The asellotan female genital system has become differentiated into an oviduct and a separate vaginalike structure, the "cuticular organ" (Forsman 1944; Veuille 1978b; Wilson 1986b). Here a more descriptive name for this structure is adopted: *spermathecal duct* (fig. 13.4 E–F). In taxa where details are available (Veuille 1978b; Wilson 1986a,b), the spermathecal duct consists of a long thin-walled cuticular tube open externally and terminating internally at the spermatheca. The latter structure is a small, sometimes multilayered sac inside the proximal oviduct. In some Asellota (Aselloidea, Stenetrioidea), the spermatheca also may be covered with a thin layer of cuticle, but it is uncovered in others (Janiroidea). In most janiroidean Asellota, this structure opens on the dorsal surface (fig. 13.4F), not at the oopore (Forsman 1944; Wolff 1962; Lincoln & Boxshall 1983; Lincoln 1985), hence its original designation *dorsal cuticular organ*. In the other asellotan superfamilies, where insemination once was thought to take place through the oopore (Ridley 1983), the spermathecal duct opens ventrally, is adjacent to the oopore (fig. 13.4E), and is contained in the tissues of the oviduct.

The spermathecal duct has been found only in the Asellota (fig. 13.4E–F), although a potential precursor state, a cuticular lining in the oviduct, occurs in other isopods (e.g., the valviferan *Idotea* and the phreatoicid *Colubotelson;* fig. 13.5). The homologies between the cuticular lining in the oviduct of the nonasellotans and the asellotan spermathecal duct, however, are not clear, and no intermediate condition has yet been observed. The primitive state within the Asellota seems to be the direct association of the two ducts: the spermathecal duct is buried in the tissues of the oviduct, and the openings for both are ventral and adjacent. Copulation, therefore, occurs primitively at the oopore, with the sperm being shunted into the spermathecal duct in the Asellota.

Within the Janiroidea asellotans, the dorsal position of the spermathecal duct appears to be the consequence of disassociation of the two ducts and migration of the spermathecal duct dorsally (Wilson 1986b). While some janiroideans (like *Munna* and *Santia*) retain a ventral opening, an anterolateral position occurs in the Janiridae (fig. 13.6), and the duct is completely dorsal in others. Examples include the Dendrotiidae (Lincoln & Boxshall 1983), the Haploniscidae (Wolff 1962, Lincoln 1985), and the Ischnomesidae (fig. 13.6). During copulation, the male stylet on the pleopod II endopod is inserted directly into the female's dorsal spermathecal duct (Veuille 1978a). Why the duct moved dorsally within the Janiroidea is not clear, although selection for copulatory efficiency has been proposed (Wilson 1986b). All other asellotes that have been examined in detail have accessory structures around the oopore to help guide the male's appendix masculina, such as the stylet receptacle of *Pseudojanira* or the oopore pocket and the folds around the opening of the spermathecal duct in *Asellus* or *Stenetrium*

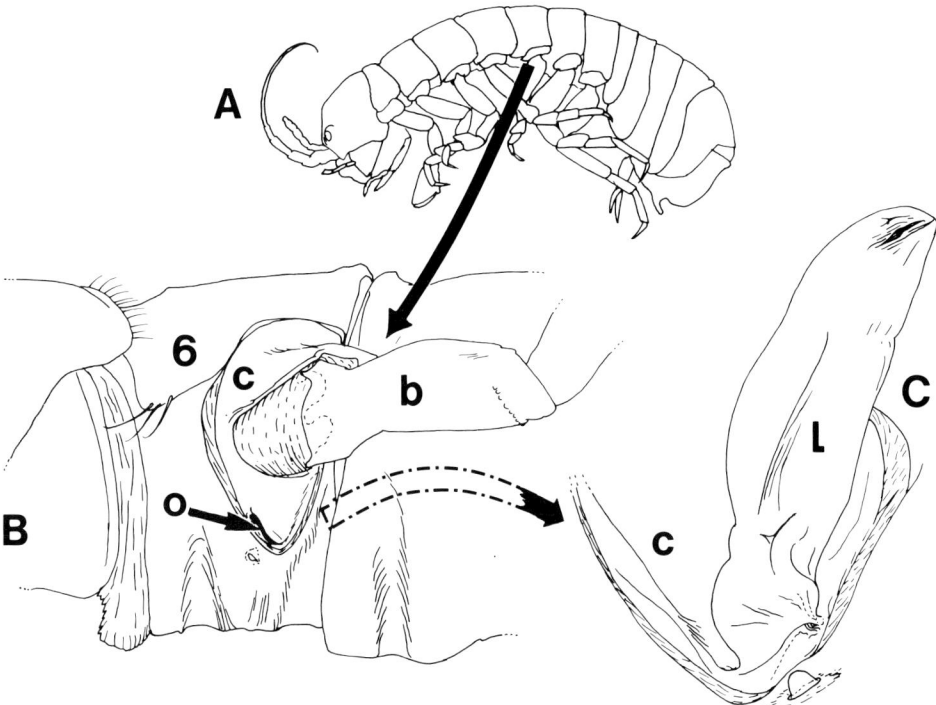

FIGURE 13.5. Cuticular genital structures in a female phreatoicidean isopod (*Colubotelson*), showing the position of the oopore and a cuticular lining of the oviduct. A. Lateral view of a preparatory female. B. External ventrolateral view of the sixth thoracic somite (pereonite 5), showing oopore (small arrow) on coxa. C. Internal medial view of thoracopod coxa in specimen prepared with potassium hydroxide (soft tissues removed), showing cuticular lining of the oviduct. b, basis of sixth thoracopod (pereopod 5); c. coxa of sixth thoracopod; l, lining of oviduct; o (and small arrow), oopore; 6, sixth thoracic somite. Large arrows indicate sources of enlargements in B and C, respectively.

(Wilson 1986b). In contrast, the Janiroidea simply have an opening for the stylet to penetrate (fig. 13.6). Once the stylet is inserted directly into the female's spermathecal duct without the intervention of other structures, perhaps the pore of the spermathecal duct was no longer under selection pressure to remain near the oopore.

MATING SYSTEMS AND DEEP-SEA ISOPODS

The success of the Janiroidea in the deep sea (Kussakin 1973; Hessler, Wilson & Thistle 1979; Hessler & Wilson 1983) may be related to the mating system found in these animals. Despite the tremendous morphological diversity of deep-sea isopods (ibid.), the Janiroidea are defined by their nearly constant sexual morphology (Wilson 1987). This is in decided contrast to the copulatory morphology

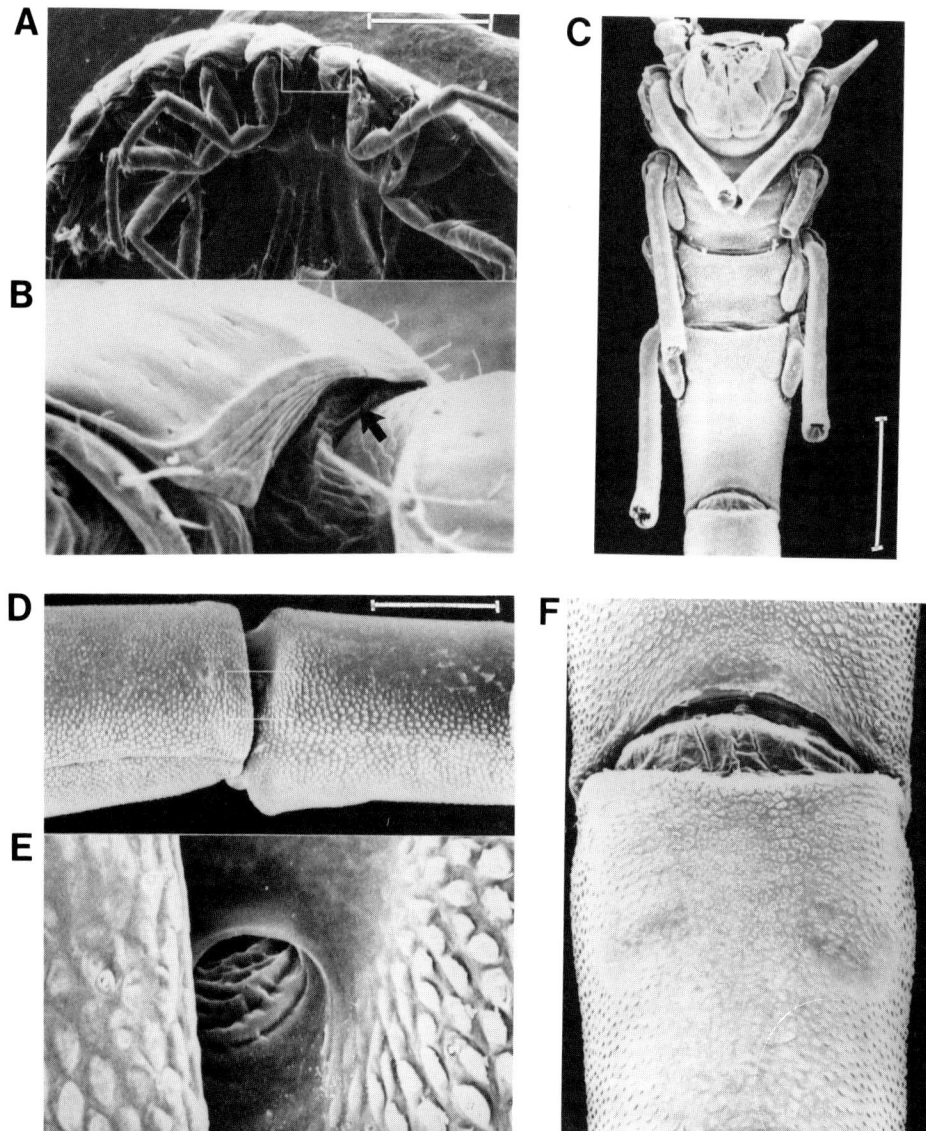

FIGURE 13.6. External pore of the spermathecal duct in two Janiroidean Isopods. A–B. *Ianiropsis* preparatory female. A. Lateral view, small box indicating position of enlargement. Scale bar = 0.5 mm. B. Junction between fifth and sixth thoracic somites, arrow indicating position of spermathecal pore, enlarged 8 times. C–F. *Ischnomesus*, preparatory female. C. Ventral view of anterior part of body. Scale bar = 1.0 mm. D. Dorsolateral view of junction between fifth and sixth thoracic somites, small box indicating position of enlargement. E. Spermathecal pore, enlarged 8 times. F. Unopen oopores on anterior part of thoracic somite 6, enlarged 3.6 times from C.

seen in the remainder of the Asellota, where considerable variability is seen in both the male and female systems (Magniez 1974; Wilson 1987). Consequently, the mating system may be the key adaptation that has led to their success. I propose that, in addition to having the adaptable habits of detritivory (common to many small Crustacea), the mating system of these isopods is ideal for the low populations densities necessary for success in the deep sea (Wilson & Hessler 1987). Because mate location may be a serious barrier to species continuation in low population densities, the ability to mate over a longer time period than in most Crustacea could have important selective benefits. If females can be available for mating during several instars, instead of only during the molt to brooding condition, then the probability of a male finding a female during her receptive period is substantially increased. The complex janiroidean male pleopods, which are mechanically constrained for injecting sperm into a tiny opening in the female's back, may be the adaptation that makes this possible. By decreasing the size of the spermathecal duct, the female can keep the duct open longer without risking invasion by parasites. Increasing the probability that a species can mate successfully means that they can have populations at lower densities than a crustacean that must mate during a short, restricted period in the female's molt cycle. Consequently, the Janiroidea may have been preadapted for deep-sea low population densities. When these isopods colonized this environment, they prospered and became among the most abundant crustacean taxa there. This hypothesis needs further development, although the existing facts are compelling.

FUTURE RESEARCH

Many important questions about isopod genitalia remain to be answered. Outstanding descriptive problems include the detailed morphology of the female genitalia and how the male organs interlink with the female to transfer the sperm. Furthermore, the diversity of isopod genitalia should be incorporated into a general framework so that their functional morphology and phylogenetic patterns are better understood. Finally, a theme interwoven throughout this paper is that the ubiquity and diversity of the isopods may be a result of key adaptations in their reproductive systems. While these ideas are not yet on firm ground scientifically, I hope that they will stimulate more research on isopod mating systems.

Acknowledgments

I thank Ray Bauer, the organizer of this symposium volume, for inviting me to participate. His invitation encouraged me to begin a research program that has resulted in this preliminary review of isopod genitalia and an NSF grant for further study. Moreover, I thank the editors and two outside referees for reviewing and editing my sometimes obscure prose. This research was partially supported by NSF grants BSR86-04573 and BSR88-18448.

Literature Cited

Bowman, T. E. & L. G. Abele. 1982. Classification of the Recent Crustacea. In L. G. Abele, ed., *The Biology of Crustacea.* Vol. 1, *Systematics, the Fossil Record, and Biogeography,* pp. 1–27. New York: Academic Press.

Brusca, R. C. 1984. Phylogeny, evolution and biogeography of the marine isopod subfamily Idoteinae (Crustacea: Isopoda: Idoteidae). *San Diego Soc. Nat. Hist. Trans.* 20:99–134.

Cotelli, F., M. Ferraguti, G. Lanzavecchia & C. L. L. Donin. 1976. The spermatozoon of Peracarida. I. The spermatozoon of terrestrial isopods. *J. Ultrastruct. Res.* 55:378–390.

Forsman, B. 1944. Beobachtungen uber *Jaera albifrons* Leach an der schwedischen Weskuste. *Ark. Zool.* 35A:1–33.

Hessler, R. R. 1982. The structural morphology of walking mechanisms in eumalacostracan crustaceans. *Philos. Trans. R. Soc. Lond. B Biol. Sci.* 296:245–298.

Hessler, R. R. & J.-O. Strömberg. 1989. Observations on the behavior of janiroidean isopods, with special reference to genera that are found in the deep sea. *Sarsia* 74:145–159.

Hessler, R. R. & G. Wilson. 1983. The origin and biogeography of malacostracan crustaceans in the deep sea. In R. W. Sims, J. H. Price & P. E. S. Whalley, eds., *Evolution, Time, and Space: The Emergence of the Biosphere,* Syst. Assoc. Spec. Publ. 23:227–254.

Hessler, R. R., G. Wilson & D. Thistle. 1979. The deep-sea isopods: A biogeographic and phylogenetic overview. *Sarsia* 64:67–75.

Hiraiwa, Y. K. 1935. Studies on a bopyrid, *Epipenaeon japonica* Thielemann. III. Development and life-cycle, with special reference to the sex differentiation in the bopyrid. *J. Sci. Hiroshima Univ.,* series, B div. 1 *(Zool.)* 4:101–141.

Klapow, L. A. 1970. Ovoviviparity in the genus *Excirolana* (Crustacea: Isopoda). *J. Zool. Lond.* 162:359–369.

Kussakin, O. G. 1973. Peculiarites of the geographical and vertical distribution of marine isopods and the problem of deep-sea fauna origin. *Mar. Biol. (Berl.)* 23:19–34.

Kussakin, O. G. 1979. *Marine and Saltwater Isopoda of the Cold and Temperate Waters of the Northern Hemisphere. I. Suborder Flabellifera.* Leningrad: Akad. Nauk. (In Russian.)

Lincoln, R. J. 1985. Deep-sea asellote isopods of the north-east Atlantic: The family Haploniscidae. *J. Nat. Hist.* 19:655–695.

Lincoln, R. J. & G. A. Boxshall. 1983. Deep-sea asellote isopods of the north-east Atlantic: The family Dendrotionidae and some new ectoparasitic copepods. *J. Linn. Soc. Lond. Zool.* 79:297–318.

Luxmore, R. A. 1982. The reproductive biology of some serolid isopods from the Antarctic. *Polar Biol.* 1:3–11.

MacMurrich, J. P. 1895. Embryology of the Isopod Crustacea. *J. Morphol.* 11:63–146.

Maercks, H. H. 1931. Sexualbiologische studien an *Asellus aquaticus* L. *Abt. Allegemeine Zool. Phys. Tier* 48:399–507.

Magniez, G. 1974. Données faunistiques et ecologiques sur les Stenasellidae. *Int. J. Speleol.* 6:1–180.

Mead, F. 1976. La place de l'accouplement dans le cycle de reproduction des Isopodes terrestres (Oniscidea). *Crustaceana (Leiden)* 31:27–41.

Menzies, R. J. 1954. The comparative biology of reproduction in the wood-boring isopod crustacean *Limnoria Bull. Mus. Comp. Zool. Harv. Univ.* 112:362–388.

Michel, W. C. 1986. Contact chemoreception and mate recognition by an Antarctic crustacean. *Chem. Sens.* 11:638–639.

Monod, T. 1926. Les Gnathiidae: Essai monographique. *Mém. Soc. Sci. Nat. Phys. Maroc Zool.* 13:1–667.

Nicholls, G. E. 1943. The Phreatoicidea. Part I. The Amphisopidae. *Pap. Proc. R. Soc. Tasman.* 1942:1–145.

Pires, A. M. S. 1987. *Potiicoara brasiliensis:* A new genus and species of Spelaeogriphacea (Crustacea: Peracarida) from Brazil with a phylogenetic analysis of the Peracarida. *J. Nat. Hist.* 21:225–238.

Radu, V. G. & C. Craciun. 1972. Ultrastructure du segment terminal du canal déférent chez *Porcellio scaber* Latr. *Rev. Roum. Biol., Ser. Zool.* 17:167–173.

Raup, D. M. 1986. Biological extinction in earth history. *Science (Washington, D.C.)* 231:1528–1533.

Reger, J. F., P. W. Itaya & M. E. Fitzgerald. 1979. A thin-section and freeze-fracture study on membrane specializations in spermatozoa of the isopod *Armadillium vulgare. J. Ultrastruct. Res.* 67:180–193.

Ridley, M. 1983. *The Explanation of Organic Diversity: The Comparative Method and Adaptations for Mating.* Oxford: Clarendon Press.

Sars, G. O. 1899. Isopoda. *Crustacea of Norway*, 2:1–270. Bergen, Norway: Bergen Museum.

Schram, F. R. 1970. Isopod from the Pennsylvanian of Illinois. *Science (Washington, D.C.)* 169:854–855.

Schram, F. R. 1974. Paleozoic Peracarida of North America. *Fieldiana Geol.* 33:95–124.

Sheppard, E. 1957. Isopod Crustacea. Part II. The suborder Valvivera, families Idotheidae, Pseudidotheidae and Xenarcturidae, Fam. N. with a supplement to Isopoda Crustacea. Part I. The family Serolidae. *Discovery Rep.*, no. 29:141–198.

Shuster, S. 1987. Alternative reproductive behaviors: Three discrete male morphs in *Paracerceis sculpta*, an intertidal isopod from the northern Gulf of California. *J. Crustacean Biol.* 7: 318–327.

Van Lieshout, S. E. N. 1983. Calabozoidea, a new suborder of stygobiont Isopoda, discovered in Venezuela. *Bijdr. Dierkd.* 53:165–177.

Veuille, M. 1978a. Biologie de la reproduction chez *Jaera* (Isopode Asellote) I. Structure et fonctionnement des pieces copulatrices males. *Cah. Biol. Mar.* 19:299–308.

Veuille, M. 1978b. Biologie de la reproduction chez *Jaera* (Isopode Asellote) II. Evolution des organes reproducteurs femelles. *Cah. Biol. Mar.* 19:385–395.

Veuille, M. 1980. Sexual behaviour and evolution of sexual dimorphism in body size in *Jaera* (Isopoda Asellota). *Biol. J. Linn. Soc.* 13:89–100.

Wägele, J.-W. 1982. A new hypogean *Cyathura* from New Caledonia (Crustacea, Isopoda, Anthuridea). *Bull. Zool. Mus. Univ. Amst.* 8:189–197.

Wägele, J.-W. 1983. On the origin of the Microcerberidae (Crustacea: Isopoda). *Z. Zool. Syst. Evolutionsforsch.* 21:249–262.

Wägele, J.-W. 1988. Aspects of the life-cycle of the Antarctic fish parasite *Gnathia calva* Vanhöffen, 1914 (Crustacea: Isopoda). *Polar Biol.* 8:287–291.

Wägele, J.-W. 1989. Evolution und Phylogenetisches system der Isopoda. Stand der Forschung und neue Erkenntnisse. *Zoologica* 140:1–262.

Wilson, G. 1986a. Pseudojaniridae (Crustacea: Isopoda), a new family for *Pseudojanira stenetrioides* Barnard, 1925, a species intermediate between the asellote superfamilies Stenetrioidea and Janiroidea. *Proc. Biol. Soc. Wash.* 99:350–358.

Wilson, G. 1986b. The evolution of the female cuticular organ in the Asellota (Crustacea, Isopoda). *J. Morphol.* 190:297–305.

Wilson, G. 1987. The road to the Janiroidea: The comparative morphology and evolution of the asellote Isopoda crustaceans. *Z. Zool. Syst. Evolutionsforsch.* 25:257–280.

Wilson, G. & R. Hessler. 1987. Speciation in the deep sea. *Annu. Rev. Ecol. Syst.* 18:185–207.

Wilson, G., D. Thistle & R. Hessler. 1976. The Plakarthriidae (Isopoda; Flabellifera): Déjà vu. *Zool. J. Linn. Soc.* 58:331–343.

Wirth, U. 1984. Die Struktur der Metazoen-Spermien und ihre Bedeutung für die Phylogenetik. *Verh. Naturwiss. Ver. Hambg.* 27:295–362.

Wolff, T. 1962. The systematics and biology of bathyal and abyssal Isopoda Asellota. *Galathea Rep.* 6:1–320.

FOURTEEN
Functional Morphology of Spermatophores and Sperm Transfer in Calanoid Copepods

PAMELA I. BLADES-ECKELBARGER

Harbor Branch Oceanographic Institution, Fort Pierce, Florida.

Abstract

Spermatophores produced by most calanoid copepods are simple, tube-shaped flasks that contain the spermatozoa and seminal secretions. They adhere to the external cuticle of the female by means of a cementlike substance present on the tapered, open end of the flask. Other calanoids have evolved spermatophore flasks connected to highly complex, chitinlike coupling plates. These have a specific configuration that corresponds to the morphology of the conspecific female's urosome, thus assuring proper placement. Spermatophore transfer is accomplished by a ritualized and precise mating behavior controlled by pheromonal attraction and structurally modified copulatory appendages. The distinctive morphologies of the spermatophores, copulatory appendages, and female genitalia, in combination with species-specific mating behavior, function as controlling factors of reproductive isolation in calanoid copepods.

SPERMATOPHORES PRODUCED by copepods are typically tube-shaped flasks that contain spermatozoa and seminal secretions. In calanoid copepods, a single mature spermatophore is stored within the male's metasome until copulation, at which time it is extruded and attached to the external surface of the female genital segment with the aid of specific male appendages. The spermatophoric contents empty into internal chitin-lined spermathecal sacs where they are stored until the oocytes are spawned.

Published information concerning spermatophore morphologies and mating behaviors in calanoids is very limited relative to the large number of species in this group. This paper will combine information from the literature with new observations on the functional morphology of spermatophores, copulatory appendages, and female genitalia of some marine calanoid copepods. The significance of these reproductive structures and mating behavior in establishing reproductive isolating barriers will be discussed.

MORPHOLOGY OF REPRODUCTIVE STRUCTURES

Copulatory Appendages

In numerous calanoid families (e.g., Pontellidae, Centropagidae, Temoridae, Metridiidae, Pseudodiaptomidae, Lucicutiidae, Heterorhhabdidae, Augaptilidae, Arietillidae, Candaciidae, Bathypontiidae, Acartiidae, and Tortanidae [Brodskii 1967a,b]), the antennule (A1) of the male (generally the right one) is geniculate and prehensile, modified with one or two hinged joints that enable the antennule to fold back upon itself (fig. 14.1). Segments on either side of the hinge are lined with cuticular teeth and equipped with sensory setae. Intergeneric differences include the relative location of the hinges along the length of the antennule, extent of development of cuticular teeth, and number of sensory setae or aesthetascs. This appendage is used for the initial capture of the female (fig. 14.9A).

Calanoid copepods typically have five pairs of thoracic swimming legs. Pairs 1–4 are biramous, symmetrical, and in the majority of species essentially identical. However, in many calanoid families the fifth pair of swimming legs (P5) on both the male and female rarely resembles P 1–4 and its structure is used often as a taxonomic feature of males. The female P5 is symmetrical but usually reduced or absent. This appendage has been observed in females of *Labidocera aestiva* to function in the removal of the empty spermatophore (Blades & Youngbluth 1979).

The male P5 is asymmetrical to varying degrees, but often is strongly modified to hold the female and transfer the spermatophore. Structural modifications of the male's P5 range from a slight reduction or thickening of spines and setae to the development of a chela with powerful musculature on one leg of the pair (fig. 14.2A). Development of this clawlike structure is notable in some calanoid families such as the Pontellidae and Centropagidae. In *Labidocera aestiva*, for example, the right leg is uniramous and the terminal segment is modified into a large chela (fig. 14.2A). The chela is equipped with several spines, presumably sensory in function (Blades & Youngbluth 1980). The left leg is biramous, con-

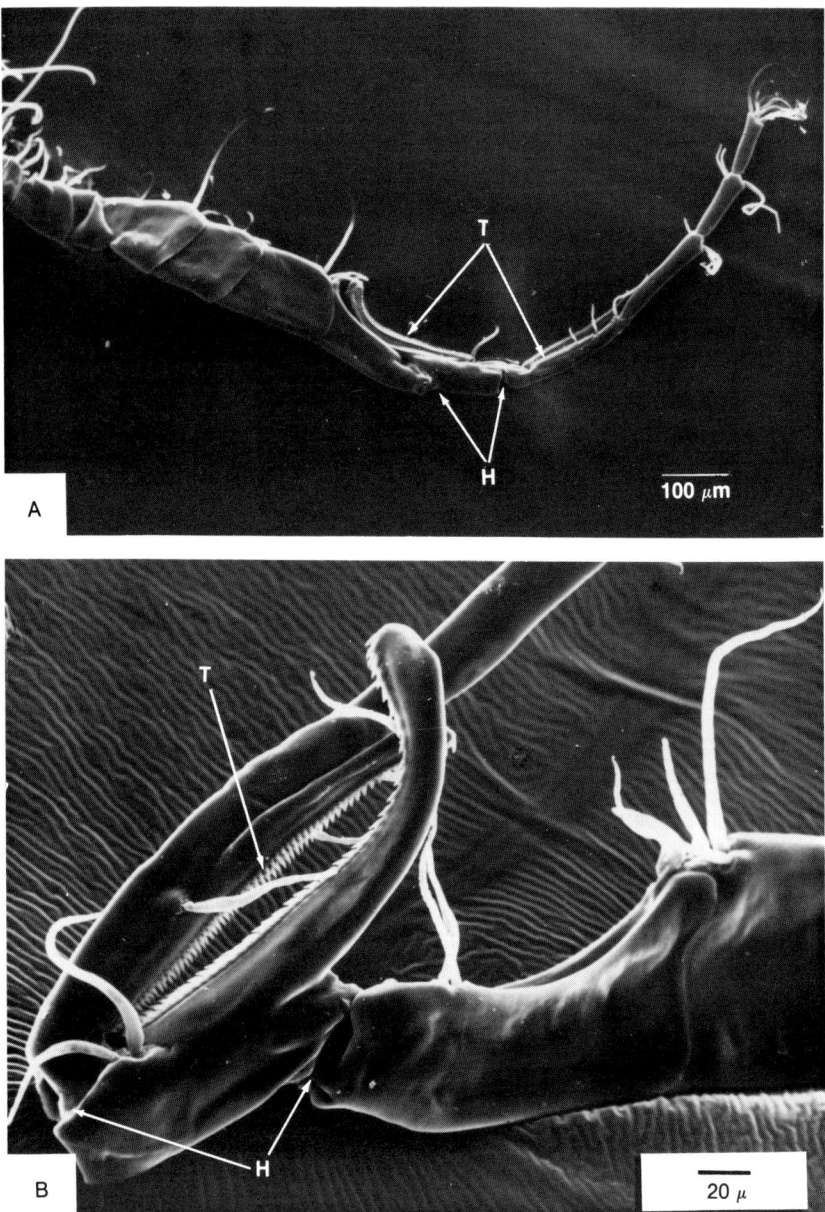

FIGURE 14.1. Distal segments of geniculate antennule from male *Labidocera aestiva*. A. Antennule open. B. Antennule closed. H, hinges; T, toothed margins. Note numerous hairs. Figure 14.1B from Blades and Youngbluth 1979.

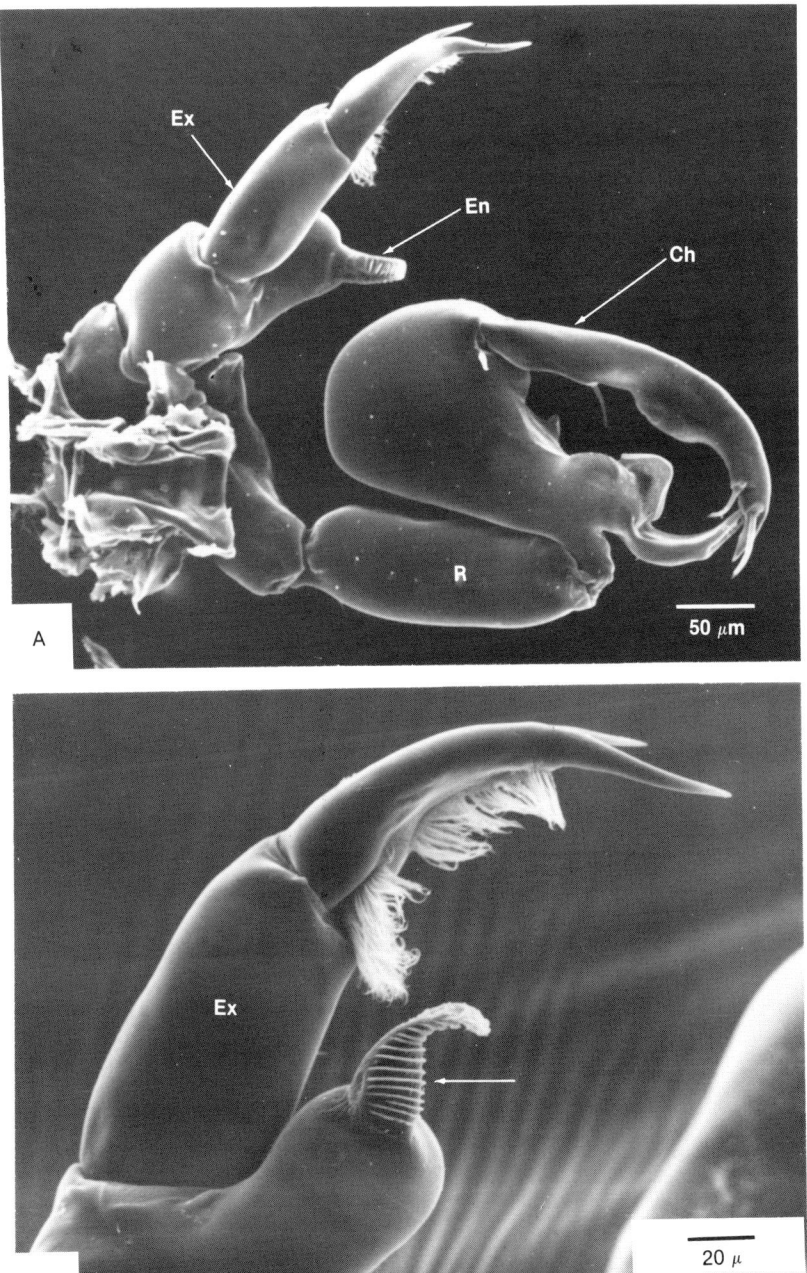

FIGURE 14.2. A. P5 from male *Labidocera aestiva* showing large chela (Ch) of right leg (R). En, endopod; Ex, exopod; L, left leg. B. Closer view of left leg showing details of exopod and corrugated surface on distal tip of endopod (unlabeled arrow). Both figures from Blades and Youngbluth 1979.

sisting of a unisegmented endopod and a double-segmented exopod (fig. 14.2). The distal section of the exopod terminates in a bifurcation of one short and one long spine and has two distinct patches of long setae located along the medial margin (fig. 14.2B). The distal acuminate portion of the endopod has a distinctly corrugated surface (fig. 14.2B). The specific functions of this pair of legs have been described by Blades and Youngbluth (1979, 1980). The chela holds the female's urosome in the copulatory position (Figs. 14.9B–D). The left leg has a dual role: it functions in the tactile examination of the female's genital field and "cleaning" of the spermatophore attachment site by the corrugated endopod, and it transfers the spermatophore that adheres to the exopodal hairs.

In some calanoid genera, the chelate structure of the male's fifth leg is much less pronounced, but this appendage may assume other unusual forms. For example, the P5 on the male *Euchaeta* is characterized by its length (nearly that of the metasome) and the complex morphology of the left leg's distal segments that function to hold the spermatophore (fig. 14.3; Park 1975, 1978; Hopkins, Mauchline & McLusky 1978; Ferrari & Dojiri 1987). Variable morphological features of the left P5 include: (1) a serrated lamelliform process off segment 2 that may be flat and bladelike (*E. marina* [figs. 14.3A, B] or partially toothed and dagger-shaped (*E. norvegica*); (2) a digitiform process off segment 2 of variable size and length that may be reduced and tapered (*E. marina* [fig. 14.3B]; *E. antarctica*), but more often is club-shaped with a corrugated surface (*E. norvegica* [fig. 14.3C]); (3) an assortment of short stiff spines and tufts of thickened setae along the medial margin of segment 3 (figs. 14.3B,C). At the base of the third exopodal segment of the left leg in *E. marina* is a small, thumblike process that appears to be "fleshy" in texture and covered with papillae (fig. 14.3B). This process was also noted in *E. rimana*, *E. marinella*, and *E. indica* by Bradford (1974), who termed it a "thin-skinned lobe." Male *Euchaeta* are frequently observed in plankton collections holding a spermatophore between the digitiform process and the lamelliform structure (fig. 14.3C).

A most unusual P5 is present in the male calanid *Undinula vulgaris* (fig. 14.4). The right leg is greatly reduced, consisting of a tiny 3-segmented endopod and a larger 3-segmented exopod. In contrast, the left leg is extremely long; when it is extended, its length equals nearly that of the metasome (fig. 14.4A). The leg hinges at two locations, enabling the male to carry it folded against the metasome. The last two segments each terminate in long stout spines, and together they form a shape that may be considered clawlike (figs. 14.4A,B). The last segment is very complex, composed of a large knob with a partially corrugated surface (fig. 14.4C). Extending from this knob is another elongate process that appears to be composed of soft, flexible tissue (figs. 14.4B, C, D). The length of this fleshy protuberance is covered with papillae, and its broadened tip is covered with numerous fine setae. A short, pointed digitiform process projects off one

FIGURE 14.3. A. Left (L) and right (R) legs of P5 from male *Euchaeta marina* extending along length of urosome. Lp, lamelliform process. B. *E. marina*, higher magnification, terminal segments of left leg showing serrated lamelliform process (Lp) and digitiform process off segment 2; and thumblike process (Tp) off base of segment 3. C. *E. norvegica*, male P5. Spermatophore (*) held between club-shaped, digitiform process (Dp) and lamelliform process. Note corrugated surface of digitiform process.

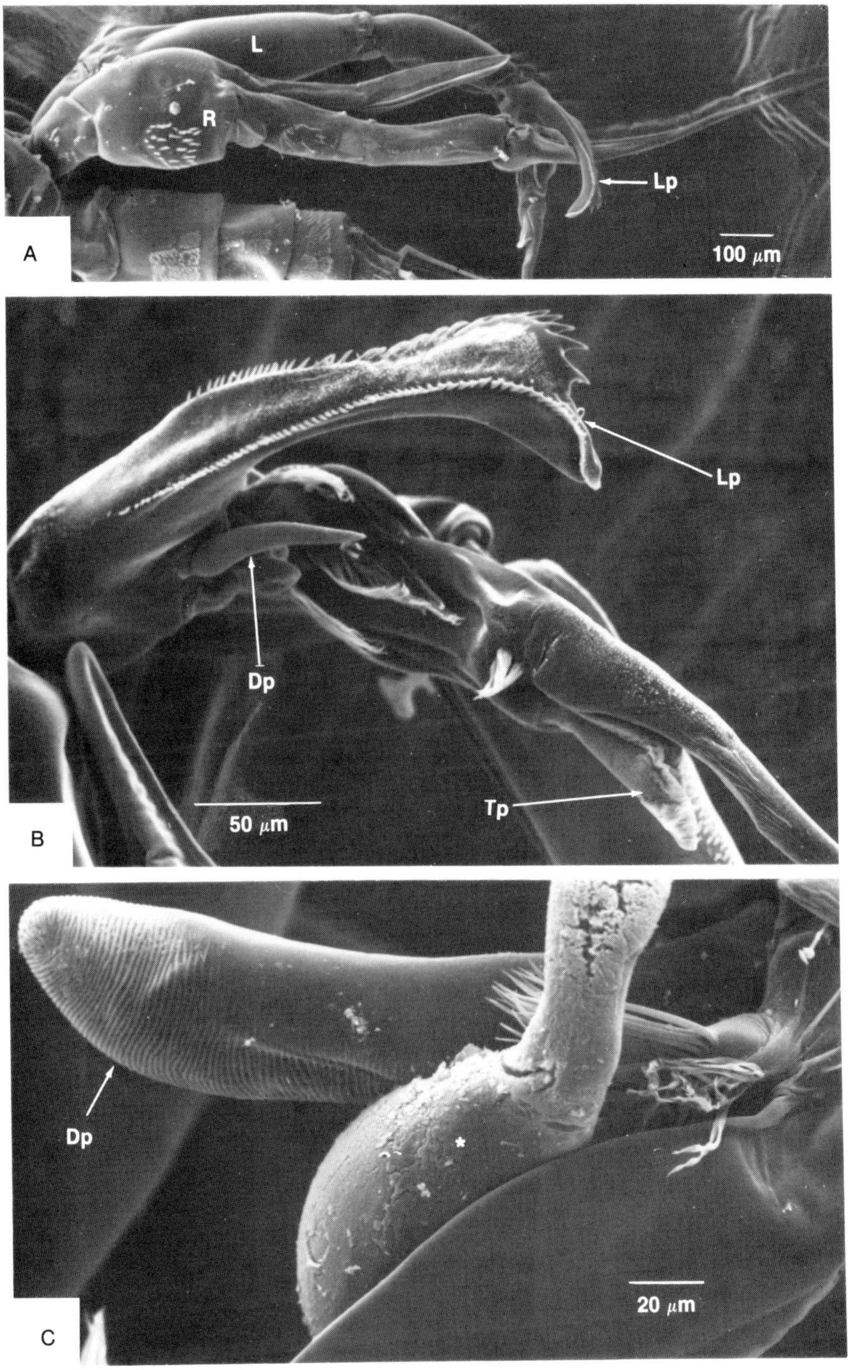

Spermatophore Morphology in Calanoid Copepods 251

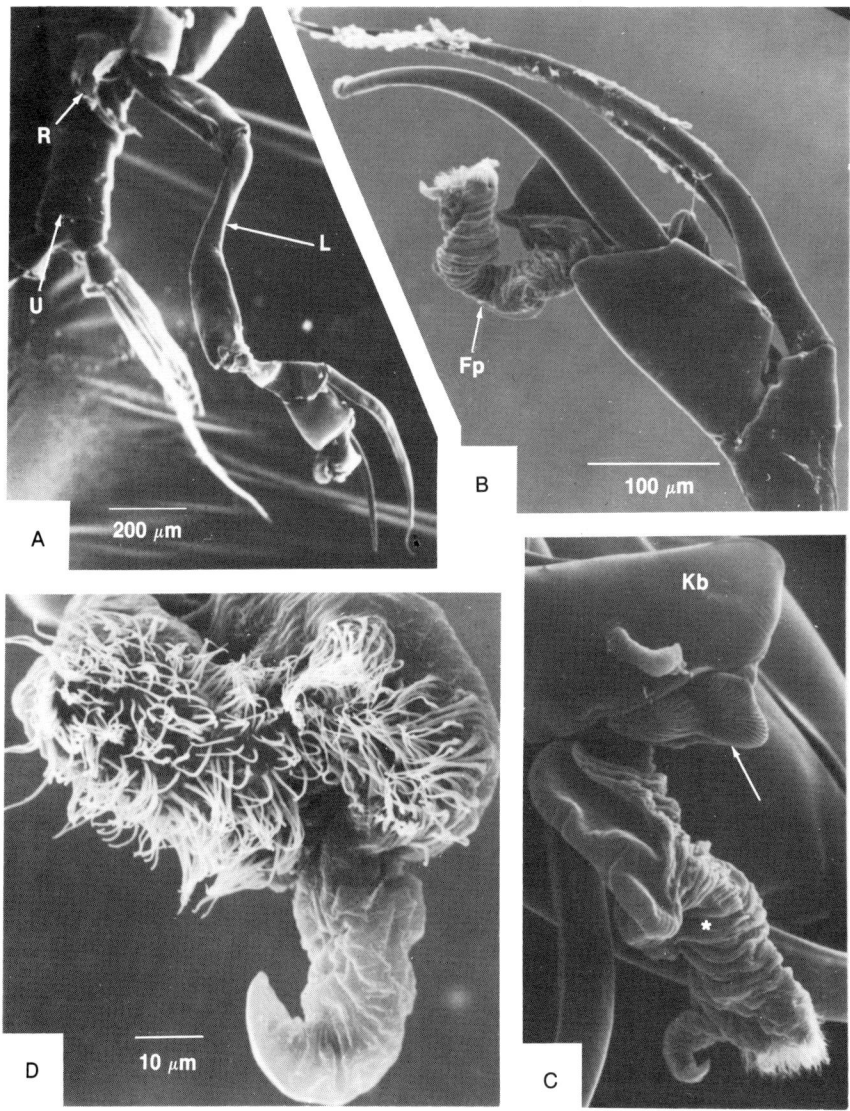

FIGURE 14.4. P5 of *Undinula vulgaris* male. A. Left leg (L) extended. R, right leg; U, urosome. B. Distal segments of left P5 showing clawlike shape and flexible protuberance (Fp). C. Last segment of left P5 showing knob (Kb) with corrugated surface (unlabeled arrow) and flexible protuberance (*). Scale bar = 30 μm. D. Distal tip of flexible protuberance showing hairs and thumblike projection.

edge. The manner in which the male *U. vulgaris* uses this leg during spermatophore transfer remains unknown.

Female Genitalia

In female calanoids, the external and internal morphology of the genital region are notably variable and often used as taxonomic characters (Williams 1972;

Park 1975, 1978). The opening of the genital tract, the genital pore, is positioned on the ventral surface of the first urosomal segment, the genital segment (figs. 14.5, 14.10c). The genital pore in most genera is covered by a flaplike structure of variable morphology, referred to as the genital plate or operculum, that articulates along its anterior margin (figs. 14.5B, 14.10c). The cuticular area surrounding the genital pore, the genital field, may be smooth (figs. 14.5A,D) or adorned to various degrees with cuticular spines, hairs, and glandular pores (fig. 14.10A,C). In some species, (e.g., Euchaetidae) the morphology of the genital field is very complex and species-specific, exhibiting paired lateral "flanges" and an assortment of grooves and cuticular infoldings (fig. 14.5c and Geptner 1968).

Internally, the paired oviducts terminate at separate gonopores that lie just beneath the genital pore. The gonopores typically merge at a common atrium, although in a few species, e.g., *Scolecethrix danae* (fig. 14.5D), the gonopores open to the exterior separately, a condition that supposedly is primitive (G. Boxshall, personal communication). Paired, chitin-lined spermathecal pouches, usually referred to as spermathecae or seminal receptacles, are also present beneath the genital opening on either side of the genital plate. Depending on the species, the spermathecae merge with the oviducts separately or open into the common atrium.

Spermatophores

Spermatophores of calanoid copepods are simple, tube-shaped flasks, sometimes referred to as the spermatophore flask or spermatophore proper (figs. 14.6A,B, 14.7A,B,D, 14.10A, 14.11A, 14.12c), that store the spermatozoa and associated seminal secretions. Toward its open end, the flask narrows into a spermatophore neck of variable length (figs. 14.6A, 14.7A,D). In the majority of calanoids, the spermatophore adheres to the female by a cementlike secretion present on the outside of the spermatophore neck or by secretions extruded from the spermatophore itself (figs. 14.6A, 14.10A,B, 14.12c). However, in some calanoids, most notably members of the Pontellidae and Centropagidae, the spermatophore flask is connected to one or more chitinlike plates (figs. 14.6B,C, 14.7A,B,D, 14.11A). These plates, termed coupling plates, the coupling apparatus, or coupler (after Koppler, Heberer 1932), carry an adhesive secretion that, in combination with their unique shape, secures the spermatophore to the urosome of the female. The morphology of the coupling plates, therefore, is unique in each species and corresponds to the external morphology of the conspecific female's urosome and genital region. This has been called a "key-and-lock" relationship (Fleminger 1967; Lee 1972), implying that the coupler assures attachment of the spermatophore at a precise location on a conspecific female (figs. 14.6B,C, 14.7A,B). The configuration of the coupling apparatus within a single genus may be highly variable, as reported by Fleminger (1967, 1975, 1979) for the pontellid *Labidocera*. For example, the plates of the coupler form simple ventral shields in *L. aestiva* (figs. 14.6B, 14.7D), whereas they encapsulate the urosome totally in *L. scotti* (fig. 14.6c). Even more elaborate examples were reported for *L. barbadiensis* (Fleminger & Moore 1977) and *L. barbudae* (Fleminger 1979), in which the spermatophores are connected to several coupling plates arranged in a complex

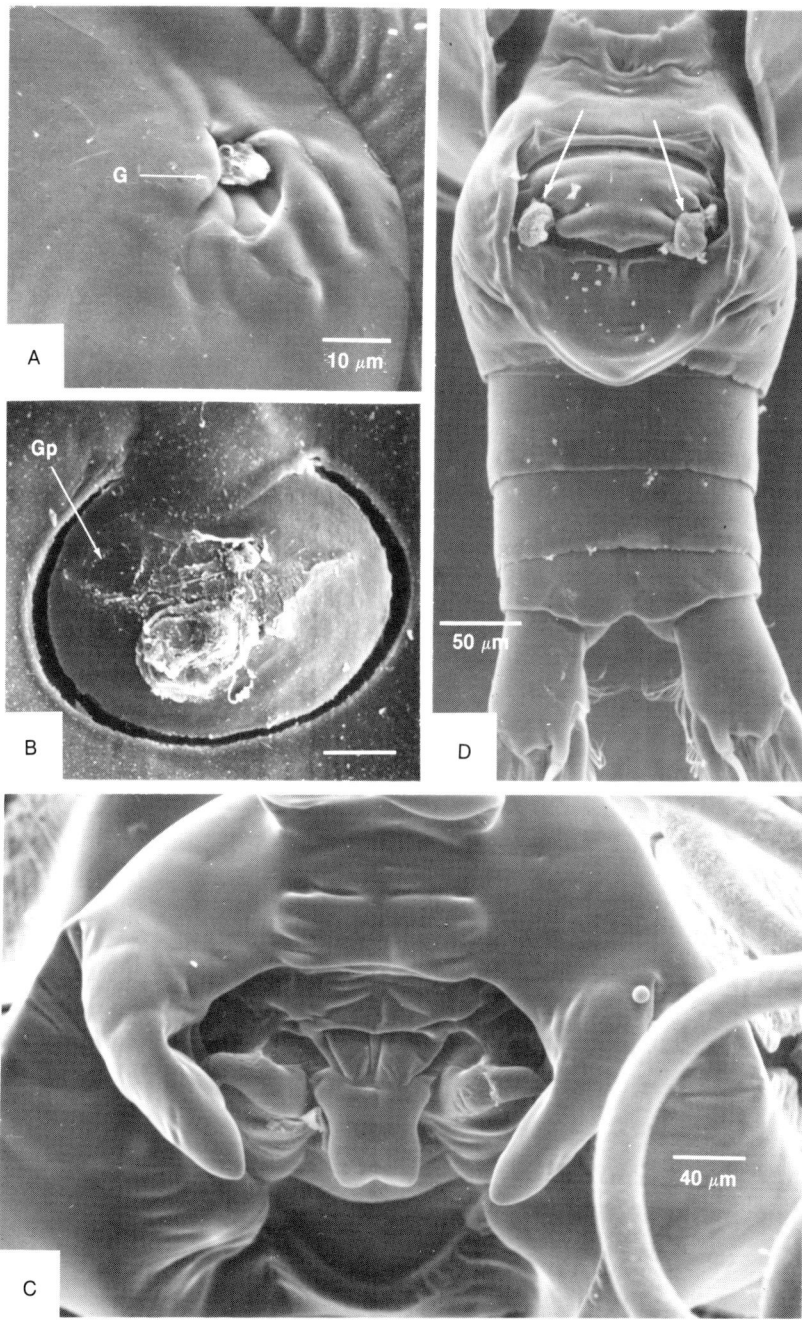

FIGURE 14.5. Morphological diversity of female genital pores (G). A. *Pleuromamma gracilis*. B. *P. abdominalis*. Gp, genital plate. Scale bar = 20 μm. C. *Euchaeta antarctica*. D. *Scolecethrix danae*. Unlabeled arrows indicate two separate genital openings, both with remnants of spermatophore contents.

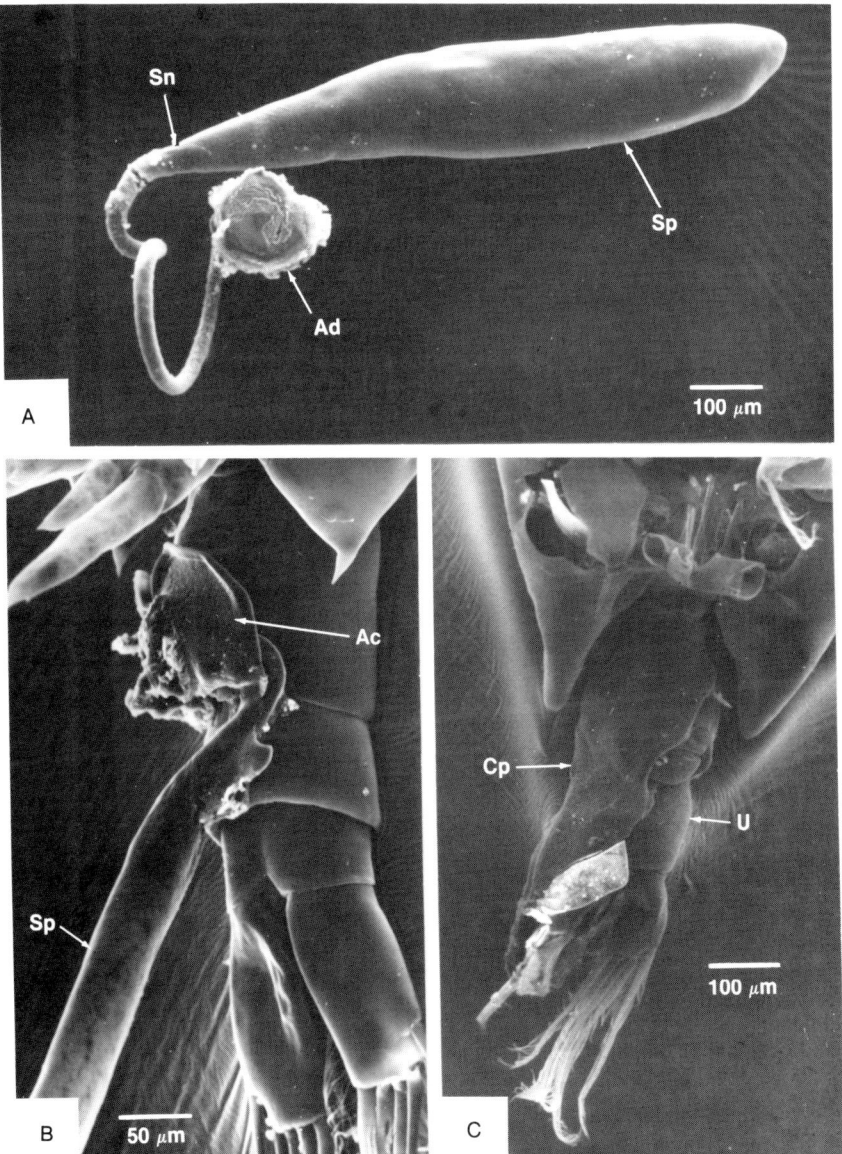

FIGURE 14.6. Spermatophore morphology. A. Simple spermatophore from *Euchaeta norvegica*. Ad, attachment disc; Sn, spermatophore neck; Sp, spermatophore proper. B. *Labidocera aestiva* female, urosome with spermatophore attached to ventral surface; simple coupler showing anterior coupling plate (Ac) covering genital pore. Sp, spermatophore proper. C. *L. scotti* female urosome (U) with spermatophore attached. Note encapsulating coupler (Cp).

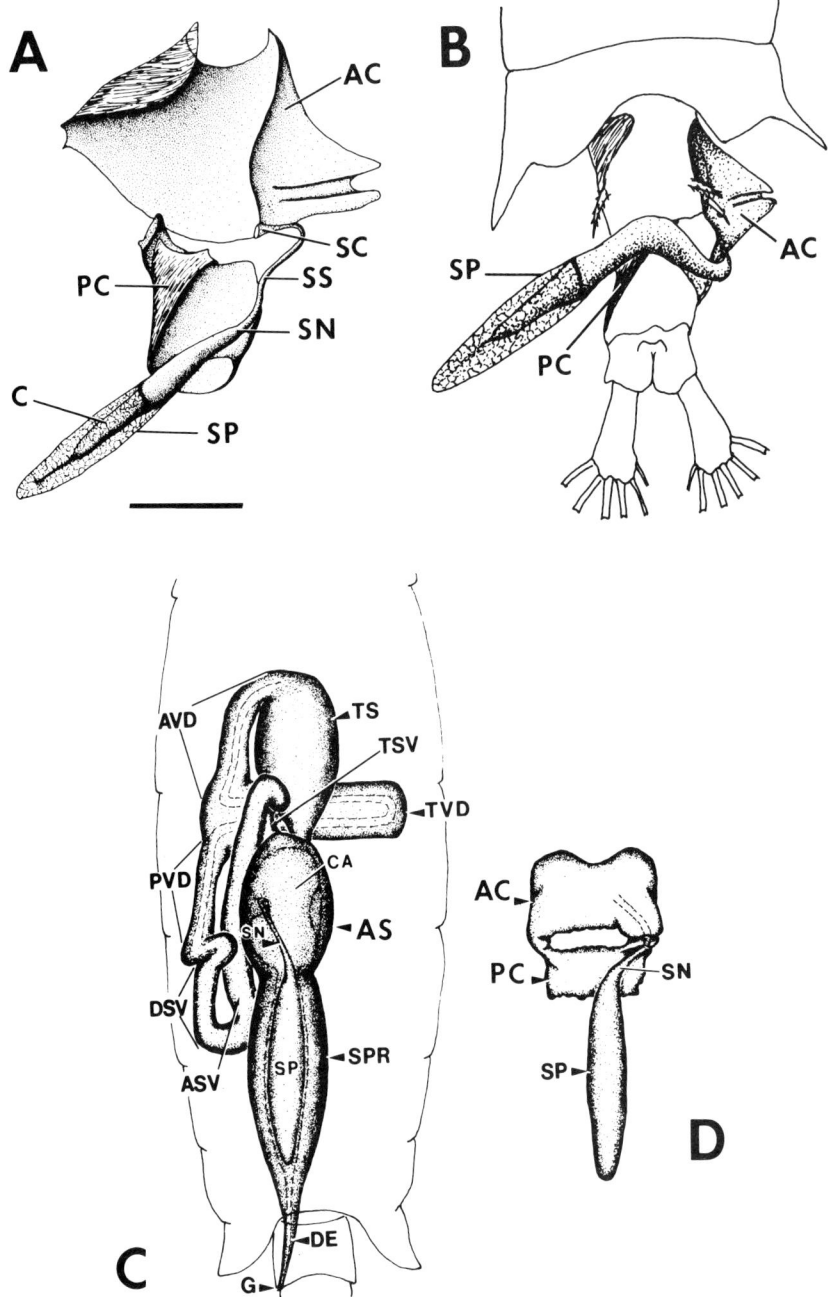

manner around the urosome, with the dorsal plate sending a long extension over the dorsal surface of the cephalothorax.

Male Reproductive System and Spermatophore Formation

The spermatophore is produced entirely within the male reproductive system. The morphology of the male genital system and the process of spermatophore formation have been studied previously for a few calanoid copepods using light microscopy (Heberer 1932; Park 1966). Ultrastructural studies on copepods that produce simple spermatophores are available for *Calanus finmarchicus* (Raymont et al. 1974) and *Euchaeta norvegica* (Hopkins 1978). Ultrastructural details of the more complex system in *Labidocera aestiva*, which produces a spermatophore with coupling plates, are found in Blades and Youngbluth (1981) (fig. 14.7c,D).

The reproductive system of male calanoids contains organs comparable in function to the male accessory glands of other invertebrates (see Adiyodi & Adiyodi 1975 for review) where the secretions produced by the glands contribute to the internal and external components of the spermatophore. The reproductive tract of male calanoids consists of a single testis and a long, sinuous genital duct that terminates at a gonopore on the first urosomal segment (fig. 14.7c). The genital duct is morphologically divided into the vas deferens (ductus deferens), seminal vesicle, spermatophore sac, and chitin-lined ductus ejaculatorius. As immature spermatozoa pass from the testis through these regions, the various components of the seminal fluid, spermatophore wall, and coupling apparatus (if present) are produced and secreted into the lumen.

In general, the first part of the vas deferens produces secretions that compose the seminal fluid, often referred to as the core secretion, that surrounds the spermatozoa inside the spermatophore. These secretions take on the form of flocculent material and granules of varying size and density. The last part of the vas deferens continues to produce seminal secretions as well as material that forms the surrounding spermatophore wall. The seminal vesicle is much less glandular yet may produce additional material for the spermatophore wall.

The large, elongate, and highly glandular spermatophore sac is divided into two morphologically distinct but contiguous regions, the anterior spermatophore sac (="former" or "molder" of Heberer 1932) and the posterior sac proper

FIGURE 14.7. A. Spermatophore complex of *Centropages typicus*. Scale bar = 100 μm. B. Diagramatic illustration, dorsal view of properly positioned spermatophore on urosome of *C. typicus* female (not drawn to scale). C. Diagrammatic illustrations of male reproductive system, dorsal view. D. Spermatophore complex of *Labidocera aestiva*. AC, anterior coupling plate; ASV, ascending seminal vesicle; AVD, anterior vas deferens; C, core secretion; CA, coupling apparatus (inside lumen of anterior spermatophore sac [AS]). DE, ductus ejaculatorius; DSV, descending seminal vesicle; G, gonopore; PC, posterior coupling plate; PVD, posterior vas deferens; SC, spermatophore cup; SN, spermatophore neck; SP, spermatophore proper; SPR, spermatophore sac proper; SS, spermatophore stalk; TVD, transverse vas deferens; TS, testis; TSV, terminal part of seminal vesicle. Figures 14.7A and 14.7B from Blades 1977, 14.7c and 14.7D from Blades-Eckelbarger 1986.

(fig. 14.7c). Depending on the species, simple adhesive secretions or more complex plates of a coupling apparatus are produced within the anterior spermatophore sac and connected to the neck region of a mature spermatophore flask that lies in the lumen of the sac proper. The size and cellular composition of the anterior spermatophore sac apparently are related to the final configuration and complexity of the coupling apparatus, if present. For example, *Euchaeta norvegica* produces a spermatophore without coupling plates (Hopkins 1978). Consequently, the anterior spermatophore sac is represented by a simple gland that secretes an adhesive substance onto the spermatophore neck. The next level of complexity is observed in *Undinula vulgaris* (Blades-Eckelbarger personal observation), in which the anterior spermatophore sac is composed of 4 to 6 morphologically distinct secretion types. Favorable sections through this region (fig. 14.8A) reveal that once released into the lumen, these secretions form a large, indistinct mass that adheres to the spermatophore neck (fig. 14.12c).

With respect to calanoids that form spermatophores associated with couplers, *Pleuromamma abdominalis* produces a small cup-shaped coupler within a much reduced anterior spermatophore sac composed of 4 to 6 different types of secretion (fig. 14.8B; Heberer 1932; Blades-Eckelbarger, personal observation). In contrast, this glandular region in the pontellids *Epilabidocera amphitrites* (Park 1966) and *Labidocera aestiva* (Blades & Youngbluth 1981) produces 7 to 8 different secretion types that are molded by the shape of the lumen into the highly complex plates of the coupling apparatus (figs. 14.6B, 14.7D, 14.8C).

MATING BEHAVIOR

Reports of copulatory behavior in calanoid copepods as observed on living specimens (Gauld 1957; Jacobs 1961; Katona 1975; Blades 1977; Blades & Youngbluth 1979; Jacoby & Youngbluth 1983), in addition to other descriptions of reproductive morphology and spermatophore attachment sites (see Vaupel Klein 1982 for review), provide convincing evidence that the mating behavior of calanoid copepods is precise, ritualized, and species-specific. In general, the mating behavior of calanoids appears to follow the same sequence of events: (1) attraction of the male to the female, (2) capture of the female by the male, (3) movement into the copulatory position, (4) spermatophore transfer and attachment, (5) release of the female, and (6) discharge of the spermatophoric contents.

Upon receiving chemical stimulation, probably by perception of pheromones emitted by the female (Katona 1973; Griffiths & Frost 1976), the male performs a distinctively different swimming behavior often referred to as "searching

FIGURE 14.8. Cellular diversity of the anterior spermatophore sac (AS) outlined region, (light micrographs, 1 μm thick sections). A. *Undinula vulgaris*, sagittal section. Note adhesive mass (*) in lumen connected to partial section of spermatophore proper (Sp). B. *Pleuromamma abdominalis*, sagittal section. Note cup-shaped coupler (C) in lumen (L). L', lumen of spermatophore sac proper (SPR). C. *Labidocera aestiva*, frontal section. Note shape of lumen that "molds" anterior coupling plate (AC) and posterior coupling plate (PC).

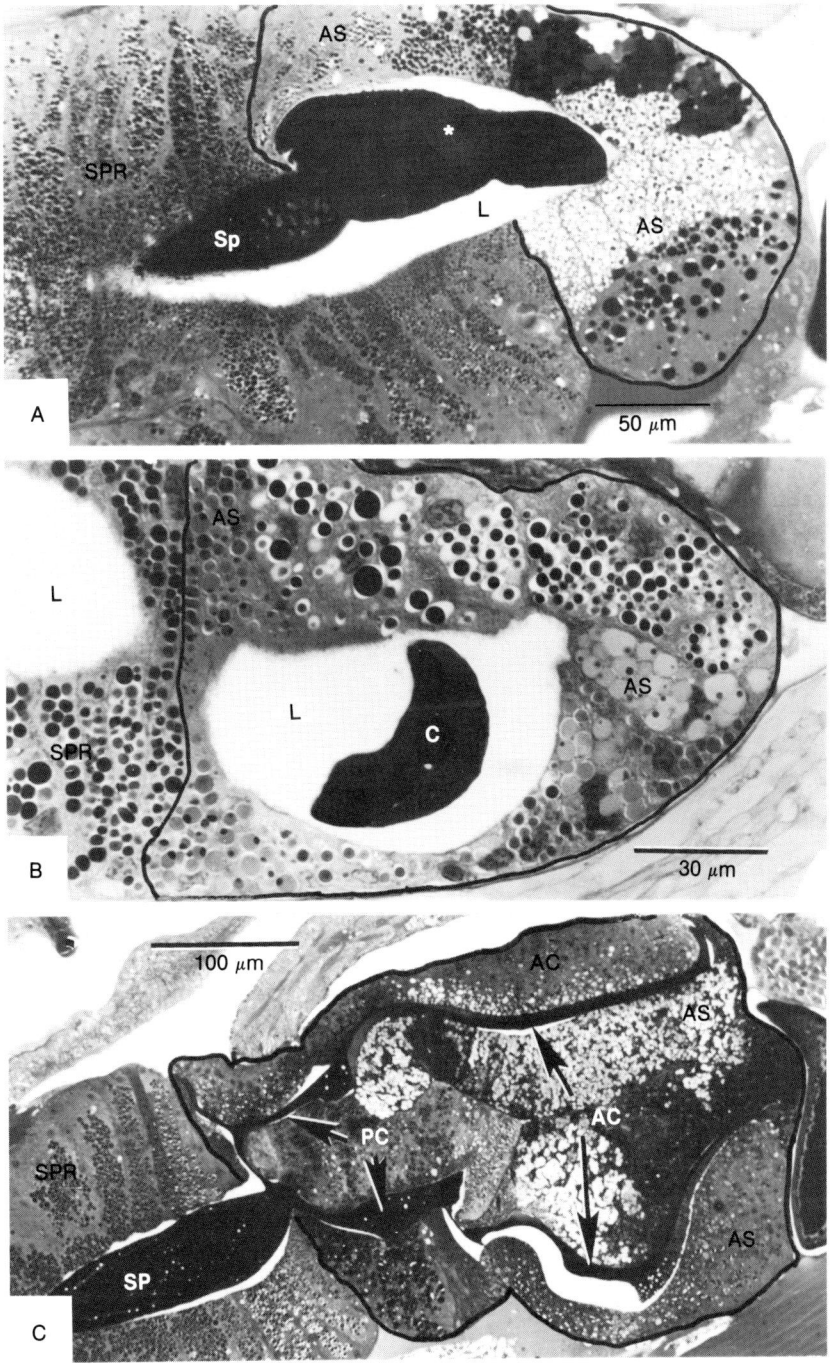

movements" or "mate-seeking behavior" (Parker 1902; Jacobs 1961; Katona 1973, 1975; Jacoby & Youngbluth 1983). The male's erratic swimming movements bring him in close proximity to a potential mate, where he may receive further chemical or mechanical cues from the swimming wake of the female (Strickler & Bal 1973). The male uses the geniculate antennule to grasp the female by her caudal rami or caudal setae (fig. 14.9A). During this initial physical contact, specific hairs on the male's antennule (fig. 14.1B) may be stimulated when it closes, thereby signaling the male to move into the copulatory position (Blades & Youngbluth 1980).

In the copulatory position (fig. 14.9B), the male secures the female with the chelate part of his fifth leg and releases the antennular hold. Now the mating pair lie in the same plane with heads facing in opposite directions. The male extrudes a spermatophore (fig. 14.9C) and, using the other part of his P5, attaches it to the female (fig. 14.9D). The male then releases the female, and the spermatophore discharges. The above is a generalized description of the mating behavior in calanoids using *Labidocera aestiva* as an example (Blades & Youngbluth 1979). Reported variations in this behavioral sequence involve (1) timing between and within phases, (2) additional tactile inspection of the female by the male and, (3) the moment of spermatophore extrusion relative to attachment (see "Importance of Reproductive Morphology in Copepod Speciation").

SPERMATOPHORE ATTACHMENT AND DISCHARGE

The development of spermatophores in calanoid copepods provides an efficient mechanism by which the aflagellate and immobile spermatozoa can be transferred to the female. The spermatophore is attached to the external surface of the female but may or may not be in direct contact with the genital pore. In the latter case, the sperm must exit the spermatophore and be transported over the cuticle of the female to the spermathecal sacs.

In those calanoid species in which the spermatophore is attached to the genital opening, referred to as direct placement (Hopkins & Machin 1977, Ferrari 1978), the spermatophoric contents empty almost completely into the spermathecal sacs. Direct placement of a simple spermatophore is illustrated here in *Acartia tonsa* (fig. 14.10A), (Hammer 1978; Blades-Eckelbarger personal observation). The attachment sites of spermatophores with coupling plates (figs. 14.6B,C, 14.7A,B, 14.10B) may also be considered as direct placements because the configuration of the coupling apparatus conforms to the morphology of the female's urosome and, when attached properly, brings the open end of the spermatophore flask in close proximity to the genital pore (Heberer 1932; Fleminger & Tan

FIGURE 14.9. Mating behavior of *Labidocera aestiva*. A. Initial capture of female by male. Arrow indicates grasp with geniculate antennule. B. Copulatory position, ventral view. C. Spermatophore extrusion, dorsal view of male showing left leg exopod holding coupling plate (arrow) of fully extruded spermatophore. D. Spermatophore attachment. Ch, chelate P5 of male holding female's urosome; SP, spermatophore proper. From Blades & Youngbluth 1979.

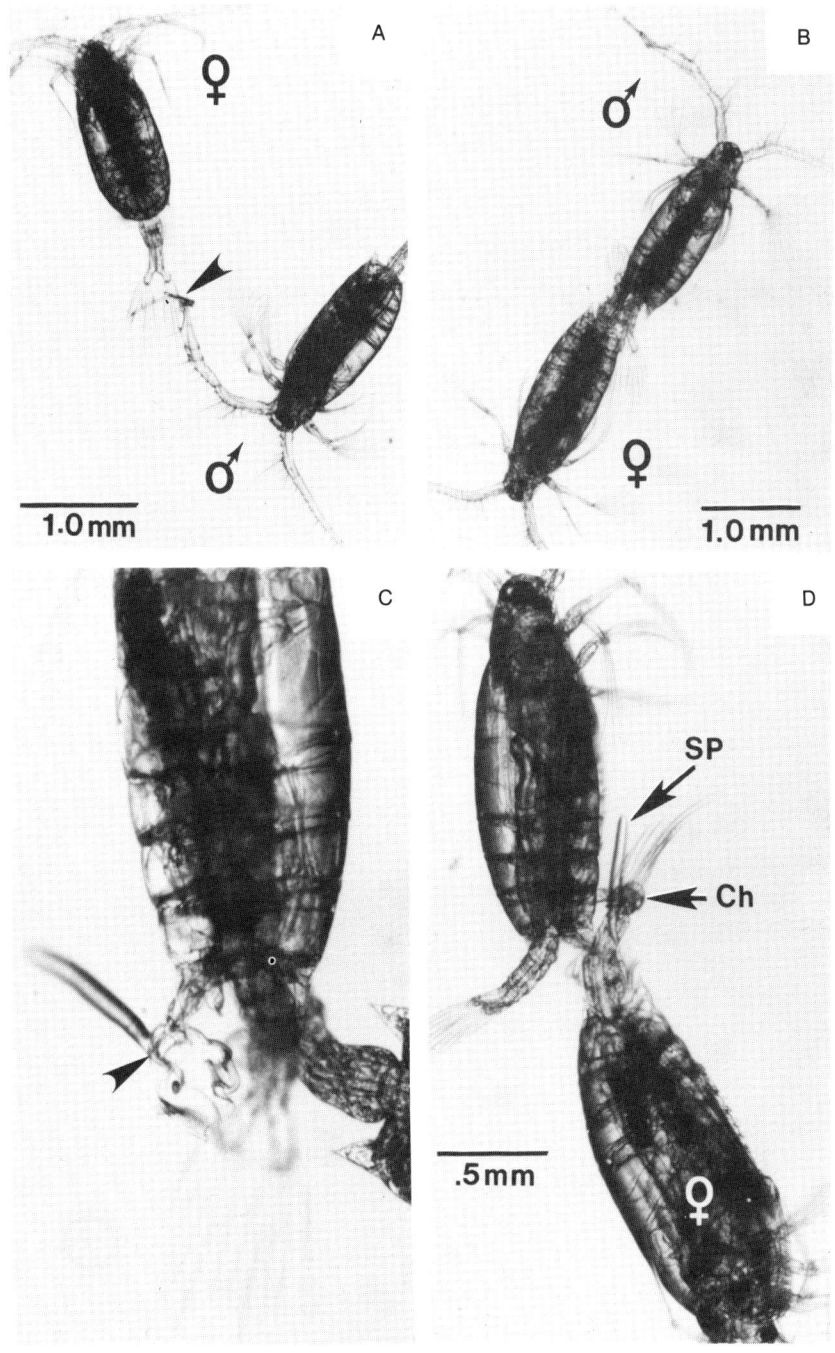

1966; Fleminger 1967, 1975; Lee 1972; Fleminger & Moore 1977; Blades 1977; Blades & Youngbluth 1979). In *Labidocera aestiva* the spermatophore attachment site corresponds to a distinct field of cuticular pores or "pit-pores" that are present on the second urosomal segment of the female (figs. 14.10c,d). TEM examination revealed that these pores connect to secretory cells that may produce a substance that dissolves the adhesive holding the spermatophore to the female (Blades & Youngbluth 1979).

In *Euchaeta norvegica*, which possesses a simple spermatophore without coupling plates, two common sites of attachment, direct and nondirect, have been described (Hopkins & Machin 1977; Ferrari 1978). Ferrari and Dojiri (1987) reported also that both "correct" and "alternate" attachment sites are common in *E. antarctica*. Furthermore, these authors found two morphologically distinct spermatophores attached to the two different areas on the genital segment of the females. They suggested that the males may be able to purposefully produce either type of spermatophore depending upon where it will be attached.

Hopkins and Machin (1977) observed with *Euchaeta norvegica* that when the spermatophore was attached directly over the female genital cavity the contents of the spermatophore emptied into the spermathecae. However, if the attachment site was a distance from the genital area, the initial contents of the spermatophore flowed onto the surrounding cuticle to form a circular mass of cementlike material, which they referred to as the attachment disc (fig. 14.6a). In approximately 30 percent of the nondirect spermatophores observed on *E. norvegica* a narrow "fertilization tube" (fig. 14.2c–f) extended from the attachment disc to the female's genital opening. Transport of the sperm to the female's genital opening, therefore, appears to be accomplished by the physical properties of the seminal secretions within the spermatophore. The granules and flocculent material that compose the core secretion solidify on contact with seawater and form a tube (fig. 14.12f) through which the liquid interior carrying the sperm continues to pass en route to the female's gonopore (Blades 1977; Hopkins & Machin 1977; Hopkins 1978).

An example of extensive fertilization tube formation is observed in *Undinula vulgaris*, where the simple spermatophore is attached to the right dorsolateral surface of the last thoracic segment (fig. 14.12c). A fertilization tube extends from the adhesive mass and traverses a distinct field of cuticular pores (figs. 14.12a–d) to what appears to be an attachment disc. From the attachment disc, the tube passes between the junction of the metasome and urosome to ultimately connect with the genital opening. TEM observations of the area beneath the pore field revealed an extensive glandular tissue (Blades-Eckelbarger personal observation). The function of the secretory product contained within the associated cells remains unknown.

FIGURE 14.10. A. Simple spermatophore (SP) attached directly to genital pore (G) of *Acartia tonsa* female. B. Spermatophore with simple coupling plate (C) covering genital opening of *Anomalocera ornata* female. C. Genital segment and second urosomal segment of female *Labidocera aestiva* showing genital plate (Gp) and field of pit-pores (PP). Note hairs and spines over genital field. D. Higher magnification of pit-pores. Figures 14.10c and 14.10d from Blades & Youngbluth 1979.

Spermatophore Morphology in Calanoid Copepods 263

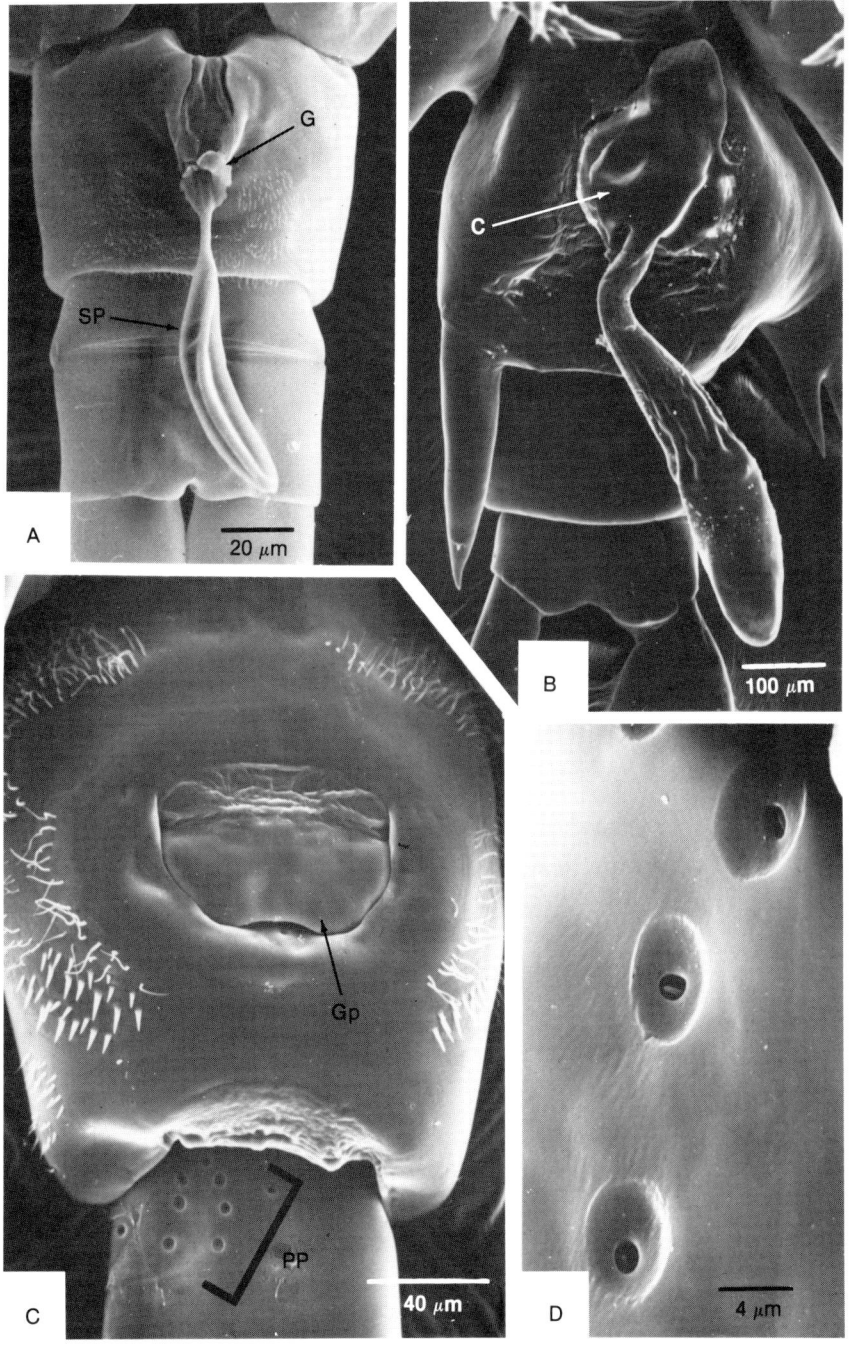

Discharge of the spermatophoric contents is accomplished by means of mechanical or hydrostatic pressure resulting from water uptake and consequent swelling of modified cells located immediately within the wall of the spermatophore flask (fig. 14.11). These cells were originally called swelling spermatozoa ("Quellspermatozoen" or "Q-sperm") by Heberer (1932), later referred to as foamlike bodies in *Euchaeta norvegica* (Hopkins 1978), and more recently identified as modified spermatozoa in *Labidocera aestiva* (Blades & Youngbluth 1981). In *L. aestiva*, these modified cells are present in the distal half of the spermatophore (fig. 14.11A), whereas in *Undinula vulgaris* they extend throughout the length of the spermatophore (fig. 14.11B).

The mechanism by which the fertilization tube flows in the correct direction to the gonopore is still unclear. Chemical attraction has been discounted because the spermatophore and its secretions are of acellular nature and incapable of "sensing" chemical gradients (Vaupel Klein 1982). The present observations of *Undinula vulgaris*, along with those on *Euchaeta norvegica* (Hopkins & Machin 1977, Ferrari 1978), indicate that these tubes follow a distinct trail to the female genital area. Ferrari (1978) and Vaupel Klein (1982) suggested that general body shape in combination with mechanical stimuli from the configuration and cuticular structures on the female genital region would control the path of growth of the tube while it was forming.

The quantity of core secretion is notably variable among genera and may correspond to the distance traveled by the fertilization tube to the female's genital pore. Favorable sections through the seminal vesicle or spermatophore proper of a male calanoid can provide a simple means by which relative quantities of core secretion can be compared. This type of morphological information may be used to determine direct or nondirect attachment sites for species on which spermatophores are not easily observed. For example, the spermatophores of *Undinula vulgaris* (Blades-Eckelbarger personal observation) and *Euchaeta norvegica* (Hopkins 1978) contain large quantities of seminal secretions that can form long fertilization tubes. In contrast, the spermatophore of *Pleuromamma abdominalis* contains very little core secretion and most of its volume is filled with spermatozoa (Heberer 1932, 1937; Blades-Eckelbarger personal observation). Although the site of spermatophore attachment has not been reported for *Pleuromamma*, the lack of abundant core secretion suggests that the spermatophore is attached directly to the female's genital opening.

IMPORTANCE OF REPRODUCTIVE MORPHOLOGY IN COPEPOD SPECIATION

Interspecific reproductive isolation in calanoid copepods is considered to be the result of a combination of the following factors: (1) differences in mate recognition due to species-specific pheromones (chemical), (2) morphological diversity

FIGURE 14.11. Spermatophore discharge. A. Partially discharged spermatophore of *Labidocera aestiva* showing coupling apparatus (CP), spermatophore proper (SP), and swelling cells (SW). B. Partially discharged spermatophore of *Undinula vulgaris*. Note swelling cells (SW) extend length of spermatophore proper. CR, core secretion; FT, fertilization tube. C. Fully discharged spermatophore of *Centropages furcatus*.

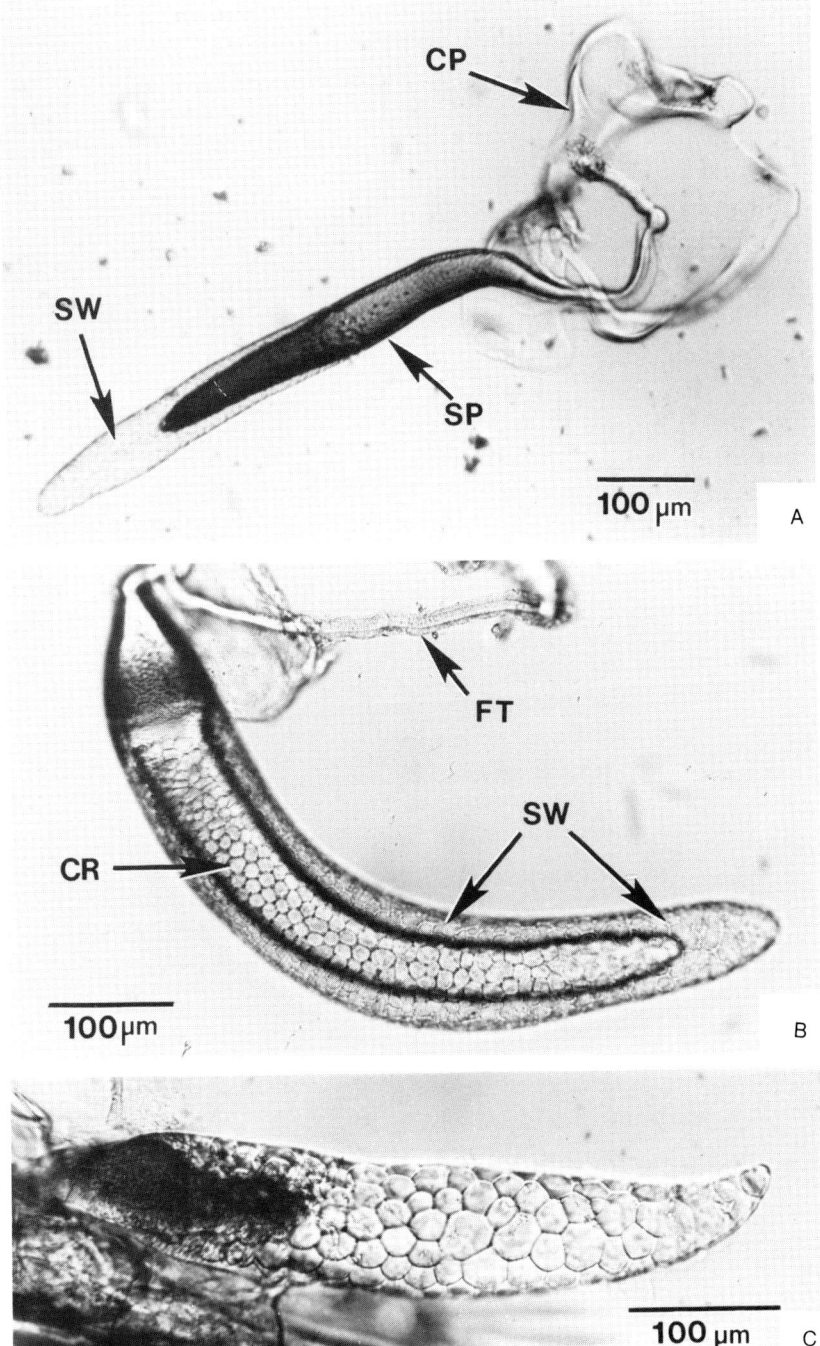

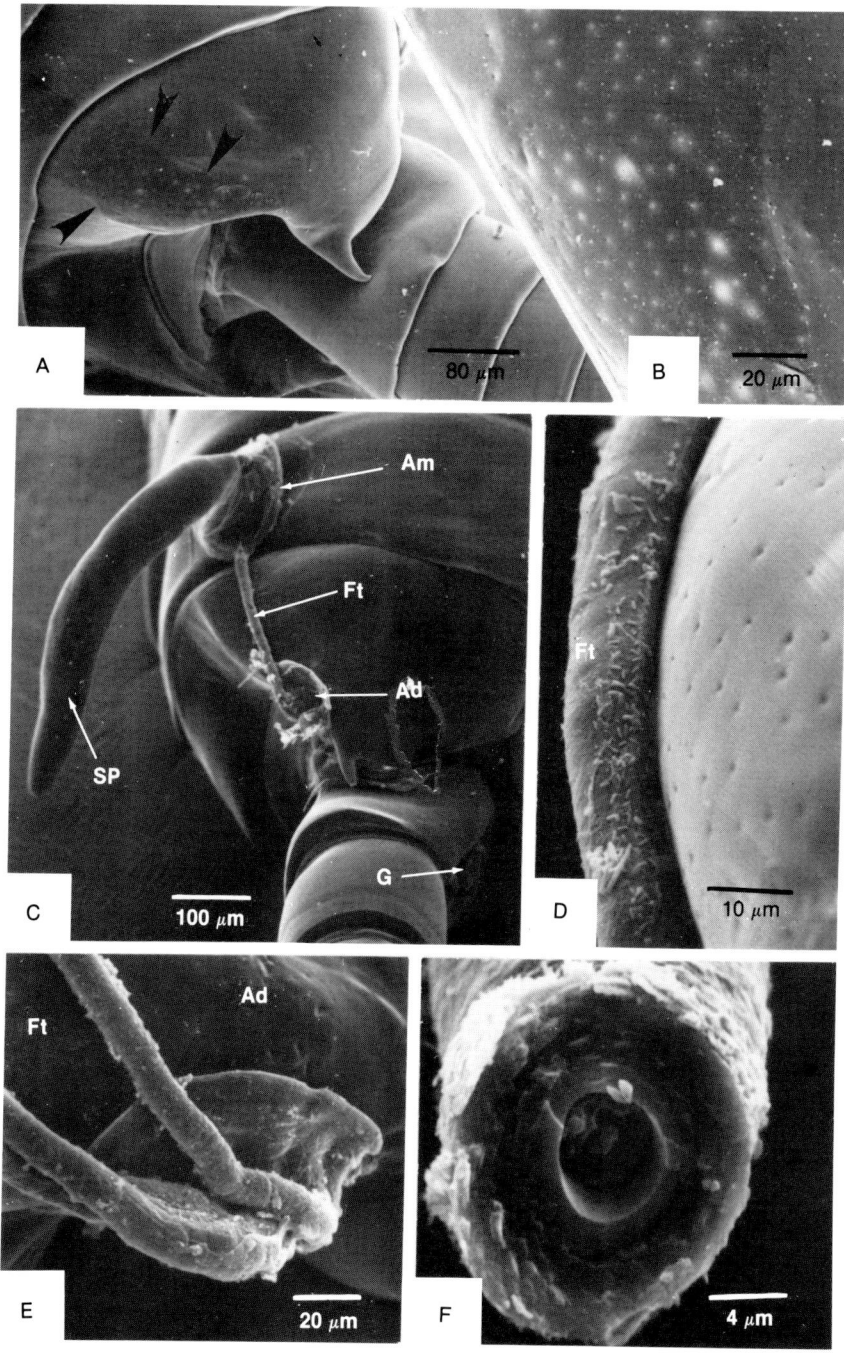

in the primary and secondary sexual structures (mechanical) and (3) deviations in mating behavioral patterns (Fleminger 1967, 1975; Frost & Fleminger 1968; Fleminger & Hulsemann 1974; Vaupel Klein 1982; Jacoby & Youngbluth 1983). These barriers to hybridization may be expressed at varying stages of the mating encounter depending on the species and degree of geographical overlap between sympatric species. The present review defines premating barriers in calanoids as those that are expressed prior to spermatophore extrusion, whereas postcopulation or prezygotic barriers are those that occur after spermatophore extrusion.

Chemical premating barriers have been recognized in some calanoids during the initial attraction or search phase, where the male acts upon pheromonal signals from the female (Parker 1902; Jacobs 1961; Katona 1973; Griffiths & Frost 1976; Jacoby & Youngbluth 1983). Jacoby and Youngbluth (1983) found chemical cues to be specific among three species of *Pseudodiaptomus*. When attempting to cross heterospecific pairs, the authors noted that males performed fewer searches and fewer copulations.

Species-specific morphological differences, which can act at either the premating or prezygotic level, are most obvious with respect to the male fifth leg, female genital area, and spermatophores. A critical step in the mating encounter involves the exact coupling of the male's P5 with the female's urosome, which is controlled by the morphology of these parts (see review by Vaupel Klein 1982). The male's grip on the female must be firm, and it must be positioned precisely to enable the male to place the spermatophore correctly. Such precision has been observed in *Centropages typicus* (Blades 1977), *Labidocera aestiva* (Blades & Youngbluth 1979), and *Pseudodiaptomus* spp. (Jacoby & Youngbluth 1983). The male may receive mechanical signals through sensory hairs or spines on the clasping part of the P5 (Blades & Youngbluth 1980), confirming that he has captured a conspecific female and that his fifth leg is grasping the right area. Furthermore, tactile inspection of the female genital region by the male has been described for *L. aestiva* in which the male performs a deliberate "stroking behavior" with the left fifth leg over the spermatophore attachment site prior to extrusion of the spermatophore. Such behavior may inform the male of an existing spermatophore on the site and may prepare the site for attachment of his spermatophore (Blades & Youngbluth 1979).

Literature on calanoids that produce spermatophores with complex coupling plates, i.e., members of the Pontellidae (Fleminger 1967, 1975; Blades & Youngbluth 1979) and *Centropages* spp. (Lee 1972; Blades 1977), strongly indicates that a species-specific "key-and-lock" mechanism is formed by the morphology of the

FIGURE 14.12. A–E. Spermatophore attachment site on *Undinula vulgaris*. A. Site corresponds to distinct field of pores (within arrowheads) on right dorsolateral surface of last thoracic segment. B. Higher magnification of pores. C. spermatophore attached to this site. Note fertilization tube (Ft) traversing pore field to attachment disc (Ad). G, genital opening of female. D. Higher magnification of fertilization tube crossing pore field. E. Fertilization tubes from three spermatophores joining at one attachment disc. F. Transverse break in fertilization tube from *Euchaeta norvegica* spermatophore showing solidified flocculent and granular components of seminal secretion surrounding central canal through which liquid core secretion and sperm flow.

coupling plates, the conspecific female's genital configuration, and the specific movements of the male while holding and orienting to the female. Due to the structural specificity of the coupler, the spermatophore of one species will not accurately "fit" the urosome of a heterospecific female. Futhermore, the degree of divergence of these morphological characters appears to be closely related to the extent of geographical overlap (Fleminger 1967, 1975). Therefore, hybridizing barriers in species with elaborate coupling plates may be more effective during the copulation stage rather than the premating search and capture stage (Vaupel Klein 1982).

The "key-and-lock" mechanism is not applicable to those species that produce simple spermatophores lacking couplers. Some studies suggest that males of these species are capable of quickly and accurately attaching the simple spermatophore directly onto the genital opening of the female (Gauld 1957; Katona 1975; Hammer 1978; Jacoby & Youngbluth 1983). Indications are that these species have become more efficient in mate recognition and/or the actual act of placement of the spermatophore, a result perhaps of highly evolved pheromonal or mechanical cues. Additional observations of other calanoids with simple spermatophores are needed to determine the dominant character or characters that may function as a block to hybridization.

Less obvious morphological differences may be found with respect to body proportion and total body size. These characters also have been cited as playing a role in reproductive isolation (Fleminger 1967; Lawson 1977). In addition, species-specific patterns of cuticular setae and pores may provide important chemical and mechanical cues in those calanoid species that have not developed highly modified secondary sexual characters (Fleminger 1975; Fleminger & Hulsemann 1974; Blades & Youngbluth 1979; Fleminger 1986).

Behavioral reproductive barriers may be expressed in combination with chemical and mechanical cues during one or more phases of the mating encounter: (1) timing between and within phases, as observed for three species of *Pseudodiaptomus* by Jacoby and Youngbluth (1983), (2) additional tactile inspection of the female by the male, as noted in the pontellid *Labidocera aestiva* by Blades and Youngbluth (1979), and (3) the moment of spermatophore extrusion relative to attachment. For example, in *L. aestiva* and three species of *Pseudodiaptomus* (Blades & Youngbluth 1979; Jacoby & Youngbluth 1983), the male extrudes the spermatophore only after attaining a firm grasp with the modified P5. In *Centropages typicus* the male extrudes the spermatophore prior to this, while holding the female with his antennule (Blades 1977), whereas in *Eurytemora affinis* and *Euchaeta* spp. the male swims about with a spermatophore gripped in the P5 before making physical contact with a female (Katona 1975; Hopkins, Mauchline & McLusky 1978; Ferrari & Dojiri 1987; Blades-Eckelbarger, personal observation). Once the male extrudes a spermatophore from the spermatophore sac, there remains the chance that he may drop it or fail to attach it correctly to a conspecific female and thus risk wasting sperm. Therefore, premating barriers functioning prior to spermatophore extrusion are considered to be the most efficient (Mayr 1963) serving to conserve energy and gametes.

Acknowledgments

This symposium paper is dedicated to the memory of Dr. Abraham Fleminger, a friend and colleague, whose energy, enthusiasm, and warm encouragement greatly influenced my early interests in copepod reproduction. I would like to thank Dr. Geoff Boxshall of the British Museum of Natural History and Dr. Taisoo Park of Texas A&M University for helpful consultation on various aspects of the review and Dr. Petra Sierwald of the Delaware Museum of Natural History for providing translations of German papers. This paper is Contribution No. 698 of the Harbor Branch Oceanographic Institution.

Literature Cited

Adiyodi, K. G. & R. G. Adiyodi. 1975. Morphology and cytology of the accessory sex glands in invertebrates. In G. H. Bourne & J. F. Danielli, eds., *International Review of Cytology* 43:353–399. New York: Academic Press.

Blades, P. I. 1977. Mating behavior of *Centropages typicus* (Copepoda: Calanoida). *Mar. Biol. (Berl.)* 40:47–64.

Blades, P. I. & M. J. Youngbluth. 1979. Mating behavior of *Labidocera aestiva* (Copepoda: Calanoida). *Mar. Biol. (Berl.)* 51:339–355.

Blades, P. I. & M. J. Youngbluth. 1980. Morphological, physiological, and behavioral aspects of mating in calanoid copepods. In W. C. Kerfoot, ed., *Evolution and Ecology of Zooplankton Communities*, pp. 39–51. Hanover, New Hampshire: University Press of New England.

Blades, P. I. & M. J. Youngbluth. 1981. Ultrastructure of the male reproductive system and spermatophore formation in *Labidocera aestiva* (Crust.: Copepoda). *Zoomorphology (Berl.)* 99:1–21.

Blades-Eckelbarger, P. I. 1986. Aspects of internal anatomy and reproduction in the Copepoda. *Syllogeus* 58:26–50.

Bradford, J. M. 1974. *Euchaeta marina* (Prestandrea) (Copepoda, Calanoida) and two closely related new species from the Pacific. *Pac. Sci.* 28:159–169.

Brodskii, K. A. 1967a. *Calanoida of the Far Eastern Seas and Polar Basin of the USSR*. Moskva-Leningrad: Isdatel'stvo Akademii Nauk SSSR. (Note: Translation from Russian by A. Mercado in 1950. Israel Program for Scientific Translations).

Brodskii, K. A. 1967b. Types of genitalia and heterogeneity in the genus *Calanus* (Copepoda). *Dokl. Akad. Nauk SSSR* 176:222–225.

Ferrari, F. 1978. Spermatophore placement in the copepod *Euchaeta norvegica* Boeck 1872 from the deepwater dumpsite 106. *Proc. Biol. Soc. Wash.* 91:509–521.

Ferrari, F. & M. Dojiri. 1987. The calanoid copepod *Euchaeta antarctica* from southern ocean Atlantic sector midwest trawls, with observations on spermatophore dimorphism. *J. Crustacean Biol.* 7:458–480.

Fleminger, A. 1967. Taxonomy, distribution, and polymorphism in the *Labidocera jollae* group with remarks on evolution within the group (Copepoda: Calanoida). *Proc. U.S. Natl. Mus.* 120:1–61.

Fleminger, A. 1975. Geographical distribution and morphological divergence in American coastal-zone planktonic copepods of the genus *Labidocera*. In L. E. Cronin, ed., *Estuarine Research. Vol. 1, Chemistry, Biology and the Estuarine System*, pp. 392–419. New York: Academic Press.

Fleminger, A. 1979. *Labidocera* (Copepoda, Calanoida): New and poorly known Caribbean species with a key to species in the Western Atlantic. *Bull. Mar. Sci.* 29:170–190.

Fleminger, A. 1986. The pleistocene equatorial barrier between the Indian and Pacific oceans and a likely cause for Wallace's line. *UNESCO Tech. Pap. Mar. Sci.* 49:84–97.

Fleminger, A. & K. Hulsemann. 1974. Systematics and distribution of the four sibling species comprising the genus *Pontellina* Dana (Copepoda, Calanoida). *U.S. Natl. Mar. Fish. Serv. Fish. Bull.* 72:63–120.

Fleminger, A. & E. Moore. 1977. Two new species of *Labidocera* (Copepoda, Calanoida) from the Western North Atlantic region. *Bull. Mar. Sci.* 27:520–529.

Fleminger, A. & E. Tan. 1966. The *Labidocera mirabilis* species group (Copepoda, Calanoida) with descriptions of a new Bahamian species. *Crustaceana (Leiden)* 11:291–301.

Frost, B. & A. Fleminger. 1968. A revision of the genus *Clausocalanus* (Copepoda: Calanoida) with remarks on distributional patterns in diagnostic characters. *Bull. Scripps Inst. Oceanogr. Univ. Calif.* 12:1–235.

Gauld, D. T. 1957. Copulation in calanoid copepods. *Nature (Lond.)* 180:510.

Geptner, M. V. 1968. Structure and taxonomic significance of the genital complex in copepods of the family Euchaetidae (Calanoida). *Oceanology* 8:543–552.

Griffiths, A. M. & B. W. Frost. 1976. Chemical communication in the marine planktonic copepods *Calanus pacificus* and *Pseudocalanus* sp. *Crustaceana (Leiden)* 30:1–8.

Hammer, R. M. 1978. Scanning electron microscope study of the spermatophore of *Acartia tonsa*. (Copepoda: Calanoida). *Trans. Am. Microsc. Soc.* 97:386–389.

Heberer, G. 1932. Untersuchungen uber Bau und Funktion der Genitalorgane der Copepoden. I. Der mannliche Geitalapparat der calanoiden Copepoden. *Z. Mikrosk. Anat. Forsch. (Leipz.)* 31:250–424.

Heberer, G. 1937. Weitere Ergebniss euber Bildung und Bau der Spermatophoren und Spermatophorenkoppelapparate bei calanoiden Copepoden. *Verh. Dtsch. Zool. Ges.* 39:86–93.

Hopkins, C. C. E. 1978. The male genital system and spermatophore production and function in *Euchaeta norvegica* Boeck (Copepoda: Calanoida). *J. Exp. Mar. Biol. Ecol.* 35:197–231.

Hopkins, C. C. E. & D. Machin. 1977. Patterns of spermatophore distribution and placement in *Euchaeta norvegica* (Copepoda: Calanoida) *J. Mar. Biol. Assoc. U. K.* 57:113–131.

Hopkins, C. C. E., J. Mauchline & D. S. McLusky. 1978. Structure and function of the fifth pair of pleopods of male *Euchaeta norvegica* (Copepoda: Calanoida). *J. Mar. Biol. Assoc. U.K.* 58:631–637.

Jacobs, J. 1961. Laboratory cultivation of the marine copepod *Pseudodiaptomus coronatus* Williams. *Limnol. Oceanogr.* 6:443–446.

Jacoby, C. A. & M. J. Youngbluth. 1983. Mating behavior in three species of *Pseudodiaptomus* (Copepoda: Calanoida). *Mar. Biol. (Berl.)* 76:77–86.

Katona, S. K. 1973. Evidence for sex pheromones in planktonic Copepods. *Limnol. Oceanogr.* 18:574–583.

Katona, S. K. 1975. Copulation in the copepod *Eurytemora affinis* (Poppe, 1880). *Crustaceana (Leiden)* 28:90–94.

Lawson, T. J. 1977. Community interactions and zoogeography of the Indian Ocean Candaciidae (Copepoda: Calanoida). *Mar. Biol. (Berl.)* 43:71–92.

Lee, C. M. 1972. Structure and function of the spermatophore and its coupling device in the Centropagidae (Copepoda: Calanoida). *Bull. Mar. Ecol.* 8:1–20.

Mayr, E. 1963. *Animal Species and Evolution*. Cambridge, Massachusetts: Harvard University Press.

Park, T. S. 1966. The biology of a calanoid copepod, *Epilabidocera amphitrites* McMurrich. *Cellule* 66:1–251.

Park, T. S. 1975. Calanoid copepods of the family Euchaetidae from the Gulf of Mexico and Western Caribbean Sea. *Smithson. Contrib. Zool.* 196:1–26.

Park, T. S. 1978. Calanoid copepods (Aetideidae and Euchaetidae) from Antarctic and sub Antarctic waters. *Antarct. Res. Ser.* 27:91–290.

Parker, G. H. 1902. The reactions of copepods to various stimuli and the bearing of this on daily depth migrations. *Bull. U.S. Fish Comm.* 21:103–123.

Raymont, J. E. G., S. Krishnaswamy, M. A. Woodhouse & R. L. Griffin. 1974. Studies on the fine structure of Copepoda. Observations on *Calanus finmarchicus* (Gunnerus). *Proc. R. Soc. Lond. B Biol. Sci.* 185:409–424.

Strickler, R. & A. K. Bal. 1973. Setae of the first antennae of the copepod *Cyclops scutifer* (Sars): Their structure and importance. *Proc. Natl. Acad. Sci. U.S.A.* 70:2656–2659.

Vaupel Klein, J. C. von. 1982. A taxonomic review of the genus *Euchirella* Giesbrecht, 1888 (Copepoda, Calanoida). II. The type-species, *Euchirella messinensis* (Claus, 1863). A. The female of *F. typica*. *Zool. Verh. (Leiden)* 198:1–131.

Williams, R. 1972. A further character for identification of the adult females of *Calanus finmarchicus*, *C. helgolandicus*, and *C. glacialis* (Crustacea: Copepoda). *Bull. Mar. Ecol.* 8:53–60.

FIFTEEN
The Reproductive Biology of Two Species of Remipedes

JILL YAGER

*Biology Department,
Antioch College,
Yellow Springs, Ohio.*

Abstract

Remipedes are an unusual group of troglobitic crustaceans that live exclusively in submerged caves. The reproductive biology of two species of remipedes, *Speleonectes benjamini* and *Godzilliognomus frondosus,* is presented. These species were found to be simultaneous hermaphrodites. Various degrees of oogenesis and spermatogenesis were documented in each individual. The occurrence of mature oocytes was associated with mature sperm. The female reproductive system begins in the posterior portion of the head and extends to the gonopore on the protopod of the seventh trunk appendage. The male system extends from the posterior portion of the seventh trunk appendage to the gonopore on the fourteenth trunk appendage. Emphasis is placed on the male reproductive system, and ultrastructural details of the sperm and spermatophore are given. The sperm have an ovoid nucleus and a flagellum with a 9+2 microtubular arrangement. In the posterior portion of the vas deferens the sperm are packaged into spermatophores with a distinct shape. The sperm nuclei are in the proximal end of the spermatophore, and the flagella extend into the distal portion. The spermatophore covering and surrounding flocculent material are PAS-positive.

REMIPEDES ARE primitive crustaceans that were first described about a decade ago from anchialine caves in the northern Bahamas (Yager 1981). Anchialine caves are water-filled caves with inland surface openings and subsurface connections to the sea. The caves typically have a distinct density interface beneath which the water is low in dissolved oxygen, usually less than 1 ppm. Remipedes appear to be confined to this hypoxic environment (Yager 1987a, b, 1989b).

Speleonectes benjamini and *Godzilliognomus frondosus* are common to anchialine caves of the northern Bahamas (Yager 1987a, 1989b). *G. frondosus* is the smallest remipede species to date. Specimens average about 7 mm in length and have a maximum number of 16 trunk appendages. Adults of *S. benjamini* average 14 mm in length, and the maximum number of trunk segments is 27. Both species are simultaneous hermaphrodites. Oogenesis and spermatogenesis leading to mature gametes were observed in the same individual.

The female reproductive system extends from the posterior portion of the head to the seventh trunk appendage. The male system extends from trunk segments 7–14. A diagrammatic sketch of a remipede indicating approximate positions of the female and male reproductive systems is given in figure 1.

The following is a review of what has been published about remipede reproduction anatomy to date.

Schram (1986) described the male reproductive system as extending from the head posteriorly and lying ventral to the gut. This was corrected by Itô and Schram (1988), who described the testes of *Lasionectes entrichoma* Yager and Schram (part of species name, not a reference to that paper) as originating in the posterior portion of the seventh trunk segment. They stated that the testes originate ventral to the gut. After examination of many specimens of several species, I am unable to confirm the origination of the testes in this position. In the material examined, the testes were always found dorsal or dorsolateral to the gut.

Gonadal material seen as a strand leading from the second maxilla to approximately the fourteenth trunk segment was noted in the initial description of *Speleonectes lucayensis* (Yager 1981). The presence of sperm was mentioned but not detailed in the species description of *S. benjamini* (Yager 1987a). Itô and Schram (1988) described a second pair of gonopores located on the seventh trunk appendages associated with the female reproductive system. They described the female gonopore and offered a general description of the reproductive system of *Lasionectes entrichoma*. Ovaries with immature oocytes and paired oviducts were noted for *S. benjamini*. Itô and Schram (1988) confirmed that remipedes are hermaphroditic but were uncertain whether they were simultaneous or sequential hermaphrodites. Yager (1989a) described the reproductive system of *S. benjamini* and confirmed that the species is a simultaneous hermaphrodite, finding evidence of both spermiogenesis and oogenesis in the same individual.

The male gonopore has been illustrated with scanning electron micrographs for *Lasionectes entrichoma* (Schram, Yager & Emerson 1986; Itô & Schram 1988), *Speleonectes benjamini* (Yager 1987a, 1989a), *Cryptocorynetes haptodiscus* (Yager 1987a), and *Pleomothra apletocheles* (Yager 1989b).

Reproductive Biology of Remipedes **273**

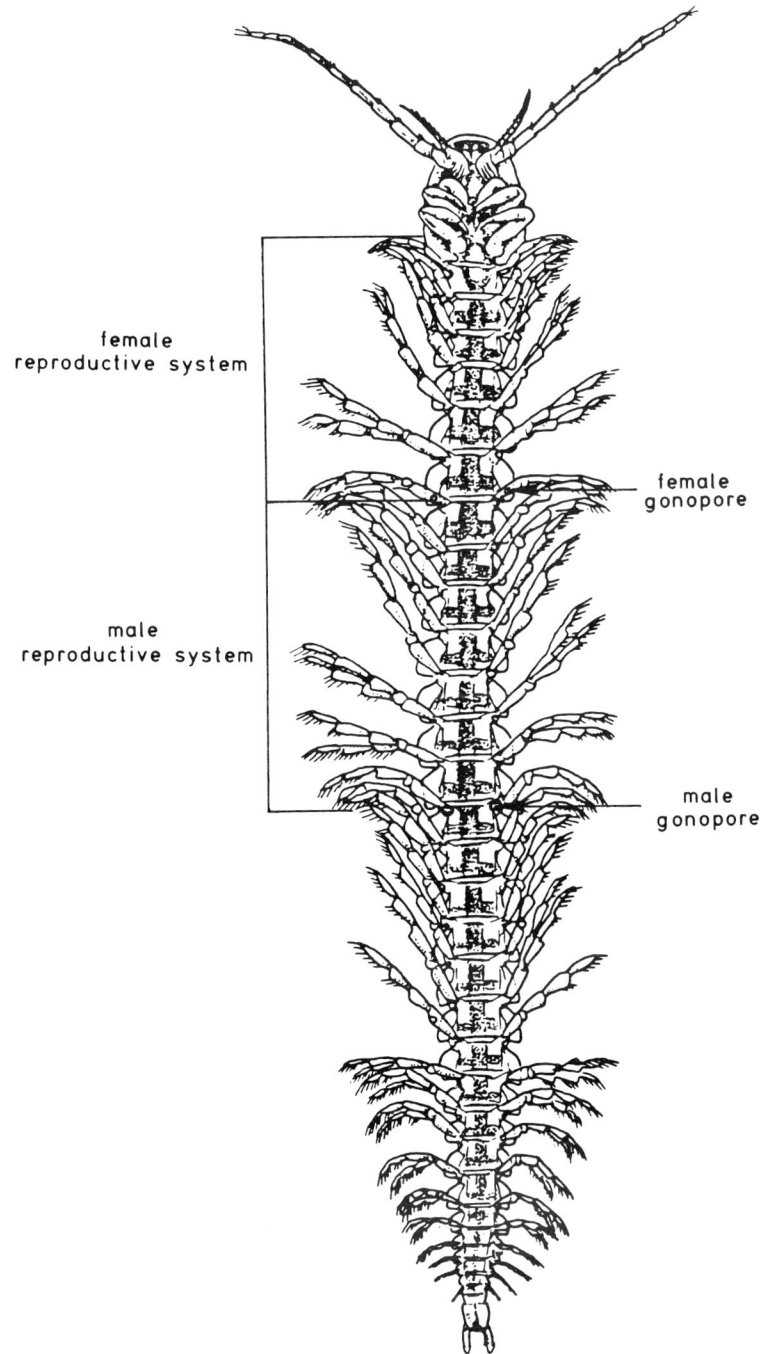

FIGURE 15.1. Diagrammatic remipede showing locations of the male and female reproductive systems, ventral view. Modified from Yager and Schram 1986.

274 *Jill Yager*

THE FEMALE REPRODUCTIVE SYSTEM

The ovary of *Speleonectes benjamini* originates in the head at about the level of the maxilliped. The broad organ lies dorsal to the midgut and ventral to the simple circulatory vessel (fig. 15.2A). By about the posterior portion of the first trunk segment the organ is separated into distinct paired ovaries. The medial margins of the ovaries lie close together. In *S. benjamini* the germinal zone appears at the anteromedial portion of the single ovary. Oogonia and immature oocytes can be observed (fig. 15.3). In *S. benjamini*, nuclei of the smallest oogonia measure approximately 8 μm in diameter. Oogonia of *Godzilliognomus frondosus* are about 9 μm in diameter. Nongerminal squamous epithelial cells with flat-

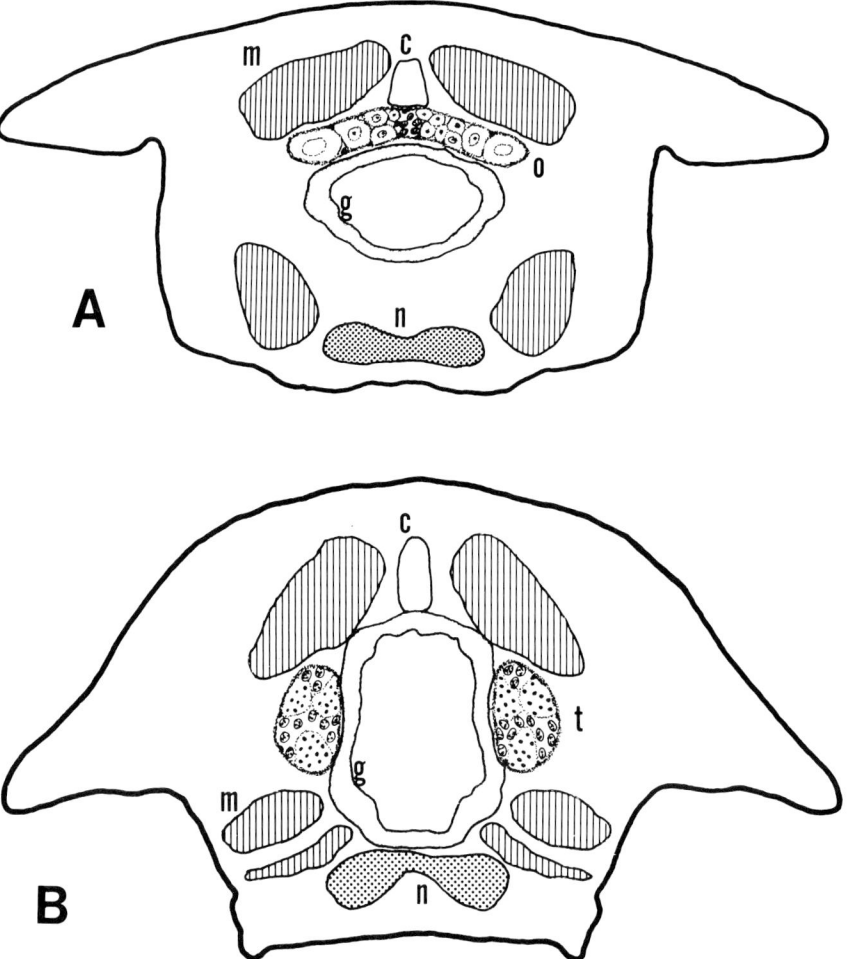

FIGURE 15.2. A. Cross section of the female reproductive system of a remipede. B. Cross section of the male reproductive system. From Yager 1989a. c, dorsal circulatory vessel; g, midgut; m, muscle; n, ventral nerve cord; o, ovary; t, testis.

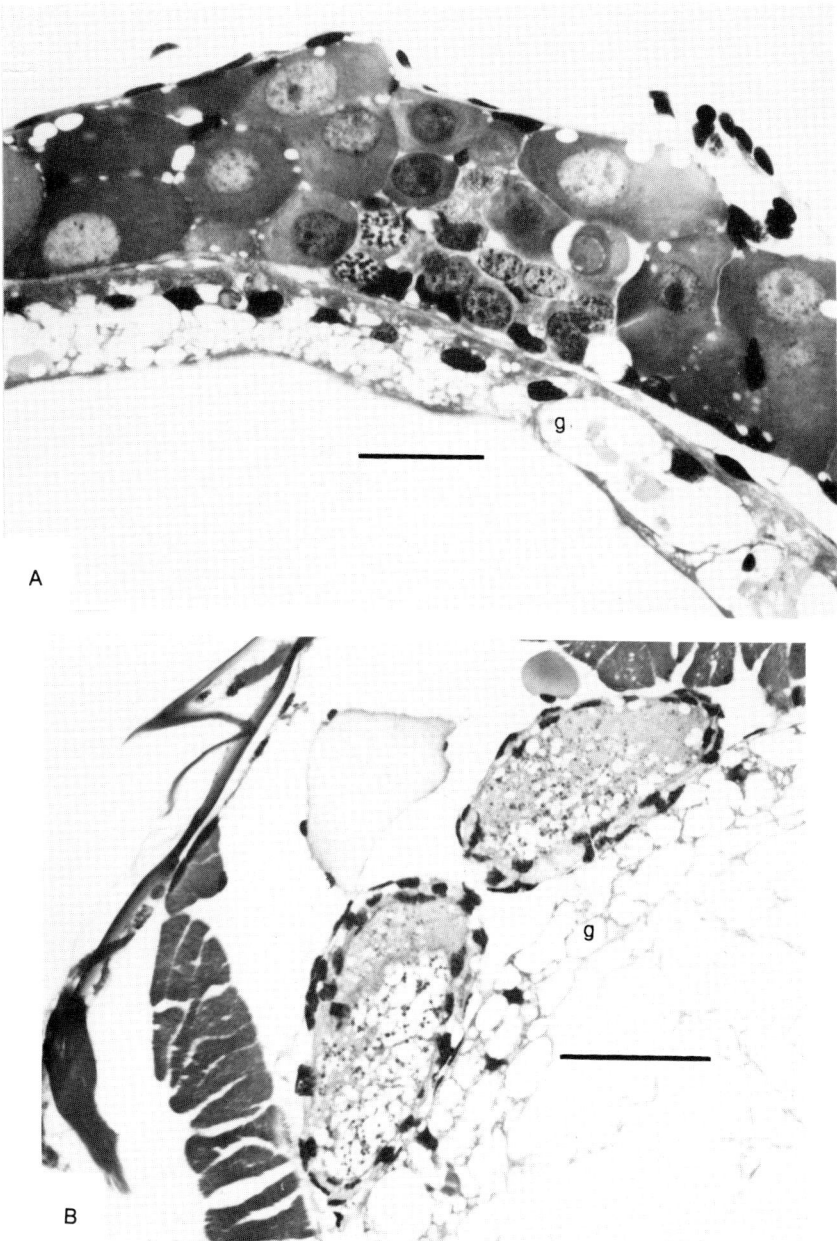

FIGURE 15.3. A. Ovary, *Speleonectes benjamini*, with medial zone of differentiation. g, midgut. Scale bar = 25 μm. B. Oocytes with double layer of cells, *Godzilliognomus frondosus*. g, midgut. Scale bar = 75 μm.

tened nuclei surround the ovaries. Because of insufficient ultrastructural details to confirm various degrees of development, the oocytes will be termed primary, intermediate, and mature based on nuclear or cytoplasmic size.

Primary oocytes of *Speleonectes benjamini* (fig. 15.3A) are small and have a finely granular ooplasm. The cells are elliptical and range in size from about 20–30 × 25–38 µm. The large nucleus measures about 12–15 µm in diameter. A large nucleolus with a diameter of 4–5 µm was present, as well as several smaller nucleoli. Primary oocytes were found in the ovary at about trunk segment 2. Young oocytes of *Godzilliognomus frondosus* were present in trunk segment 1. They were round and about 65 µm in diameter with a nuclear diameter of 30 µm. Primary oocytes have been found in individuals collected throughout the year.

Intermediate oocytes were distinguished from primary or immature oocytes by the presence of small, uniform granules in the cytoplasm. In both species, the intermediate oocytes were found in the paired ovaries at trunk segment 2. Oocytes of *Speleonectes benjamini* are elliptical, about 56 × 88 µm, with a nucleus of about 16 µm and a nucleolar diameter of 6 µm. Intermediate oocytes of *Godzilliognomus frondosus* are elliptical and measure about 100 × 180 µm. The nucleus is about 41 µm and the nucleolus about 10 µm.

The largest cross-section diameters of mature oocytes of *Speleonectes benjamini* were found in trunk segments 3–5. The size ranged from about 75 × 88 µm to 100 × 160 µm. The nuclear diameter was from 35–50 µm, and the large nucleolus was about 13 µm in diameter. The largest mature oocytes for *Godzilliognomus frondosus* ranged from approximately 150 × 225 µm to 280 × 350 µm. The nuclear diameter was about 66 µm and the nucleolus 15 µm. The granular cytoplasmic inclusions in the mature oocytes of *G. frondosus* were large globules or droplets of varying sizes, from 6–10 µm in diameter. Composition of these droplets was not determined.

In sagittal sections the ovaries can be seen enlarging distally as the eggs develop. Oocytes are positioned along the ovary with the smallest (youngest) oocytes anteriorly and the largest (mature) oocytes proceeding posteriorly. The smallest oocytes are observed at the beginning of the ovary between about the level of the maxilliped and the first trunk segment. Mature oocytes have been found in the ovaries beginning at about trunk segment three and extending no farther posteriorly than about the fifth trunk segment. These mature oocytes have been observed only in the ovaries dorsal to the midgut. The large eggs appear to be few in number, with no more than five observed in sagittal sections of *Speleonectes benjamini*.

Histological differentiation between ovaries and oviduct was not apparent. In the two species examined there is evidence of differentiation or maturation of the oocytes after the ovary splits into the two lateral components. At this point, I refer to the paired "oviducts" of Itô and Schram (1988) as paired ovaries. Oocytes with a double layer of epithelial cells (fig. 15.3B) appear posteriorly in what may be the oviducts or the transitional tissue leading to them. However, at the light microscope level the tubes containing the double layered oocytes appear to lack the secretory and muscle cells commonly found in oviducts. The oviducts may be poorly developed until the time of oviposition.

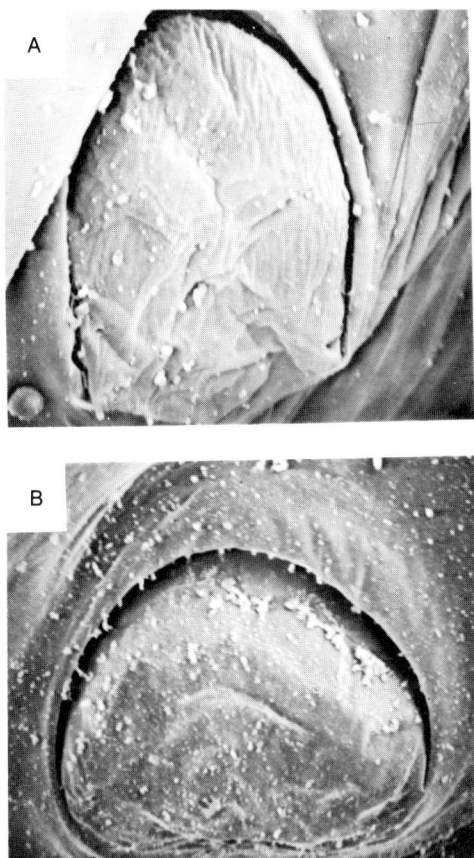

FIGURE 15.4. A. Female gonopore, *Godzilliognomus frondosus*, 724x, width of gonopore 67 μm. B. Female gonopore, *Godzillius robustus*, 311x, width of gonopore 166 μm.

The female gonopore (fig. 15.4) is located on the posterior base of the protopod of the seventh trunk appendage. The gonopore is semicircular in shape and similar in all species. Beneath a chitinous rim is a semicircular pad. The entire complex is slightly raised on papilliform. The width of the gonopore of *Speleonectes benjamini* is about 72 μm, that of *Godzilliognomus frondosus* (fig. 15.4A), measures approximately 166 μm. Examination of sectioned material from several species has revealed little microscopic detail of this area. No mature oocytes or sperm have been seen at or near the female gonopore.

THE MALE REPRODUCTIVE SYSTEM

The testes (fig. 15.2B) originate in the posterior portion of the seventh trunk segment and extend to about the tenth. They are paired organs that lie dorsolateral to the midgut. The outer surface is covered with thin squamous epithelium.

Clusters (cohorts) of cells at various stages of spermatogenesis can be observed in the testis of *Godzilliognomus frondosus* (fig. 15.5A). The cells within each group appear to be in the same stage of differentiation. Based on the nucleus, several cell types can be distinguished. Spermatogonia have nuclei that are about 10–11 μm in diameter with several nucleoli present. A few small, scattered clumps of condensed chromatin appear throughout the nucleus as well as along the edge of the nuclear envelope. The nucleus of an undetermined cell type is about 11–12 μm, with a homogeneous distribution of dense chromatin. Several spermatocyte stages are present. The nucleus of primary spermatocytes is about 9–10 μm in diameter and more dense and has clumps of condensed chromatin equally throughout. The nuclear diameter of secondary spermatocytes is smaller, about 5–6 μ in diameter, and is mostly condensed, darkly staining chromatin throughout. Mitochondria are present in the granular cytoplasm of all cell types observed.

The testes of *Speleonectes benjamini* (fig. 15.5B) produce young spermatids with a round nucleus about 4–5 μm in diameter (Yager 1989a). Young round spermatids of *Godzilliognomus frondosus* also have a diameter of about 4–5 μm. A flagellum with a distinct 9+2 microtubule arrangement was observed in young spermatids of *G. frondosus*. Associated with the developing gametes are accessory cells located between the clusters of differentiating cells. The accessory cells have irregularly shaped nuclei with small scattered clumps of condensed chromatin, and the cytoplasm is less dense than that of the gametes. These cells are known as Sertoli cells in vertebrates and are called accessory, nutritive, nurse, follicle, or support cells in invertebrates. They are thought to play a role in the nutrition and mechanical support of the germ cells (Blades-Eckelbarger & Youngbluth 1982).

At about trunk segment 10 there is a transition between testes and vas deferens. The squamous epithelial cells become cuboidal. In this area the young, round spermatids of *Speleonectes benjamini* and *Godzilliognomus frondosus* elongate and become elliptical with a tapered end.

The paired, coiled vasa deferentia are continous with the testes and extend lateral to the midgut from about trunk segments 10–14. At the fourteenth trunk segment they pass ventrolaterally to the gonopore. Cuboidal epithelial cells (fig. 15.6A) surround the lumen of the vas deferens. The cells have a large, round to ovoid nucleus with a diameter of about 6–8 μm. The cells secrete a homogeneous, PAS-positive material into the lumen of the vas deferens. From about trunk segment 10–12 the lumen contains elongate spermatids (fig. 15.6A) measuring about 8–10 μm long and 3–5 μm wide at the head end. The tail end is tapered, and flagella are visible. The spermatid nuclei are Feulgen-positive and PAS-negative. The acrosome is eosinophilic and PAS-negative. As the spermatids mature and move toward the posterior portion of the vas deferens, they are bound into distinctly shaped spermatophores (fig. 15.6B). At about trunk segment 12 in *Speleonectes benjamini* the epithelial cells that line the lumen lose their characteristic thick cuboidal shape and the cells become unequal in size. PAS-positive secretory products are present in these cells and released into the lumen. The lumen of the vas deferens increases in size to hold the spermatophores and a large volume of homogeneous, PAS-positive, moderately electron

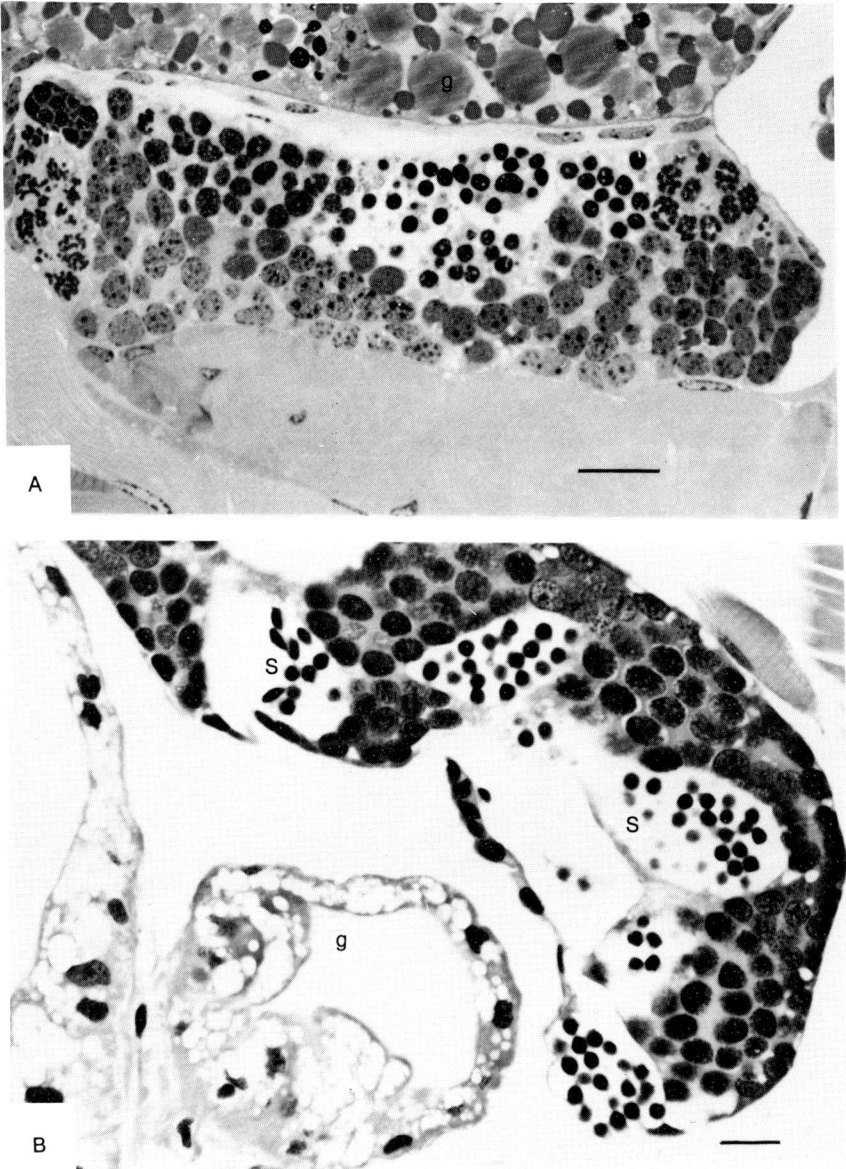

FIGURE 15.5. A. Testis, *Godzilliognomus frondosus*. Several groups of differentiating cells are apparent. B. Testis, *Speleonectes benjamini*. g, gut; s, spermatids. Scale bar = 20 μm.

dense flocculent material. As indicated by the positive PAS reaction, the materials are probable mucopolysaccharides. Spermatophores were found in specimens of *S. benjamini* collected in March, June, July, and September. Spermatophores were recorded from *Godzilliognomus frondosus* in specimens collected in March and July.

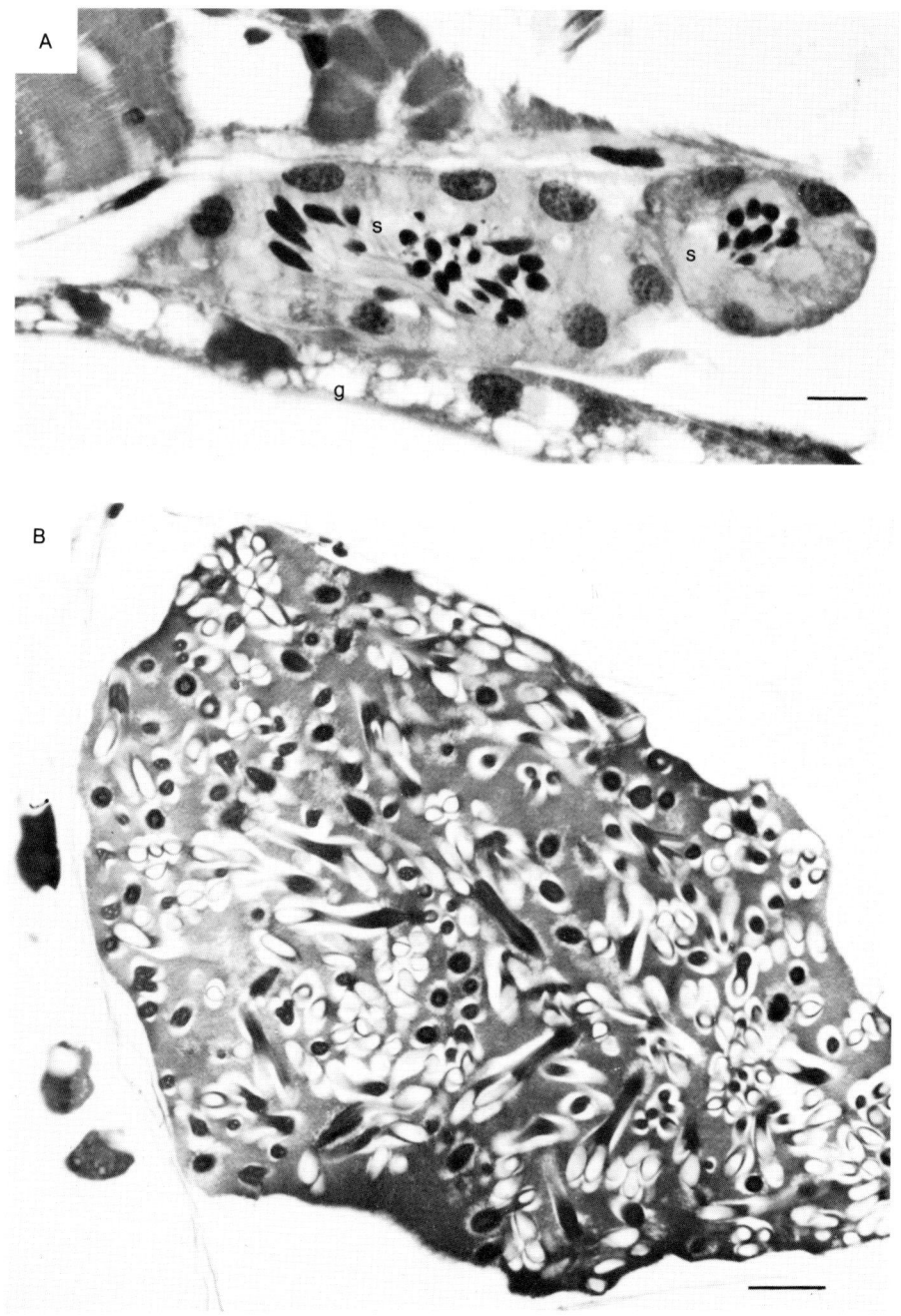

FIGURE 15.6 Vas deferens with elliptical spermatids, *Speleonectes benjamini*. e, epithelial cell; g, gut; s, spermatids. Scale bar = 8 μm. B. Vas deferens with spermatophores, *Speleonectes benjamini*, PAS reaction. Scale bar = 18 μm.

SPERMATOPHORE ULTRASTRUCTURE

The spermatophore of *Speleonectes benjamini* (fig. 15.7) is about 38 μm in length, is PAS-positive, and lacks a discernible membrane (Yager 1989a). The sperm nuclei are in separate extensions at one end of the spermatophore, and their flagella extend into the single distal end. The distal portion of the spermatophore is covered with small, cylindrical, villilike projections (fig. 15.8). Around the small distal projections the spermatophore shows differential staining by lead citrate and uranyl acetate (fig. 15.8). Also, a more positive response to the PAS reaction adjacent to the projections is evident (fig. 15.6B). Profiles of cross sections indicate from three to six flagella in the spermatophores of *S. benjamini*. The spermatophores of *Godzilliognomus frondosus* are similar in appearance to those of *S. benjamini*. However, they are shorter, approximately 22 μm in length, and appear to have only one or two sperm packaged into the proximal end. Flagella have been observed extending into the distal end of the spermatophores of *G. frondosus*, and the similar villilike extensions are present.

The mature sperm of *Speleonectes benjamini* have an ovoid nucleus, cup-shaped acrosome and flagellum with a 9+2 microtubular arrangement. The large nucleus measures about 8–9 μm by 5 μm. The chromatin is highly condensed and fibrillar in texture. The cup-shaped acrosome (see figs. 15.7, 15.9B) has granular bands in the subacrosomal space. What looks like an acrosomal filament or rod extends through the length of the nucleus. Centrioles have not been observed. Several mitochondria are present in the cell. In *S. benjamini* the flagellum appears to loop around the nucleus before extending into the distal end of the spermatophores (fig. 15.9A). Although motility has not been confirmed, the arrangement of cellular ultrastructure suggests that the sperm are capable of movement. A spermatophore with a flagellum extending out of the distal end can be seen in figure 15.9B.

The male gonopore (fig. 15.10) is a protuberance appearing at the posterior base of the protopod on the fourteenth trunk appendage. Gonopore anatomy differs among species. The gonopore of *Pleomothra apletocheles* is a simple bulbous structure (fig. 15.10A). The gonopore of *Godzilliognomus frondosus* is located on the underside of a large lateral extension of the protopod (fig. 15.10B).

Associated with the male gonopore is an oblong genital plate (figs. 15.10B, 15.11). The plate is located posterior to the gonopore between trunk segments 14 and 15. The plate measures approximately 150×200 μm in *Godzilliognomus frondosus*. The surface is covered with many small openings. In *Speleonectes lucayensis* and *S. tulumensis* small, hollow-tipped sensilla-like structures emerge from the openings (fig. 15.11c). In *G. frondosus* and *S. benjamini* the openings lack sensilla-like structures and may be secretory in nature (see fig. 15.11A, B). It is possible, however that the structures are present but retracted into the openings. Internally the plate is associated with columnar epithelial cells. The function of this plate is unknown. If it is secretory in nature it might form a kind of glue for egg or spermatophore attachment. However, every remipede specimen collected to date has been examined and no eggs or spermatophores have been found attached outside the body. One alternative is that the plate serves in a sensory capacity during mating.

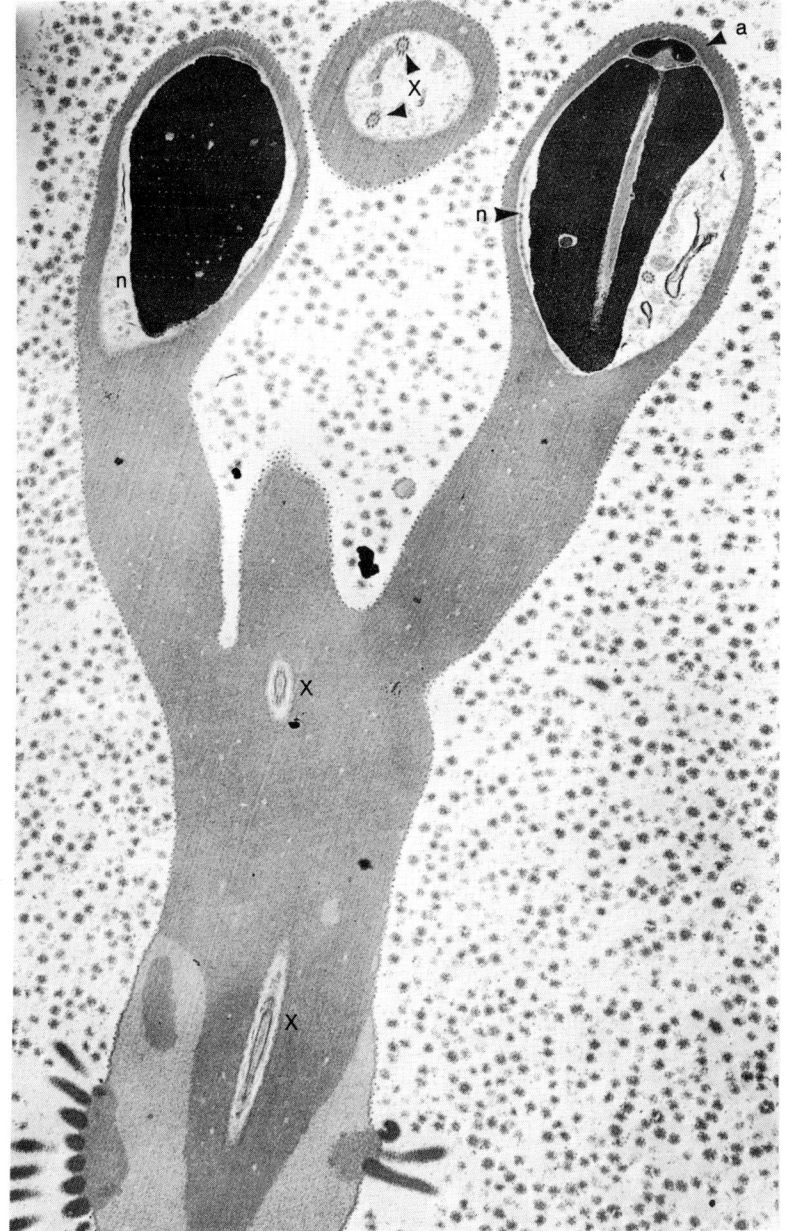

FIGURE 15.7. Spermatophore, *Speleonectes benjamini*. a, acrosome; n, nucleus; x, axoneme of flagellum. Total length about 38 μm. The spermatophore is surrounded by moderately electron dense material. From Yager 1989a.

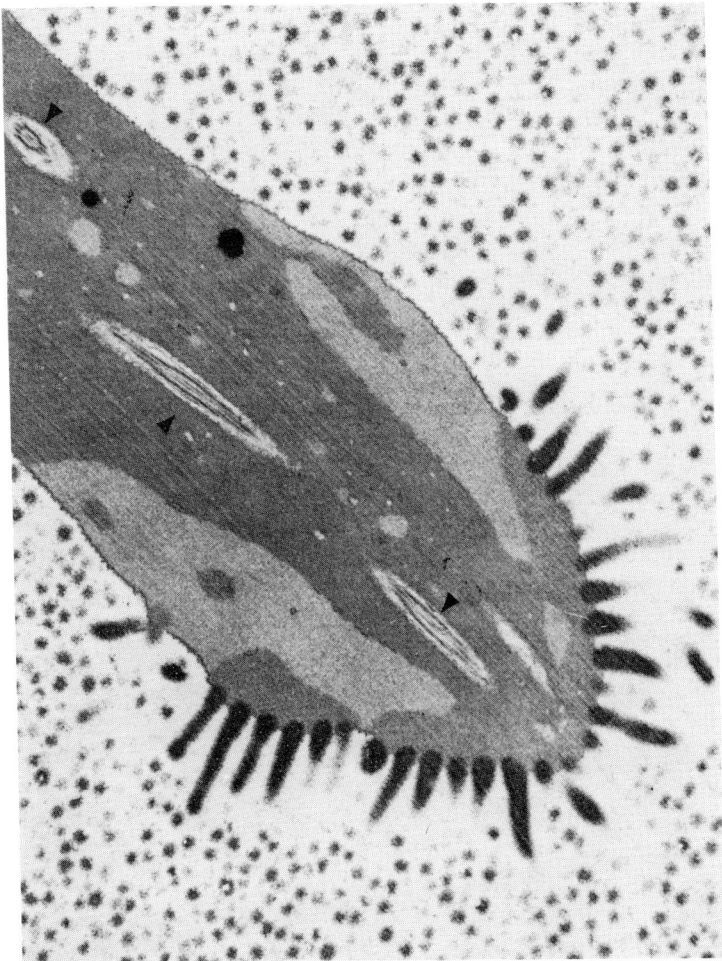

FIGURE 15.8. Spermatophore with axonemes of flagella in distal end, *Speleonectes benjamini*. Arrows indicate axonemes.

COMPARISONS OF CRUSTACEAN SPERM AND SPERMATOPHORE MORPHOLOGY

The basic or generalized sperm model includes four features: an ovoid nucleus, acrosome, mitochondria, and flagella with a 9+2 microtubular arrangement. This basic sperm is widely distributed in the phylum Arthropoda and is considered conservative (Baccetti 1970, 1979: Franzen 1970, 1987). Any loss of one or more of the four structures probably constitutes an irreversible evolutionary step (Brown 1970). The Branchiopoda, Cephalocarida, Ostracoda, Copepoda, and Malacostraca possess varied and/or aberrant sperm that lack flagella. Five crustacean groups are known to possess flagellate sperm: the Branchiura, Cirripedia, Mystacocarida, Ascothoracida, and Remipedia (Grygier 1981: Yager 1989a). The Branchiura, Cirripedia, and Mystacocarida have flagellated sperm that deviate

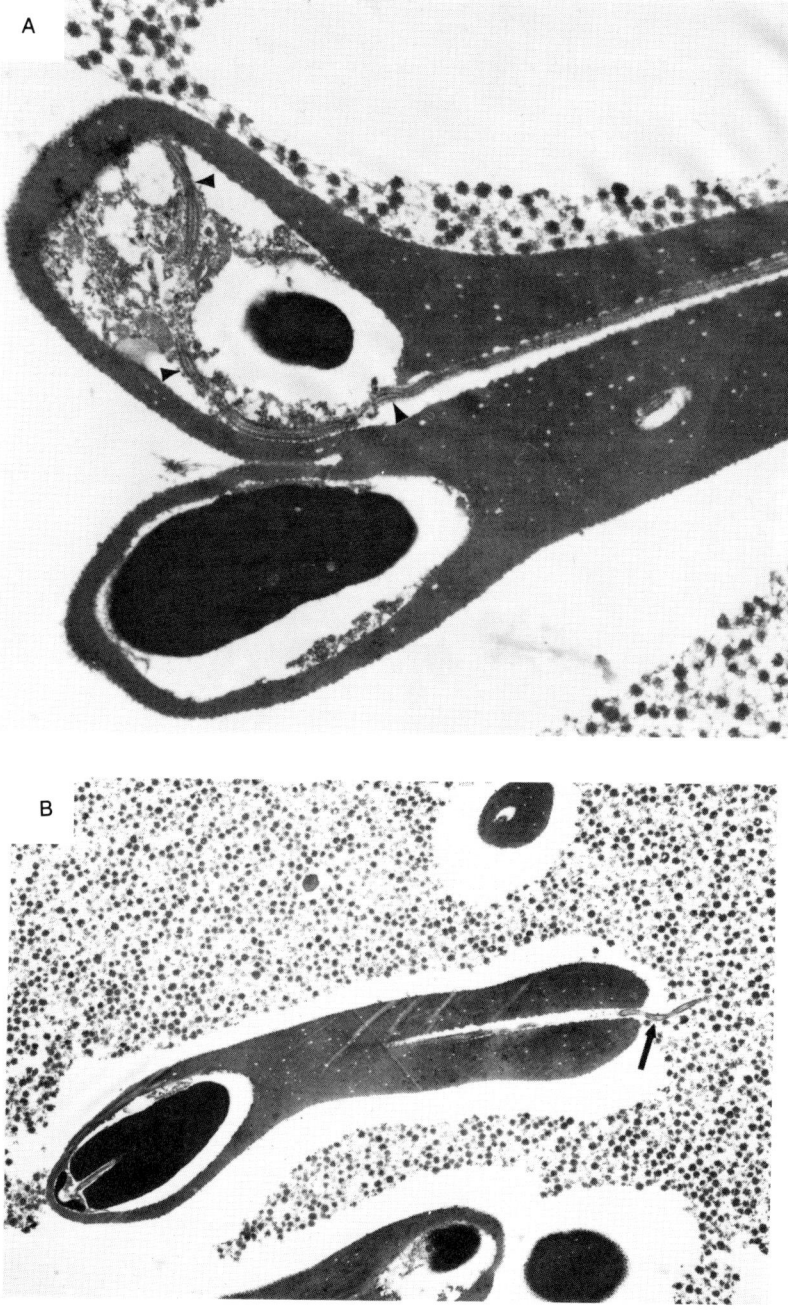

FIGURE 15.9. A. Spermatophore, *Speleonectes benjamini*. Arrows mark axoneme of flagellum looping around the nucleus before extending distally. B. Spermatophore, *Speleonectes benjamini*. Arrow indicates flagellum extending out of the distal end.

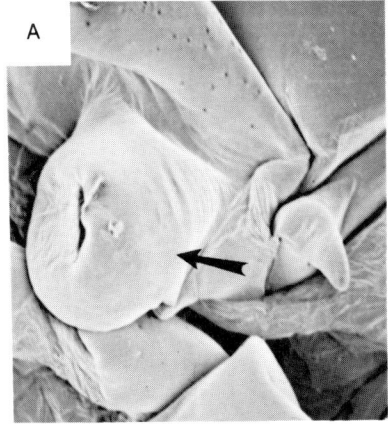 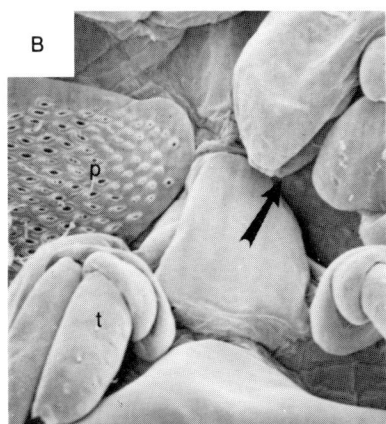

FIGURE 15.10. A. Male gonopore, *Pleomothra apletocheles*, 142x. Arrow indicates bulbous gonopore. B. Male gonopore complex, *Godzilliognomus frondosus*, 233x. Fourteenth trunk appendage of one side removed to expose genital plate found anterior to trunk appendage 15. Arrow indicates location of gonopore underneath triangular extension of protopod of fourteenth trunk appendage. p, genital plate; t, trunk appendage 15.

in some way from the conventional sperm model (see reviews by Grygier 1981: Pochon-Masson 1983; Adiyodi 1985). Grygier (1981, 1982) described the sperm of Ascothoracida as having the most generalized sperm morphology known in Crustacea. Although generalized in other aspects, the ascothoracid sperm is characterized by a cylindrical nucleus. The mature sperm of *Speleonectes benjamini* seem to resemble the basic sperm model with ovoid nucleus, acrosome, and simple flagellum (Yager 1989a). The remipede sperm contain several mitochondria, but a distinct arrangement of them is not evident. Two centrioles are found in the midpiece of most invertebrate sperm; however, they have not yet been observed in remipede sperm. The ultrastructure of the flagella of *Speleonectes benjamini* and *Godzilliognomus frondosus* shows a 9+2 microtubular arrangement. Motility of remipede sperm has not been confirmed. However, the morphology observed is consistent with that of other motile sperm.

The structure of the sperm of *Speleonectes benjamini* and *Godzilliognomus frondosus* does not appear to diverge from the stereotyped conventional sperm pattern. The data suggest that the remipedes may have the most generalized sperm morphology found in the Crustacea. The sperm of the horseshoe crab *Limulus* (Subphylum Chelicerata, Class Merostomata) is considered to be the most generalized sperm morphology of the arthropods (Fahrenbach 1973; Baccetti 1979). The sperm of *S. benjamini* appears to more closely resemble that of *Limulus* than the sperm of any other crustacean (Yager 1989a).

The production of spermatophores has been documented in a number of crustaceans, including decapods, copepods, mystacocarids, and remipedes. Although they exhibit various shapes and sizes, spermatophores generally consist of a mass of sperm encapsulated by one or more layers of acellular materials. Histochemical observations indicate that the main components of spermato-

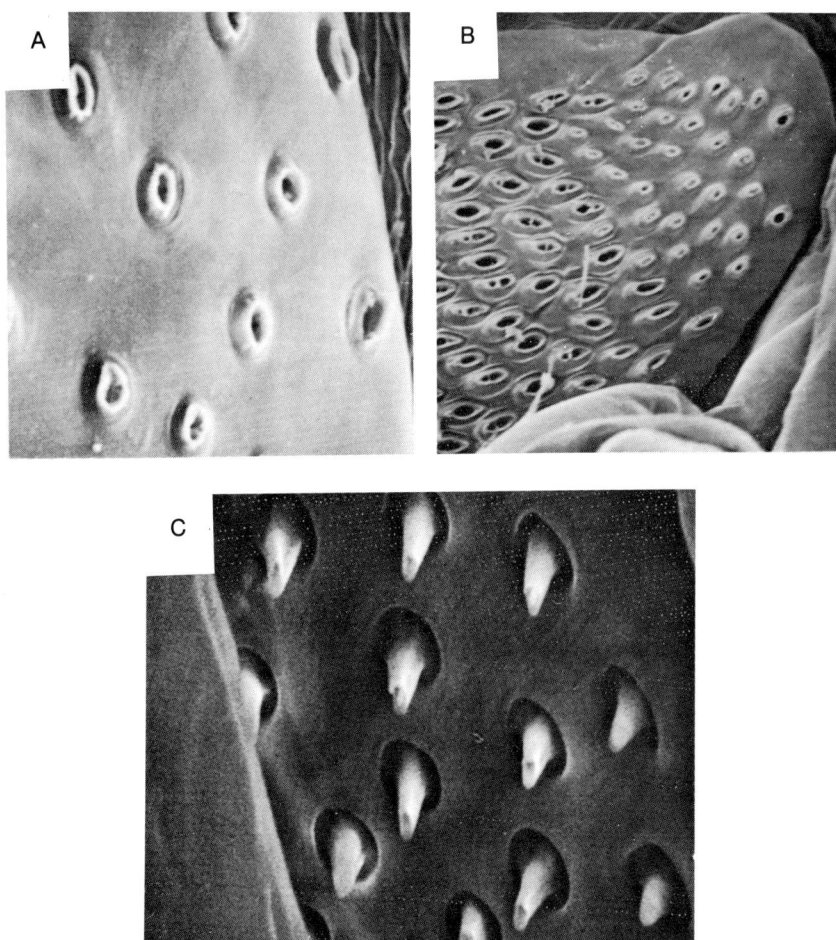

FIGURE 15.11. A. Genital plate, *Speleonectes benjamini*, 1710x. From Yager 1989a. B. Genital plate, *Godzilliognomus frondosus*, 450x. C. Genital plate, *Speleonectes lucayensis*, 3360x.

phore layers are mucopolysaccharides (Kooda-Cisco & Talbot 1982; Dudenhausen & Talbot 1983; Subramoniam 1984; Dougherty & Harris 1986; Yager 1989a).

The primary spermatophore layer directly surrounds and consolidates the sperm into a mass. Approximately 100 sperm are grouped into ovoid spermatophores in macruran decapods (Hinsch & McKnight 1988). In mystacocarids two sperm cells are encapsulated into a distinctly shaped spermatophore (Brown & Metz 1967). The two remipede species examined have from 2–6 sperm cells bound in a distinct shape. An intermediate or middle layer surrounds the encapsulated sperm and consists of extracellular secretions produced by the vas deferens. In most decapods a further acellular layer usually surrounds the entire spermatophoric mass and the whole structure is extruded. The extruded spermatophore typically takes on a distinct shape (see review of decapod spermatophores in Dudenhausen & Talbot 1983). It is not known if remipedes surround their spermatophores and flocculent matrix with an outer acellular wall and extrude a single structure. To date, no evidence of an outer wall has been observed in remipede material. The spermatophores and surrounding flocculent material have been found in the vas deferens as well as in the protopod of the fourteenth trunk appendage in what may be a seminal vesicle.

REPRODUCTIVE BEHAVIOR—HYPOTHESES

Remipedes have not been observed mating. They lack copulatory processes found in other crustaceans, such as appendix masculinae or penes. Itô and Schram (1988) hypothesized that the female gonopore pad is depressed inward by the male gonopore complex during copulation, allowing eggs to be released while being fertilized. My observations suggest that there is an alternative hypothesis to direct copulation.

The female gonopore of *Godzilliognomus frondosus* is about 67 μm wide and 75 μm in depth. The male gonopore complex is approximately 130 μm wide. Measurements of the gonopores of several species indicate that for the most part, the male gonopore complex is wider than the female gonopore opening by almost half. The gonopore complex of *G. frondosus* is on the underside of a large triangular extension of the protopod (fig. 15.10B). This flaplike extension would have to be lifted to expose the actual gonopore. Thus the ability to push the cuticular pad of the female gonopore inward by insertion of the male gonopore complex may not be anatomically possible. It is more likely that spermatophores are extruded through the male gonopore and transferred directly to the female gonopore. There they would be placed either inside the female gonopore in a seminal receptacle or onto the surface of the female gonopore area. Internal fertilization would seem more likely, since each spermatophore bears only a few spermatozoa. The mode of remipede spermatophore transfer is unknown.

DEVELOPMENT

To date, nothing is known about remipede development. Examination of material from plankton tows taken in remipede habitat has not revealed the presence of larval stages. Although there is no direct evidence, the data suggest that remipedes have direct development. Even the smallest juvenile collected to date has more than 14 trunk segments and hence both male and female gonopores. Remipedes appear to produce only a few large oocytes, another possible indication of direct development.

Based on the number and shape of trunk segments, total body length, and biomass I have determined three age classes in remipedes. Juveniles and subadults of several species have been collected. The juvenile and subadults have rounded tergites, while tergites of adults are shorter and wider, having a subrectangular appearance. Juveniles are basically miniature copies of the adults but about one-third or less as long. Subadults are more than half the length of adults, but they still retain the rounded tergites. Table 15.1 gives size and segment number comparisons for all apparent adults and for the smallest individuals collected to date.

TABLE 15.1. Comparisons of trunk segment number and length of largest adult with smallest juvenile or subadult found for all species to date.

Species	Trunk Segment Number adult/juvenile	Total Body Length adult/juvenile
Speleonectes benjamini	27/20	18.0 mm/ 4.9 mm
Speleonectes lucayensis	32/16	24.0 mm/ 4.0 mm
Speleonectes ondinae	25/20[a]	16.7 mm/10.1 mm
Speleonectes tulumensis	38/17	30.2 mm/ 7.2 mm
Cryptocorynetes haptodiscus	31/28[a]	17.6 mm/ 9.7 mm
Lasionectes entrichoma	32/16	32.8 mm/ 9.0 mm
Godzillius robustus	29/26[a]	45.1 mm/25.7 mm
Godzilliognomus frondosus	16/15	9.3 mm/ 3.7 mm
Pleomothra apletocheles	25/16	17.1 mm/ 5.2 mm

[a] Subadult specimen.

SUMMARY

The reproduction biology of two species of remipedes, *Speleonectes benjamini* and *Godzilliognomus frondosus*, was examined. Both are simultaneous hermaphrodites. The female system extends from the posterior portion of the head to the female gonopore at the seventh trunk appendage. The gonopore is similar in all species examined and consists of a semicircular opening at the posterior base of the protopod. In *S. benjamini* and *G. frondosus* all stages of oogenesis have been observed.

The male reproductive system extends from the posterior portion of the seventh trunk segment to the male gonopore on the fourteenth trunk appendage. The testes are paired organs that lie dorsolateral to the midgut and terminate at about the tenth trunk segment. In the transition between testes and vas deferens, round spermatids become elliptical with a tapered end. The paired, coiled vas deferens begins at about trunk segment 10 and terminates at the male gonopore. In the vas deferens the sperm become encapsulated into distinctly shaped spermatophores. Mature sperm have an ovoid nucleus and a flagellum with a $9+2$ microtubular arrangement.

Much remains to be learned about remipede reproduction. Specimens of most species are rare. After extensive fieldwork, fewer than 20 individuals have been collected for five of the nine species. Further investigations are in progress that I hope will clarify many of the unknowns about the reproductive biology of the remipedes.

Acknowledgments

The research for this manuscript was carried out as partial fulfillment of my doctoral dissertation at Old Dominion University. The Department of Biological Sciences and the Electron Microscopy Laboratory supported the laboratory research. I thank Keith A. Carson and especially Ralph W. Stevens III for their helpful discussions during the course of this work.

Literature Cited

Adiyodi, R. G. 1985. Reproduction and its control. In D. M. Skinner, ed., *The Biology of Crustacea*, 9:147–215. New York: Academic Press.

Baccetti, B. 1970. The spermatozoon of Arthropoda. IX. The sperm cell as an index of Arthropod phylogenesis. In B. Baccetti, ed., *Comparative Spermatology*, pp. 29–46. Rome: Accademia Nazionale dei Lincei; New York: Academic Press.

Baccetti, B. 1979. Ultrastructure of sperm and its bearing on arthropod phylogeny. In A. P. Gupta, ed., *Arthropod Phylogeny*, pp. 609–644. New York: Van Nostrand Reinhold.

Blades-Eckelbarger, P. I. & M. J. Youngbluth. 1982. Ultrastructure of the male reproductive system and spermatophore formation in *Labidocera aestiva* (Crust.: Copepoda). *Zoomorphology (Berl.)* 99:1–21.

Brown, G. G. 1970. Some comparative aspects of selected crustacean spermatozoa and crustacean phylogeny. In B. Baccetti, ed., *Comparative Spermatology*, pp. 183–203. New York: Academic Press.

Brown, G. G. & C. B. Metz. 1967. Ultrastructural studies on the spermatozoa of two primitive crustaceans, *Hutchinsoniella macracantha* and *Derocheilocaris typicus*. *Z. Zellforsch. Mikrosk. Tech.* 80:78–92.

Dougherty, W. J., M. M. Dougherty & S. G. Harris. 1986. Ultrastructural and histochemical observations on electroejaculated spermatophores of the palaemonid shrimp, *Macrobrachium rosenbergii*. *Tissue & Cell* 18:709–724.

Dudenhausen, E. E. & P. Talbot. 1983. An ultrastructural comparison of soft and hardened spermatophores from the crayfish *Pacifistacus leniusculus* Dana. *Can. J. Zool.* 61:182–194.

Fahrenbach, W. H. 1973. Spermiogenesis in the horseshoe crab, *Limulus polyphemus*. *J. Morphol.* 140:31–52.

Franzen, A. 1970. Phylogenetic aspects of the morphology of spermatozoa and spermiogenesis. In B. Baccetti, ed., *Comparative Spermatology*, pp 29–46. New York: Academic Press.

Franzen, A. 1987. Spermatogenesis. In A. C. Geise, J. S. Pearse & V. B. Pearse, eds., *Reproduction of Marine Invertebrates*, 9:1–47. Palo Alto and Pacific Grove, Calif.: Blackwell/Boxwood Press.

Grygier, M. J. 1981. Sperm of the ascothoracican parasite *Dendrogaster*, the most primitive found in Crustacea. *Invert. Reprod. Devel.* 3:65–73.

Grygier, M. J. 1982. Sperm morphology in Ascothoracida (Crustacea: Maxillopoda): Confirmation of generalized nature and phylogenetic importance. *Invert. Reprod. Devel.* 4:323–332.

Hinsch, G. W. & C. E. McKnight. 1988. The vas deferens of the Spanish lobster, *Scyllarus chacei*. *Invert. Reprod. Devel.* 13:267–280.

Itô, T. & F. R. Schram. 1988. Gonopores and the reproductive system of nectiopodan Remipedia. *J. Crustacean Biol.* 8:250–253.

Kooda-Cisco, M. J. & P. Talbot. 1982. A structural analysis of the freshly extruded spermatophore from the lobster, *Homarus americanus*. *J. Morphol.* 172:193–207.

Pochon-Masson, J. 1983. Arthropoda—Crustacea. In K. G. Adiyodi and R. G. Adiyodi, eds., *Reproductive Biology of Invertebrates*, 2:407–449. Chichester: Wiley.

Schram, F. R. 1986. *Crustacea*. New York: Oxford University Press.

Schram, F. R., J. Yager & M. J. Emerson. 1986. Remipedia. Part 1. Systematics. *San Diego Soc. Nat. Hist. Mem.* 15:1–60.

Subramoniam, T. 1984. Spermatophore formation in two intertidal crabs *Albunea symnista* and *Emerita asiatica* (Decapoda: Anomura). *Biol. Bull. (Woods Hole)* 166:78–95.

Yager, J. 1981. Remipedia, a new class of Crustacea from a marine cave in the Bahamas. *J. Crustacean Biol.* 1:328–333.

Yager, J. 1987a. *Cryptocorynetes haptodiscus*, new genus, new species, and *Speleonectes benjamini*, new species, of remipede crustaceans from anchialine caves in the Bahamas, with remarks on distribution and ecology. *Proc. Biol. Soc. Wash.* 100:302–320.

Yager, J. 1987b. *Speleonectes tulumensis* n. sp. (Crustacea: Remipedia) from two anchialine cenotes of the Yucatan Peninsula, Mexico. *Stygologia* 3:160–166.

Yager, J. 1989a. The male reproductive system, sperm, and spermatophores of the primitive, hermaphroditic, remipede crustacean *Speleonectes benjamini*. *Invert. Reprod. Devel.* 15:75–81.

Yager, J. 1989b. *Pleomothra apletocheles* and *Godzilliognomus frondosus*, two new genera of remipede crustaceans (Godzilliidae) from anchialine caves in the Bahamas. *Bull. Mar. Sci.* 44:1195–1206.

Yager, J. & F. R. Schram. 1986. *Lasionectes entrichoma*, new genus, new species (Crustacea: Remipedia) from anchialine caves in the Turks and Caicos, British West Indies. *Proc. Biol. Soc. Wash.* 99:65–70.

SIXTEEN
Structure and Chemical Content of the Spermatophores and Seminal Fluid of Reptantian Decapods

GERTRUDE W. HINSCH

*Department of Biology,
University of South Florida,
Tampa.*

Abstract

Reptantian crustacean sperm appear to become embedded in a matrix of testicular or extreme anterior vas deferens origin. As sperm enter the anterior vas deferens, additional component(s) that form the wall of the sperm bundle are added.

In brachyurans, these bundles become surrounded by seminal fluids secreted by the middle and posterior regions of the vas deferens. Sperm bundles may be stored for some time within the male tract and are transferred to the seminal receptacles of the females at the time of copulation. The sperm may be stored for long periods of time in the seminal receptacles prior to oviposition and fertilization.

In lobsters, crayfishes, and anomurans, the sperm become aggregated in sperm bundles in the form of a highly convoluted ribbon or in individual bundles attached to peduncles arising from the spermatophore wall. Several layers of mucopolysaccharide materials are deposited around these sperm bundles to form the spermatophore. At mating, the spermatophores are transferred to the ventral surface of the female. The spermatophores may be internalized or remain attached to the exoskeleton until the time of oviposition. The role that the mucopolysaccharide acellular materials of the seminal fluids and spermatophore walls play in sperm protection, survival, or nutrition is little known or understood.

THE BIOLOGY of the male crustacean reproductive tract has received relatively little attention. The formation and the morphology of reptantian spermatophores have been described at the light microscope level by many (reviewed in Dudenhausen & Talbot 1983). However, until only recently few ultrastructural studies of the vas deferens and spermatophores have been reported (Hinsch & Walker 1974; Kooda-Cisco & Talbot 1982, Dudenhausen & Talbot 1983; Martin, Herzig & Narimatsu 1987; Beninger et al. 1988; Hinsch & McKnight 1988; Hinsch 1988; Talbot & Beach 1989). This presentation will review the works published previously as well as include original observations of the author on additional species. Tissues for such work were fixed as indicated in earlier works of the author.

Considerable morphological variation exists among representatives of the reptant decapods, and members of the various groups have evolved different strategies of transferring sperm from the male to the female. Among the brachyurans, sperm masses surrounded only by a seminal fluid are transferred from the male during copulation to an internal site, the seminal receptacle of the female. In contrast, among the anomurans and "macrurans" several layers of acellular materials form an elaborate spermatophore that surrounds the sperm bundles. These spermatophores are transferred and attached to the ventral surface of the exoskeleton of the female at the time of mating. In some cases, the female then internalizes the spermatophore and may retain the sperm for several years (Bumpus 1891).

SPERM MASS

Decapod sperm become embedded in a matrix of testicular origin before beginning their passage to the vas deferens. This matrix appears to be mucopolysaccharide in nature when stained with PAS and varies in quantity and appearance when viewed with the electron microscope depending upon the species involved (figs. 16.1A,B, 16.2A,C, 16.5D, 16.6A,C,D, 16.7A, 16.10E).

Spermatophore formation occurs in the vasa deferentia of males of decapod crustaceans. In general, synthesis of the walls of the spermatophores occurs in the various regions of the anterior vas deferens. The epithelium of this region generally consists of tall columnar cells containing cytoplasmic organelles associated with synthetic activity. The secretion products may be exocytosed to the lumen of the vas deferens (*Libinia* [Hinsch & Walker 1974], *Homarus* [Kooda-Cisco & Talbot 1982]), or the apical ends of the cells may bleb off into the lumen by apocrine secretion (*Cherax*, [Talbot & Beach 1989], *Scyllarides* [Hinsch personal observation]).

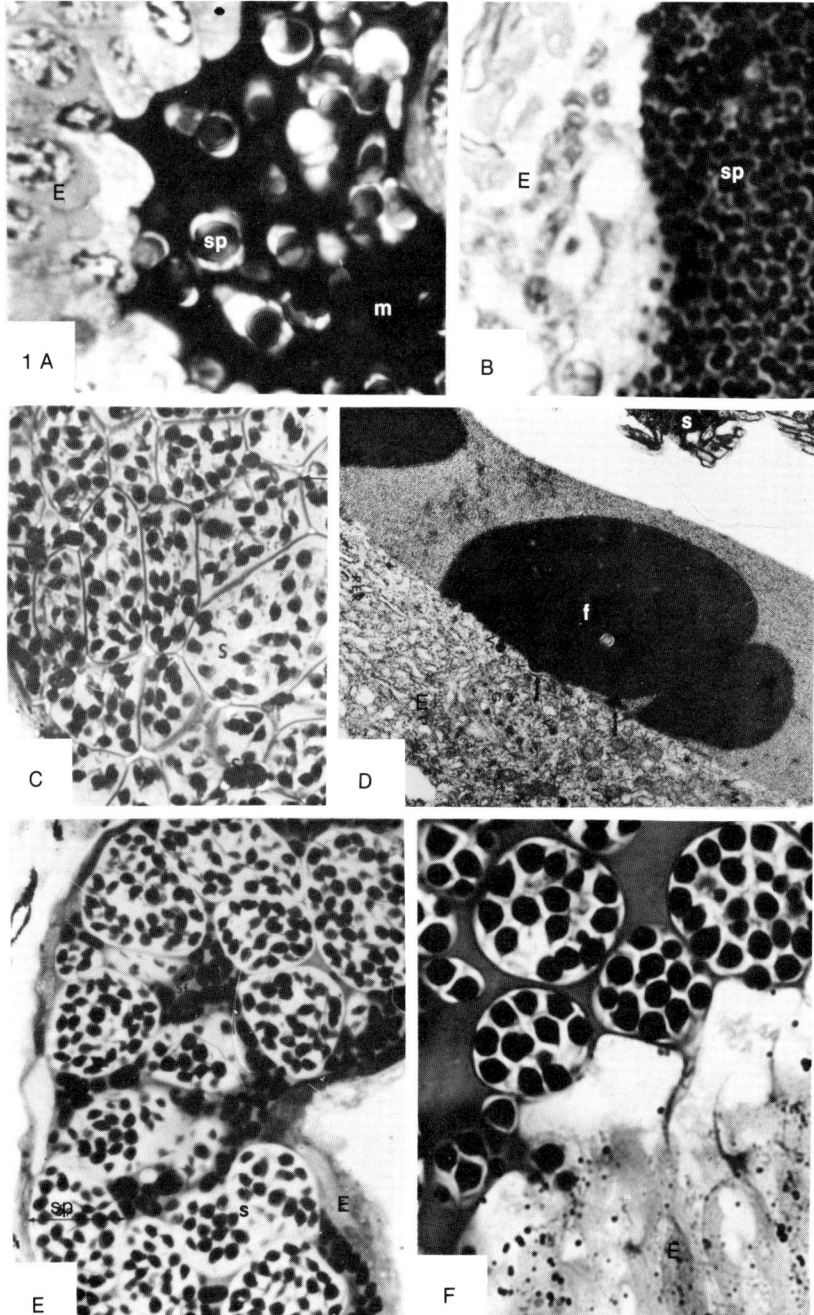

FIGURE 16.1. A. Sperm (sp) of *Uca pugilator* surrounded by matrix (m) in the sperm duct of the testis. E, epithelium. X 900. B. Sperm (sp) of *Callinectes sapidus* in the sperm duct. X 400. C. Sperm (s) becoming organized into discrete aggregates in the anterior vas deferens of *Uca*. X 400. D. While the sperm may be clumped into discrete aggregates, seminal fluids (f) secreted by the epithelium (E) of the vas deferens do not form a definite wall around the sperm (s) and matrix initially. X 10000. E–F. Light micrographs of the sperm masses of *Uca* and surrounding seminal fluids (f). E, X 700; F, X 850.

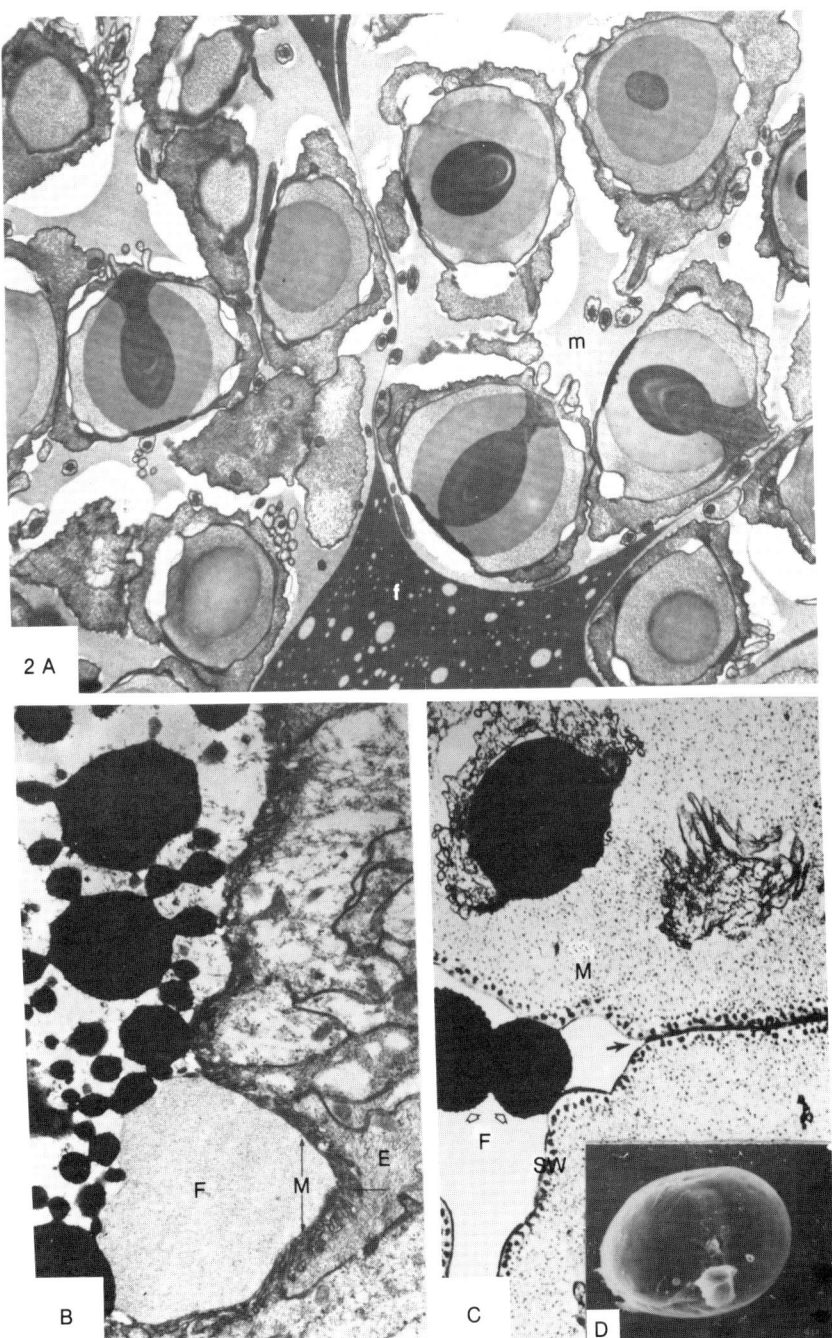

FIGURE 16.2. A. TEM of aggregates of sperm of *Libinia emarginata* embedded in matrix (m) surrounded by seminal fluid (f) but still lacking a discrete wall to the sperm mass. X 4900. B. Seminal fluid (F) and globules in the lumen of the vas deferens of *Uca*. M, microvilli; E, epithelium; X 8200. C. Materials from the seminal fluids condense to produce the wall of the sperm mass in *Uca*. In this crab, the wall appears as a thin electron dense layer with numerous small electron globules to the interior. X 10000. D. Typical brachyuran sperm mass.

The epithelium appears to decrease in height along the length of the vas deferens. The posteriormost regions of the brachyuran vas deferens appear to secrete seminal fluid(s) to serve in the storage of mature seminal products until copulation occurs.

STRUCTURE OF THE SPERMATOPHORES

Brachyurans

As the sperm surrounded with matrix materials secreted in the testis or efferent duct move posteriorly into the vas deferens, they become aggregated into discrete masses. These sperm masses of the brachyurans become surrounded by secretion products from the anterior vas deferens (figs. 16.1C–F, 16.2A–C). This product condenses into a layer around the sperm and matrix to form the sperm mass wall. In many species, e.g., in the genera *Libinia* and *Ovalipes*, this wall is smooth-surfaced and formed of a single layer (Hinsch 1986; figs. 16.1F, 16.2C,D, 16.3A). In the golden crab, *Chaceon fenneri* (Hinsch 1988b; Manning & Holthuis 1989), *Uca* (Becker 1983), and *Scylla serrata* (Uma and Subramoniam 1979), the sperm mass wall appears to have two discrete layers (fig. 16.3B). In the snow crab, *Chionoecetes opilio*, the surface of the sperm mass is convoluted (Beninger et al. 1988).

At the time of copulation, the sperm masses together with other secretions of the vas deferens are transferred to the seminal receptacles of the females. In *Libinia emarginata*, free sperm appear in the lumen of the seminal receptacles of the females shortly after copulation (fig. 16.4A). In *Ovalipes ocellatus* the droplets of seminal fluid appear to have coalesced and fused with the walls of the sperm masses, forming a compact sperm plug (fig. 16.4B,C); Hinsch 1986).

Anomurans

In the land hermit crab *Coenobita clypeatus* the sperm become associated with fibrillar materials while still in the testis (figs. 16.7A–C, 16.8A, 16.B). These fibrils as well as matrix materials come to surround the individual sperm and separate them one from the other (figs. 16.7D, 16.8A). Within the vas deferens, these materials condense into aggregates of varying geometric patterns and are grouped into individual masses of sperm by walls produced in the anterior vas deferens (figs. 16.8C,D).

Similarly, in the red crab, *Pleuroncodes planipes*, the sperm are surrounded by aggregates and matrix of testicular origin around which a wall of material secreted by the epithelium of the anterior vas deferens forms (fig. 16.5A).

As the sperm masses of some anomurans move posteriorly in the vas deferens, the secretion product of the anterior vas deferens may become shaped by the contraction of the walls of the vas deferens. An individual sperm-filled mass becomes attached by a peduncle to the surrounding secretions (fig. 16.6A; Matthews 1953). These secretions of varying appearance and quantity are derived from the more posterior regions of the vas deferens. They are added progressively around the sperm masses to form the definitive spermatophore mass that

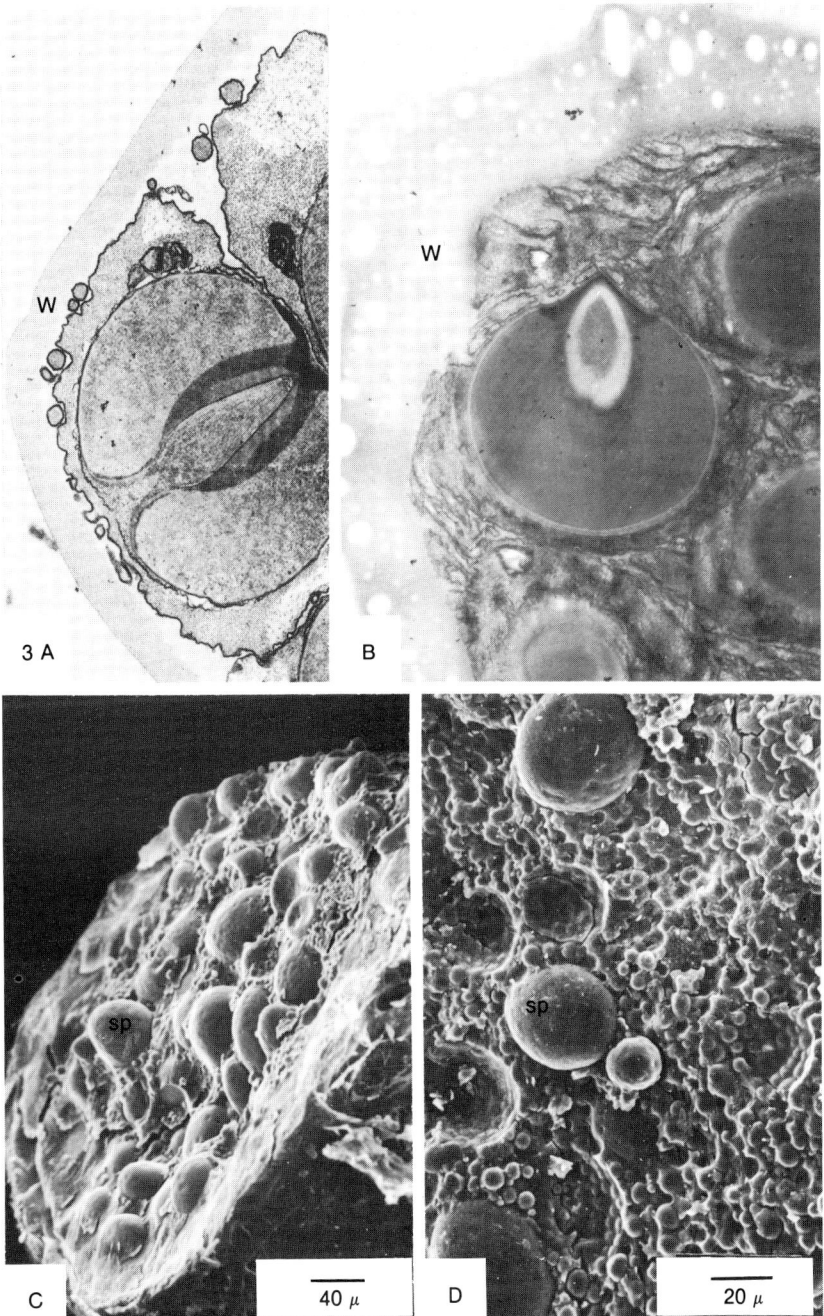

FIGURE 16.3. A. Section through the sperm mass of *Libinia emarginata* showing the wall (w). X 14200. B. Section through the sperm mass of *Chaceon fenneri* showing that in this instance the wall (w) is composed of two layers, with the outermost layer being pierced with numerous vesicles. X 10800. C. SEM of fractured surface of the vas deferens of *C. fenneri* showing the sperm masses (sp) and surrounding seminal fluids. D. SEM showing the sperm masses (sp) and surrounding seminal fluids of *Chaceon*.

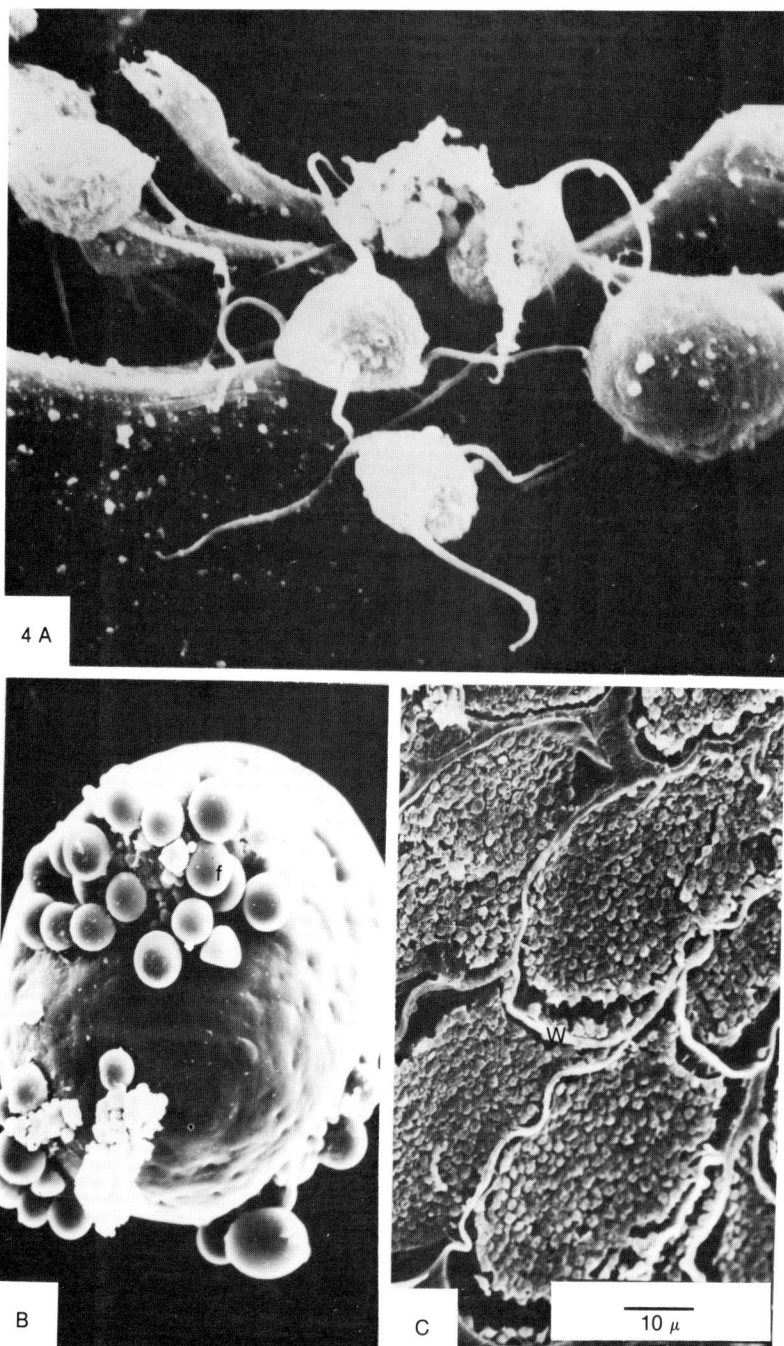

FIGURE 16.4. A. *Libinia emarginata* sperm taken from the seminal receptacle of a recently mated female. X 3500. B. Brachyuran sperm mass and associated seminal fluid droplets (f). X 2400. C. Sperm from the seminal receptacle of a recently mated *Ovalipes ocellatus* female. In this case, the wall (w) of the sperm masses appear to have fused with the droplets of seminal fluid, forming a mass of sperm packets.

is to be transferred to the female at the time of copulation (figs. 16.5A–D, 16.6A–C, 16.9A).

"Macrurans" (Lobsters and Crayfishes)

As in the brachyurans and anomurans, the macruran sperm become surrounded by a matrix of testicular origin before becoming enclosed in a wall of the sperm cord secreted by the anterior vas deferens (figs. 16.10D,E, 16.11E).

As the sperm cord moves posteriorly in the vas deferens the progressive deposition of the definitive layers of the spermatophore mass become apparent. The number and appearance of layers that surround the sperm cord vary from species to species (*Homarus* [Kooda-Cisco & Talbot 1982], *Pacifastacus* [Dudenhausen & Talbot 1983], *Enoplometopus* [Haley 1984], *Panulirus* [Martin, Herzig & Narimatsu 1987], *Scyllarus* [Hinsch & McKnight 1988], *Cherax* [Talbot & Beach 1989]). As in the anomurans, the shape of the sperm cord varies among species. In some forms, as the sperm cord passes posteriorly, contractions of the wall of the vas deferens (Matthews 1956a,b) appear to break the sperm cord up into individual units that resemble the sperm masses of the "macruran" *Scyllarus* (Hinsch & McKnight 1988) or result in a tightly coiled tube embedded in spermatophore matrix as in *Panulirus* (Martin, Herzig & Narimatsu 1987) and *Cherax* (Talbot & Beach 1989). Changes in the layers of the spermatophore have been observed between softened and hardened spermatophores (Dudenhausen & Talbot 1983; Martin, Herzig & Narimatsu 1987).

CHEMICAL COMPOSITION

Relatively few studies have delved into the nature of the spermatophore structure. Most studies have indicated that the materials comprising the numerous layers associated with these structures are PAS (periodic acid Schiff)–positive. Studies of spermatophores of several decapods have indicated that these noncellular materials are predominantly neutral and acidic mucopolysaccharides (Uma & Subramoniam 1979; Subramoniam 1984; Radha & Subramoniam 1985; Sasikala & Subramoniam 1987). In *Scylla*, the outer wall of the sperm mass is chitinous and the inner wall is nonchitinous (Uma & Subramoniam 1979). Spaulding (1942) and King (1948) have reported the presence of chitin in spermatophores as well.

The role of the spermatophore components and seminal fluids in sperm protection once they have been transferred to the seminal receptacle in brachyurans or the ventral surface of the females in anomurans and macrurans is unknown. Among the brachyurans, the walls may remain intact for some time within the seminal receptacles as in *Ovalipes* (Hinsch 1986) and *Chaceon* (Hinsch 1988a) or disappear after insemination as in *Libinia* (Hinsch 1986).

The factors that bring about the change in seminal fluids to initiate their fusion with the sperm mass wall in *Ovalipes* or the dissolution of the wall in *Libinia* are not known. What role if any the cells of the seminal receptacles play in sperm storage and nutrition has not been investigated. It has been reported

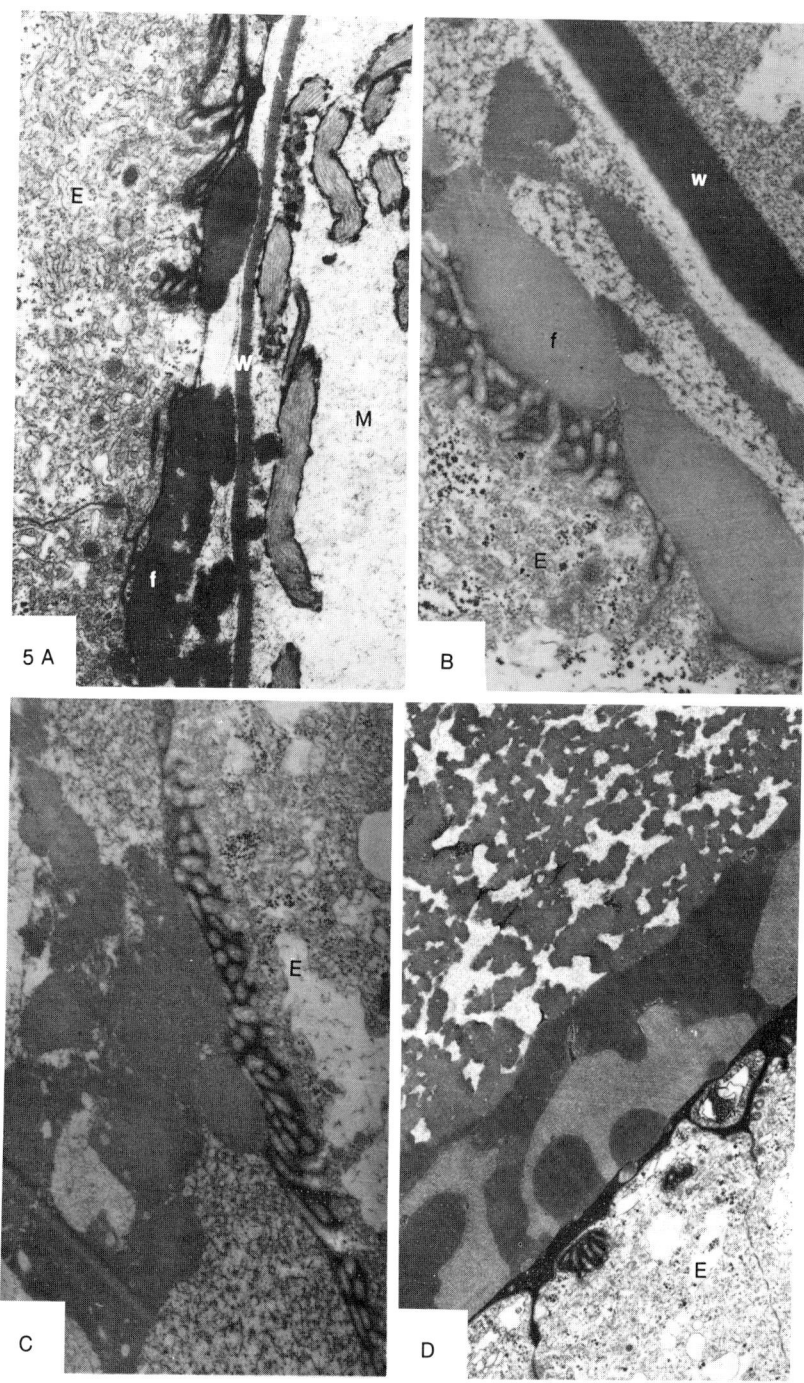

(Anilkumar & Adiyodi 1977) that the cells of the seminal receptacle of *Parathelphusa* change with the ovarian cycle. Might these changes affect the sperm stored in the seminal receptacle as well?

Jeyalectumie and Subramoniam (1987) reported that seminal secretions containing spermatophores of *Parathelphusa* have large quantities of free carbohydrates, proteins, and lipids. They suggest that the spermatozoa in the posterior vas deferens may use fatty substances for oxidative metabolism. Further, the presence of lactate dehydrogenase isozymes in reproductive tissues may undergo anaerobic metabolism. This poses an interesting question, as reptantian sperm, being supposedly nonmotile, lack normal mitochondria and their enzymes for oxidative phosphorlyation (Pearson & Walker 1975).

In anomurans and "macurans" the spermatophore is transferred to the ventral surface of the female. Here it may persist for some time or it may be internalized by the female (Matthews 1954). Freshly extruded spermatophores may be very sticky (e.g., *Scyllarides nodifer*) and have been observed adhering to aquarium walls (Hinsch personal observation). In some forms newly deposited spermatophores slowly harden and darken (*Panulirus* [Martin, Herzig & Narimatsu 1987]). In *Pacifastacus* the outer part of the spermatophore is lost after binding, except at the site of attachment to the female, and the inner layers persist (Dudenhausen & Talbot 1983).

The physical and chemical basis of these changes has not been studied. The structural makeup of spermatophores appears to change with transference to the female (Dudenhausen & Talbot 1983; Martin, Herzig & Narimatsu 1987). These changes may include phenolic tanning (Malek & Bawab 1971). However, Uma and Subramoniam (1979) were unable to identify a protein capable of phenolic tanning or an enzyme that could initiate such a tanning process. It has been suggested that hardening of the spermatophore may result from calcification (Berry 1970; Subramoniam 1984).

CONCLUSIONS

While several descriptions of sperm masses and spermatophore structure have appeared in recent years, we still know little about the exact chemical nature of these structures. The walls of the sperm masses and spermatophores are performed primarily in the anterior vas deferens. In most decapods, the middle and posterior vas deferens serve as sites of storage of the seminal product. Some

FIGURE 16.5. A. Section of anterior region of the vas deferens of *Pleuroncodes planipes* showing the matrix (m) materials surrounding the sperm arms. The wall (w) of the sperm mass and the formation of the first secretions (f) to form the mass of the spermatophore appear to the left. E, epithelium of vas deferens; X 11000. B. More posteriorly in the vas deferens of *P. planipes* additional secretions appear between the epithelium (E) and sperm mass wall (w). X 12900. C. Still more posteriorly, an electron dense secretion product is visible between the microvilli of the epithelial cells. Several materials of varying electron density have accumulated in the lumen of the vas deferens. X 18000. D. Various patterns appear as the secretion products within the lumen aggregate to form the spermatophore. X 9000.

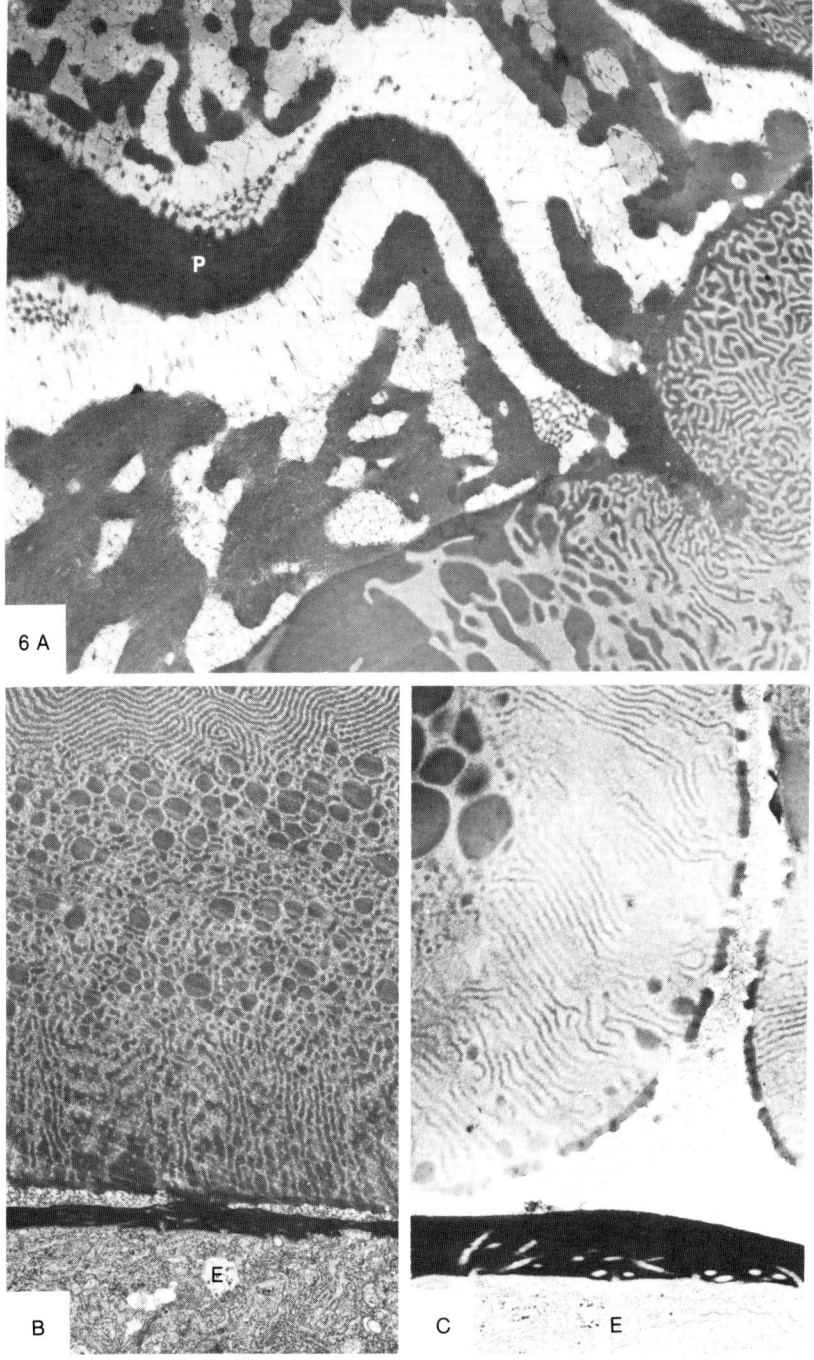

FIGURE 16.6. A. In species such as *Pleuroncodes planipes* that have a pedunculate type of spermatophore, the peduncle (p), which supports the sperm mass, is composed of material that appears to be identical to the wall. This is in turn surrounded by several different types of materials and is attached at the base of the spermatophore wall. X 9600. B–C. Patterning of the spermatophore wall materials within the posterior vas deferens. A final layer of very electron dense secretion product lies adjacent to the surface of the epithelium (e). B, X 8000; C, X 12000.

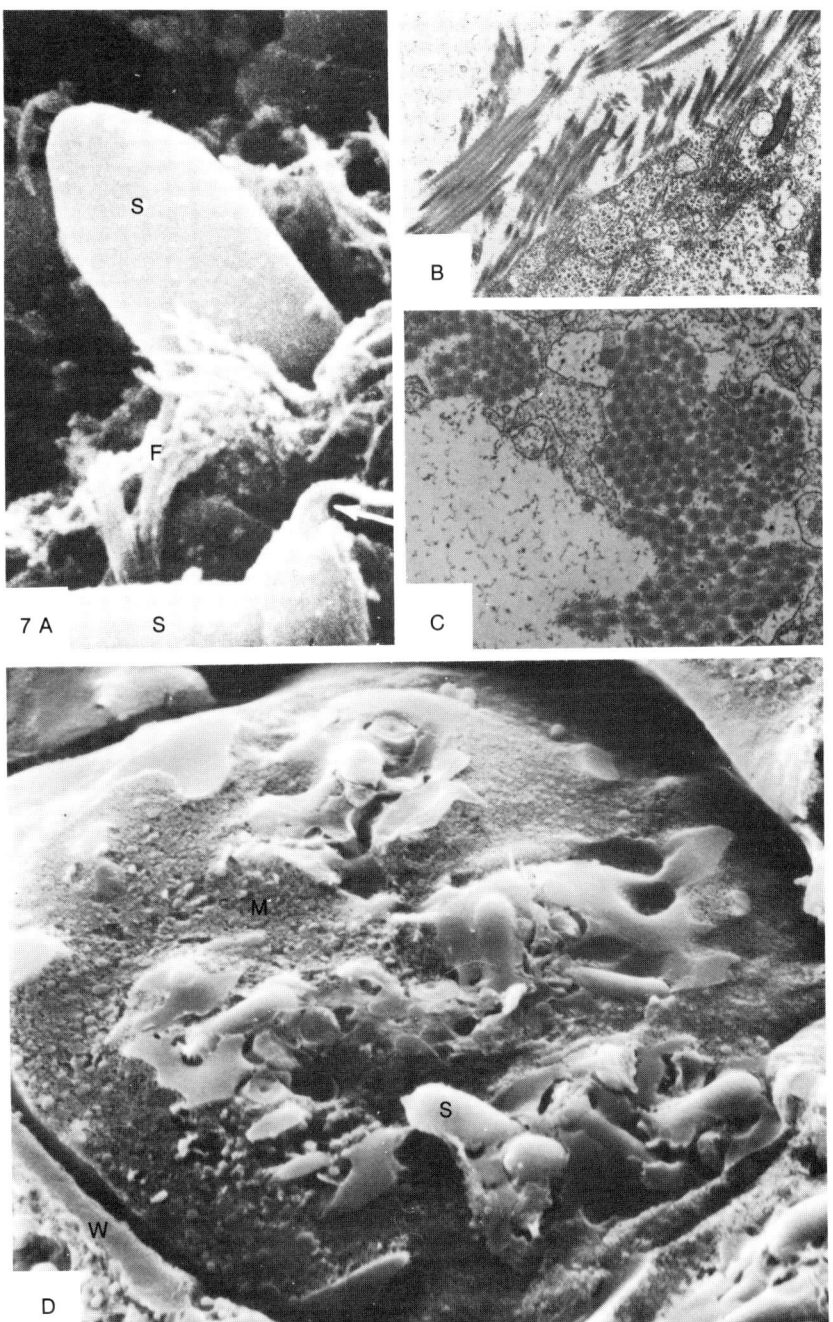

FIGURE 16.7. A. Sperm (s) of *Coenobita clypeatus* surrounded by fibrils (f) within ampulla in vas deferens. X 7500. B. Fibrils in sperm duct. X 7000. C. Cross sections of the fibrils. X 20000. D. SEM of sperm ampulla of *C. clypeatus* showing sperm (s) embedded in a matrix (m) within the sperm mass wall (w). X 2500.

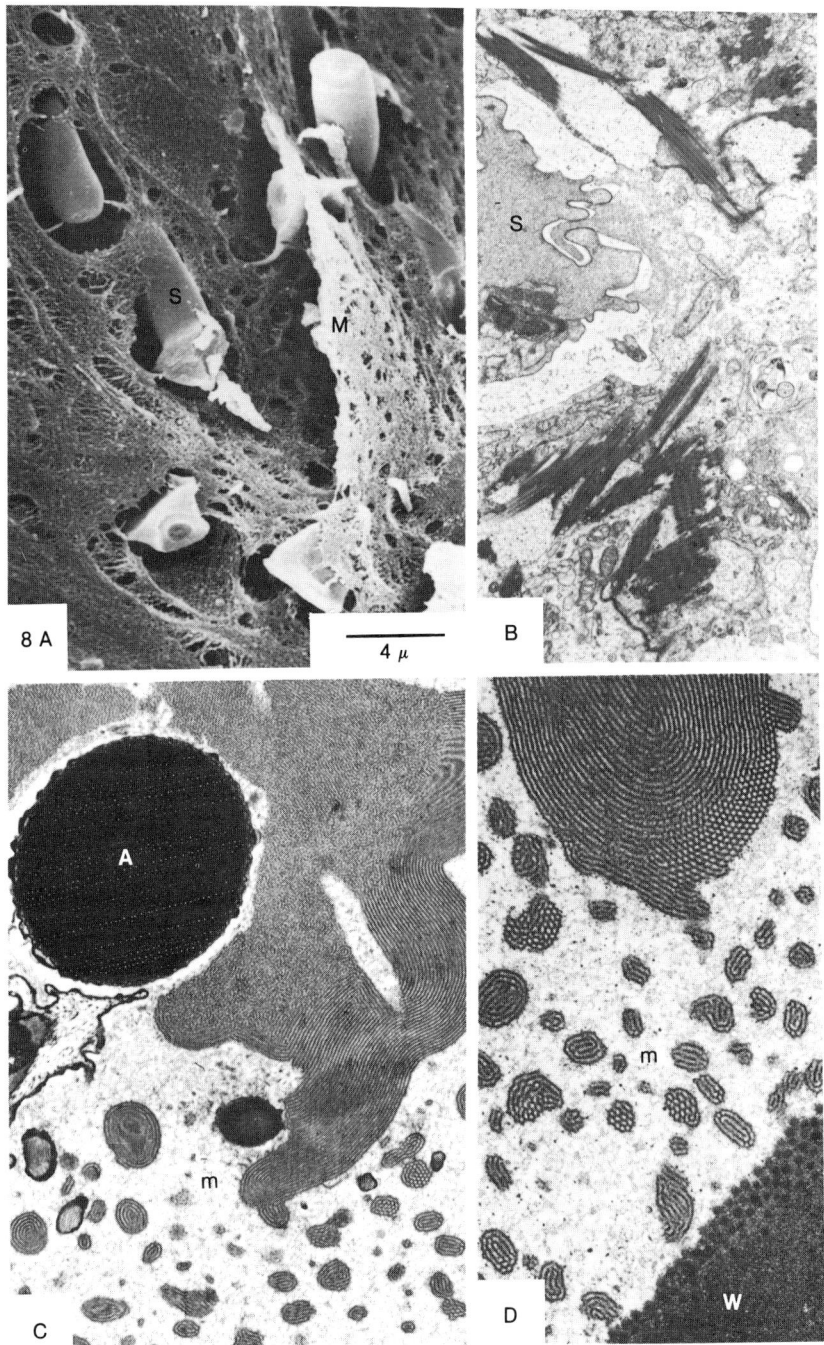

FIGURE 16.8. A. *Coenobita clypeatus* sperm embedded in matrix within the sperm mass. B. TEM of a sperm (s) in the testis of *C. clypeatus* showing the fibrils in close proximity. X 6300. C–D. TEMs of interior of the sperm mass from the anterior vas deferens showing the condensation of matrix materials (m) that surround the *C. clypeatus* sperm. A, acrosome; W, wall of sperm mass; C, X 10200; D, X 16,700.

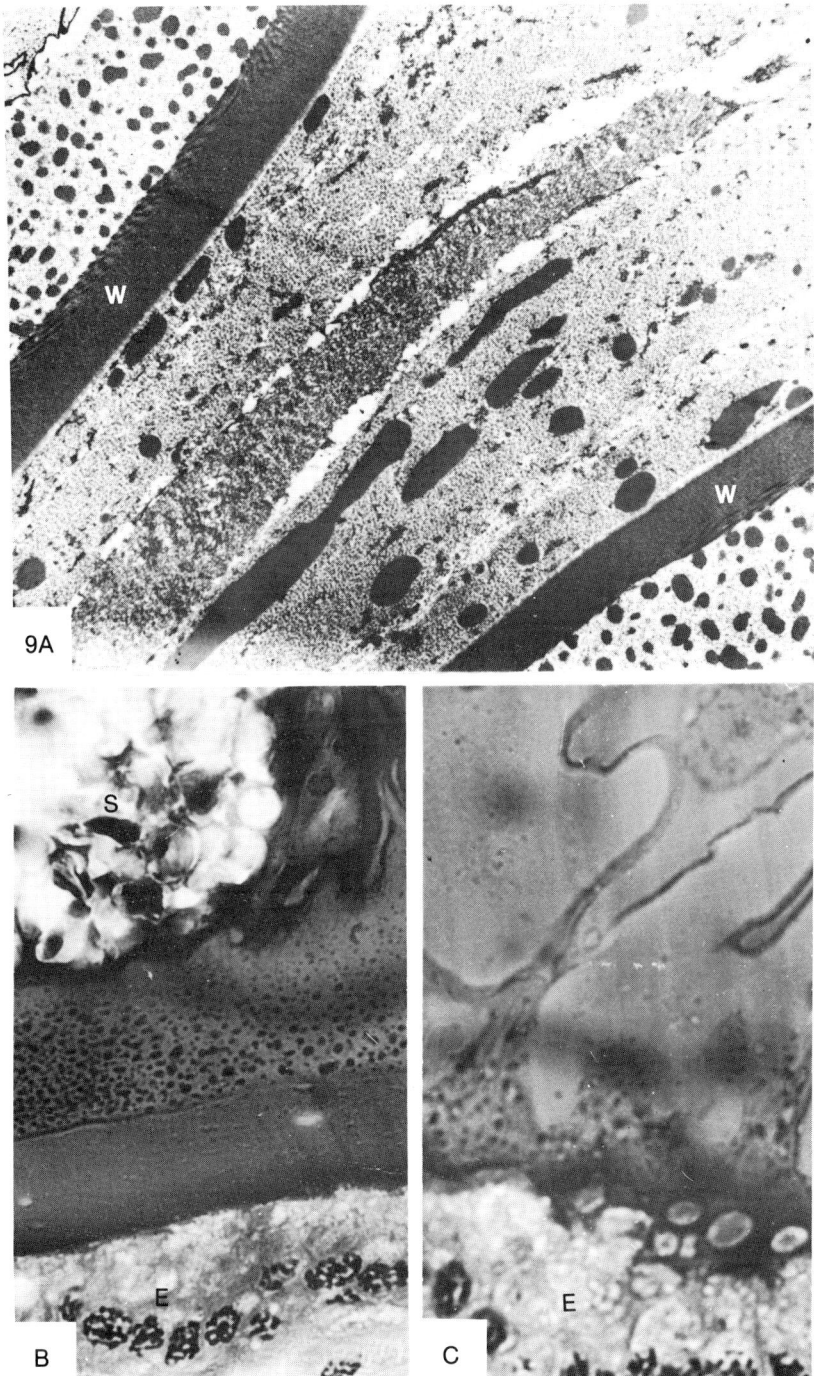

FIGURE 16.9. A. Varied secretions that become assembled between the walls (w) of two adjacent sperm (s) masses to form the spermatophore in *Coenobita clypeatus*. X 6300. B–C. Light micrographs of the two sides of the spermatophore of *Scyllarides nodifer;* E, epithelium. X 460.

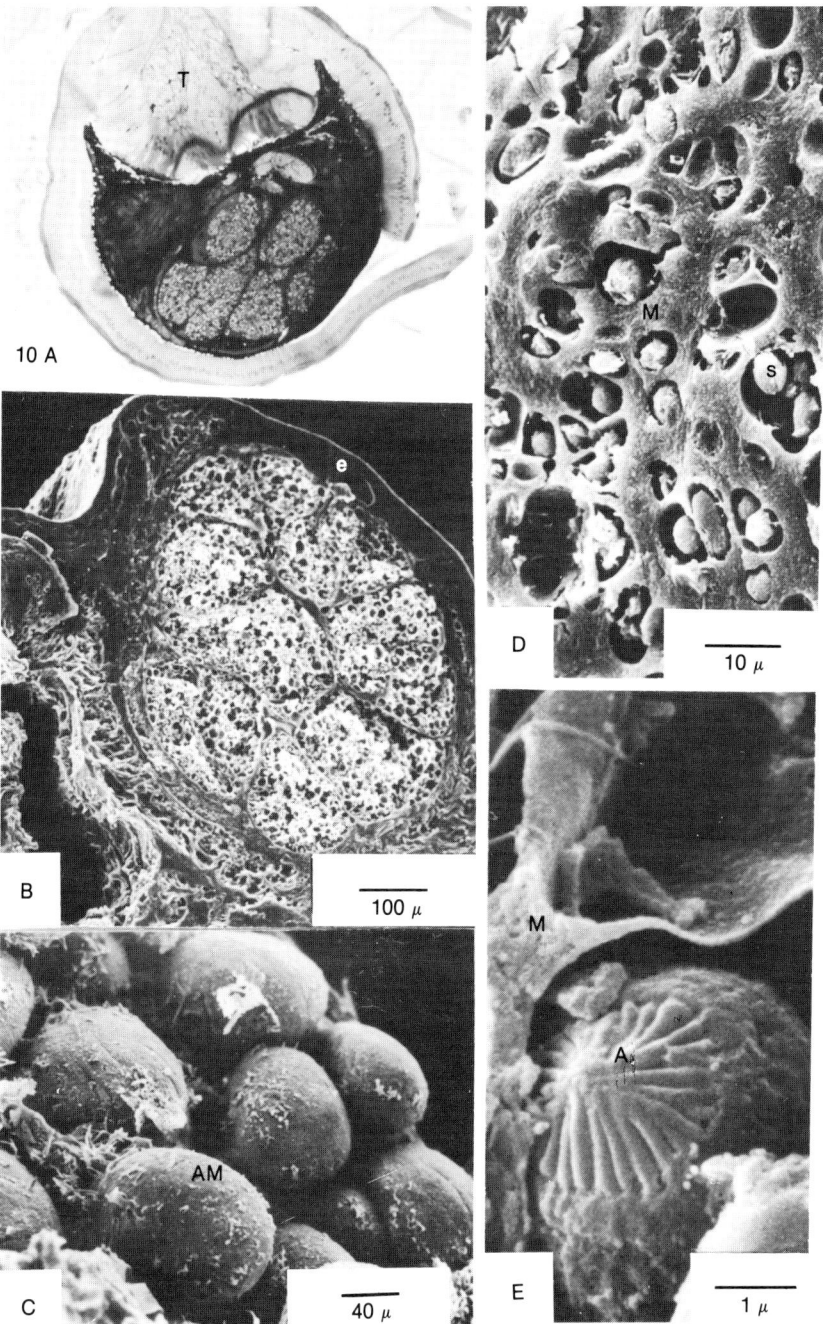

FIGURE 16.10. A. Thick section of a portion of the vas deferens of *Scyllarides nodifer* showing sperm masses surrounded by spermatophore wall materials. T, typhosole; X 200. B. SEM of anterior vas deferens of *S. nodifer* showing sperm masses. w, wall of mass; e, epithelium. C. SEM of sperm mass (am) within the vas deferens. D. Sperm (s) embedded within the matrix of sperm mass. E. SEM of matrix (m) that surrounds the sperm of *S. nodifer*. A, acrosome.

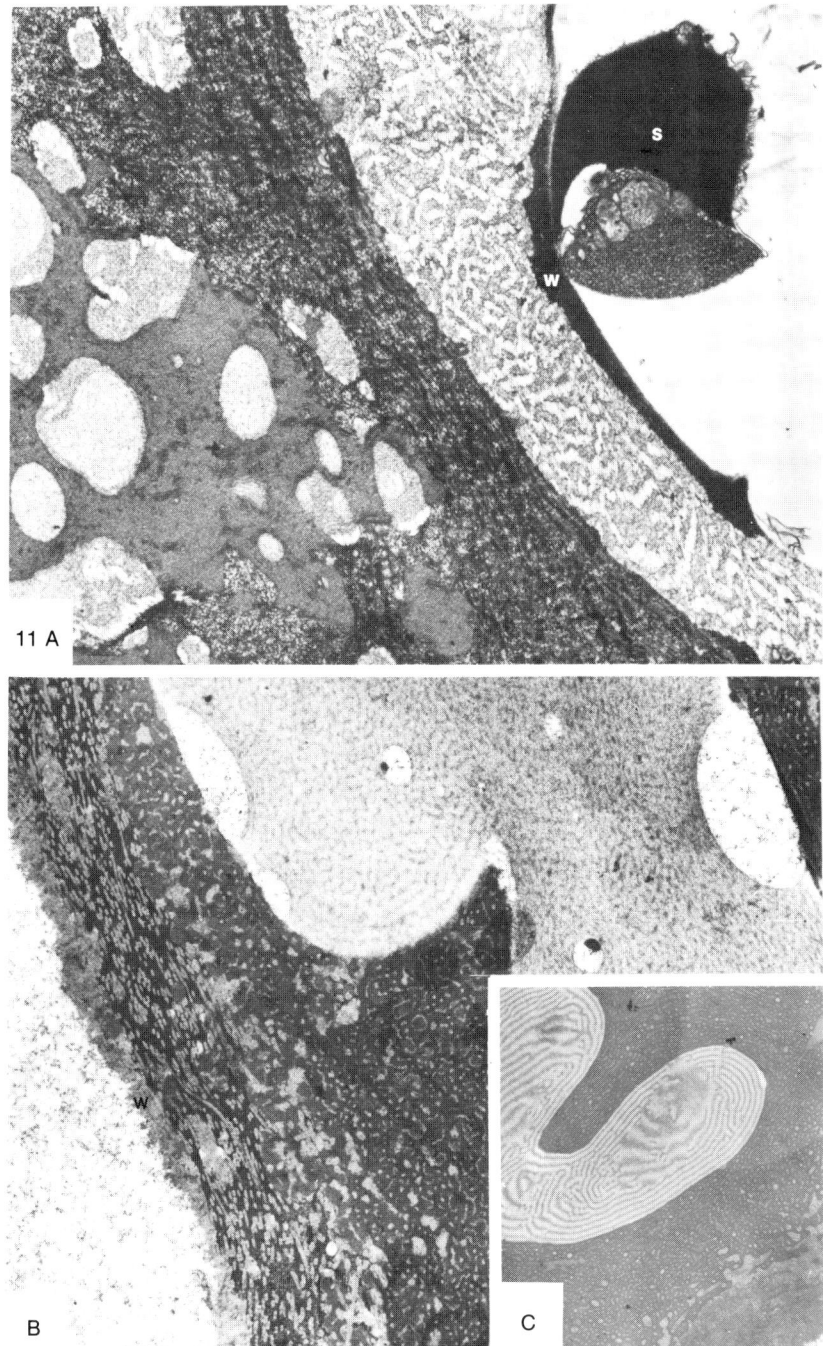

FIGURE 16.11. A. The wall (w) of the sperm mass (s) in *S. nodifer* appears electron-dense and becomes surrounded by materials of varying electron density that were secreted by the epithelial cells of the vas deferens. X 8500. B–C. As the spermatophoric mass passes posteriorly in the vas deferens, the materials of varying electron density become organized into several layers; X 9900.

seminal components may be formed in these regions as well. Ultrastructural studies indicate that the materials forming the walls of the sperm mass as well as the seminal fluids and/or spermatophore walls appear sequentially as they are deposited along the length of the vas deferens. It is presumed that one of the major roles of the layers secreted around the sperm is that of protection. Additionally, it has been suggested that the acid mucopolysaccharides of the spermatophores might act as a cementing agent or to prevent dehydration or as an antimicrobial agent (Sasikala & Subramoniam 1987). We do not know what role, if any, these materials may play in maintenance and nutrition of the sperm. We need to determine how sperm remain viable over long periods of time either within the seminal receptacles or attached to the ventral surface of decapod females.

Literature Cited

Anilkumar, G. & K. G. Adiyodi. 1977. Spermatheca of the freshwater crab, *Paratelphusa hydrodromous* (Herbst) in relation to the ovarian cycle. In K. G. Adiyodi & R. G. Adiyodi, eds., *Advances in Invertebrate Reproduction*, 1: 269–274. Kerala, India: Peralam-Kenoth.

Becker, J. 1983. Microscopic structure of the male reproductive tract and analysis for steroid in the fiddler crab. *Uca pugilator*. M.S. thesis, University of South Florida.

Beninger, P. G., R. W. Elner, T. P. Foyle & P. H. Odense. 1988. Functional anatomy of the male reproductive system and the female spermatheca in the snow crab *Chionoectes opilio* (O. Fabricius) (Decapoda: Majidae) and a hypothesis for fertilization. *J. Crustacean Biol.* 8:322–332.

Berry, P. F. 1970. Mating behavior, oviposition and fertilization in the spiny lobster *Panulirus homarus* (Linneaus). *Oceanogr. Res. Inst. (Durban) Invest. Rep.* 24:1–16.

Bumpus, H. C. 1891. The embryology of the American lobster. *J. Morphol.* 5:215–262.

Dudenhausen, E. E. & P. Talbot. 1983. An ultrastructural comparison of soft and hardened spermatophores from the crayfish *Pacifastacus leniusculus* Dana. *Can. J. Zool.* 61:182–194.

Haley, S. R. 1984. Spermatogenesis and spermatophore production in the Hawaiian red lobster *Enoplometopus occidentalis* (Randall) (Crustacea, Nephropidae). *J. Morphol.* 180:181–193.

Hinsch, G. W. 1986. A comparison of sperm morphologies, transfer and sperm mass storage between two species of crab, *Ovalipes ocellatus* and *Libinia emarginata*. *Int. J. Invert. Reprod. Dev.* 10:79–87.

Hinsch, G. W. 1988a. Morphology of the reproductive tract and seasonality of reproduction in the golden crab *Geryon fenneri* from the eastern Gulf of Mexico. *J. Crustacean Biol.* 8:254–261.

Hinsch, G. W. 1988b. Ultrastructure of the sperm and spermatophores of the golden crab *Geryon fenneri* and a closely related species, the red crab *G. quinquedens*, from the eastern Gulf of Mexico. *J. Crustacean Biol.* 8:340–345.

Hinsch, G. W. & C. E. McKnight. 1988. The vas deferens of the Spanish lobster, *Scyllarus chacei*. *Int. J. Invert. Reprod. Dev.* 13:267–280.

Hinsch, G. W. & M. H. Walker. 1974. The vas deferens of the spider crab, *Libinia emarginata*. *J. Morphol.* 143:193–207.

Jeyalectumie, C. & T. Subramoniam. 1987. Biochemical composition of seminal secretions with special reference to LDH activity in the reproductive tissues of the field crab, *Paratelphus hydrodromous* (Herbst). *Exp. Biol. (Berl.)* 46:231–236.

King, J. E. 1948. A study of the reproductive organs of the common marine shrimp, *Penaeus setiferus* (Linnaeus). *Biol. Bull. (Woods Hole)* 94:244–262.

Kooda-Cisco, M. J. & P. Talbot. 1982. A structural analysis of the freshly extruded spermatophore from the lobster, *Homarus americanus*. *J. Morphol.* 172:193–207.

Malek, S. R. A. & F. M. Bawab. 1971. Tanning in the spermatophore of a crustacean *(Penaeus triculcatus)*. *Experientia* (Basal) 27:1098.

Manning, R. B. & L. B. Holthuis. 1989. Two new genera and nine new species of Geryonid crabs (Crustacea, Decapoda, Geryonidae). *Proc. Biol. Soc. Wash.* 102:50–77.

Martin, G. G., C. Herzig & G. Narimatsu. 1987. Fine structure and histochemistry of the freshly extruded and hardened spermatophore of the spiny lobster, *Panulirus interruptus*. *J. Morphol.* 192:237–246.

Matthews, D. C. 1953. The development of the pedunculate spermatophore of a hermit crab, *Dardanus asper* (De Haan). *Pac. Sci.* 7:255–66.

Matthews, D. C. 1954. The development of the spermatophoric mass of the rock lobster, *Parribacus antarcticus* (Lund). *Pac. Sci.* 8:28–33.

Matthews, D. C. 1956a. The probable method of fertilization in terrestrial hermit crabs based on a comparative study of spermatophores. *Pac. Sci.* 10:303–309.

Matthews, D. C. 1956b. Further evidences of anomuran nonpedunculate spermatophores. *Pac. Sci.* 11:380–385.

Pearson, P. J. & M. H. Walker. 1975. Alterations of cytochrome C oxidase activity during spermiogenesis in *Carcinus maenas*. *Cell Tissue Res.* 164:401–410.

Radha, T. & T. Subramoniam. 1985. Origin and nature of the spermatophoric mass of the spiny lobster *Panulirus homarus*. *Mar. Biol.* (Berl.) 86:13–19.

Sasikala, S. L. & T. Subramoniam. 1987. On the occurrence of acid mucopolysaccharides in the spermatophores of two marine prawns, *Penaeus indicus* (Milne-Edwards) and *Metapenaeus monoceros* (Fabricius) (Crustacea: Macrura). *J. Exp. Mar. Biol. Ecol.* 113:145–153.

Spaulding, J. F. 1942. The nature and formation of the spermatophore and sperm plug in *Carcinus maenas*. *O. J. Microsc. Sci.* 83:399–422.

Subramoniam, T. 1984. Spermatophore formation in two intertidal crabs, *Albunea symnista* and *Emerita asiatica* (Decapoda: Anomura), *Biol. Bull. (Woods Hole)* 166:78–95.

Talbot, P. & D. Beach. 1989. Role of the vas deferens in the formation of the spermatophore of the crayfish *(Cherax)*. *J. Crustacean Biol.* 9:9–24.

Uma, K. & T. Subramoniam. 1979. Histochemical characteristics of spermatophore layers of *Scylla serrata* (Forskal) (Decapoda: Portunidae). *Int. J. Invert. Reprod.* 1:31–40.

SEVENTEEN
Chemical Composition of Spermatophores in Decapod Crustaceans

T. SUBRAMONIAM

Department of Zoology,
University of Madras,
India.

Abstract

The spermatozoa of decapod crustaceans are aflagellate and nonmotile. In general, sperm transfer occurs by means of spermatophores. Recent histochemical studies indicate the predominance of various mucopolysaccharides in spermatophores. These substances protect the delicate spermatozoa from damage caused by desiccation and microbial infection during their epizoic storage on the female body prior to fertilization in many groups. The seminal plasma enclosed within the spermatophore may provide energy-yielding substances for sperm maintenance during prolonged storage. Histochemical characterization of spermatophore layers has yielded information regarding spermatophore hardening after attachment to the female body. Studies on the biochemical composition of the seminal plasma of brachyuran crabs reveal that their plasma is rich in organic substances necessary for sperm metabolism. The possibility of cryopreservation of spermatophores and seminal plasma of decapods, required in artificial insemination, is discussed.

THE NONMOTILE spermatozoa of decapod crustaceans are enclosed in packets called spermatophores for transmission to females during mating. Spermatophores exhibit gross morphological variations among different decapod groups. In brachyuran crabs spermatophores are vesiculate or spherical; in anomuran crabs, they are generally pedunculate, consisting of an ampoule with sperm mass, a peduncle, and a gelatinous pedestal. In "macrurans," the spermatophores are tubular with surrounding accessory mucoid secretions, as in lobsters and crayfish. Spermatophores may contain accessory structures such as attachment wings in penaeid shrimps. Spermatophores, usually with a distinctive wall, are molded into characteristically complex shapes by changes in the contour of the lumen of the vas deferens as well as by contraction of the ejaculatory duct musculature.

Histological studies on the origin and form of spermatophores of decapod crustaceans are many (see Mann 1984 for review). Recently several ultrastructural investigations also have been made to trace the origin of spermatophoric components in crabs, lobsters, and crayfish (Hinsch & Walker 1974; Kooda-Cisco & Talbot 1982; Dudenhausen & Talbot 1983). The present paper deals with the chemical composition of the spermatophores of decapod crustaceans in an attempt to correlate functional significance with respect to sperm storage, spermatophore attachment, and dehiscence.

BRACHYURA

The simplest type of spermatophore is found in brachyuran crabs. In extreme conditions, a distinct spermatophore may be lacking and hence the spermatozoa are transported in the seminal plasma, as reported in the Hawaiian crab *Ranina ranina* (Ryan 1984). However, in other crabs a well-defined spermatophore is always present. In the mud crab *Scylla serrata*, each spermatophore is enveloped by two layers, an outer thick and an inner thin layer, the latter being confluent with the inner sperm mass (Uma & Subramoniam 1979). These two layers show histochemical variation; further, the chemical identity of the inner layer with the sperm mass substance suggests their common origin from the vas deferens epithelium. Interestingly, chitin, a structural polysaccharide so commonly found in arthropod exoskeleton (Richards 1951), is histochemically demonstrable in the outer layer. The chitinous nature of the spermatophore wall has also been indicated in another crab, *Carcinus maenas* (Spalding 1942), and in the harpacticoid copepod *Tisbe holothuriae* (Pochon-Masson & Gharagozlou–Van Ginneken 1977). In addition to chitin, the outer layer of the *S. serrata* spermatophore is also rich in sulphated acid mucopolysaccharide (SAMP). Neither chitin nor SAMP is present in the inner layer; here the acid mucopolysaccharide (AMP) is rich in carboxyl groups. Protein components of the two layers appear to be rich in tryptophanyl reactive sites, but tyrosyl and other phenolic substances are wanting. Protein stabilization by way of phenolic tanning in these layers may be ruled out, as they fail to give positive reaction to the enzyme phenol oxidase, which is essential for the oxidation of phenolic substances into crosslinking quinones. Similarly, stabilization by -S-S bonding is also wanting in these layers, as inferred from the absence of sulphur-containing amino acids in the spermatophore layers. Such a chemical composition of the spermatophore layers of *S. serrata* has a bearing on their physical properties.

Permeability studies conducted on the spermatophore envelopes of *Scylla serrata* suggest that they are permeable to low molecular weight vital dyes (Uma & Subramoniam 1979). Furthermore, when spermatophores are treated with tap water, they swell and form a conelike projection that ultimately breaks up to release spermatozoa. However, the outer layer is resistant to alkali and acid treatment. Histochemical studies by the above authors have indicated that the sperm mass substance inside the spermatophore is also rich in AMPs. The liquid-imbibing property of AMPs may be responsible for the absorption of low molecular weight substances from the spermathecal fluid resulting in the dehiscence of the spermatophores at the time of spawning (Ezhilarasi & Subramoniam 1982).

ANOMURA

Anomuran spermatophores are generally pedunculate and are structurally species-specific. In different hermit crabs, the spermatophoric components include an ampoule, peduncle, and pedestal. Conversely, the two sand crabs *Emerita asiatica* and *Albunea symnista* produce a spermatophoric mass, which is fastened to the ventral sternum of the females during mating (Subramoniam 1984). In spite of the morphological diversity of anomuran spermatophores, only the spermatophores of the sand crabs *E. asiatica* and *A. symnista* have received attention in regard to chemical composition (Subramoniam 1984). That study revealed that mucopolysaccharide (MP) complexed with protein is the main component of the spermatophore. In *A. symnista*, the fully formed spermatophoric mass consists of three components: the tubular spermatophore, a basal gelatinous cord, and the protective gelatinous matrix. The spermatophore wall is mainly composed of a neutral MP, although this component does not include chitin, as shown by the negative chitosan test. The binding substance of the spermatozoa within the spermatophore stains metachromatically with toluidine blue, suggesting strongly acidic groups in the mucus. These acidic groups are in the form of sulphated polyanions, whereas the spermatozoa gave a diastase labile PAS (Periodic Acid Schiff's)–positive reaction in the axial structure of the acrosomal vesicle, suggesting the occurrence of glycogen. Such a differential histochemical property between the sperm-binding substance and the spermatozoa is clearly observed in the Alcian Blue–PAS technique and with aldehyde fuchsin. The basal gelatinous cord contains again a neutral MP conjugated to a protein that is rich in basic and aromatic groups. Interestingly, the gelatinous cord gave a positive reaction to catechol incubation indicating the enzyme phenolase. The protective gelatinous matrix is also rich in AMP, but the acidic property is mainly due to carboxyl groups, as the metachromatic reactions with toluidine blue are obtained in high pH. The proteins of the gelatinous matrix contain tryptophanyl groups, but tyrosine and other phenolic compounds are absent.

In *Emerita asiatica,* the spermatophoric mass consists of two types of pedunculate spermatophores, one of truncated cone shape, the other of tumbler shape. The peduncles of both types are connected ventrally to filaments that run along the entire length of the spermatophoric ribbon. The whole structure is embedded in a gelatinous matrix that binds the spermatophore tightly. Histochemically, the spermatophoric mass of *E. asiatica* is similar to *Albunea symnista,* especially in its MP content. However, the gelatinous matrix of the *E. asiatica* spermatophoric mass contains yet another MP species that contains vicinyl hydroxyl groups as well as acidic groups, staining blue with Alcian blue. Such an MP heterogeneity in the various spermatophoric components of the two sand crabs may be correlated to their protective as well as structural functions (Jeanloz 1970; Montgomery 1970). Thus the highly acidic groups of *A. symnista*'s gelatinous matrix might bind inorganic ions such as calcium from the seawater to produce "hardening" of the deposited spermatophore into a puttylike mass.

"MACRURA" (LOBSTERS AND CRAYFISH)

Tubular, nonpedunculate spermatophores are produced by several species of lobsters and crayfishes. In the spiny lobster *Panulirus penicillatus,* Mathews (1951) described a complex spermatophore consisting of a puttylike matrix surrounding the highly coiled, continuous spermatophoric tube enclosing the agglutinated spermatozoa. Unlike the American lobster, *Homarus americanus,* which receives and stores the spermatophore in a seminal receptacle during mating, the spermatophoric masses of Palinuridae are carried externally by the female. Berry (1970) recognized 3 distinct horizontal matrix layers in the deposited spermatophoric mass of *Panulirus homarus.* They are: (1) an outer crustlike layer, termed the protective matrix, (2) a middle layer, bearing the spermatophore, and (3) a basal spongy layer, termed the adhesive matrix. The protective matrix is composed of minute granules that are strongly eosinophilic and a few agarophilic granules, suggesting the possible presence of calcium salts. The spermatophoric matrix is more granular and differs from the protective matrix only in its consistency. The agarophilic granules are more numerous in the middle layer. The adhesive matrix tends to have a striated appearance but is strongly eosinophilic on its basal margin. This basic pattern of the spermatophoric mass has been shown to be present in many other species belonging to the Palinuridae, although a few deviations, such as the absence of a distinct spermatophoric wall, as in *Panulirus angulatus* and *Linuparus trigonus,* do occur (Berry & Heydorn 1970).

Spermatophores of several lobster species are known to undergo hardening and blackening on exposure to seawater. However, the mechanism of hardening has long been controversial, mainly because of paucity of information on the chemical nature of the constituent layers of lobster spermatophores. Recent ultrastructural studies on lobster and crayfish spermatophores have endeavored to resolve the problem by examining the structural transformation the constituent layers undergo upon deposition to the sternum of the females (Kooda-Cisco & Talbot 1982; Dudenhausen & Talbot 1983; Martin, Herzig & Narimatsu 1987;

Talbot & Beach 1989). In *Homarus americanus*, the inner mass of randomly oriented spermatozoa supported by a matrix of secretory material is surrounded by three investment layers: (1) a primary spermatophoric layer that is amorphous and PAS positive, (2) an intermediate layer containing PAS positive granules, and (3) an outer layer comprised of small filaments and a flocculent material that imparts stickiness to the freshly extruded spermatophore. The *Homarus* spermatophore does not undergo hardening in sea water, but spermatophoric components are expected to play a major role in sperm maintenance during prolonged storage inside the receptacles of the females. A recent histochemical investigation on the spermatophoric mass of the spiny lobster *Panulirus homarus* has revealed that MPs form the main component (Radha & Subramoniam 1985). The wall of the spermatophoric tube consists of neutral MPs, whereas the sperm mass as well as the protective gelatinous matrix are rich in AMPs. However, isolation of AMP following the procedure of Rahemtulla and Lovtrup (1974) has yielded a single fraction of chondroitin sulphate on agarose gel electrophoresis. By virtue of their high viscosity and viscoelastic properties (Pigman & Horton 1970), the chondroitin sulphate can provide elasticity and resistance to compression of spermatophores. The AMPs may also have an important role in maintaining a microenvironment within the spermatophoric tube for prolonged sperm survival in the spiny lobsters.

The behavior of the spermatophoric mass in seawater also varies among lobsters. A correlation between the nature of deposited spermatophores and the time of ovulation in relation to spermatophore deposition has been suggested by Berry and Heydorn (1970) for palinurid lobsters. The observation that MPs of the protective matrix of *Panulirus homarus* are rich in polyanionic groups may indicate possible calcium binding to these groups resulting in calcification (Radha & Subramoniam 1985). A recent ultrastructural and histochemical study on spermatophores of the spiny lobster *Panulirus interruptus* also throws some light on the changes related to the hardening process (Martin, Herzig & Narimatsu 1987). In this lobster, the freshly extruded spermatophore is white, soft, and sticky on all surfaces. The highly coiled sperm tube lies near the surface of the spermatophore's foot region by which it becomes attached to the exoskeleton of the female. The matrix surrounding the sperm tube is composed of small granules embedded in a loose weave of filaments. However, after attachment of the spermatophore to the female, the outer surface (cap) hardens and darkens, whereas the granules found in the foot region of the matrix layer dissolve to form a thick layer characterized by vertical striations. The matrix layer between the cap and the foot region undergoes a process of degranulation. No such structural change occurs in the sperm tube. Histochemically, the primary layer enclosing the sperm tube is both PAS and Alcian blue positive, whereas the matrix layer stains uniformly for PAS and is rich in tyrosine. This layer also is positive to dihydroxy phenyl alanine oxidase, whereas the darkened cap of the spermatophore bleaches in hydrogen peroxide. All this may indicate the possible occurrence of phenolic tanning in the spermatophore. The blackening may also be due to melanistic pigmentation. Interestingly, in a sand lobster, *Thenus orientalis*, the gelatinous protective matrix is shown to contain elastic fibrils (Shyamasundari & Rao 1986). In *Homarus americanus*, Kooda-Cisco and Talbot (1982) described the ultrastructural organization of collagenlike fibers dispersed throughout the sperm-bearing matrix.

The spermatophore of freshwater crayfish is similar to that of palinurid lobsters. In *Pacifastacus leniusculus*, the wall of the unextruded spermatophore is composed of three concentric layers: a thin, primary spermatophoric layer that surrounds the inner sperm mass, a thick middle layer composed primarily of electron-dense spherical granules, and a thick outer globular layer (Dudenhausen & Talbot 1983). On extrusion and hardening, the sticky outer global layer transforms into a fibrillar thickened ridge anchoring the oval end of the spermatophore to the female body. However, the middle layer is responsible for the hardening of the spermatophore for prolonged storage by virtue of the modification of the outer portion into a reticulated structure. This structural organization of the spermatophore envelope also occurs in two other Australian crayfish species, *Cherax tenuimanus* and *C. albidus* (Beach & Talbot 1987).

PENAEOIDEAN AND CARIDEAN SHRIMPS

Because of their economic importance, the reproductive biology of shrimps, particularly the sperm transfer mechanism and artificial insemination for hybridization, has received considerable attention recently (Primaveera 1985). Penaeoid shrimp produce spermatophores of varying complexities and deposit them in or on the female thelycum. Females with open thelyca generally receive a spermatophore with complicated accessory structures such as wings, whereas in closed thelycal forms the males produce simpler spermatophoric masses (Burkenroad 1934, 1936; Bauer 1986). Conversely, caridean shrimp produce relatively simple spermatophores, although deposition is external (Bauer 1976; Chow 1982; Dougherty, Dougherty & Harris 1986).

Malek and Bawab (1971, 1974a,b) traced the origin of spermatophores with special reference to the chemical nature of constituent layers and their mode of hardening in a penaeid prawn, *Penaeus kerathurus*. The precursor materials, originating in the glandular epithelium of the vas deferens, form concentric layers over the main body enclosing the sperm mass substances, in addition to the formation of the accessory structure or "wing." Further molding of the spermatophore occurs in the ampoule region by additional secretion. As many as seven spermatophoric layers have been distinguished in this prawn.

In the freshwater prawn *Macrobrachium rosenbergii*, the spermatophore consists of a sperm mass surrounded by two matrices composed of basophilic and eosinophilic materials (Chow 1982). This pod-shaped spermatophore, as redescribed by Dougherty, Dougherty, and Harris (1986), consists of a lateral sperm mass, a medial mucus mass, and a noncellular capsule that sticks to the surface of the sternum of the females. At the ultrastructural level, the capsule material primarily consists of tightly packed fine fibrils that gave positive reactions for glycoprotein and AMPs. The fibrillar capsule layer is subdivided into an outer region containing large globules of nonsulphated glycosamino glycans and an inner region containing many small membraneless granules of glycoprotein and acidic glycosamine glycan composition. The packing of the outer capsular subdivision forms a reticulum with electron-lucent interstices, or canaliculi, permitting water imbibition by the spermatophore.

The matrix material of the sperm mass resembles the canaliculated reticulum and globules of the outer capsule subdivision. The threads found in the sperm mass are suggested to play a role in the orientation of the unistellate spermatozoa within the spermatophore (Dougherty 1987). The median mucus mass is similar to the capsule and sperm mass substance in structural composition and chemical nature. Unlike the penaeid prawns, where the spermatophoric components are formed in a sequential manner from a variety of secretory materials emanating from the glandular epithelial cells (Malek & Bawab 1974 a,b), the spermatophore is not preformed in carideans, but the secretions are poured into the ampullar region. The spermatophore proper is formed upon extrusion (Bauer 1976). Hence it is likely that the distinctive components of the *Macrobrachium rosenbergii* spermatophore are formed from a mixture of seminal secretions stored within the ampoule.

Histochemical properties of decapod spermatophores, reviewed above, indicate the predominance of MPs. Further characterization has been made on the spermatophores of two marine prawns, *Penaeus indicus* and *Metapenaeus monoceros*, employing agarose gel electrophoresis (AGE) as well as quantification of total AMPs (Sasikala & Subramoniam 1987). The *P. indicus* spermatophore consists of a sperm mass enclosed within the sperm sac to which is attached the foliacious wing. In *M. monoceros,* the freshly extruded spermatophore consists of a sperm mass and milky white crystalline structures. AGE of the isolated and purified AMPs of spermatophores of both prawns revealed two fractions corresponding to chondroitin sulphate C and hyaluronic acid. Further, the sperm sac contained the maximum quantity of AMPs while the "wing" contained the least in *P. indicus*. In *M. monoceros*, the AMPs content of the sperm mass and crystalline structures was similar. High AMP content in the sperm sac may have functional implications, enabling sperm release by imbibition of water (Uma & Subramoniam 1979). The wing structure of the spermatophore is composed of gelatinous protein that tends to swell and dissolve in water. However, the AMP content of the wing may aid in the adhesion and attachment of the spermatophores beneath thelycal plates of the females. The spermatophores are retained within the thelycum of the females until fertilization. The occurrence of large quantities of AMPs in spermatophores might keep the spermatozoa viable in a fluid medium until fertilization. Again, the antimicrobial activity of sulphated AMPs such as hyaluronic acid and chondroitin sulphate may protect the delicate sperm cells when exposed to seawater, both during their transfer to the thelycum and during their release for fertilization.

SIGNIFICANCE OF SPERMATOPHORE HARDENING

In spite of the extensive literature available on spermatophore formation in a wide variety of crustacean species, information pertaining to its chemical composition is scanty. In many crustacean species where the spermatophores are externally deposited and carried by the females until fertilization, the spermatophore has been reported to undergo hardening. A recurrent mode of hardening in arthropod structures is phenolic tanning and chitinization (Richards 1951). Fryer (1960) was the first to provide histochemical evidence for phenolic tanning of a crustacean spermatophore in the branchiuran *Dolops ranarum*. In this species, the spermatophores are not chitinous but proteinaceous in composition. Using various solubility tests and histochemical reagents, Fryer suggested the occurrence of quinone tanning. In this respect, the spermatophore is comparable to insect ootheca rather than to the cuticle, because the spermatophoric wall lacked chitin. A noteworthy observation is that the color of the spermatophore differed from that of most such quinone-tanned proteins. The spermatophore also remains soft and pliable. The glandular tissue of the spermatophoric vesicle secretes the aromatic material that tanned the protein secreted by the spermatophoric gland. Using histochemical tests, Malek and Bawab (1971) characterized the chemical composition of the spermatophore layers of the prawn *Penaeus kerathurus* with special reference to the possible occurrence of phenolic tanning. Three layers (II, IV, and V) are of lipoprotein nature, the protein of which is rich in phenolic groups, owing to the presence of tyrosine. No free phenol could be histochemically detected, but the enzyme phenolase was demonstrated by "Nadi" reagent as well as the catechol incubation test. Using the aminophenols derived from tyrosyl residues of the protein, Malek and Bawab suggested that the existing phenolase could promote active oxidation resulting in the formation of cross-linking quinones. Although the tanned layers failed to darken, as the tanned cuticle does, the layers showed pronounced changes in affinity for stains and in isoelectric point, and they developed resistance to acid treatment. Malek and Bawab indicated that the spermatophore of *P. kerathurus* owes its hardness not to the mere exposure to seawater (Heldt 1938) but to a definite enzymatically catalyzed chemical transformation. Evidence for the occurrence of phenolic tanning in spermatophores has also been indicated in the sand crab *Albunea* (Subramoniam 1984) and the lobster *Panulirus interruptus* (Martin, Herzig & Narimatsu 1987).

An interesting feature of the spermatophore tanning process is the "softness" and colorless nature of the tanned layers. Although authors who have described such a hardening process have not offered any explanation regarding the actual mechanism of protein stabilization, a plausible explanation is found in the work of Anderson (1971); Anderson demonstrated a new type of quinone tanning that results in the formatioin of a colorless sclerotin, which Brunet and Kent (1955) earlier termed as "white sclerotins." This process (also known as β-sclerotization), involving side-chain linking of the quinones with the tyrosyl residues of the tanning protein, conferred softness and resilience to the structures. However, it is important to characterize the type of phenols involved in the sclerotization as well as to determine the properties of the phenol oxidase before arriving at such conclusions for spermatophore tanning. Melanization, a product of quinone polymer, has also been reported by Martin, Herzig & Narimatsu (1987) in the lobster *Panulirus interruptus*.

While Anderson's side-chain sclerotizatioin may also explain spermatophoric tanning, there may be another function of the phenolic compounds of crustacean spermatophores. Brunet (1980) argues that some of the sclerotizing agents (phenols) may be involved in quite different antibiotic or antioxidant processes. Spermatophores deposited onto the female body are exposed to microbial attack in the water medium. In the spiny lobster *Panulirus interruptus*, Martin, Herzig, and Narimatsu (1987) reported bacterial colonization on the hardened spermatophore surface, but bacteria were not found to penetrate into the deeper layers of the spermatophores, probably because of the antibacterial activity of the phenolic substances present in it. Clearly, a detailed investigation is necessary before we make more speculations on the biological significance of phenolic substances in the spermatophores of decapod crustaceans. But a hypothesis may be made on the correlation between the nature of spermatophore layer stabilization and the type of fertilization in the Crustacea: when both fertilization and spermatophore deposition are external, spermatophore layers may be impermeable to seawater; when insemination is internal through true copulation (e.g., the Brachyura), the spermatophore layer may be permeable to low molecular weight substances in order to bring about dehiscence and sperm release, as indicated in an experimental study on *Scylla serrata* spermatophores (Uma & Subramoniam 1979). It should be noted here that the female spiny lobsters are known to use their powerful chelae to remove the protective matrix and gouge the spermatophoric tube open to release the spermatozoa at the time of fertilization (Fielder 1964; Berry 1969).

Chitinized structures are widespread in arthropod species. In decapod crustaceans, the epithelial cells of brachyuran oviducts are known to secrete a chitin intima, which is discarded at every molt (Hartnoll 1979). Therefore, it is reasonable to assume that the vas deferens epithelial cells of brachyuran crabs are also capable of secreting chitinous materials. Although some workers have suggested the occurrence of chitinous materials in seminal secretions and spermatophores (Spalding 1942; King 1948), the occurrence of chitin in the polymerized form has only been histochemically demonstrated in the *Scylla serrata* spermatophore (Uma & Subramoniam 1979). In a harpacticoid copepod, Gharagozlou–Van Ginneken (1978) reported a chitin/proteinlike lamellar pattern in the spermatophore wall. This author, however, could not demonstrate the occurrence of chitin histochemically. Evidently, there is a paucity of biochemical information regarding the occurrence of chitinous materials in the seminal secretions of crustaceans. Our recent observations on the glycosidase activity on the substrate O-Nitrophenyl-2-Acetamide-2-Deoxy-β-D-glucopyranoside suggest the involvement of N-acetyl glucosidase in the metabolism of N-acetyl glucosamine for the synthesis of chitin. Nevertheless, the presence of chitin synthetase is required for the polymerization of chitin; biochemical work in this regard is highly desirable.

BIOCHEMICAL COMPOSITION OF SPERMATOPHORES AND SEMINAL PLASMA

Data on the biochemical composition of crustacean spermatophores are surprisingly sparse, but some information in this respect is available for cirripede seminal plasma (Barnes 1962a, b; Barnes & Blackstock 1974). The seminal plasma of *Balanus balanus* is highly proteinaceous and rich in amino acids but low in lipids. Unlike mammals, cirripede semen is devoid of fructose but is high in glycogen, glucose, and a small quantity of lactic acid. The low oxygen uptake of *B. balanus* spermatozoa together with low motility and ATP content may suggest glycolysis under anaerobic conditions for energy production in sperm metabolism. Conversely, spermatozoa of insects such as the honey bee *Apis* readily oxidize not only glycolytic compounds in the seminal plasma, but also utilize, under aerobic conditions, intracellular compounds such as phospholipids as an energy source for sperm motility (Blum, Glowska & Taber 1962; Taber 1977).

Biochemical information concerning sperm metabolism in cirripede seminal plasma cannot be extrapolated to decapod crustaceans, as the latter possess nonmotile spermatozoa enclosed in spermatophores. Whereas macruran and anomuran decapods produce spermatophores only for sperm transfer, the brachyuran crabs produce spermatophores that are carried in a fluid medium of seminal plasma. Females receive copious quantities of seminal plasma along with spermatophores at every mating and store them in spermatheca, pending ovulation. Therefore, survival and maintenance of spermatozoa within spermatophores during their transit and storage in the females require an efficient metabolic machinery. A recent study on the spermatophores and seminal plasma of the field crab *Parathelphusa hydrodromous* has revealed that they are rich in protein, free carbohydrates, and lipids (Jeyalectumie & Subramoniam 1987). The occurrence of considerable quantities of carbohydrates in spermatophores corroborates earlier findings that sperm cells of many decapod crustaceans are endowed with glycogen storage (Pochon-Masson & Gharagozlou–Van Ginneken 1977). A quantitative assay of the enzyme lactate dehydrogenase (LDH) has also revealed its high activity within the spermatophore of *P. hydrodromous*. This observation is significant because it shows, as in mammals, an equally high free carbohydrate content in the seminal plasma. Electrophoretic analysis of the isoenzyme pattern of spermatophore LDH reveals further interesting results. At least six fractions of LDH are resolved, and the most conspicuous of them is a fraction homologous to the LDH_x of mammalian spermatozoa, having an electrophoretic mobility between LDH_3 and LDH_4. Furthermore, LDH_x of the spermatophores seems more related to M_4 (LDH_5) (muscle type) than to H_4 (LDH_1) (heart type) (Clausen & Ovlisen 1965). This comparison may suggest that the spermatozoa in spermatophores may undergo anaerobic metabolism utilizing glycogen-derived glycosyl units or free sugars. These studies, though preliminary, may indicate a similarity in sperm metabolism between Crustacea and mammals.

In *Scylla serrata*, the formation and maturation of the spermatophore is accompanied by certain biochemical changes (Jeyalectumie 1989). Organic components such as protein, lipid, and carbohydrate significantly increase with spermatophore maturation, as does LDH activity. On the contrary, water content declines. In the giant octopus *Octopus dofleini*, Mann, Martin, and Thiersch (1981) reported that the spermatophoric plasma is a fluid salt solution; during maturation along the male reproductive tract, inorganic ions are replaced by organic substrates. Such a transformation in seminal plasma and spermatophore fluids of *S. serrata* and other crabs could occur, but data are scanty. Such a nutrient rich seminal plasma may help in the maintenance of spermatozoa during their prolonged storage within female spermathecae, as the interval between mating and the first ovarian maturation, and hence ovulation, is rather long (Ezhilarasi & Subramoniam 1982). Jeyalectumie (1989) reports that, of all the organic substrates, only the carbohydrates (both free and bound) undergo a sharp decline in the spermatheca. A concomitantly high activity of LDH enzyme in spermatophores during this period again suggests that anaerobic glycolysis may be the chief form of energy production during sperm storage.

CONCLUSIONS

Spermatophores play a major role in sperm transfer and storage of decapod crustaceans. They not only serve to protect the spermatozoa during transmission to females but also serve the function of providing energy rich substrates for prolonged sperm storage in the females. However, information on the chemical nature of decapod spermatophores is still inadequate. Our understanding of their chemical composition comes mainly from recent histochemical investigations. MPs complexed with protein form the main components of spermatophores. Crustacean spermatophores could serve as a good model system for histochemical investigation of MPs in view of the occurrence of a variety of MPs in spermatophore structures (table 17.1). More biochemical studies are also desirable to determine spermatophore functions in sperm protection and survival during prolonged storage. In recent years, several attempts have been made to cryopreserve spermatophores and sperm cells of various decapod crustaceans in an attempt to formulate artificial insemination techniques (Chow, Taki & Ogasawara 1985; Ishida, Talbot & Kooda-Cisco 1986; Anchordoguy et al. 1988). A

TABLE 17.1. Summary of the characterization of mucopolysaccharides of the spermatophoric components of *Albunea symnista* and *Emerita asiatica*. From Subramoniam 1984.

Spermatophoric Components	*Chemical Nature*	*Origin*
Albunea symnista		
1. Sperm mass substance	Sulphated AMP	PVD
2. Spermatophore wall	Neutral MP	PVD
3. Gelatinous cord	Neutral MP	DVD, Ventral epithelium
4. Gelatinous matrix	Carboxylated AMP	DVD, Dorsal epithelium, especially typhlosole
Emerita asiatica		
1. Sperm mass substance	Sulphated AMP	PVD
2. Spermatophore wall		
Inner layer	Carboxylated AMP	PVD
Outer layer	Neutral MP	DVD, Ventral epithelium
3. Peduncle/gelatinous cord	Neutral MP	DVD, Ventral epithelium
4. Gelatinous matrix	Periodate reactive AMP	DVD, Dorsal epithelium, especially typhlosole

AMP = acid mucopolysaccharides; MP = mucopolysaccharides; DVD = distal vas deferens; PVD = proximal vas deferens.

knowledge of spermatophore and seminal plasma chemical composition could be of immense value in understanding the cryopreservation capacity of sperm cells. A recent investigation on the biochemical composition of spermatophores and seminal plasma of *Scylla serrata*, cryopreserved at different temperatures, indicated that compounds such as carbohydrates showed a drastic decline during storage at $-4°C$. However, no such change in substrate or enzyme occurs when stored in liquid nitrogen at $-196°C$ (Jeyalectumie & Subramoniam 1989). This study further indicates the importance of seminal secretions in sperm metabolism during their storage in the female reproductive tract. Clearly, more information regarding chemical composition of spermatophore is needed for a better understanding of their functions in several respects, such as spermatophore hardening, sperm maintenance, nutrition, and protection.

Acknowledgments

Work on *Scylla serrata* was supported by a grant from the Department of Science and Technology, Government of India (22(8P-8)/STP-II). I am also thankful to Drs. P. Ramaswamy, K. Nellaiappan, and C. Jeyalectumie for discussion on the manuscript.

Literature Cited

Anchordoguy, T. H., J. Crowe, J. Griffin & W. H. Clark, Jr. 1988. Cryopreservation of sperm from the marine shrimp *Sicyonia ingentis*. *Cryobiology* 25:238–243.

Anderson, S. O. 1971. Phenolic compounds isolated from insect hard cuticle and their relationship to the sclerotization process. *Insect Biochem.* 1:157–170.

Barnes, H. 1962a. The composition of the seminal plasma of *Balanus balanus*. *J. Exp. Biol.* 39:345–351.

Barnes, H. 1962b. The oxygen uptake and metabolism of *Balanus balanus* spermatozoa. *J. Exp. Biol.* 39:353–358.

Barnes, H. & J. Blackstock. 1974. Biochemical composition of the seminal plasma of the cirripede *Balanus balanus* (L) with particular respect to free amino acids and proteins. *J. Exp. Mar. Biol. Ecol.* 16:47–85.

Bauer, R. T. 1976. Mating behaviour and spermatophore transfer in the shrimp, *Heptacarpus pictus* (Stimpson) (Decapoda: Caridea: Hippolytidae). *J. Nat. Hist.* 10:415–440.

Bauer, R. T. 1986. Phylogenetic trends in sperm transfer and storage complexity in decapod crustaceans. *J. Crustacean Biol.* 6:313–325.

Beach, D. & P. Talbot. 1987. Ultrastructural composition of sperm from the crayfishes *Cherax terminarus* and *Cherax albidus*. *J. Crustacean Biol.* 7:201–218.

Berry, P. F. 1969. Occurrence of an external spermatophoric mass in the spiny lobster, *Palinurus gilchristi*. *Crustaceana (Leiden)* 17:223–224.

Berry, P. F. 1970. Mating behaviour, oviposition and fertilizatioin in the spiny lobster, *Panulirus homarus* (Linnaeus). *Oceangr. Res. Inst. (Durban) (Invest. Rep.)* 24:1–16.

Berry, P. F. & A. E. F. Heydorn. 1970. A comparison of the spermatophoric masses and mechanism of fertilization in South African spiny lobsters (Palinuridae). *Oceangr. Res. Inst. (Durban) Invest. Rep.* 25:1–18.

Blum, R. S., Z. Glowska & S. T. Taber. 1962. Chemistry of the drone honey bee reproductive system II carbohydrates in the reproductive organs and semen. *Ann. Entomol. Soc. Am.* 55:135–139.

Brunet, P. C. J. 1980. The metabolism of the aromatic amino acids concerned in the cross-linking of insect cuticle. *Insect Biochem.* 10:467–500.

Brunet, P. C. J. & P. W. Kent. 1955. Observations on the mechanism of a tanning reaction in *Periplaneta* and *Blatta*. *Proc. R. Soc. Lond. B Biol. Sci.* 144:259–274.

Burkenroad, M. D. 1934. The Penaeidea of Louisiana with a discussion on their world relationships. *Bull. Am. Mus. Nat. Hist.* 68:61–143.

Burkenroad, M. D. 1936. The Aristaeinae, Solenocerinae, and the pelagic Penaeidea of the Bingham collection. Materials for the revision of the oceanic Penaeidae. *Bull. Bingham Oceanogr. Collect. Yale Univ.* 5:1–151.

Chow, S. 1982. Male reproductive system and fertilization of the Palaemonid shrimp, *Macrobrachium rosenbergii*. *Bull. Jpn. Soc. Sci. Fish.* 48:177–183.

Chow, S., Y. Taki & Y. Ogasawara. 1985. Cryopreservation of spermatophore of the freshwater shrimp, *Macrobrachium rosenbergii*. *Biol. Bull (Woods Hole)* 168:471–475.

Clausen, J. & B. Ovlisen. 1965. Lactate dehydrogenase isoenzymes of human semen. *Biochem. J.* 97:513–517.

Doughtery, W. J. 1987. Oriented spermatozoa in the spermatophore of the Palaemonid shrimp, *Macrobrachium rosenbergii*. *Tissue & Cell* 19:145–162.

Dougherty, W. J., M. M. Dougherty & S. G. Harris. 1986. Ultrastructural and histochemical observations on electroejaculated spermatophores of the Palaemonid shrimp, *Macrobrachium rosenbergii*. *Tissue & Cell* 18:709–724.

Dudenhausen, E. E. & P. Talbot. 1983. An ultrastructural comparison of soft and hard spermatophores from the crayfish, *Pacifastacus leniusculus* Dana. *Can. J. Zool.* 61:182–194.

Ezhilarasi, S. & T. Subramoniam. 1982. Spermathecal activity and ovarian development in *Scylla serrata* (Forskal) (Decapoda: Portunidae), In T. Subramoniam & Sudha Varadarajan, eds., *Progress in Invertebrate Reproduction and Aquaculture*, pp. 77–88. Madras: New Century Press.

Fielder, D. R. 1964. The process of fertilization in the spiny lobster, *Jasus lalandei* (H. Milne Edwards). *Trans. R. Soc. S. Aust.* 88:161–166.

Fryer, G. 1960. The spermatophores of *Dolops ranarum* (Crustacea, Branchiura), their structure, formation and transfer. *Q. J. Microsc. Sci.* 95:205–215.

Gharagozlou–Van Ginneken, I. D. 1978. Secretion and organization of the stratified wall of spermatophore in some copepods: Ultrastructure and cytochemistry. *Cytobiologie* 1:231–243.

Hartnoll, R. G. 1979. The phyletic implications of spermathecal structure in the Raninidae (Decapoda: Brachyura). *J. Zool. (Lond.)* 187:75–83.

Heldt, J. H. 1938. La reproduction chez les crustacés decapods de la famille des pénéides. *Ann. Inst. Oceanogr. Monaco* 18:31–206.

Hinsch, G. W. & M. H. Walker. 1974. The vas deferens of the spider crab, *Libinia emarginata*. *J. Morphol.* 143:1–20.

Ishida, T., P. Talbot & M. Kooda-Cisco. 1986. Technique for the long-term storage of lobster *(Homarus)* spermatophores. *Gamete Res.* 14:183–195.

Jeanloz, R. W. 1970. Mucopolysaccharides of higher animals. In W. Pigman & D. Horton, eds., *The Carbohydrates*, 2B:590–615. New York: Academic Press.

Jeyalectumie, C. 1989. Biochemical investigations on the reproductive tissues and cryopreservation of seminal secretions of a brachyuran crab *Scylla serrata* (Forskal) Decapoda: Portunidae. Ph.D. dissertation, University of Madras.

Jeyalectumie, C. & T. Subramoniam. 1987. Biochemical composition of seminal secretions with special reference to LDH activity in the reproductive tissues of the field crab, *Paratelphusa hydrodromous* (Herbst). *Exp. Biol. (Berlin)* 46:231–236.

Jeyalectumie, C. & T. Subramoniam. 1989. Cryopreservation of spermatophores and seminal plasma of the edible crab, *Scylla serrata*. *Biol. Bull (Woods Hole)* 177:247–253.

King, J. E. 1948. A study of the reproductive organs of the common marine shrimp, *Penaeus setiferus* (Linnaeus). *Biol. Bull. (Woods Hole)* 94:244–262.

Kooda-Cisco, M. J. & P. Talbot. 1982. A structural analysis of the freshly extruded spermatophore from the lobster, *Homarus americanus*. *J. Morphol.* 172:193–207.

Malek, S. R. A. & F. M. Bawab. 1971. Tanning in the spermatophore of a crustacean *(Penaeus testiculatus)*. *Experientia (Basel)* 27:1098.

Malek, S. R. A. & F. M. Bawab. 1974a. The formation of spermatophore in *Penaeus kerathurus* (Forskal, 1775) (Decapoda, Penaeidae). I. The initial formation of sperm mass. *Crustaceana (Leiden)* 26:273–285.

Malek, S. R. A. & F. M. Bawab. 1974b. The formation of spermatophore in *Penaeus kerathurus* (Forskal, 1775) (Decapoda, Penaeidae). II. The deposition of the layers of the body and of the wing. *Crustaceana (Leiden)* 27:72–83.

Mann, T. 1984. *Spermatophores*. Berlin: Springer-Verlag.

Mann, T., A. W. Martin & J. B. Thiersch. 1981. Changes in the spermatophoric plasma during spermatophore development and during the spermatophoric reaction in the giant octopus of the North Pacific, *Octopus dofleini martini*. *Mar. Biol. (Berl.)* 63:121–127.

Martin, G. G., C. Herzig & G. Narimatsu. 1987. Fine structure and histochemistry of the partly extruded and hardened spermatophore of the spiny lobster, *Panulirus interruptus*. *J. Morphol.* 192:237–246.

Mathews, D. C. 1951. The origin, development and nature of the spermatophoric mass of the spiny lobster, *Panulirus penicillatus* (Oliver). *Pac. Sci.* 5:359–371.

Mathews, D. C. 1954. The development of spermatophoric masses of the rock lobster, *Parribacus antarcticus* (Lund). *Pac. Sci.* 8:28–34.

Montgomery, R. 1970. Glycoproteins. In W. Pigman & D. Horton, eds., *The Carbohydrates*, 2B:628–710. New York: Academic Press.

Pigman, W. & D. Horton. 1970. *The Carbohydrates*, vol. 2B. New York: Academic Press.

Pochon-Masson, J. & I. D. Gharagozlou–Van Ginneken. 1977. Peculiar features of the sperm of an Harpacticoid copepod, *Tisbe holothuriae*. In K. G. Adiyodi & R. G. Adiyodi, eds., *Advances in Invertebrate Reproduction*, 1:61–75. Kerala, India: Peralam-Kennoth.

Primaveera, J. H. 1985. A review of maturation and reproduction in closed thelycum penaeids. *Proceedings of the 1st International Conference on the Culture of Penaeid prawns/shrimps*, pp. 47–64. Illoilo City, Philippines.

Radha, T. & T. Subramoniam. 1985. Origin and nature of the spermatophoric mass of the spiny lobster, *Panulirus homarus*. *Mar. Biol. (Berl.)* 86:13–19.

Rahemtulla, F. & S. Lovtrup. 1974. The comparative biochemistry of invertebrate mucopolysaccharides. I: Methods: Platyhelminthes. *Comp. Biochem. Physiol.* 49B:631–637.

Richards, A. G. 1951. *The Integument of Arthropods*. Minneapolis: Minnesota University Press.

Ryan, E. P. 1984. Spermatogenesis and the male reproductive system in the Hawaiian crab *Ranina ranina* L. In U. Engels, ed., *Advances in Invertebrate Reproduction*, p. 629. New York: Elsevier.

Sasikala, S. L. & T. Subramoniam. 1987. On the occurrence of acid mucopolysaccharides in the spermatophores of two marine prawns, *Penaeus indicus* (Milne Edwards) and *Metapenaeus monoceros* (Fabricius) (Crustacea: Macrura). *J. Exp. Mar. Biol. Ecol.* 113:145–153.

Shyamasundari, K. & K. H. Rao. 1986. Histochemical investigations of the male reproductive system of the sand lobster, *Thenus orientalis* (Lund): In M. F. Thompson et al., eds., *Indian Ocean-Biology of Benthic Marine Organisms*, pp. 48–54. New Delhi: Oxford and IBH.

Spalding, J. F. 1942. The nature and formation of the spermatophore and sperm plug in *Carcinus maenas*. *Q. J. Microsc. Sci.* 83:399–422.

Subramoniam, T. 1984. Spermatophore formation in two intertidal anomuran crabs, *Albunea symnista* and *Emerita asiatica* (Decapoda: Anomura). *Biol. Bull. (Woods Hole)* 166:78–95.

Taber, S. 1977. Semen of *Apis mellifera*: Fertility, chemical and physical characteristics. In K. G. Adiyodi & R. G. Adiyodi, eds., *Advances in Invertebrate Reproduction*, 219–252. Kerala, India: Peralam-Kennoth.

Talbot, P. & D. Beach. 1989. Role of the vas deferens in the formation of the spermatophore of the crayfish (cherax). *J. Crustacean Biol.* 9:9–24.

Uma, K. & T. Subramoniam. 1979. Histochemical characteristics of spermatophore layers of *Scylla serrata* (Decapoda: Portunidae). *Int. J. Invertebr. Reprod.* 1:31–41.

EIGHTEEN
Morphological Diversity of Decapod Spermatozoa

BRUCE E. FELGENHAUER

*Department of Biology,
University of Southwestern Louisiana,
Lafayette,
and*

LAWRENCE G. ABELE

*Department of Biological Science,
Florida State University,
Tallahassee.*

Abstract

The ultrastructural features of decapod spermatozoa are reviewed on the basis of some original data and published literature with reference to the possible value of spermatozoan morphology in phylogenetic studies. Decapod spermatozoa are rather unusual in being nonmotile and aflagellate. The considerable variation that they exhibit within both the Caridea and the Reptantia appears to be phylogenetically irrelevant because it is not concordant with phylogenies based on morphological or molecular characters. However, so few data are available that it is premature to conclude that spermatozoa yield no information relevant to decapod crustacean phylogeny.

EARLY LITERATURE on decapod spermatozoa described these cells as bizarre, in part because details of ultrastructure were misunderstood and in part because these cells are very different from typical animal spermatozoa, which basically consist of a nucleus, an acrosome, and a flagellum. In contrast, most crustacean spermatozoa, and all decapod spermatozoa, are aflagellate and nonmotile.

Although there are early scattered references to decapod spermatozoa (e.g., Siebold 1836; fig. 24), Grobben (1878) was the first to attempt a synthesis when he examined representative species of all major groups. He later (1906) suggested a classification of the decapods based on spermatozoa and in 1934 provided a detailed analysis of the phylogenetic position of the Caridea based on a comparison of spermatozoa. At the same time, Koltzoff (1906) summarized all information on decapod spermatozoa known to that date and related spermatozoan types to phylogeny according to a system of imprecisely defined descriptive terms (p. 424; see comments by Nichols 1909). Although his written descriptions were somewhat confusing, his drawings were at least recognizable, particularly those of the unistellate spermatozoa of *Lysmata seticaudata* and *Sicyonia carinata*. Spitschakoff (1909) provided drawings and descriptions of caridean spermatozoa, focusing on the palaemonid *Leander adspersus,* and compared them to those of *Pasiphaea sivado* and *Galathea squamifera*. Fasten (1914) summarized earlier literature concerning spermiogenesis, including most major references to decapod spermatozoa. Nath (1942) also provided a detailed and critical review of the literature and examined a large number of taxa himself. Although Nath was critical of the quality of earlier work, there are questions concerning his own work. For example, he described and illustrated the spermatozoa of *Penaeus indicus* as lacking "rays or pseudopodia," although every other species of *Penaeus* has a distinct spike. Similar comments could be made about some of his other descriptions, but caution is warranted, because his 1937 paper on *Macrobrachium lamarrei* (as *Palaemon*) is accurate and variation of spermatozoan morphology occurs within genera (see Wingstrand 1978 and "Discussion"). He developed an "evolutionary plan" of the decapods based on the stage of spermiogenesis at which the Golgi material disappears. Another evolutionary plan based on organelles is that of Wielgus (1973), who based his phylogenetic study of decapod spermatozoa on the number of centrioles, considering one primitive and two advanced. Neither of these approaches seems to have much promise.

Two difficulties plagued these and other early works: the technical limitations of the light microscope and the absence of any cytochemical tests that would allow comparative inferences. Thus work prior to about 1960 provides limited information relative to today's standards. In the early 1960s the situation changed when workers began to use the electron microscope (e.g. Yasuzumi 1960; Moses 1961a, b) to investigate spermatozoan ultrastructure as well as to characterize cytochemically the nucleus and acrosomal complex. Shortly thereafter Pochon-Masson (1968a, b) provided detailed comparisons of four major groups: carideans *(Crangon crangon)*, astacideans *(Homarus gammarus, Astacus astacus)*, anomurans *(Pagurus bernhardus)*, and brachyurans *(Carcinus maenas)*. She concluded that, despite great differences in general form, all decapod spermatozoa conform to the same fundamental plan of organization. Subsequent studies, especially by Hinsch (1969, 1971, 1973, 1980, 1986, 1988; McKnight & Hinsch 1986), Clark and coworkers (Clark et al. 1973; Lu, Clark & Franklin 1973; Kleve, Yudin & Clark 1980; Lynn & Clark 1983a, b; Shigekawa & Clark 1986; Griffin et al. 1987), and Talbot and coworkers (Talbot & Summers 1978; Talbot & Chanmanon 1980a, b; Dudenhausen & Talbot 1979, 1982, 1983; Beach & Talbot 1987) established the standards and comparative basis for modern work.

Our interest here is to provide a brief review of decapod spermatozoa and to evaluate the form and ultrastructure of these cells as possible characters in phylogenetic studies. Table 18.1 is a list of the species whose spermatozoa have been described. Items on the list were chosen by the following criteria. We have cited mostly references that used electron microscopy, but we have included literature based on light microscopy if the drawings illustrate information not available elsewhere. For example, Grobben (1906) published the only available information on the Pasiphaeidae, a group that is important in comparing caridean spermatozoa, particularly those of the caridean family Rhynchocinetidae. Except for comparative purposes, we have not cited early literature if a more recent reference provides better information. One difficulty with this criterion is the case of Burkenroad (1981), who commented negatively on the value of spermatozoa in phylogeny. He gave a very brief description of the spermatozoa of several species, e.g., *Lucifer faxoni*, *Trachypenaeus byrdi*, *Parapenaeus longirostris*, and *Stenopus* cf. *scutellatus*, stating also that certain atyids lack a unistellate spermatozoan. Although his comments in general agree with our findings (see our descriptions of *Stenopus hispidus*), the atyid spermatozoa we have examined (those of *Atya margaritacea* and *Typhlatya rogersi*) are unistellate, and it is possible that his reference to "reptant-like" penaeid spermatozoa is in error. However, we have cited his work because no other author has dealt with these taxa. Similarly, for comparative purposes, we have included in table 18.1 some of the species studied by Koltzoff (1906) and Nath (1942). We have used the currently accepted taxonomy rather than that used by the original authors.

MATERIALS AND METHODS

Spermatozoa examined with the scanning electron microscope (SEM) in this investigation were freed from the distal portion of the vas deferens (ejaculatory bulb) with a Potter-Elvehjem® tissue homogenizer. Suspended live sperm were dropped onto plastic coverslips treated briefly with a 0.1 percent solution of polylysine (after Mazia, Schatten & Winfield 1975) for twenty minutes. The attached cells were fixed in 2.5 percent glutaraldehyde in 0.1 M sodium cacodylate (pH 7.4) for three hours at room temperature. The cells were then postfixed in 1 percent osmium tetroxide for two hours. After dehydration in a graded ethanol series, the cells were critical-point dried in CO_2, sputter-coated with approximately 20 nm of gold-palladium, and examined in a JEOL 840 scanning electron microscope at accelerating voltages of 5–20 kV.

For transmission electron microscopy (TEM), the distal portion of the vas deferens was carefully dissected from fresh specimens and fixed in 2.5 percent glutaraldehyde in 0.1 M sodium cacodylate buffer (marine forms) and 0.1 M sodium phosphate (pH 7.4) (freshwater forms) for three hours at room temperature and postfixed in 1 percent osmium tetroxide for two hours. After dehydration through a graded series of acetone, the material was infiltrated with Spurr's low-viscosity resin (Spurr 1969). Thin sections were cut with a Diatome® diamond knife and collected on carbon-stabilized, 200 μm, thin-bar copper grids and stained sequentially for 40 min in 7 percent uranyl acetate and 15 min in aqueous lead citrate. Electron micrographs were taken on a JEOL CX1200 transmission electron microscope at 80 kV.

COMPARATIVE ULTRASTRUCTURAL FEATURES OF DECAPOD SPERMATOZOA

Suborder Dendrobranchiata

The spermatozoa of the Dendrobranchiata are rather uniform. Most of the relevant ultrastructural details are available for some members of the Penaeidae and Sicyoniidae (table 18.1). However, as noted earlier, Burkenroad (1934, 1981) made some brief comments on several dendrobranchiate taxa (e.g., *Lucifer*, *Xiphopenaeus*) without publishing any formal descriptions.

In general, the dendrobranchiate spermatozoan is a highly polarized cell consisting of three relatively distinct anatomical divisions: the cell body, cap region, and spike (fig. 18.1c). The cell body is an elongate sphere housing an uncondensed nucleus. The nucleus is not membrane bound and is extremely fibrillar (fig. 18.1D; also see Kleve, Yudin & Clark 1980; fig. 2). Prominent myelinlike configurations or membrane-bound vesicles are also commonly encountered near the margins of the cell body (fig. 18.1D); these may be formed from invaginations of the plasma membrane.

The cap region morphology varies among species. Kleve, Yudin, and Clark (1980) described a structurally complex cap region for *Sicyonia ingentis* with at least seven morphologically distinct components essentially making up an acrosomal complex (= subacrosome and acrosomal vesicle). Griffin et al. (1987) demonstrated an acrosomal reaction in *S. ingentis* by using egg components (egg water). Further details of the acrosome of dendrobranchiates have been presented by Shigekawa & Clark (1986), who included an excellent discussion of the decapod acrosome. The cap region of *Penaeus setiferus* (fig. 18.1D) is not as morphologically complex as the highly organized one of *S. ingentis*. Only two regions in the cap can be distinguished by differences in electron density.

Spike morphology is very simple in all of the dendrobranchiates investigated to date. Kleve, Yudin, and Clark (1980) described two major components, a limiting membrane and an internal spike component. Felgenhauer, Abele, and Kim (1988) described the spike of *Penaeus setiferus* as a simple unit composed of an amorphous, electron dense material with no fibers but exhibiting periodic cross-striations like those seen in many caridean shrimp species. Both Kleve, Yudin, and Clark (1980) and Felgenhauer, Abele, and Kim (1988) reported a curious spiral appearance to the spike (fig. 18.1D; arrow).

Infraorder Caridea

Although more than 22 families of caridean shrimp are recognized, relatively few studies exist on the ultrastructure of their spermatozoa. The spermatozoa of this heterogeneous group are mostly uniform in their basic morphology. The

TABLE 18.1. List of Malacostraca whose spermatozoa have been described.

Taxon / Species	Reference	Taxon / Species	Reference
Hoplocarida		Palaemonidae	
Gonodactylus bredini	this study	*Macrobrachium lamarrei*	Nath 1937
Decapoda		*M. rosenbergii*	Lynn & Clark 1983a,b
Suborder Dendrobranchiata			Dougherty et al. 1986
Penaeidae			Dougherty 1987
Hymenopenaeus robustus	Burkenroad 1981	*Palaemon elegans*	Pochon-Masson 1969
Parapenaeus longirostris	Burkenroad 1934, 1981	*P. serratus*	Papathanassiou & King 1984
Penaeus aztecus	Clark et al. 1973	*Palaemonetes kadiakensis*	Felgenhauer et al. 1988
P. indicus	Nath 1942	*P. paludosus*	Koehler 1979
P. japonicus	Ogawa & Kakuda 1987	Alpheidae	
P. setiferus	Felgenhauer et al. 1988	*Alpheus glaber*	Grobben 1878
Trachypenaeus byrdi	Burkenroad 1981	*Athanas nitescens*	Grobben 1878
T. similis	Burkenroad 1934	Hippolytidae	
Sicyoniidae		*Heptacarpus pictus*	Bauer 1976
Sicyonia brevirostris	Brown et al. 1977	*Hippolyte zostericola*	this study
S. carinata	Koltzoff 1906	*Lysmata seticaudata*	Koltzoff 1906
S. ingentis	Kleve et al. 1980	*Thor manningi*	Bauer 1986
	Shigekawa & Clark 1986	Pandalidae	
	Anchordoguy et al. 1988	*Pandalus kessleri*	Kashiwagi et al. 1972
	Griffin et al. 1988	*P. nipponensis*	Tamura 1950
Sergestidae		*P. platyceros*	Hoffman 1972
Lucifer faxoni	Burkenroad 1981	Crangonidae	
Suborder Pleocyemata		*Crangon crangon*	Arsenault et al. 1979
Infraorder Caridea		*C. septemspinosa*	Pochon-Masson 1968b
Procarididae		Infraorder Stenopodidea	
Procaris ascensionis	Felgenhauer et al. 1988	Stenopodidae	
Rhynchocinetidae		*Stenopus hispidus*	this study
Rhynchocinetes typus	Dupré & Barros 1983	*S.* cf. *scutellus*	Burkenroad 1981
Pasiphaeidae		Infraorder Astacidea	
Pasiphaea sivado	Grobben 1906	Nephropidae	
		Homarus americanus	Talbot & Chanmanon 1980a,b

Taxon	Reference
H. gammarus	Pochon-Masson 1968b
Nephrops norvegicus	Chevaillier & Maillet 1965
Astacidae	
Astacus astacus	Pochon-Masson 1968b
	Lopez-Camps et al. 1981
Pacifastacus leniusculus	Dudenhausen & Talbot 1982
	Dudenhausen & Talbot 1983
Cambaridae	
Procambarus clarkii	Moses 1961a,b
P. leonensis	this study
Parastacidae	
Cherax albidus	Beach & Talbot 1987
C. tenuimanus	Beach & Talbot 1987
Infraorder Thalassinoidea	
Enoplometopidae	
Enoplometopus occidentalis	Haley 1986
Upogebiidae	
Upogebia pusilla	Koltzoff 1906
Infraorder Palinura	
Palinuridae	
Panulirus argus	Brown et al. 1977
	Talbot & Summers 1978
P. guttatus	Talbot & Summers 1978
P. polyphagus	Nath 1942
Scyllaridae	
Scyllarus arctus	Koltzoff 1906
S. chacei	McKnight & Hinsch 1986
	Hinsch & McKnight 1988
Infraorder Anomura	
Galatheidae	
Galathea squamifera	Koltzoff 1906
Munida rugosa	Koltzoff 1906
Hippidae	
Emerita talpoida	Nichols 1909
	Barker & Austin 1963
Paguridae	
Pagurus bernhardus	Chevaillier 1967
	Pochon-Masson 1968a
P. punctulatus	Nath 1942
Diogenidae	
Clibanarius longitarsus	Nath 1942
C. nathi	Nath 1942
C. vittatus	this study
Coenobitidae	
Coenobita clypeatus	Hinsch 1980
C. rugosus	Nath 1942
Infraorder Brachyura	
Dromiidae	
Dromidia antillensis	this study
Dromia personata	Koltzoff 1906
Dorippidae	
Dorippe lanata	Koltzoff 1906
Leucosiidae	
Ilia nucleus	Koltzoff 1906
Iliacantha subglobosa	this study
Calappidae	
Calappa granulata	Koltzoff 1906
Matuta lunaris	Nath 1942
M. planipes	Nath 1942
Majidae	
Chionoecetes opilio	Beninger et al. 1988
Herbstia condyliata	Koltzoff 1906
Inachus dorsettensis	Koltzoff 1906
Libinia dubia	Hinsch 1973
L. emarginata	Hinsch 1969, 1973, 1986
Macrocoeloma trispinosum	Hinsch 1973
Maja verrucosa	Koltzoff 1906
Mithrax sp.	Hinsch 1973
Pitho lherminieri	Hinsch 1973

TABLE 18.1. List of Malacostraca whose spermatozoa have been described. Continued

Taxon Species	Reference	Taxon Species	Reference
Podochela gracilipes	Hinsch 1973	Xanthidae	
P. riisei	Hinsch 1973	*Eriphia smithii*	Nath 1942
Stenorhynchus seticornis	Hinsch 1973	*E. verrucosa*	Koltzoff 1906
Parthenopidae		*Etisus laevimanus*	Nath 1942
Heterocrypta granulata	Hinsch 1973	*Eurypanopeus depressus*	this study
Parthenope serratus	Hinsch 1973	*Eurytium limosum*	this study
Cancridae		*Leptodius hydrophilus*	Nath 1942
Cancer borealis	Langreth 1969	*L. sanguineus*	Nath 1942
C. irroratus	Langreth 1969	*Menippe mercenaria*	Bindford 1913
C. magister	Langreth 1969	*M. rumphii*	Nath 1942
C. productus	Langreth 1969	*Pseudozius caystrus*	Nath 1942
Geryonidae		Pinnotheridae	
Geryon fenneri	Hinsch 1988	*Pinnixa* sp.	Reger 1970
G. quinquedens	Hinsch 1988	*Pinnotheres pinnotheres*	Koltzoff 1906
Portunidae		Grapsidae	
Callinectes sapidus	Brown 1966	*Eriocheir sinensis*	Yasuzumi 1960
	this study		Du et al. 1987a,b
Carcinus maenas	Chevaillier 1966, 1967, 1969	*Grapsus grapsus*	Nath 1942
	Pochon-Masson 1968a	*Metopograpsus messor*	Nath 1942
	Goudeau 1982	*Sesarma quadratum*	Nath 1942
	Reger et al. 1984	*S. reticulatum*	this study
Charybdis cruciata	Nath 1942	*S. taeniolatum*	Nath 1942
C. merguiensis	Nath 1942	Parathelphusidae	
Macropipus arcuatus	Koltzoff 1906	*Parathelphusa spinigera*	Nath 1932
M. corrugatus	Koltzoff 1906	Ocypodidae	
Ovalipes ocellatus	Hinsch 1986	*Macrophthalmus pectinipes*	Nath 1942
P. pelagicus	Nath 1942	*Ocypode ceratophthalma*	Nath 1942
P. sanguinolentus	Nath 1942	*O. rotundata*	Nath 1942
Scylla serrata	Nath 1942	*Uca annulipes*	Nath 1942
Thalamita crenata	Nath 1942	*U. lactea*	Nath 1942
		U. urvillei	Nath 1942

spermatozoan is a polarized, unistellate cell that resembles, in gross morphology, that of the Dendrobranchiata (compare figs. 18.1A, B to Figs. 18.1C, D). The mature spermatozoan, in most species, resembles a "thumbtack" or "button" and consists of a cell body, a cap region, and a spike (fig. 18.1A) within the cell body. The cytoplasmic layer between the nucleus and the plasmalemma may contain numerous vesicular bodies (Fig. 18.1B), which are particularly common in the family Palaemonidae (for examples see Koehler 1979; Papathanassiou & King 1984; Dougherty 1987). Koehler (1979) suggested that these intranuclear vesicles play an acrosomal role in fertilization. Most other organelles are absent in mature spermatozoa, although mitochondria are present in developing spermatid stages (Koehler 1979).

The cap region of carideans is not nearly as structurally organized as that of dendrobranchiates (e.g., *Sicyonia ingentis*). A distinct acrosomal complex is absent (fig. 18.1B) in all caridean shrimp examined to date. However, Arsenault, Clattenburg, and Odense (1979) considered the electron-dense region of the cap and spike to be a proacrosome in spermatids and an acrosome in mature spermatozoa. The most prominent structures within the cap are anastomosing fibrils, which extend into the spike (see fig. 18.2E). In addition in spermatids centrioles are present in the central portion of the cap (Lynn & Clark 1983a; Koehler 1979; Arsenault, Clattenburg, and Odense 1979; Felgenhauer, Abele & Kim 1988). Presumably the centrioles in this region organize the fibrils that extend into the spike. These radial fibrils (= radial spines) are particularly prominent in *Rhynchocinetes typus* (see Dupré & Barros 1983: fig. 13) and *Pasiphaea sivado* (see Grobben 1906).

The spikes of the carideans sectioned to date can be divided into two types: solid ones containing closely packed radial fibrils that exhibit periodic cross-striations (e.g. *Palaemonetes* spp. [Koehler 1979; Lynn & Clark 1983a; Felgenhauer, Abele & Kim 1988]) and tube like ones with distinct electron-dense walls containing anastomosing radial fibrils (e.g. *Rhynchocinetes typus* [Dupré & Barros 1983], *Procaris ascensionis* [Felgenhauer, Abele & Kim 1988]). In *Rhynchocinetes typus*, Dupré and Barros (1983) found a cross-striation with an 18.5 nm periodicity, whereas Felgenhauer, Abele, and Kim (1988) did not report any periodicity in the spike fibrils of *P. ascensionis*.

Infraorder Stenopodidea

Few details concerning the ultrastructure of this taxonomically problematic group are available. Burkenroad (1981), as already noted, described the spermatozoa of *Stenopus* cf. *scutellus* as superficially resembling that of the mantis shrimp *Gonodactylus bredinii* (compare figs. 18.4c and 18.4E). He described it simply as "a slightly flattened spheroid, apparently lacking appendages, with a small refractile body in the middle of one side of the disc." This description is quite accurate, as our TEM investigations of *Stenopus hispidus* reveal the spermatozoan to be a simple elliptical cell (approximately 7–10 μm diameter) with a prominent lamellar body located to one side against the plasma membrane (fig. 18.4c). This lamellar body is similar to that flanking the complex acrosome in spermatozoa of several species of brachyuran crabs, such as *Callinectes sapi-*

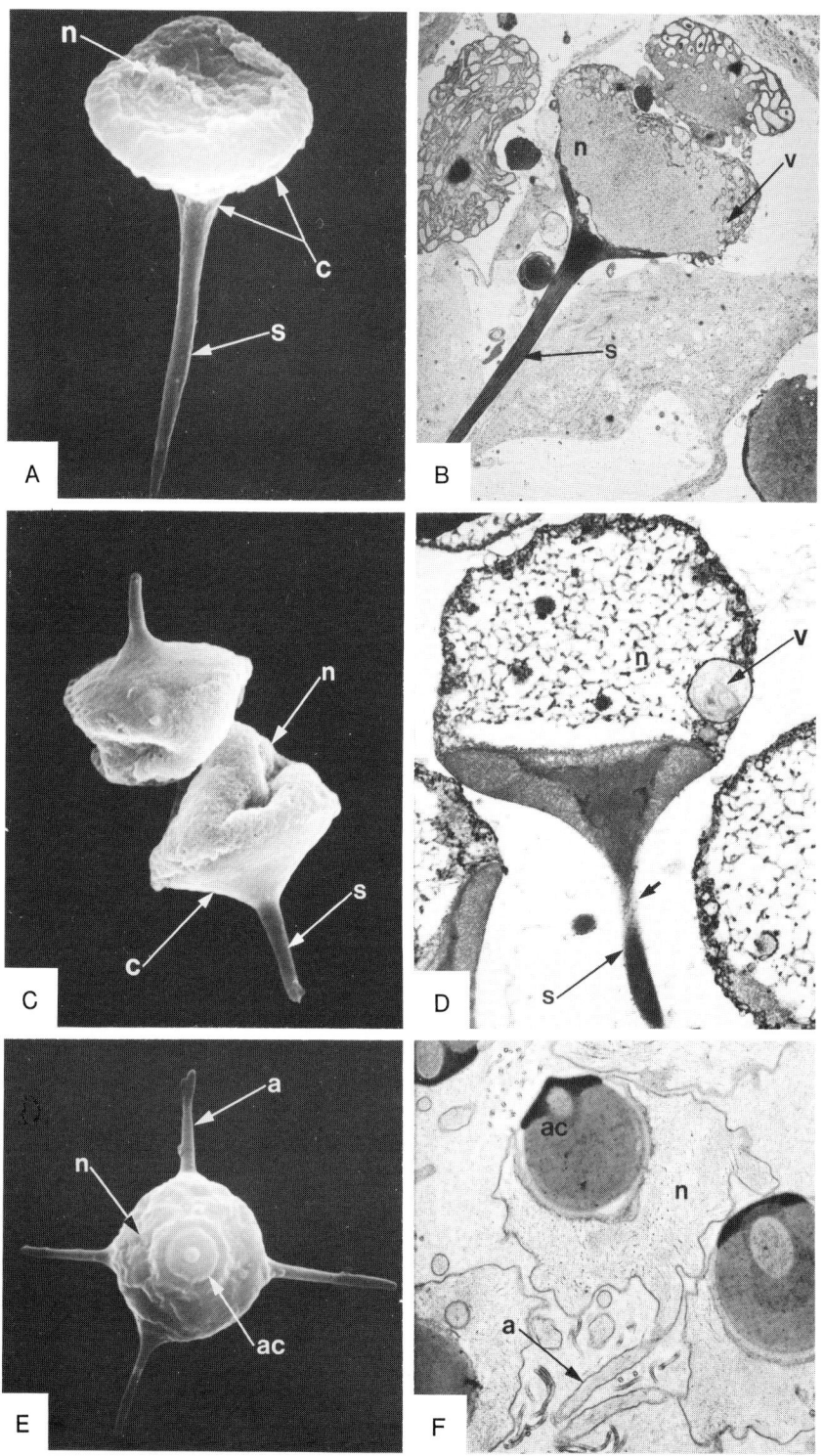

dus (figs 18.2A, B). No distinct acrosomal region or stellate appendages (radial arms), common features of reptant spermatozoa, are present (e.g. fig. 18.1F).

Although we have examined more than a dozen males of *Stenopus hispidus*, including six field-captured males that were paired with females, we are not convinced that this elliptical spermatozoan (figs. 18.4C, D) represents the mature cell simply because spermatozoa of all other decapods have some sort of radiating structures, either single spikes or arms as in the Reptantia. In addition, the absence of any type of acrosome or acrosomal complex is perplexing.

Reptantia

The reptant decapods consist of the infraorders Astacidea, Thalassinidea, Palinura, Anomura, and Brachyura (Bowman & Abele 1982). Space does not allow us to describe in detail the spermatozoa of each group that has been studied (but see table 18.1 for references). We describe here the general ultrastructural features of reptant spermatozoa and comment on the variation we have noted within the infraroders.

The reptant spermatozoan is a multistellate cell with radiating appendages (= arms or spikes) extending from the main cell body (figs. 18.1E, 18.2C). These stellate appendages are not homologous to the unistellate spike found in natant decapods (Talbot & Summers 1978; Hinsch 1986), and their function in motility or sperm-egg attachment is unknown. The origin, morphology, and number of radial arms vary among decapod species. The arms of some species appear to originate as extensions of the nucleus, whereas those of other species appear to be extensions of the cytoplasm (e.g., Hinsch 1986). Microtubules, presumably providing support, are present in the arms of some species (astacideans, palinurans, and some brachyurans) and absent in others (many brachyurans; compare figs. 18.2E and 18.3F). Spermatozoa of a few species (e.g., *Libinia emarginata*) contain both microtubules and chromatin in the arms; the arms of other taxa *(Homarus, Nephrops, Panulirus)* contain microtubules but lack chromatin, and the arms of *Ovalipes ocellatus* (see Hinsch 1986) lack microtubules but contain chromatin (Hinsch 1969; Langreth 1969; Talbot & Summers 1978; Lopez-Camps et al. 1981; and compare figs. 18.1F and 18.2D of this study). The number of arms or spikes can vary from 20 or more in *Procambarus leonensis* to 3 robust arms in *Iliacantha subglobosa* (compare figs. 18.3D, E, F to fig. 18.2C).

We are aware of only two papers dealing with thalassanidean spermatozoa: Koltzoff's (1906) brief comments on *Upogebia pusilla* and Haley's (1986) substantial paper on *Enoplometopus occidentalis*.

FIGURE 18.1. Decapod spermatozoan types. A. Unistellate spermatozoan of *Palaemonetes kadiakensis*. c, cap; n, nucleus; s, spike; by SEM; X 4000. B. Unistellate spermatozoan of *P. kadiakensis*. n, nucleus; s, spike; v, vesicle; by TEM; X 3640. C. Unistellate spermatozoan of *Penaeus setiferus*. n, nucleus; s, spike; by SEM; X 4000. D. Unistellate spermatozoan of *P. setiferus*. n, nucleus; s, spike; v, vesicle; arrow indicates twist in the spike; by TEM; X 3800. E. Multistellate spermatozoan of *Eurytium limosum*. a, arms; acp, acrosomal cap; n, nucleus; by SEM; X 3800. F. Multistellate spermatozoan of *Callinectes sapidus*. a, arms; ac, acrosome; n, nucleus; by TEM; X 4100.

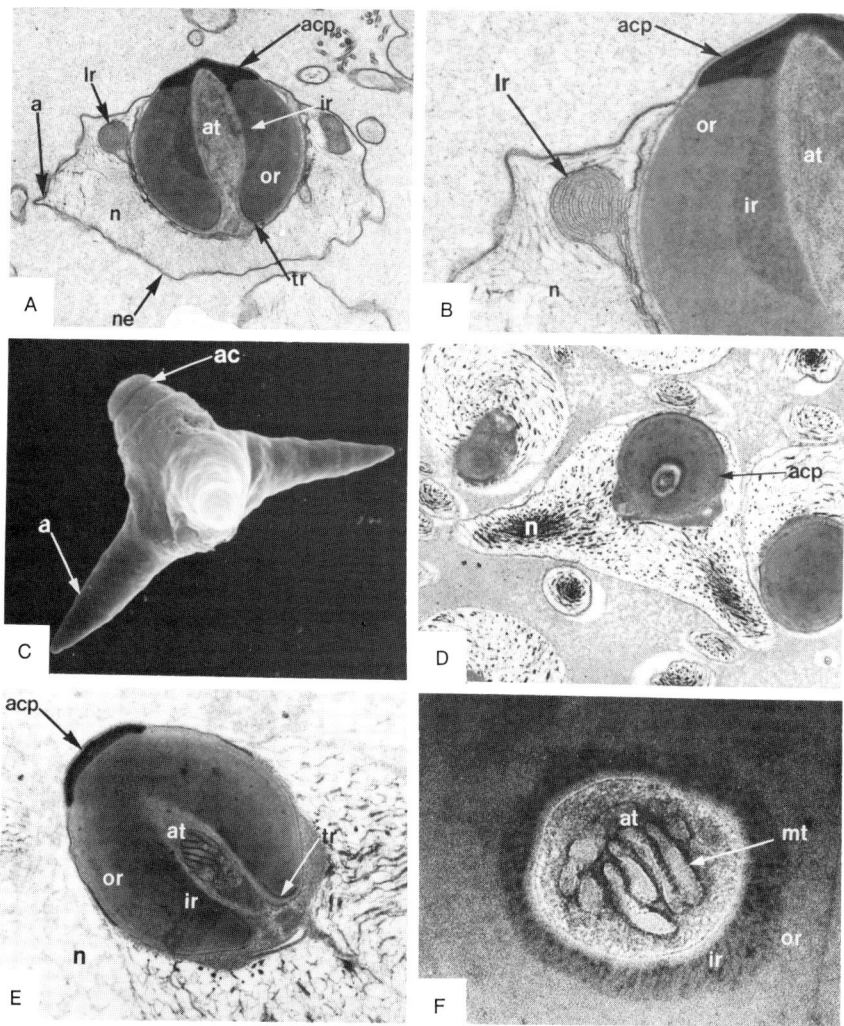

FIGURE 18.2. Reptantian spermatozoa. A. Multistellate spermatozoan of *Callinectes sapidus*. a, arm; acp, acrosomal cap; at, acrosomal tubule; ir, inner region of acrosomal vesicle; lr, lamellar region; n, nucleus; ne, nuclear envelope; or, outer region of acrosomal vesicle; tr, thickened ring; by TEM; X 3000. B. Close-up of acrosomal region of *C. sapidus*. acp, acrosomal cap; at, acrosomal tubule; ir, inner region of acrosomal vesicle; lr, lamellar region; n, nucleus; or, outer region of acrosomal vesicle; by TEM; X 10,000. C. Multistellate spermatozoan of *Iliacantha subglobosa*. a, arm; ac, acrosome; by SEM; X 9000. D. Ultrastructural features of the spermatozoan of *I. subglobosa*. acp, acrosomal cap; n, nucleus; by TEM; X 9500. E. Close-up of acrosomal region of *I. subglobosa*. acp, acrosomal cap; at, acrosomal tubule; ir, inner region of acrosomal vesicle; n, nucleus; or, outer region of acrosomal vesicle; tr, thickened ring (indicated by white arrow); by TEM; X 28,000. F. Cross section through the acrosomal tubule of *I. subglobosa* showing microtubules and bilayered condition of the acrosomal vesicle. at, acrosomal tubule; ir, inner region of acrosomal vesicle; mt, microtubule; or, outer region of acrosomal vesicle; by TEM; X 80,000.

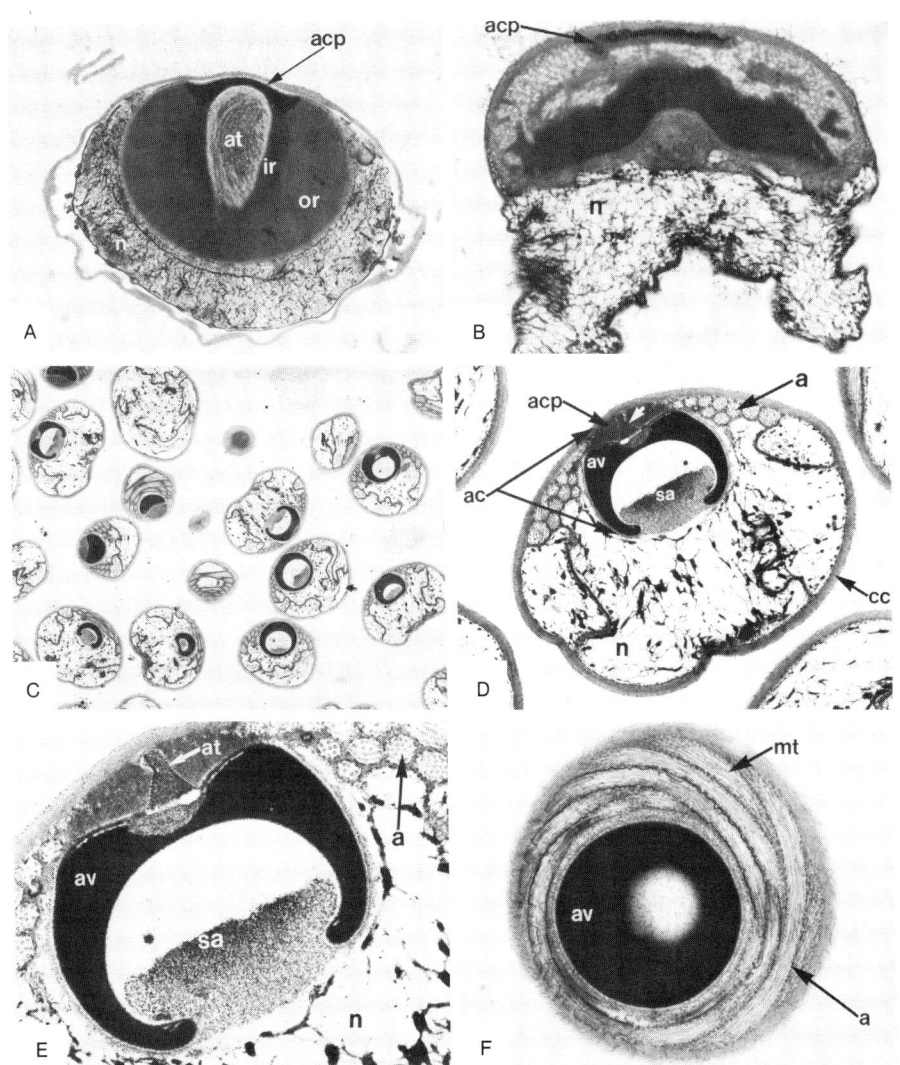

FIGURE 18.3. Reptantian spermatozoa. A. Spermatozoan of *Sesarma reticulatum*. acp, acrosomal cap; n, nucleus; by TEM; X 15,000. B. Multistellate spermatozoan of *Dromidia antillensis*. acp, acrosomal cap; at, acrosomal tubule; ir, inner region of acrosomal vesicle; n, nucleus; or, outer region of acrosomal vesicle; by TEM; X 14,000. C. Spermatozoa of *Procambarus leonensis* within spermatophore. By TEM; X 3000. D. Ultrastructural details of spermatozoan of *P. leonensis*. a, arms (= spikes); ac, acrosome; acp, acrosomal cap; at, acrosomal tubule (indicated by white arrow); av, acrosomal vesicle; cc, cell capsule; n, nucleus; sa, subacrosomal region; by TEM; X 18,000. E. Close-up of acrosomal vesicle of *P. leonensis*. a, arms (= spikes); av, acrosomal vesicle; n, nucleus; sa, subacrosomal region; note microtubules supporting arms. White arrow points to acrosomal tubule; by TEM; X 36,000. F. Section through anterior region showing arms (= spikes) supported by microtubules. a, arms; av, acrosomal vesicle; mt, microtubules; by TEM; X 19,000.

The acrosome is a highly complex structure throughout the reptant decapods (e.g., figs. 18.2A, B, E, 18.3A, E). Usually surrounded by the nucleus (figs. 18.2A, D), it consists of a highly refringent (electron dense) acrosomal vesicle (fig. 18.2A) that houses the acrosomal tubule (figs. 18.2A, B) (= perforatorium of stomatopods) in the undischarged condition. The steps in the acrosomal reaction result essentially in cell eversion (i.e. the spermatozoan is turned inside out; Brown 1966; Talbot & Chanmanon 1980b; Goudeau 1982; and others) and subsequent ejection of the nucleus into the cavity formed by the everted acrosome (Talbot & Chanmanon 1980b). The acrosome structure is relatively constant within each infraorder. The acrosomal complex is extremely uniform among the brachyurans that have been studied (e.g. Brown 1966; Goudeau 1982; Hinsch 1986; figs. 18.2A, E, 18.3B). The acrosomal vesicle has a characteristic bilayered appearance, and the layers have been termed the inner and outer regions of the vesicle (fig. 18.2B). The acrosomal tubule is located within the acrosomal vesicle and contains filaments or microtubules depending on the species (figs. 18.2E, F). A lamellar region flanking the acrosome is common in many species (figs. 18.2A, B). For comparison, we sectioned spermatozoa of the primitive brachyuran *Dromidia antillensis* (Dromiidae) and the more advanced *Callinectes sapidus* (Portunidae). The ultrastructure of the acrosomal complex is nearly identical in the two species (compare fig. 18.2A with fig. 18.3A). The unusual spermatozoan (fig. 18.3B) of *D. antillensis* is more reminiscent of those of scyllarid lobsters (see McKnight and Hinsch 1986) than those of brachyuran crabs. In contrast, the ultrastructure of the acrosomal complex of *Sesarma reticulatum* is nearly identical to that of *C. sapidus* (compare fig. 18.2A with 18.3A). Most of the brachyuran species we examined share this belayered acrosomal vesicle (see table 18.1).

The acrosome of the astacoid astacideans differs the most from the basic reptant plan of organization. In this group the acrosomal vesicle is horseshoe shaped (figs. 18.3D, E) in all freshwater crayfish species described to date (e.g. Moses 1961a, b; Lopez-Camps et al. 1981; Dudenhausen & Talbot 1982; Beach & Talbot 1987). The vesicle has a distinct crystalline nature produced by periodically arranged parallel laminae (fig. 18.3E). No study to date has mentioned an acrosomal tubule or perforatorium within the astacoid Astacidea. Dudenhausen & Talbot (1983) and Beach & Talbot (1987) have commented on the difficulty of discerning details of astacoid spermatozoa because of fixation problems and also on the absence of any information concerning the mechanics of the acrosomal reaction. Our investigation of *Procambarus leonensis* has disclosed a prominent acrosomal tubule just below the acrosomal cap (figs. 18.3D, E; white arrow). As in other astacoids investigated, an amorphous subacrosomal accumulation is located just beneath the acrosomal vesicle (figs. 18.3D, E). In addition to the differences in the acrosomal complex, it is worth noting that the cell membrane is extremely thick in astacoid spermatozoa and forms what has been termed a cell capsule (figs. 18.3C, D). This cell capsule is similar in appearance to the cell coat that surrounds the plasma membrane in stomatopods (Cotelli & Lora Lamia Dorin 1983; fig. 18.4D, E) and syncarids (Jespersen 1983).

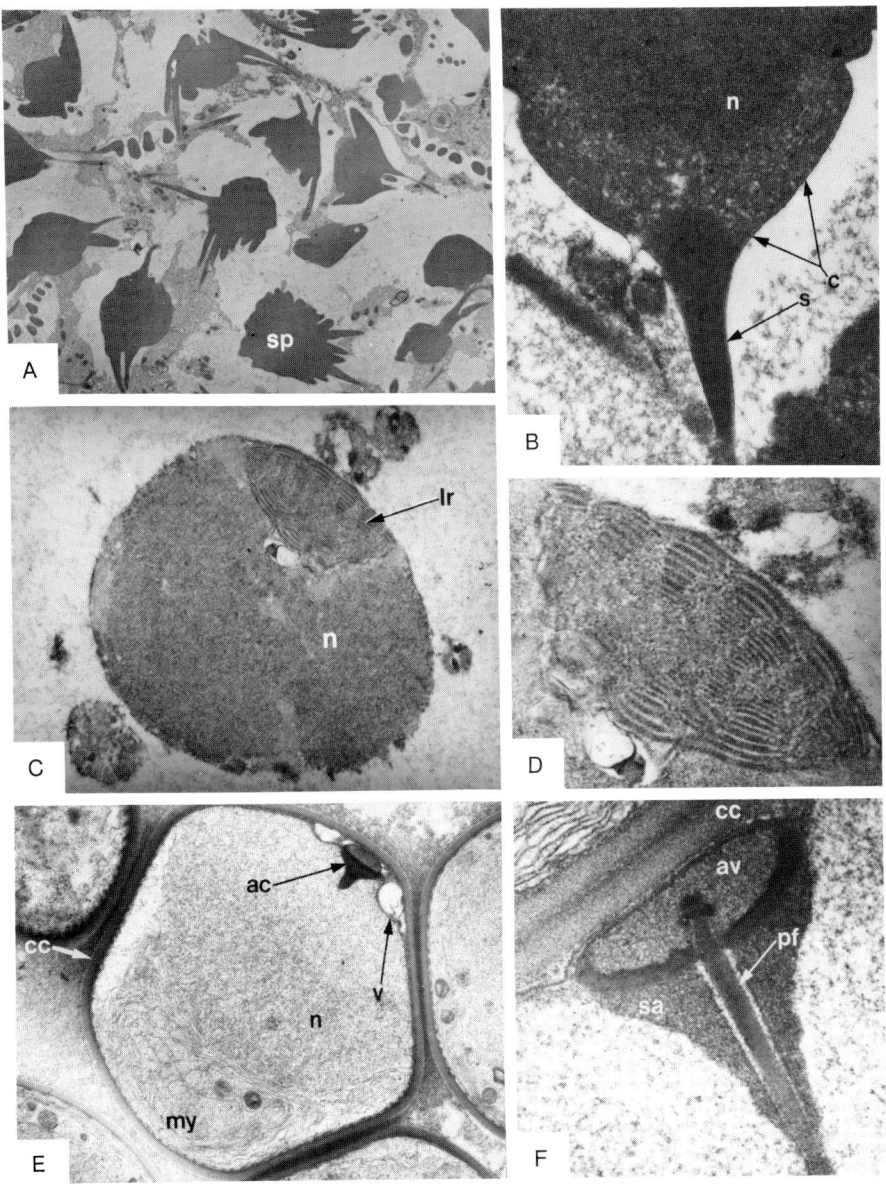

FIGURE 18.4. Diversity of crustacean spermatozoa. A. Spermatozoa of the caridean shrimp *Hippolyte zostericola*. sp, spermatozoan; by TEM; X 7200. B. Close-up of cap region and spike of the spermatozoan of *H. zostericola*; note the difference between the two regions. c, cap; n, nucleus; s, spike; by TEM; X 22,000. C. Spherical spermatozoan of *Stenopus hispidus*. lr, lamellar region; n, nucleus; by TEM; X 10,000. D. Close-up of lamellar region pictured in C; by TEM; X 20,000. E. Spermatozoan of *Gonodactylus bredinii* ac, acrosome; cc, cell capsule; my, myelin figures; n, nucleus; v, vesicles; by TEM; X 6500. F. Close-up of acrosome of *G. bredinii* av, acrosomal vesicle; cc, cell capsule; pf, perforatorium; by TEM; X 20,000.

DISCUSSION

Early classifications of the Decapoda (reviewed by Felgenhauer & Abele 1983) recognized a "Natantia" group consisting of penaeids and their relatives, carideans, and stenopodideans and a "Reptantia" group consisting of the remaining decapods. The first discussions of decapod spermatozoa tended to follow this classification, recognizing a unistellate natantian form and a multistellate reptantian form (see Balss 1957). Modern classifications and phylogenies (see Abele & Felgenhauer 1986; Schram 1986) recognize a fundamental division of the Decapoda into two suborders, the Dendrobranchiata (e.g. *Penaeus, Sicyonia*) and the Pleocyemata (e.g., *Macrobrachium, Stenopus, Callinectes*). Within the Pleocyemata a number of infraorders are recognized, which are sometimes divided into three major "groups" the Caridea (including the Procarididae), the Stenopodidea, and the remaining decapods (placed in a heterogeneous "reptant" assemblage). This division into three groups is supported by morphological (Abele & Felgenhauer 1986) and molecular (18S rRNA) data (Kim & Abele 1990) and offers a framework within which spermatozoan form and ultrastructure can be examined. If spermatozoa reflect phylogeny, major differences should exist between dendrobranchiate spermatozoa and those of the remaining decapods. Similarly, at least three types of spermatozoa should exist within the Pleocyemata, representing the Caridea, the Stenopodidea, and the reptants.

The morphology of decapod spermatozoa does not conform exactly to the pattern suggested by other morphological features. The available information on dendrobranchiates reveals a common plan (the comments of Burkenroad 1981 notwithstanding), but there is great diversity within both the Caridea and other infraorders. For example, the spermatozoa of the carideans *Procaris ascensionis* (see Felgenhauer, Abele & Kim 1988), *Macrobrachium rosenbergii* (see Lynn & Clark 1983a), *Rhynchocinetes typus* (see Dupré & Barros 1983), and *Hippolyte zostericola* (this report; figs. 18.4A, B) all differ from each other in what appear to be significant features, i.e., the form of the spike, presence of radial fibrils, and general shape. However, it must be remembered that there are 22 families of caridean shrimp, and although some information is available on 8 of these (table 18.1), only a few species in three families have been examined in detail.

There is also substantial diversity of spermatozoan form within the reptant decapods, even within a relatively homogeneous group such as the Palinuridea. In *Scyllarus chacei* the radial arms are of cytoplasmic origin (McKnight & Hinsch 1986), whereas in *Panulirus* the arms are of nuclear origin (Talbot & Summers 1978) and contain microtubules.

It would be premature, however, to conclude that spermatozoa have no phylogenetic value in decapod crustaceans for two reasons: (1) a relatively small number of species has been examined, and (2) studies on other crustacean groups suggest that spermatozoan morphology can be quite diverse yet still yield clues to relationships. Examples of the value of spermatozoa in phylogeny can be drawn from the work of Wingstrand (1972, 1978, 1988). In the first of three papers (1972), a comparison of the spermatozoa of a pentastomid and a branchiuran crustacean revealed so many shared details that Wingstrand concluded the parasitic pentastomids are not an independent phylum but are highly modified crustaceans closely related to fish lice, a conclusion recently supported by molecular data (Abele, Kim & Felgenhauer 1989). In 1978, Wingstrand examined spermatozoa of 71 species of branchiopod crustaceans. Despite an incredible diversity of sizes and shapes among the spermatozoa, even within a single genus, he was able to demonstrate that spermatozoan morphology provides important phylogenetic information. From these and other studies it appears that firm generalizations regarding spermatozoan morphology and phylogeny are not possible; a taxon-by-taxon assessment is necessary because among some taxa (pentastomid-branchiuran) morphology is highly conservative, whereas in others (the branchiopod genus *Moina*) spermatozoa have undergone a radiation at the genus level. At this time, there does not appear to be sufficient published information to warrant such a comparative study of the decapods.

Acknowledgments

The authors wish to thank Sandra Silvers, Kim Riddle, and Tom Fellers of the Florida State University Electron Microscopy Center for their skillful assistance with the electron microscopes. Special thanks must be given to Travis Boline for typing and to Anne Thistle for editing the manuscript. The authors also wish to extend their thanks to Drs. Raymond Bauer and Joel Martin for organizing and inviting us to participate in the symposium entitled "Sex Attraction, Mating Behavior and Insemination in the Crustacea" held at the American Society of Zoologists meetings (1988) in San Francisco, California.

While the present article was in proof these papers have appeared and contain valuable information on decapod spermatozoa: Jamieson, B. G. M. 1989. A comparison of the spermatozoa of *Oratosquilla stephensoni* and *Squilla mantis* (Crustacea, Stomatopoda) with comments on the phylogeny of the Malacostraca. *Zoologica Scr.* 18:509–517. Jamieson, Barrie G. M. 1989. Ultrastructural comparison of the spermatozoa of *Ranina ranina* (Oxystomata) and of other crabs exemplified by *Portunus pelagicus* (Brachygnata) (Crustacea, Brachyura). *Zoomorphology.* 109:103–111. Jamieson, B. G. M. 1989. The ultrastructure of the spermatozoa of four species of xanthid crabs (Crustacea, Brachyura, Xanthidae). *J. Submicrosc. Cytol. Pathol.* 21(3): 579–584. Jamieson, B. G. M. 1990. The ultrastructure of the spermatozoa of *Petalomera lateralis* (Gray) Crustacea, Brachyura, Dromiacea) and its phylogenetic significance. *Invertebrate Reproduction and Development.* 17:(1)39–45. Jamieson, B. G. M. and C. C. Tudge. 1990. Dorippids are Heterotemata: Evidence from ultrastructure of the spermatozoa of *Neodorippe astuta* (Dorippidae) and *Portunus pelagicus* (Portunidae) Brachyura: Decapoda. *Marine Biology,* vol. 105, in press.

Literature Cited

Abele, L. G. & B. E. Felgenhauer. 1986. Phylogenetic and phenetic relationships among the lower Decapoda. *J. Crustacean Biol.* 6:385–400.

Abele, L. G., W. Kim & B. E. Felgenhauer. 1989. Molecular evidence for inclusion of the phylum Pentastomida in the Crustacea. *Mol. Biol. Evol.* 6:685–691.

Anchorduguy, T., J. H. Crowe, F. J. Griffin & W. H. Clark, Jr. 1988. Cryopreservation of sperm from the marine shrimp *Sicyonia ingentis*. *Cryobiology* 25:238–243.

Arsenault, A. L., R. E. Clattenburg & P. H Odense. 1979. Further observations on spermiogenesis in the shrimp, *Crangon septemspinosa*. A mechanism for cytoplasmic reduction. *Can. J. Zool.* 58:497–506.

Balss, H. 1957. Decapoda. Part 11. Stammesgeschichte. In H. G. Bronn, ed., *Klassen und Ordnungen des Tierreichs*, Fünfter Band, Abteilung 1, Buch 7, Lieferung 12:1797–1821. Leipzig: Akademische Verlagsgesellschaft, Geest & Portig K.-G.

Barker, K. R. & C. R. Austin. 1963. Sperm morphology of *Emerita talpoida*. *Biol. Bull. (Woods Hole)* 125:361–362.

Bauer, R. T. 1976. Mating behavior and spermatophore transfer in the shrimp *Heptacarpus pictus* (Stimpson) (Decapoda: Caridea: Hippolytidae). *J. Nat. Hist.* 10:415–440.

Bauer, R. T. 1986. Sex change and life history pattern in the shrimp *Thor manningi* (Decapoda: Caridea): A novel case of partial protandric hermaphroditism. *Biol. Bull. (Woods Hole)* 170:11–31.

Beach, D. & P. Talbot. 1987. Ultrastructural comparison of sperm from the crayfishes *Cherax tenuimanus* and *Cherax albidus*. *J. Crustacean Biol.* 7:205–218.

Beninger, P. G., R. W. Elner, T. P. Foyle & P. H. Odense. 1988. Functional anatomy of the male reproductive system and the female spermatheca in the snow crab *Chionoecetes opilio* (O. Fabricius) (Decapoda: Majidae) and a hypothesis for fertilization. *J. Crustacean Biol.* 8322–332.

Bindford, R. 1913. The germ cells and the process of fertilization in the crab, *Menippe mercenaria*. *J. Morphol.* 24:147–204.

Bowman, T. E. & L. G. Abele. 1982. Classification of the Recent Crustacea. In D. E. Bliss and L. G. Abele, eds., *The Biology of the Crustacea*. Vol. 1, *Systematics, the Fossil Record, and Biogeography*, pp. 1–27. New York: Academic Press.

Brown, A., Jr., P. Talbot, R. G. Summers & W. H. Clark, Jr. 1977. Comparative analysis of decapod sperm. *J. Cell Biol.* 75:170A.

Brown, G. G. 1966. Ultrastructural studies of sperm morphology and sperm-egg interaction in the decapod *Callinectes sapidus*. *J. Ultrastruct. Res.* 14:425–440.

Burkenroad, M. D. 1934. The Penaeidae of Louisiana with a discussion of their world relationships. *Bull. Am. Mus. Nat. Hist.* 68:61–143.

Burkenroad, M. D. 1981. The higher taxonomy and evolution of Decapoda (Crustacea). *Trans. San Diego Soc. Nat. Hist.* 19:251–268.

Chevaillier, P. 1966. Contribution à l'étude du complexe DNA-histone dans la spermatozoïde du pagure *Eupagurus bernhardus* L. (Crustacé, Décapode). *J. Microsc. (Paris)* 5:739–758.

Chevaillier, P. 1967. Nouvelles observations sur la structure des fibres intra-nucléaires du spermatozoïde du pagure, *Eupagurus bernhardus* (Crustacé Décapode). *J. Microsc. (Paris)* 6:853–856.

Chevaillier, P. 1969. Évolution des constituants nucléaires et cytoplasmiques au cours de la spermatogénèse chez les Crustacés Décapodes. *C. R. Acad. Sci. Paris* 269:2251–2253.

Chevaillier, P. & P. L. Maillet. 1965. Structure fine et constitution cytochimique du spermatozoïde de la langoustine *Nephrops norvegicus* (Crustacés, Décapodes). *J. Microsc. (Paris)* 4:679–700.

Clark, W. H., Jr., P. Talbot, R. A. Neal, C. R. Mock & B. R. Salser. 1973. In vitro fertilization with non-motile spermatozoa of the brown shrimp *Penaeus aztecus*. *Mar. Biol. (Berl.)* 22:353–354.

Cotelli, F. & C. Lora Lamia Dorin. 1983. Ultrastructure of the spermatozoon of *Squilla mantis*. *Acta Zool. (Stockh.)* 64:131–137.

Dougherty, W. J. 1987. Oriented spermatozoa in the spermatophore of the palaemonid shrimp, *Macrobrachium rosenbergii*. *Tissue & Cell* 19:145–152.

Dougherty, W. J., M. M. Dougherty & S. G. Harris. 1986. Ultrastructural and histochemical observations on electroejaculated spermatophores of the palaemonid shrimp, *Macrobrachium rosenbergii*. *Tissue & Cell* 18:709–724.

Du, N., W. Lai & L. Xue. 1987a. Acrosome reaction of the sperm in the Chinese mitten-handed crab, *Eriocheir sinensis* (Crustacea, Decapoda). *Acta Zool. Sin.* 33:8–13.

Du, N., W. Lai & L. Xue. 1987b. Studies on the sperm of Chinese mitten-handed crab, *Eriocheir sinensis* (Crustacean, Decapoda). 1. The morphology and ultrastructure of mature sperm. *Oceanol. Limnol. Sin.* 18:119–125.
Dudenhausen, E. E. & P. Talbot. 1979. Spermiogenesis in the crayfish *Pacifastacus leniusculus. J. Cell Biol.* 83:225a.
Dudenhausen, E. E. & P. Talbot. 1982. An ultrastructural analysis of mature sperm from the crayfish *Pacifastacus leniusculus* Dana. *Int. J. Invertebr. Reprod.* 5:149–159.
Dudenhausen, E. E. & P. Talbot. 1983. An ultrastructural comparison of soft and hardened spermatophores from the crayfish *Pacifastacus leniusculus* Dana. *Can. J. Zool.* 61:182–194.
Dupré, E. & C. Barros. 1983. Fine structure of the mature spermatozoon of *Rhynchocinetes typus*, Crustacea Decapoda. *Gamete Res.* 7:1–18.
Fasten, N. 1914. Spermatogenesis of the American crayfish, *Cambarus virilis* and *Cambarus immunis* (?), with special reference to synapsis and the chromatoid bodies. *J. Morphol.* 25:587–649.
Felgenhauer, B. E. & L. G. Abele. 1983. Phylogenetic relationships among shrimp-like decapods (Penaeoidea, Caridea, Stenopodidea). In F. R. Schram, ed., *Crustacean Phylogeny*, pp. 291–311. Rotterdam: Balkema.
Felgenhauer, B., L. G. Abele & W. Kim. 1988. Reproductive morphology of the anchialine shrimp *Procaris ascensionis* (Decapoda: Procarididae). *J. Crustacean Biol.* 8:333–339.
Goudeau, M. 1982. Fertilization in a crab. I. Early events in the ovary, and cytological aspects of the acrosome reaction and gamete contacts. *Tissue & Cell* 14:97–111.
Griffin, F. J., W. H. Clark, Jr., J. H. Crowe & L. M. Crowe. 1987. Intracellular pH decreases during the in vitro induction of the acrosome reaction in the sperm of *Sicyonia ingentis. Biol. Bull. (Woods Hole)* 173:311–323.
Griffin, F. J., K. Shigekawa & W. H. Clark, Jr. 1988. Formation and structure of the acrosomal filament in the sperm of *Sicyonia ingentis. J. Exp. Biol.* 246:94–102.
Grobben, K. 1878. Beiträge zur Kenntnis der männlichen Geschlechtsorgane der Dekapoden. *Arb. aus dem Zool. Inst. der Univ. Wien und der Zool. Station in Triest* 1:57–150.
Grobben, K. 1906. Zur Kenntnis der Dekapodenspermien. *Arb. aus dem Zool. Inst. der Univ. Wien und der Zool. Station in Triest* 16:399–406.
Grobben, K. 1934. Die Dekapodenspermien und die Stellung der Eucyphidea (Garneelen) im Stammbaum der dekapoden Crustaceen. *Akad. wiss. Wien Abt.* 143:285–292.
Haley, S. R. 1986. Ultrastructure of spermatogenesis in the Hawaiian red lobster, *Enoplometopus occidentalis* (Randall). *J. Morphol.* 190:81–92.
Hinsch, G. W. 1969. Microtubules in the sperm of the spider crab, *Libinia emarginata* L. *J. Ultrastruct. Res.* 29:525–534.
Hinsch, G. W. 1972. Penetration of the oocyte envelope by spermatozoa in the spider crab. *J. Ultrastruct. Res.* 35:86–97.
Hinsch, G. W. 1973. Sperm structure of Oxyrhyncha. *Can. J. Zool.* 51:421–426.
Hinsch, G. W. 1980. Spermiogenesis in *Coenobita clypeatus.* I. Sperm structure. *Int. J. Invertebr. Reprod.* 2:189–198.
Hinsch, G. W. 1986. A comparison of sperm morphologies, transfer and sperm mass storage between two species of crab, *Ovalipes ocellatus* and *Libinia emarginata. Int. J. Invertebr. Reprod. Dev.* 10:79–87.
Hinsch, G. W. 1988. Ultrastructure of the sperm and spermatophores of the golden crab *Geryon fenneri* and a closely related species, the red crab *Geryon quinquedens* from the eastern Gulf of Mexico. *J. Crustacean Biol.* 8:340–345.
Hinsch, G. W. & C. E. McKnight. 1988. The vas deferens of the Spanish lobster, *Scyllarus chacei. Int. J. Invertebr. Reprod. Dev.* 13:267–280.
Hoffman, D. L. 1972. The development of the ovotestis and copulatory organs in a population of protandric shrimp, *Pandalus platyceros* Brandt from Lopez Sound, Washington. *Biol. Bull. (Woods Hole)* 142:251–270.
Jespersen, Å. 1983. Spermiogenesis in *Anaspides tasmaniae* (Thomson) (Crustacea, Malacostraca, Syncarida). *Acta Zool. (Stockh.)* 64:39–46.
Kashiwagi, M., J. Okawa, K. Hakoishi & C. Iida. 1972. [Title in Japanese.] *Aquaculture* 20:109–118.
Kim, W. & L. G. Abele. 1990. Molecular phylogeny of selected decapod crustaceans based on 18S rRNA nucleotide sequences. *J. Crustacean Biol.* 10:1–13.
Kleve, M. G., A. I. Yudin & W. H. Clark, Jr. 1980. Fine structure of the unistellate sperm of the shrimp, *Sicyonia ingentis* (Natantia). *Tissue & Cell* 12:29–45.
Koehler, L. D. 1979. A unique case of cytodifferentiation: Spermiogenesis of the prawn, *Palaemonetes paludosus. J. Ultrastruct. Res.* 69:109–120.
Koltzoff, N. K. 1906. Studien über die Gestalt der Zelle. I. Untersuchungen über die Spermien

der Decapoden, als Einleitung in das Problem der Zellengestalt. *Arch. Mikrobiol. Anat.* 67:364–572.

Langreth, S. G. 1969. Spermiogenesis in *Cancer* crabs. *J. Cell Biol.* 43:575–603.

Lopez-Camps, J., R. Bargallo, M. G. Bozzo, M. Durfort & R. Fontarnau. 1981. The spermatogenesis of crustaceans. VII. Review of spermatozoon of the crayfish *Astacus astacus* (Malacostraca, Decapoda, Macrura, Reptantia). *Gamete Res.* 4:65–82.

Lu, C. S., W. J. Clark & L. E. Franklin. 1973. Spermatogenesis of the decapod, *Penaeus setiferus*. *J. Cell Biol.* 59:202A.

Lynn, J. W. & W. H. Clark, Jr. 1983a. The fine structure of the mature sperm of the freshwater prawn, *Macrobrachium rosenbergii*. *Biol. Bull. (Woods Hole)* 164:459–470.

Lynn, J. W. & W. H. Clark, Jr. 1983b. A morphological examination of sperm-egg interaction in the freshwater prawn, *Macrobrachium rosenbergii*. *Biol. Bull. (Woods Hole)* 164:446–458.

McKnight, C. E. & G. W. Hinsch. 1986. Sperm maturation and ultrastructure in *Scyllarus chacei*. *Tissue & Cell* 18:257–266.

Mazia, D., G. Schatten & S. Winfield. 1975. Adhesion of cells to surfaces coated with polylysine. *J. Cell Biol.* 66:198–200.

Moses, M. J. 1961a. Spermiogenesis in the crayfish *(Procambarus clarkii)*. I. Structural characterization of the mature sperm. *J. Biophys. Biochem. Cytol.* 9:222–228.

Moses, M. J. 1961b. Spermiogenesis in the crayfish *(Procambarus clarkii)*. II. Description of stages. *J. Biophys. Biochem. Cytol.* 10:301–333.

Nath, V. 1932. The spermatid and the sperm of the crab, *Paratelphusa spinigera*. *Q. J. Microsc. Sci.* 75:544–556.

Nath, V. 1937. Spermatogenesis of the prawn, *Palaemon lamarrei*. *J. Morphol.* 61:149–163.

Nath, V. 1942. The decapod sperm. *Transcr. Natl. Inst. Sci. India* 2:87–119.

Nichols, M. L. 1909. Comparative studies in the crustacean spermatogenesis. *J. Morphol.* 20:461–478.

Ogawa, Y. & S. Kakuda. 1987. Scanning electron microscopic observations on the spermatozoa of the prawn *Penaeus japonicus*. *Bull. Jpn. Soc. Sci. Fish.* 53:975–977.

Papathanassiou, E. & P. E. King. 1984. Ultrastructural studies on gametogenesis of the prawn *Palaemon serratus* (Pennant). II. Spermiogenesis. *Acta Zool.* (Stockh.) 65:33–40.

Pochon-Masson, J. 1968a. L'ultrastructure des spermatozoïdes vésiculaires chez les Crustacés Décapodes avant et au cours de leur dévagination expérimentale. I. Brachyoures et Anomoures. *Ann. Sci. Nat. Zool., série 12*, 10:1–100.

Pochon-Masson, J. 1968b. L'ultrastructure des spermatozoïdes vésiculaires chez les Crustacés Décapodes avant et au cours de leur dévagination expérimentale. II. Discussion et conclusions. *Ann. Sci. Nat. Zool., série 12*, 10:367–454.

Pochon-Masson, J. 1969. Infrastructure du spermatozoïde de *Palaemon elegans* (de Man) (Crustacé Décapode). *Arch. Zool. Exp. Gen.* 110:363–372.

Reger, J. F. 1970. Studies on the fine structure of spermatids and spermatozoa of the crab, *Pinnixia* sp. *J. Morphol.* 132:89–100.

Reger, J. F., F. Escaig, J. Pochon-Masson & M. E. C. Fitzgerald. 1984. Observations on crab, *Carcinus maenas*, spermatozoa following rapid-freeze and conventional fixation techniques. *J. Ultrastruct. Res.* 89:12–22.

Schram, F. R. 1986. *Crustacea*. New York and Oxford: Oxford University Press.

Shigekawa, K. & W. H. Clark, Jr. 1986. Spermiogenesis in the marine shrimp, *Sicyonia ingentis*. *Dev. Growth & Differ.* 28:95–112.

Siebold, C. T. von. 1836. Ueber die Spermatozoen der Crustaceen, Insecten, Gasteropoden und einiger anderer wirbelosen Theire. *Arch. Anat. Physiol. Wiss. Medicin* 1836:13–53.

Spitschakoff, T. 1909. Spermien und Spermiohistogenese bei Cariden. *Arch. f. Zellforsch.* 3:1–43.

Spurr, A. R. 1969. A low viscosity epoxy resin embedding medium for electron microscopy. *J. Ultrastruct. Res.* 26:31–43.

Talbot, P. & P. Chanmanon. 1980a. The structure of sperm from the lobster, *Homarus americanus*. *J. Ultrastruct. Res.* 70:275–286.

Talbot, P. & P. Chanmanon. 1980b. Morphological features of the acrosome reaction of lobster *(Homarus)* sperm and the role of the reaction in generating forward sperm movement. *J. Ultrastruct. Res.* 70:287–297.

Talbot, P. & R. G. Summers. 1978. The structure of sperm from *Panulirus*, the spiny lobster, with special regard to the acrosome. *J. Ultrastruct. Res.* 64:341–351.

Tamura, T. M. 1950. On the life history of *Pandalus nipponensis*. *Nihon Suisan Gakkai-Shi* 15:721–724.

Wielgus, E. 1973. Znaczenie filogenetyczne budowy plemników u Decapoda (The phylogenetical significance of spermatozoa in Decapoda). *Przegl. Zool.* 17:420–426.

Wingstrand, K. G. 1972. Comparative spermatology of a pentastomid, *Raillietiella hemidactyli*, and a branchiuran crustacean, *Argulus foliaceus*, with a discussion of pentastomid relationships. *K. Dan. Vidensk. Selsk. Biol. Skr.* 19:1–72.

Wingstrand, K. G. 1978. Comparative spermatology of the Crustacea Entomostraca. 1. Subclass Branchiopoda. *K. Dan. Vidensk. Selsk. Biol. Skr.* 22:3–66.

Wingstrand, K. G. 1988. Comparative spermatology of the Crustacea Entomostraca 2. Subclass Ostracoda. *K. Dan. Vidensk. Selsk. Biol. Skr.* 32:1–79.

Yasuzumi, G. 1960. Spermatogenesis in animals as revealed by electron microscopy. VII. Spermatid differentiation in the crab, *Eriocheir japonicus*. *J. Biophys. Biochem. Cytol.* 7:73.

CONTRIBUTORS

Lawrence G. Abele
Department of Biological Science
Florida State University
Tallahassee, Florida 32306
U.S.A.

David E. Aiken
Fisheries and Oceans
Biological Station
St. Andrews, New Brunswick EOG2XO
Canada

Raymond T. Bauer
Center for Crustacean Research
University of Southwestern Louisiana
Lafayette, Louisiana 70504
U.S.A.

Denton Belk
Biology Department
Our Lady of the Lake University of San Antonio
San Antonio, Texas 78285
U.S.A.

Pamela I. Blades-Eckelbarger
Darling Marine Center
University of Maine
Walpole, Maine 04573
U.S.A.

Rudolf Boddeke
Netherlands Institute for Fishery Investigations
P.B. 68, 1970 AB, IJmuiden
The Netherlands

Betty Borowsky
Osborn Laboratories of Marine Science
Boardwalk at W. 8th Street
Brooklyn, New York 11224
U.S.A.

J. R. Bosschieter
Netherlands Institute for Fishery Investigations
P.B. 68, 1970 AB, IJmuiden
The Netherlands

Roy L. Caldwell
Department of Integrative Biology
University of California
Berkeley, California 94720
U.S.A.

Anne C. Cohen
Life Sciences, Invertebrate Zoology
Natural History Museum of Los Angeles County
900 Exposition Boulevard
Los Angeles, California 90007
U.S.A.

Rudolf Diesel
Fakultät für Biologie
Universität Bielefeld
Postfach 8640, 4800 Bielefeld 1
Federal Republic of Germany

Philip J. Dunham
Department of Psychology
Dalhousie University
Halifax, Nova Scotia B3H4J1
Canada

Bruce E. Felgenhauer
Department of Biology
University of Southwestern Louisiana
Lafayette, Louisiana 70504
U.S.A.

Richard A. Gleeson
Whitney Marine Laboratory
University of Florida
9505 Ocean Shore Boulevard
St. Augustine, Florida 32086
U.S.A.

P. C. Goudswaard
Netherlands Institute for Fishery Investigations
P.B. 68, 1970 AB, IJmuiden
The Netherlands

Gertrude W. Hinsch
Department of Biology
University of South Florida
Tampa, Florida 33620
U.S.A.

Jens T. Høeg
Institute of Cell Biology and Anatomy
University of Copenhagen
Universitetsparken 15, DK 2100, Copenhagen
Denmark

Alan M. Hurshman
Department of Psychology
Dalhousie University
Halifax, Nova Scotia B3H 4JI
Canada

Joel W. Martin
Life Sciences, Invertebrate Zoology
Natural History Museum of Los Angeles County
900 Exposition Boulevard
Los Angeles, California, 90007
U.S.A.

James G. Morin
Department of Biology
University of California
Los Angeles, California, 90024
U.S.A.

Stephen M. Shuster
Department of Biological Sciences
Northern Arizona University
Flagstaff, Arizona 86011
U.S.A.

T. Subramoniam
Department of Zoology
University of Madras, Guindy Campus
Madras 600-025
India

Susan L. Waddy
Fisheries and Oceans
Biological Station
St. Andrews, New Brunswick EOG2XO
Canada

George D. F. Wilson
Marine Biology
Scripps Institution of Oceanography
La Jolla, California 92093
U.S.A.

Jill Yager
Biology Department
Antioch College
Yellow Springs, Ohio 45387
U.S.A.

SYSTEMATIC INDEX

Abudefduf, 99
Acanthonyx lunulatus, 156
Acartia tonsa, 260, 262
Acartiidae, 247
Achaeus cranchii, 156
Acropora palmata, 11
Acrothoracica, 212, 215, 223-24
Aeschronectida, 69
Akentrogonidae, 209, 216
Albunea symnista, 310-11, 317
Alpheidae, 326
Alpheus armatus, 68
Alpheus glaber, 326
Ampelisca, 35
Ampelisca abdita, 43
Ampelisca vadorum, 43
Amphipoda, 33-47, 50-65
Amphiroa, 96
Amphithoe valida, 43
Anemonia sulcata, 158
Anomalocera ornata, 262
Anomura, 294, 310-11, 327, 331
Anostraca, 111-23
Anthuridae, 234
Apis, 317

Archaeostomatopodea, 69
Arcturidae, 234, 236
Argis dentata, 168
Arietillidae, 247
Aristaeomorpha, 188, 198, 203-4
Aristaeomorpha foliacea, 188
Aristeidae, 185, 192, 198
Aristeus, 203-4
Aristeus antennatus, 188
Artemia, 112-13, 116-22
Artemia franciscana, 119
Artemiidae, 113
Artemiopsis, 115
Artemiopsis stefanssoni, 112, 115, 119
Ascothoracida, 224, 283, 285
Asellidae, 239
Aselloidea, 232, 240
Asellota, 231, 234, 236-38, 240, 243
Asellus, 236, 239-40
Astacidae, 327
Astacidea, 326, 331, 334
Astacus astacus, 323, 327
Athanas nitescens, 326

Atya margaritacea, 324
Augaptilidae, 247

Bairdioidea, 2
Balanus balanus, 316-17
Balistes, 98
Bathypontiidae, 247
Bathyporeira pelagica, 43
Bathyporeira pilosa, 43
Benthesicymidae, 192, 204
Bopyridae, 239
Boschmaella, 220
Boschmaia, 216
Brachyura, 145-60, 294, 309, 316, 327, 331
Branchinecta, 113, 116-17
Branchinecta campestris, 116
Branchinectidae, 113
Branchinella kugenumaensis, 114
Branchiopoda, 283
Branchipus, 116
Branchipus stagnalis, 119
Branchiura, 283

Calabozoidea, 230
Calanoida, 246-268
Calanus finmarchicus, 257
Calappa granulata, 151, 327
Calappidae, 151, 327
Callinectes, 336
Callinectes sapidus, 17-31, 43, 45, 59, 151-52, 292, 328-29, 331-32, 334
Cambaridae, 327
Cancer antennarius, 25
Cancer anthonyi, 25
Cancer borealis, 154, 160, 328
Cancer irroratus, 328
Cancer magister, 328
Cancer productus, 328
Cancridae, 146, 154, 160, 328
Candaciidae, 247
Candona suburbana, 4, 6
Carinus maenas, 43, 45, 151, 159, 211, 309, 323, 328
Cardisoma guanhumi, 154
Caridea, 164-181, 313-14, 323, 325-26, 336
Centropages, 267
Centropages furcatus, 264
Centropages typicus, 257, 267-68
Centropagidae, 247, 253
Cephalocarida, 283
Chaceon, 295, 297
Chaceon fenneri, 294-95
Charybdis cruciata, 328
Charybdis merguiensis, 328
Chelicerata, 285
Cherax, 291, 297
Cherax albidus, 313, 327
Cherax tenuimanus, 313, 327
Chionoecetes bairdi, 138, 153-57
Chionoecetes opilio, 151-57, 294, 327
Chirocephalidae, 113, 121
Chthamalophilidae, 209, 216-17, 219-20, 223, 225
Chthamalophilus, 220
Cirripedia, 208-25, 283
Cladocopina, 2, 4
Clibanarius longitarsus, 327
Clibanarius nathi, 327
Clibanarius vittatus, 327
Clistosaccidae, 209, 216
Clistosaccus, 215, 217-19, 221, 223, 225
Clistosaccus paguri, 214, 216-18, 221, 223
Coenobita clypeatus, 294, 301-2, 327
Coenobita rugosus, 327
Coenobitidae, 327
Colubotelson, 232, 240-41
Copepoda, 246-68, 283
Corallina, 96
Corophium, 35
Corophium arenarium, 43
Corophium insidosum, 43
Corophium volutator, 43
Corystes cassivelanus, 151
Corystidae, 151
Crangon crangon, 164-81, 323, 326
Crangon septemspinosa, 166, 326

Crangonidae, 165, 326
Cryptocorynetes haptodiscus, 272, 288
Cryptogaster, 217, 219-20
Cyathura, 234
Cylindroleberididae, 2, 3
Cylindroleberidoidea, 2
Cyphosaccus, 211, 216, 219
Cypria turneri, 6
Cypridina dentata, 5, 7
Cyprididae, 2-4
Cypridinoidea, 2
Cypridoidea, 2, 4
Cytherelloidea, 2
Cytheroidea, 2

Darwinuloidea, 2
Decapoda, 184, 290-305, 308-19, 322-37
Dendrobranchiata, 184, 325-26, 329, 336
Dendrocephalus, 114
Dendrotiidae, 240
Diogenidae, 327
Dolops ranarum, 315
Dorippe lanata, 327
Dorippidae, 327
Dromiacea, 147
Dromia personata, 148, 327
Dromidia antillensis, 327, 333-34
Dromiidae, 327, 334
Duplorbis, 217, 219-20
Dynamene, 101
Dynamenella, 239
Dynameninae, 101

Elasmopus laevis, 35
Emerita asiatica, 310-11, 317
Emerita talpoida, 327
Enoplometopidae, 327
Enoplometopus, 297
Enoplometopus occidentalis, 327, 331
Epicaridea, 231, 238
Epilabidocera amphitrites, 258
Epipenaeon, 239
Eriocheir sinensis, 328
Eriphia smithii, 328
Eriphia verrucosa, 328
Etisus laevimanus, 328
Eubrachyura, 147, 159
Eubranchipus, 113, 116
Eubranchipus bundyi, 114-15, 118
Eubranchipus dadayi, 120
Eubranchipus holmani, 112
Eubranchipus moorei, 112
Eubranchipus serratus, 114, 116-21, 123
Euchaeta, 250, 268
Euchaeta antarctica, 250, 254, 262
Euchaeta indica, 250
Euchaeta marina, 250
Euchaeta marinella, 250
Euchaeta norvegica, 250, 255, 257-58, 262, 264, 267
Euchaeta rimana, 250
Euchaetidae, 253
Eulimnogammarus, 42

Eurypanopeus depressus, 328
Eurytemora affinis, 268
Eurytium limosum, 328, 331
Excirolana, 230, 239

Facetotecta, 223
Fucus, 54

Galathea squamifera, 323, 327
Galatheidae, 327
Gammaridea, 33-47
Gammarus, 35
Gammarus duebeni, 34, 42, 45-46, 53
Gammarus fossarum, 53
Gammarus lawrencianus, 35, 45, 50-65
Gammarus mucronatus, 38
Gammarus palustris, 34-39, 41-45, 54, 60-61, 64
Gammarus pulex, 34, 53
Gammarus setosus, 35
Gecarcinidae, 154
Geryon fenneri, 152, 328
Geryonidae, 151, 328
Geryon quinquedens, 328
Gnathia calva, 230
Gnathostenetroidoidea, 232
Godzilliognomus frondosus, 272, 274-79, 281, 287-88
Godzillius robustus, 277, 288
Gonodactylidae, 73, 87
Gonodactyloidea, 68-69
Gonodactylus, 73, 77, 81-83
Gonodactylus bahiahondensis, 73
Gonodactylus bredini, 71-73, 76-78, 80, 326, 329, 335
Gonodactylus oerstedii, 76
Grapsidae, 147, 328
Grapsus grapsus, 328

Halocyprida, 2
Halocyprididae, 2
Halocypridoidea, 2
Halocyprina, 2
Haploniscidae, 240
Haptosquilla trispinosa, 83-86
Harpiosquillidae, 73
Haustorius canadensis, 43
Heptacarpus paludicola, 43
Heptacarpus pictus, 44, 326
Herbstia condyliata, 327
Heterocrypta granulata, 328
Heterorhhabdidae, 247
Heterosquilla tricarinata, 75
Heterotremata, 147
Hippidae, 327
Hippolyte zostericola, 326, 335-36
Hippolytidae, 180, 326
Homarus, 291, 297, 312, 331
Homarus americanus, 42, 45, 60, 126-42, 311-12, 326
Homarus gammarus, 323, 327
Homoloidea, 147
Hoplocarida, 69-87, 326
Hyale, 42
Hyalella, 42
Hyalella azteca, 35

Hyas araneus, 151, 156-57
Hyas coarctatus, 151, 156
Hymenopenaeus robustus, 326

Ianiropsis, 242
Ibla idiotica, 212, 223
Idotea, 240
Iliacantha subglobosa, 327, 331-32
Ilia nucleus, 151, 327
Inachus communissimus, 151-52, 154, 156
Inachus dorsettensis, 151, 327
Inachus phalangium, 148, 151-54, 156, 158
Ischnomesidae, 240
Ischnomesus, 242
Isocypridina quatuorsetae, 5
Isopoda, 91-108, 228-43

Jaera, 230-31, 236-37
Janiridae, 240
Janiroidea, 232, 234, 236-37, 239-41, 243
Jassa marmorata, 35, 43

Kentrogonida, 209, 221, 224

Labidocera, 253
Labidocera aestiva, 247-49, 253, 255, 257-58, 260, 262, 264, 267-68
Labidocera barbadiensis, 253
Labidocera barbudae, 253
Labidocera scotti, 253, 255
Lasionectes entrichoma, 272, 288
Leander adspersus, 323
Leander squilla, 181
Lembos websteri, 43
Leptanthura, 235
Leptodius hydrophilus, 328
Leptodius sanguineus, 328
Lernaeodiscidae, 209, 211-12, 216-17
Lernaeodiscus procellanae, 212, 221
Leucetta, 96, 100
Leucetta losangelensis, 93
Leucosiidae, 151, 327
Libinia, 291, 294, 297
Libinia dubia, 327
Libinia emarginata, 153-54, 156, 293-96, 327, 331
Ligiidae, 236
Limnoria, 230, 235, 239
Limnoriidae, 235, 239
Limulus, 285
Linderiellidae, 113
Linuparus trigonus, 311
Liocarcinus depurator, 148, 151
Litopenaeus, 185, 188, 191-92, 198, 200, 203-4
Lucicutiidae, 247
Lucifer, 325
Lucifer faxoni, 324, 326
Lysiosquilla, 82-83, 86-87
Lysiosquilla glabriuscula, 75, 82-83
Lysiosquilla maculata, 82
Lysiosquilla panamica, 82-83

Lysiosquilla sulcata, 75, 82-83
Lysiosquilla tredecimdentata, 82
Lysiosquillidae, 73
Lysiosquilloidea, 69
Lysmata seticaudata, 323, 326

Macrobrachium, 336
Macrobrachium lamarrei, 323, 326
Macrobrachium rosenbergii, 170, 173, 313-14, 326, 336
Macrocoeloma trispinosum, 327
Macrophthalmus pectinipes, 328
Macropipus arcuatus, 328
Macropipus corrugatus, 328
Macropodia, 154, 156
Macropodia rostrata, 151-52
Macrura, 297, 311-13
Maja squinado, 148, 151-56, 158
Maja verrucosa, 152, 154-56, 327
Majidae, 145-60, 327
Malacostraca, 234, 238, 283
Marinogammarus, 42
Matuta lunaris, 327
Matuta planipes, 327
Meiosquilla, 81
Meiosquilla dawsoni, 75, 81-82
Meiosquilla lebouri, 81
Meiosquilla swetti, 75, 81-82
Meiosquilla tricarinata, 81
Mellita, 42
Menippe mercenaria, 153, 328
Menippe rumphii, 328
Mesopenaeus, 188
Merostomata, 285
Metridiidae, 247
Metapenaeus monoceros, 314
Metopograpsus messor, 328
Microcerberidae, 230
Microdeutopus gryllotalpa, 34-37, 39-44, 46, 54
Microphrys bicornutus, 153
Mictacea, 232
Millepora complanata, 11
Mithrax sp., 327
Moina, 337
Munida rugosa, 327
Munna, 230, 240
Munnogonium waldronense, 230
Mycetomorpha, 217, 219-21, 223
Myodocopa, 2, 4, 6
Myodocopida, 2, 4
Myodocopina, 2
Mysidacea, 232
Mystacocarida, 283

Nannosquilla, 73-74
Nannosquillidae, 73
Natantia, 336
Neohaustorius biarticulatus, 43
Nephropidae, 326
Nephrops, 331
Nephrops norvegicus, 327

Octopus dofleini, 317
Ocypode ceratophthalma, 328
Ocypode rotundata, 328
Ocypodidae, 147, 151, 154, 328
Odontodactylidae, 73
Odontodactylus scyllarus, 71

Oniscidae, 235
Oniscidea, 231, 236, 238-39
Oratosquilla oratoria, 70-71, 75-77
Ostracoda, 1-14, 283
Ovalipes, 294, 297
Ovalipes ocellatus, 152, 294, 296, 328, 331

Pachygrapsus crassipes, 25
Pacifastacus, 297, 299
Pacifastacus leniusculus, 313, 327
Paguridae, 327
Pagurus bernhardus, 323, 327
Pagurus punctulatus, 327
Palaemon, 323
Palaemon elegans, 326
Palaemon serratus, 326
Palaemonetes, 181, 329
Palaemonetes kadiakensis, 326, 331
Palaemonidae, 180, 326, 329
Palaemonetes paludosus, 326
Paleostomatopoda, 69
Palinura, 327, 331
Palinuridae, 311, 327, 336
Pandalidae, 326
Pandalopsis dispar, 168
Pandalus borealis, 166, 167-68
Pandalus danae, 168
Pandalus hypsinotus, 168
Pandalus jordani, 167
Pandalus kessleri, 326
Pandalus montagui, 166-67
Pandalus nipponensis, 326
Pandalus platyceros, 168, 326
Panulirus, 297, 299, 331, 336
Panulirus angulatus, 311
Panulirus argus, 133, 327
Panulirus guttatus, 327
Panulirus homarus, 311-12
Panulirus interruptus, 312, 315-16
Panulirus pencillatus, 311
Panulirus polyphagus, 327
Paracerceis, 101, 230
Paracerceis sculpta, 91-108, 230
Paragalene longicrura, 151
Paragnathia formica, 230
Paranthuridae, 235
Parapenaeus longirostris, 324, 326
Parastacidae, 327
Parathelphusa, 299
Parathelphusa hydromous, 151, 153, 317
Parathelphusa spinigera, 328
Parathelphusidae, 151, 328
Parthenope angulifrons, 151
Parthenope serratus, 328
Parthenopidae, 151, 328
Pasiphaea sivado, 323, 326, 329
Pasiphaeidae, 324, 326
Peltogasterella, 212, 219
Peltogaster paguri, 212
Peltogastridae, 209, 211-12, 216-17
Penaeidae, 185, 193, 203-4, 325-26

Penaeoidea, 183-205, 313-14
Penaeus, 185, 188, 196, 198, 200, 203-5, 323, 326
Penaeus aztecus, 186, 188-89, 191, 193-4, 196, 198, 201, 326
Penaeus duorarum, 186, 188, 189, 191
Penaeus indicus, 191, 314, 323, 326
Penaeus japonicus, 191, 326
Penaeus kerathurus, 191, 313, 315
Penaeus monodon, 205
Penaeus setiferus, 185-86, 189, 191, 194, 198, 200, 325-26, 331
Penaeus vannamei, 200
Peracarida, 232
Periplaneta americana, 44
Philomedidae, 2-3
Phreatoicidea, 232, 234, 238
Pilumnus hirtellus, 151
Pinnixa sp., 328
Pinnotheres pinnotheres, 328
Pinnotheridae, 328
Pirusaccus, 217, 220, 223
Pisa armada, 151
Pisa tetraodon, 148, 151-54, 156
Pitho lherminieri, 327
Plakarthrium, 234
Platycopida, 2
Platycopina, 2
Pleistocantha mosely, 156
Pleocyemata, 326, 336
Pleomothra apletocheles, 272, 281, 285, 288
Pleoticus, 188
Plesiopenaeus, 198
Pleuromamma, 264
Pleuromamma abdominalis, 254, 258, 264
Pleuromamma gracilis, 254
Pleuroncodes planipes, 294, 299, 300
Podochela gracilipes, 328
Podochela riisei, 328
Podocopa, 2-4, 7
Podocopida, 2
Podocopina, 2
Podotremata, 147, 159
Polyartemiidae, 113
Polycopidae, 2
Polycopoidea, 2
Pontellidae, 247, 253
Pontoporeira affinis, 35
Porcellio, 235
Porcellio scaber, 232
Portunidae, 17-31, 18, 146, 151, 154, 160, 328, 334
Portunus pelagicus, 151, 159, 328
Portunus sanguinolentus, 43, 45, 148, 151-54, 159-60, 328
Procambarus clarkii, 327
Procambarus leonensis, 327, 331, 333-34
Procarididae, 326, 336
Procaris ascensionis, 326, 329, 336
Protosquillidae, 73

Pseudodiaptomidae, 247
Pseudodiaptomus, 267-68
Pseudojanira, 240
Pseudosquilla, 82
Pseudosquilla ciliata, 70-72, 75, 80-82
Pseudosquilla ornata, 82
Pseudosquillidae, 73
Pseudozius caystrus, 328
Pugettia producta, 155

Ranina ranina, 309
Raninoidea, 147
Remipedia, 271-88
Reptantia, 290-305, 331, 336
Rhizocephala, 208-25
Rhynchocinetes typus, 326, 329, 336
Rhynchocinetidae, 324, 326
Rutidermatidae, 2-3

Sacculina carcini, 211-13, 215
Sacculinidae, 209, 211, 216-17
Santia, 240
Sarsiellidae, 2-4
Sarsielloidea, 2
Scalpellum scalpellum, 223
Scolecethrix danae, 253-544
Scylla, 297
Scyllaridae, 327
Scyllarides, 291
Scyllarides nodifer, 299, 303-5
Scyllarus, 297
Scyllarus arctus, 326
Scyllarus chacei, 327, 336
Scylla serrata, 294, 309-10, 316-17, 319, 328
Sergestidae, 326
Sesarma quadratum, 328
Sesarma reticulatum, 328, 333-34
Sesarma taeniolatum, 328
Sicyonia, 192, 196, 198, 200-1, 203-5, 336
Sicyonia brevirostris, 191, 193-94, 197-98, 326
Sicyonia carinata, 201, 323, 326
Sicyonia ingentis, 201, 325-26, 329
Sicyonia laevigata, 201
Sicyonia parri, 201-2
Sicyoniidae, 185, 193, 204, 325-26
Sigilloidea, 2
Skogsbergia lerneri, 7-8
Solenocera, 198, 203-4
Solenocera vioscai, 188, 194, 198
Solenoceridae, 185, 192, 198
Spartina alterniflora, 36
Spelaeogriphacea, 232
Speleonectes benjamini, 272, 274-79, 280-86, 288
Speleonectes lucayensis, 272, 281, 286, 288
Speleonectes ondinae, 288
Speleonectes tulumensis, 281, 288
Sphaeroma, 239
Sphaeromatidae, 93, 101, 239
Sphaeromatinae, 101

Spinacopia sandersi, 4
Squilla aculeata, 70-75
Squilla empusa, 70
Squilla holoschista, 71-72, 81
Squilla parva, 75, 77
Squillidae, 73
Squilloidea, 68
Stenetrioidea, 232, 240
Stenetrium, 240
Stenopodidae, 326
Stenopodidea, 326, 329, 336
Stenopus, 336
Stenopus hispidus, 324, 326, 329, 331, 335
Stenopus cf. *scutellus*, 324, 326, 329
Stenorhynchus lanceolatus, 156
Stenorhynchus seticornis, 328
Stomatopoda, 67-87
Streptocephalus, 113, 116-18, 123
Streptocephalus dendyi, 113
Streptocephalus dichotomus, 119
Streptocephalus dorothae, 122-23
Streptocephalus mackini, 117, 120, 122
Sylon, 217, 221, 225
Sylon hippolytes, 218
Sylonidae, 209, 216, 218
Syncarida, 232

Talitrus saltator, 35
Tantulocarida, 224
Temoridae, 247
Thalamita crenata, 328
Thalassinidea, 331
Thalassinoidea, 327
Thamnocephalus, 117
Thaumatocyprididae, 2
Thaumatocypridoidea, 2
Thecostraca, 223-24
Thenus orientalis, 312
Thompsonia, 217-20, 223
Thompsonia cubensis, 218-19
Thompsonia japonica, 218
Thoracica, 212, 215, 223-24
Thoracotremata, 147
Thor manningi, 326
Tisbe holothuriae, 309
Tortanidae, 247
Tozeuma, 99
Trachypenaeus, 191, 196, 198, 200-1, 204
Trachypenaeus byrdi, 324, 326
Trachypenaeus fuscina, 196
Trachypenaeus similis, 191-92, 194, 196, 198, 326
Tymoloidea, 147
Typhlatya rogersi, 324

Uca, 154, 292, 293-94
Uca annulipes, 328
Uca lactea, 151, 154, 328
Uca pugilator, 292-93
Uca thayeri, 153-54
Uca urvillei, 328
Uca vocans, 154
Undinula vulgaris, 250, 252, 258, 262, 264, 267
Unipeltata, 69

Systematic Index **351**

Upogebia pusilla, 327, 331
Upogebiidae, 327

Valvifera, 231, 234, 236, 238
Vargula, 9, 13
Vargula contragula, 5, 9
Vargula graminicola, 9, 12

Vargula hilgendorfi, 4-5, 7
Vargula ignitula, 9, 11
Vargula kuna, 5, 9, 11-12
Vargula lucidella, 9, 11
Vargula micamacula, 9, 11
Vargula mizonomma, 9
Vargula noropsela, 9

Vargula norvegica, 5
Vargula psammobia, 9, 12
Vargula scintilla, 9, 11
Vargula shulmanae, 9

Xanthidae, 151, 328
Xiphopenaeus, 200, 325

SUBJECT INDEX

Ablation, 22-23, 29, 72
Accessory sex gland, 81
Acrosomal vesicle, 4
Acrosome, 231-32, 278, 281-83, 285, 302, 304, 310, 322-23, 325, 329, 331-35
Adhesive, 173, 181, 185, 188, 191, 253, 258, 262, 311
Advanced, 204
Aesthetasc, 22-24, 46, 211, 219, 221-23, 247
Alpha male, 93-95, 101-8
Amplexus, 36, 38-39, 42, 45, 112, 114, 116-17, 119-20, 122
Ancestral, 223, 225, 229, 232, 234
Anchialine, 272
Androgenic gland, 28-30, 72
Antenna, 4-5, 42, 45-46, 113, 116, 122-23, 137
Antennal appendage, 114-15, 123
Antennal gland, 20
Antennule, 4-5, 8, 22-24, 29-30, 74, 82, 167, 211-12, 214, 217-19, 221, 224, 247-48, 260, 268

Appendix interna, 184, 200
Appendix masculina, 166, 173, 181, 184, 200, 231-32, 234-36, 238-40, 287
Arthropodin, 63, 221
Artificial insemination, 127, 139, 140-42

Barnacle, 208-25
Beta male, 94-95, 101, 103, 105-6
Bioluminescence, 1-14
Brood chamber, 181, 224
Brooding, 4, 7-8, 10, 35, 43, 45, 74-76, 80, 83, 100-1, 155, 157
Brood pouch, 7, 35, 45, 51, 60-61, 64, 94, 115-16, 119, 121, 229-32, 238-39

Calceoli, 45-46
Cement gland, 71-72, 78, 81-82
Centriole, 281, 285, 323, 329
Chela, 247, 249-50
Cheliped, 18, 26, 130, 135, 137
Chemoreceptor, 22
Chitin, 297, 309-10, 315-16

Chondroitin sulphate, 312, 314
Cincinnulum, 194, 198
Clade, 3, 12-14
Clasper, 4, 6, 113
Classification, 323, 336
Competition, 10, 53, 70, 81, 86, 92, 107-8, 111-23, 146, 157
Copulation, 2, 4-10, 14, 18-19, 21-22, 34-36, 38-42, 44, 51-53, 55, 62-63, 68, 73, 77, 79-82, 84, 86, 107-8, 112-13, 116, 118-20, 122, 133, 146, 153, 155, 158-60, 165, 170, 173, 175-78, 181, 184-85, 200-3, 205, 230-32, 234, 236-37, 239-40, 250, 258, 260, 267-68, 287, 291, 294, 297, 316
Coupling apparatus, 253, 255, 257-58, 260, 262, 264, 267-68
Courtship, 1-14, 18-22, 82
Crab, 17-31, 45, 59, 68, 70, 77, 134, 138, 154, 145-60, 180, 184, 210-11, 217, 294, 309-10, 315, 329, 334
Crayfish, 184, 297, 309, 311, 313, 334

Crustecdysone, 25-26
Cryptogonochorism, 216, 220, 223
Cuticular pore, 262, 267-68
Cypris, 210-23

Development, 212, 218, 221, 287
Dihydroxy phenylalanine oxidase, 312
Dorsal cuticular organ, 240
Dwarf male, 215, 223-24, 230-31, 238

Ecdysis, 35, 157; *see also* Molt
Ecdysone, 221
Ecdysterone, 36
Egg, 3, 7, 35, 71-77, 80-83, 86-87, 112, 115, 117-22, 129, 134, 138, 142, 146-47, 155-58, 166, 170, 177, 181, 211, 215-16, 219-21, 231, 239, 276, 281, 287, 325, 331
Ejaculatory duct, 147, 166, 170-71, 173, 176-77, 179, 181, 185-86, 188-89, 191-93, 196-97, 257, 309
Electroejaculation, 131, 139-41
Embryo, 7, 55, 60-61, 94, 184, 229
Entrainment, 7-9, 11-13
Externa, 209-15, 217-21, 224
Eyestalk ligation, 27-29

Fairy shrimp, 102, 107, 111-23
Female choice, 83, 102, 107, 114, 121-23, 203
Fertilization, 68, 71, 112, 138, 146, 151, 153, 157, 159, 165, 176-77, 180-81, 219-20, 225, 231, 238, 262, 264, 267, 287, 314, 316, 329
Flagellum, 231, 278, 281-85, 288, 322
Frontal appendage, 114

Gametocyte, 168, 170
Gamma male, 94-95, 101, 103, 105-6, 108
Genitalia, 4, 183-205, 228-43, 246-68
Genital pore, 3, 18, 103, 170, 177, 253-54, 260, 262
Gland, 186, 188, 191, 217, 222, 257-58, 262
Glycoprotein, 72, 313
Glycosamino glycan, 313
Gonad, 3, 165-70, 173, 216
Gonochorism, 209, 211
Gonopod, 71, 73-74, 128, 151, 153-54, 159-60
Gonopore, 71, 74, 83-85, 115, 118, 139-41, 179, 189, 191, 200-1, 253, 257, 262, 264, 272-73, 277-78, 281, 285, 287-88

Hatching, 156-58, 219
Hemipenis, 4
Hermaphroditism, 168, 209, 215-16, 220, 223, 272, 288
Homology, 196, 203-4, 212, 220, 223-24, 234, 238

Hormone, 25, 26, 28, 36, 52, 61, 63, 72
Hyaluronic acid, 314
Hybridization, 118, 123

Insemination, 5, 35, 103-4, 107, 115, 119-20, 126-42, 159, 176-77, 179, 181, 184, 193, 196, 200-1, 203, 229, 231-32, 236, 238-40, 313, 318
Instar, 2-3, 7-9, 18
Interna, 210, 219
Intromission, 36, 74, 85, 116, 118-19, 122
Iteroparous, 101

Kentrogon, 209-12, 216, 222-23

Lactate dehydrogenase, 299, 317-18
Larva, 68, 75-77, 80, 83, 86-87, 113, 156-58, 211-12, 216, 220, 222, 287
Light organ, 7
Lobster, 42, 45, 59, 126-42, 133, 138, 297, 309, 311, 313, 315-16, 334

Male choice, 68
Manca, 94, 96-97, 229
Mantis shrimp, 67-87, 329
Mantle aperture, 209, 211-17, 219-21, 223-24
Mantle cavity, 209, 211-13, 215-17, 219-21, 223-25
Mate guarding, 50-65, 68, 79, 81, 86, 107, 108, 120, 146, 155, 230
Mating, 3, 4, 10, 18, 73, 77, 80-81, 83-86, 94-95, 102, 104-8, 111-23, 126-42, 145-60, 165, 176, 181, 201-3, 205, 230, 238, 243, 258, 260, 267-68, 281, 287, 310-11, 318
Mating plug, 10
Mating system, 3, 76, 81, 86, 92, 104, 108, 229, 241, 243
Medial process, 114
Metamorphosis, 3, 73, 210-12, 224
Microtubule, 278, 281, 285, 288, 331-34, 336
Microvillus, 299
Mitochondrion, 4, 278, 281, 283, 285, 299, 329
Molt (molting), 18, 21-22, 28, 30, 35-36, 42, 51, 54, 72, 78, 80, 83, 93-94, 103-4, 127-30, 133, 136, 138-39, 142, 147-48, 155, 158, 166, 176, 200-1, 209, 212, 221, 224, 230-31, 239, 243, 316
Monandry, 5
Monogamy, 68, 82-83, 86
Morphocline, 184, 203-4
Mucopolysaccharide, 71-72, 279, 286, 291, 297, 306, 309-14, 318

Nauplius, 184, 210-11
Nucleolus, 276, 278

Nucleus, 274, 276, 278, 281-85, 288, 322-23, 325, 329, 331-35

Ofactometer, 56, 58-60, 64
Oocyte, 150-52, 154, 166-67, 170, 173, 180-81, 218, 272, 274-77, 287
Oogenesis, 215, 272, 288
Oogonium, 274
Oopore, 231, 238, 240-42
Operculum, 146-47, 155
Outgroup, 224, 232
Ovary, 3, 55, 71-72, 77-78, 81, 83, 103, 115, 119, 128, 129, 148, 150-52, 168-70, 174, 176, 211, 215, 217, 219, 221, 231, 238, 272, 274-76
Oviducal pouch, 115
Oviduct, 71, 103, 107, 112, 115, 148-49, 151-60, 166, 170, 173-74, 177, 179-81, 219, 231, 238, 240-41, 253, 272, 276
Oviparity, 68
Oviposition, 35, 36, 68, 78, 181, 217, 225, 276
Ovisac, 112, 115, 117-19, 121-22
Ovotestis, 220
Ovulation, 35-36
Ovum, 230-31, 238

Palpation, 38, 42, 44, 51, 55-57, 60-62, 74
Parasitism, 208-25, 238, 243, 337
Parthenogenesis, 3, 119, 209, 216, 218
Penis, 3, 4, 113, 115, 119, 121, 232, 234-37, 287
Periodic Acid Schiff (PAS), 278-79, 281, 291, 297, 310, 312
Petasma, 184, 194, 197-98, 200-1, 203-5
Phenol, 315-16
Phenolase, 310, 315
Phenolic tanning, 299, 309, 312, 315
Phenoloxidase, 309
Pheromone, 6, 17-31, 34, 37-39, 42-47, 52-54, 56, 59-60, 127, 146, 221, 258, 264, 267-68
Phosphoglucomutase, 95
Phylogeny, 184, 203, 224, 323-24, 337
Pleopod, 38, 40, 42, 74, 147-48, 155, 158-60, 166-67, 176, 178, 181, 184, 197-98, 200, 229, 231-32, 234-37, 240, 243
Plesiomorphic, 223-24
Polyandry, 5
Polygyny, 5, 10, 91-108, 111-23
Polymorphism, 94-95, 99, 106, 108
Postlarva, 73, 75, 82-83
Predation, 8, 36, 69-70, 81, 83-84, 95, 98-100, 102, 104, 106, 108, 121, 158
Primitive, 204, 232, 234, 238, 240, 272
Promiscuity, 68, 77, 81-82, 86

Quinone tanning, 315

Subject Index

Receptacle, 209-19, 221, 223-25
Recruitment, 168
Reproductive isolation, 3, 264, 268
Rheotaxis, 59

Sclerotin, 315
Semelparity, 94, 100-2
Semen, 117
Seminal material (fluid), 119, 147-54, 159-60, 192, 204, 253, 257, 262, 264, 267, 291, 292-97, 299, 306, 309, 314, 316-17, 319
Seminal receptacle, 3, 18, 127-29, 133, 139, 141-42, 147-54, 156-57, 159, 184, 188, 191-94, 196-97, 200-5, 211, 216, 253, 287, 291, 294, 296-97, 299, 306, 311
Seminal vesicle, 3-4, 108, 257, 264, 287
Seminal vestibule, 71, 74
Seta, 166, 222, 247, 250, 260, 268
Settlement, 212, 216
Sex attraction, 212, 221, 223, 225
Sex change, 165-68
Sex determination, 166, 225
Sex ratio, 3, 9-11, 53, 88, 96-97, 105, 118, 138, 177
Sexual selection, 3, 8, 10-11, 13, 92, 95, 104-6, 114, 116, 119, 121-23, 205
Sexual system, 209, 219-20, 223-25
Shrimp, 43-44, 68-70, 76, 99, 138, 164-81, 183-205, 313, 325, 336
Sinus gland, 29, 72
Sister group, 218, 224
Spawning, 18, 20, 72-74, 81, 127-29, 134, 137-38, 147, 152-58, 168, 170, 177, 180-81, 213, 221, 310
Sperm (spermatozoon), 3-4, 7-8, 20, 35, 71, 73-74, 80-81, 84,
119, 103, 107, 121-22, 129, 131-34, 138-39, 142, 145-160, 166-67, 170-71, 173-77, 179-81, 184, 186, 188-89, 191-93, 196, 201, 204, 211, 215-17, 223, 229, 231-32, 234, 236-40, 243, 253, 257, 260, 262, 264, 267-68, 272, 277, 281, 283, 285-87, 288, 291, 292-94, 296-97, 299, 301-2, 306, 309-12, 314, 317-18, 322-37
Sperm competition, 68, 82, 103-4, 107-8, 145-60
Sperm duct, 292
Sperm gel, 153-54
Sperm mass, 103, 170-71, 173, 186, 188-89, 191, 193-94, 196-97, 201-2, 204, 291, 293-97, 299-306, 309-10, 313-14, 318
Sperm packet, 148-54
Sperm plug, 128, 133, 139, 148-49, 155, 159-60, 294
Sperm storage, 103, 147-49, 156, 159-60, 165, 180, 184, 192, 196-97, 203-5, 297, 309, 318
Sperm transfer, 3, 50, 147, 165, 173, 176, 181, 192, 200-1, 205, 231, 313, 317-18
Spermatheca, 18, 20, 22, 192, 231, 238, 240, 253, 260, 262, 317-18
Spermathecal duct, 237, 240-43
Spermatid, 278-80, 288, 329
Spermatocyte, 169-70, 278
Spermatogenesis (spermiogenesis), 5, 218-19, 225, 272, 278, 323
Spermatogenic island, 214-15, 217, 219-21, 223
Spermatogonium, 214-15, 217-21, 223-24, 278
Spermatophore, 4-5, 7, 10, 18, 128-29, 131-34, 138-39, 141-42, 147, 152-53, 159-60, 165-66, 170, 173, 175-77, 181, 184-86, 188-89, 191-94, 196-97, 200-1, 203-5, 231-32, 234, 247,
250, 253-55, 257-58, 260, 262, 264, 267-68, 278-88, 290-306, 308-19
Spermatophore sac, 257-58
Spermatophore transfer, 252, 258, 287
Spermatophoric mass, 180, 185-86, 193, 197, 286, 297, 305, 310-11
Spermatophoric sac, 170, 173, 179, 181
Swarming, 7
Synapomorphy, 232, 238
Systematics, 181

Taxonomy, 184, 192, 324
Terminal ampoule, 185-86
Testis, 4, 71, 145, 157, 168, 170-71, 173, 185, 212, 218, 224, 257, 272, 274, 277-79, 288, 292, 294, 302
Thelycum, 147-49, 159, 184, 188, 191-94, 196-97, 200-1, 203-4, 313-14
Trichogon, 209-13, 215-17, 222-24
Tyrosine, 310, 312, 315

Urine, 20-22, 24-26, 29, 45
Uterus, 3

Vagina, 3, 103, 148-54, 160
Vas deferens, 3, 71-72, 131, 139, 147, 151, 154, 168, 170, 174, 181, 185-86, 191-93, 232, 234, 236-37, 257, 278, 280, 286-88, 291, 292-95, 297, 299-300, 304-6, 309, 313, 316, 318, 324
Velum, 148-52
Vitelline membrane, 119
Vitellogenin, 68
Viviparity, 68
Vulva, 147-49, 152, 160

X-organ, 72

Y-tube (-maze), 37, 56, 58